CAMBRIDGE MONOGRAPHS ON MATHEMATICAL PHYSICS

RELATIVITY ON CURVED MANIFOLDS

RELATIVITY ON CURVED MANIFOLDS

F. DE FELICE

"Lagrange" Institute of Mathematical Physics
University of Turin

C. J. S. CLARKE
University of Southampton

CAMBRIDGE UNIVERSITY PRESS

Cambridge

New York   Port Chester   Melbourne   Sydney

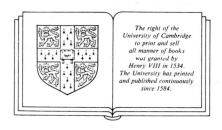

Published by the Press Syndicate of the University of Cambridge
The Pitt Building, Trumpington Street, Cambridge CB2 1RP
40 West 20th Street, New York, NY 10011, USA
10 Stamford Road, Oakleigh, Melbourne 3166, Australia

First published 1990

Printed in the United States of America

Library of Congress Cataloging-in-Publication Data

de Felice, F., 1943–
Relativity on curved manifolds / by F. de Felice and C. J. S. Clarke.
p. cm.
Includes index.
ISBN 0-521-26639-4
1. General relativity (Physics) 2. Manifolds (Mathematics)
I. Clarke, C. J. S., 1946– . II. Title
QC173.65.D43 1989
530.1′1 – dc19 88–39531
CIP

British Library Cataloging in Publication

de Felice, F., 1943–
Relativity on curved manifolds
1. Physics. Relativity. Geometric aspects
I. Title II. Clarke, C.J.S. 1946–
530.1′1

ISBN 0-521-26639-4 hard covers

CONTENTS

Preface			*xi*
	Geometry and Physics: An Overview		1
	1	Geometry	1
	2	Special relativity	4
	3	General relativity	7
	4	Local, global and infinitesimal	11
1	**The Background Manifold Structure**		14
	1.1	Topological spaces	14
	1.2	Maps	15
	1.3	Coordinate neighbourhoods	17
	1.4	Differentiable manifolds	19
	1.5	Maps of manifolds	20
	1.6	The tangent space	21
	1.7	Bases in the tangent space	24
	1.8	Transformation properties of vector components	25
	1.9	The tangent map	27
	1.10	The cotangent space	28
	1.11	Bases in the cotangent space	29
	1.12	The dual tangent map	31
	1.13	Tensors	32
	1.14	Symmetry operations on tensors	37
	1.15	The metric tensor	39
	1.16	Raising and lowering of tensor indices	47
	1.17	Alternating tensors	48
	1.18	Exterior algebra	51

1.19 Measure of lengths and the world-function 55

2 Differentiation 57
2.1 Tensor fields and congruences 57
2.2 The Lie derivative 62
2.3 The connector 68
2.4 Parallel propagation and geodesics 72
2.5 Transformation properties of the connector 77
2.6 The covariant derivative 78
2.7 Torsion and normal coordinates 82
2.8 Compatibility of the metric with the connection 84
2.9 Parallelism 85
2.10 Applications of the covariant derivative 86
2.11 The exterior derivative 88
2.12 Frobenius theorems 90
2.13 Isometries on M 99

3 The Curvature 102
3.1 The Riemann tensor 102
3.2 Symmetry properties of the Riemann tensor and the Gaussian curvature 106
3.3 Significance of a curvature tensor vanishing everywhere 109
3.4 The Ricci tensor, the curvature scalar, the Weyl tensor 111
3.5 The Bianchi identities 114
3.6 The equation of geodesic deviation 117
3.7 The covariant derivative of the world-function 120
3.8 Maximally symmetric spaces 125

4 Space-time and Tetrad Formalism 129
4.1 The space-time manifold and the physical observer 129
4.2 Construction of a tetrad 133

4.3 Relations among tetrads and the Lorentz group 135
4.4 The propagation laws for tetrads 138
4.5 The Ricci rotation coefficients 139
4.6 Differential operators related to a tetrad frame 141

5 Spinors and the Classification of the Weyl Tensor 146
5.1 Outline 146
5.2 The group SL(2, $\mathbb{C}$) 146
5.3 Lie algebras 151
5.4 Bivector algebra 155
5.5 Spinors 159
5.6 The spinor connection 166
5.7 The spinor curvature 168
5.8 The torsion case 172
5.9 Conformal spinors 174
5.10 The Weyl spinor and the Petrov classification 177

6 Coupling Between Fields and Geometry 185
6.1 Newtonian fluids 185
6.2 Generalization to special relativity 187
6.3 Coupling between fields and geometry: the field action 188
6.4 The gravitational action and the Einstein equations 191
6.5 The energy-momentum tensor of a perfect fluid 195
6.6 The energy-momentum tensor of a single particle 198
6.7 The energy-momentum tensor of the electromagnetic field 200
6.8 The energy-momentum pseudotensor 201

7 Dynamics on Curved Manifolds 206
7.1 Conservation laws 206

7.2	The equations of motion of an extended body	211
7.3	The centre-of-mass description	216
7.4	Motion of a point particle	221
7.5	Constants of motion	225
7.6	Maxwell's equations for a free electromagnetic field	227
7.7	Maxwell's equations in the presence of charges and currents	228
7.8	The radiation field	231
7.9	The light cone	236
7.10	Stationary space-times	238
7.11	The geometry of stationary null surfaces	244
8	**Geometry of Congruences**	250
8.1	Tetrad decomposition of the Riemann tensor	250
8.2	The expansion equation	255
8.3	The vorticity equation	258
8.4	The Einstein equations in tetrad form	259
8.5	The geometry of null rays	261
8.6	Singularities	267
9	**Physical Measurements in Space-time**	274
9.1	The concept of measurement	274
9.2	The measurement of time intervals and space distances	275
9.3	Measurements of angles	281
9.4	Curvature effects in the measurement of angles	282
9.5	Measurement of frequency	286
9.6	Measurement of relative velocities	287
9.7	The velocity composition law	295
9.8	Energy and momentum of a particle	297
9.9	Measurement of electric and magnetic fields	298
9.10	The properties of a fluid	300
9.11	The equations of motion of a fluid	304
9.12	The small curvature limit	306

9.13 Gravitational radiation 314

10 Spherically Symmetric Solutions 320
10.1 The spherically symmetric line element 320
10.2 The external Schwarzschild solution 322
10.3 The internal Schwarzschild solution 326
10.4 The global structure of spherically symmetric space-times 329
10.5 The extended external Schwarzschild solution 332
10.6 Penrose diagrams 338
10.7 Time-like geodesics in the external Schwarzschild solution 342
10.8 The precession of the apsidal points 347
10.9 The plunging-in observer 349
10.10 Null geodesics in the external Schwarzschild solution 351
10.11 The bending of light rays 354
10.12 The Reissner-Nordström solution 358
10.13 The extended Reissner-Nordström solution 361
10.14 Particle behaviour near the Reissner-Nordström singularity 369
10.15 Homogeneous and isotropic cosmology 372
10.16 The Friedmann solutions 378
10.17 Cosmological effects 384

11 Axially Symmetric Solutions 389
11.1 The axially symmetric line element: the canonical form 389
11.2 The Kerr solution 392
11.3 Physical interpretation of the Kerr metric 397
11.4 The space-time structure 400
11.5 Time-like geodesics 406
11.6 Rotationally induced effects 408
11.7 The angular geodesic equation 413
11.8 The equatorial circular geodesics 414
11.9 Null geodesics 418

11.10 The Kerr-Newman solution 421
11.11 The Weyl and the T-S solutions 424
11.12 Hawking radiation: an overview 426

Notation *431*

References *434*

Index *441*

PREFACE

A knowledge of general relativity is now essential to the understanding of modern physics but the power of the theory cannot be exploited fully without a detailed knowledge of its mathematical structure. The aim of this book is to set up this structure and use it to develop those applications which have been central to the growth of relativity. The book is sufficiently self-contained to be accessible to mathematicians with some knowledge of special relativity as well as to physicists.

We thank all our colleagues in our Institutions for their interest. One of us (F. de F.) wishes to express his thanks in particular to Prof. D.W. Sciama, Prof. L. Nobili, Prof. M. Calvani, Dr. R. Turolla, and Dr. S. Sonego for their help in the preparation of the manuscript and for stimulating discussions and Mr. A. Rampazzo for drawing the figures.

F. de Felice
Istituto di Fisica Matematica "J.-Louis Lagrange,"
Universitá di Torino, 10123 Torino, Italy

C.J.S. Clarke
Faculty of Mathematical Studies,
University of Southampton, Southampton, S09 5NH, U.K.

GEOMETRY AND PHYSICS: AN OVERVIEW

1 Geometry

Our theme is general relativity, seen not just as a theory of gravity, but as the expression of an approach to the world in which geometry and physics are united. The subsequent chapters will set out systematically the concepts and techniques needed; here we shall try to give an overview of the ideas that have led us to this approach, and to remind our readers of the basic facts of special relativity, with which we assume they have some acquaintance.

The laws of Euclidean geometry have since antiquity had a double status. On the one hand they were part of technology: empirical discoveries revealed when man started using the ruler and compass, which were developed and put to use in surveying and building. On the other hand they were for the Greeks examples of truths that could be discovered by pure reason, an innate capacity of the mind (as argued by Plato in the Meno, for example), which therefore did not depend for their validity on empirical measurements. This gave geometry an absolute status. Since the application of reason alone could derive these truths, they could not be other than what they were. Although, from Aristotle onwards, it was noted that there was a possibility of altering some of the axioms – notably the axiom about parallels not meeting and there being exactly one line parallel to a given line through a given point – without there being a logical contradiction, nonetheless such alterations were regarded as meaningless logical games. The fact that modifications in the axioms were manifest nonsense constituted a reductio ad absurdum proof of their absolute validity. In Kant's terminology, the laws of geometry were a priori truths which formed the necessary background conditions for the empirical world.

By the nineteenth century it was of course known that there was a sense in which non-Euclidean geometry was possible at a rather stronger level than that of a formal logical consistency: there exists what we now call a model for a non-Euclidean geometry. Namely, if one agrees that the term "straight line" denotes a great circle on a sphere, then it follows that the angles in a triangle add up to more than two right angles and parallels do not exist. But for most this was not held to violate the absolute status of Euclidean geometry. One used Euclidean geometry to prove the statement just given, and only escaped from its axioms by verbal trickery. A circle is not turned into a straight line by perversely calling it one, and it remains true that real straight lines obey the parallel axiom.

Two trends led to a shift in this consensus. First, mathematicians started to realise that non-Euclidean geometries were fun; they were interesting intellectual constructions in their own right and so a part of pure mathematics. Secondly, philosophers became increasingly suspicious of a priori truths. If an assertion was made about empirical objects, then the truth or falsity of that assertion had an empirical content. If by a "straight line" one means a real straight line drawn with a ruler or some other means then the properties of real straight lines can be examined empirically, and if it turns out that their properties are non-Euclidean (for instance, if very delicate measurements show that the angles of a triangle add up to more than two right angles), this only shows that real "straight lines" are not appropriately represented by the Euclidean ideal. And in such circumstances, we should have to enlist the help of mathematicians to provide us with a more appropriate non-Euclidean geometry.

A vital contributor to this development was Poincaré, who examined the meaning of geometrical terms in an empirical context. If we are to discover the geometry of the world, empirically then we must first decide on our procedure for making measurements. We must give a concrete specification of how we are going to draw a straight line, how we are going to measure a length and so on. In the case of length, for example, we must fix on an appropriate sort of measuring rod that is to be carried around from place to place in order to determine the ratios of different intervals in different places.

Poincaré gave the following example, which exactly illustrates the thinking that led to general relativity. Imagine a two-dimensional world confined to a disc in Euclidean space, inhabited by two-dimensional beings with measuring rods. And suppose that these beings and their rods are affected by a mysterious influence that causes them to shrink as they approach the edge of their world. Then for a certain simple shrinking law, the geometry which these creatures will determine by their empirical measurements will be non-Euclidean: the edge will be infinitely far away and, in this case, the angles of a triangle add up to less than two right angles and there are infinitely many parallels to a given line through a given point.

If we further suppose that these beings are strict empiricists, then they will conclude that their geometry is non-Euclidean. Any dissident non-empiricist among them might propose the alternative explanation that they live on a Euclidean disc affected by a shrinkage-influence. But the issue could only be decided if they could communicate with the God-like three-dimensional beings who could see their world from the outside and reveal to them that the non-empiricist view fitted better with the three-dimensional world.

For us, the evidence is that the empirical geometry of space-time is not Euclidean. The non-empirical option is always open to us: we could hold that space-time is "really" Euclidean, but shrinkage-forces distort the situation. But in the absence of a revelation from five-dimensional beings outside our world we would have no way of knowing what the "real" Euclidean geometry is; no way of knowing, for example, which, if any, of the infinitely many families of curves that could be drawn so as to obey the axioms for Euclidean straight lines, were straight lines as seen from outside our world. So in practice we have little choice but to accept that the appropriate geometry for describing our world is not a Euclidean one.

Before moving on to space-time, one qualifying remark is needed. The non-Euclidean geometry of which we have been speaking is actually still too like Euclidean geometry to be an appropriate tool for general relativity. In non-Euclidean geometry it is assumed that the elements of Euclidean geometry (points, lines, planes, distance, angle, etc.) are all defined, and that all results are to be deduced from a set of axioms that specifies how these elements intersect (such as, the axiom that there is exactly one straight line through any pair of points). What will be needed for general relativity is differential geometry, whose axioms make no statements about lines and planes as a whole, but only about their limiting behaviour when very small regions are considered. The result is not Euclidean, and it is a sort of geometry!

2 Special relativity

Special relativity is a framework for physics in which space and time have a fundamentally different structure from that which obtains in Newtonian physics, although the geometrical concepts remain basically Euclidean. General relativity is a generalisation of this framework in the same way that differential geometry is a generalisation of Euclidean geometry, and so we need to review some of the fundamental ideas of special relativity, referring the reader to one of the many available textbooks for details.

Newtonian physics rests on one of the most fascinating paradoxes in the history of science. Its key fact, which forced the break with the Aristotelian approach to dynamics, was the realisation by Galileo that a uniform movement is indistinguishable from a state of rest, from the point of view of an observer undergoing the movement. To use Galileo's own example, a person enclosed in a ship sailing steadily across a smooth sea could not perform any local experiment, confined to the cabin of the ship, which would enable him to determine whether or not he was moving through the water. This meant that acceleration, not velocity, was the prime kinetic quantity, and the laws of dynamics had to be laws determining acceleration. But the concept which enabled Newton to give the definitive form to those laws was his idea of absolute space, thought of as a direct manifestation of God. Motion was simply motion from one point in absolute space to another, and to describe it all that was necessary was to set up a Cartesian coordinate system in absolute space. The theory that enshrined the equivalence of rest and motion was based on a mathematical description in which rest and motion (relative to absolute space) were absolutely distinct.

The empiricist argument acts in Newtonian physics in the same way as in geometry. Newton's laws themselves ensure that the absolute space in terms of which they are formulated is an unobservable chimaera. All we can actually observe is not absolute place but relative place. I am in the same place now relative to the earth as I was ten minutes ago, but several thousand kilometres further on relative to the sun. With the benefit of hindsight we can say that absolute space, as an array of places that preserve their identity throughout time enabling us to say that I am now in the same place as before, is a myth. The only reliable background in which to situate events is space-time, the set of all places-at-particular-times. My writing this sentence occupies a different place-and-time as my writing the first sentence of this chapter; but I cannot meaningfully say that it occupies a different place, in an absolute sense. Thus the idea of space-time is inherent in the laws of Newtonian dynamics.

In Newtonian physics the two events just referred to can be said to occupy a different time, though not meaningfully a different place. This is because a time-ordering is a basic part of the assumed structure of Newtonian space-time. For any two events A and B it is possible to say either that A is before B, or B is before A, or they are simultaneous. Consistency in this order requires that simultaneity is an equivalence relation: space-time is divided up into equivalence classes of mutually simultaneous events, with each class constituting the universe at a given time.

In special relativity this concept of simultaneity is abandoned. Each observer has his own idea of simultaneity, but different observers disagree (just as, in Newtonian physics, each has his idea of "in the same place"). Thus in Newtonian dynamics (or better, in view of Newton's own predelictions, Galilean dynamics) there is absolute time but no absolute space, whereas in special relativity there is neither.

There remain some absolute structures, however. In both theories there exists a privileged class of observers, called inertial observers, characterised by the fact that, like Galileo's observer in the cabin of a ship, they do not feel any of the characteristic effects of acceleration. By contrast a non-inertial observer, who is necessarily accelerating relative to any inertial observer, feels apparent *inertial forces*, such as centrifugal and Coriolis forces, attributable to the acceleration. (Recall that if one is in a car that is accelerating forwards, then one feels a force pushing one backwards.)

The two different structures on space-time, Galilean and special relativistic, are expressed through the transformation laws relating the coordinates which different inertial observers place on space-time, using the same standard measuring techniques. (A coordinate system constructed by a standard procedure is called a *frame of reference* or simply a *frame*, and the coordinates of an inertial observer are called an *inertial frame*.) To a large extent these laws of transformation between inertial frames are the same. If x, y, z, t and x', y', z', t' are the three coordinates of space and one of time defined by two different inertial observers O and O', respectively, then they are related by some combination of the following (where we write $\vec{r}$ for (x, y, z) and $\vec{r}'$ for (x', y', z'), and $\vec{a}$ and b are a vector and a number depending on the two observers chosen):

(i)	a relative rotation of the space coordinates	$\vec{r} = A\vec{r}'$
(ii)	a displacement of the space coordinates	$\vec{r} = \vec{r}' + \vec{a}$
(iii)	a displacement of the time coordinates	$t = t' + b$
(iv)	a boost (see below)	

The difference comes in the form of the boost. In the Galilean case it is given by

$$t = t'$$

$$x = x' + vt'$$

$$y = y'$$

$$z = z'$$

while in the case of special relativity it is given by

$$ct = \frac{ct' + (v/c)x'}{\sqrt{1 - v^2/c^2}}$$
$$x = \frac{x' + vt'}{\sqrt{1 - v^2/c^2}}$$
$$y = y'$$
$$z = z'$$

where v is a number depending on the observers and c is an absolute constant of nature with the dimensions of velocity. We note the crucial distinction that in the Galilean case the time coordinate is unchanged, so that the observers agree on the definition of simultaneity. The above form shows that c has the role of a conversion constant (like Boltzmann's constant in thermodynamics) needed to relate the quantities t and x which have different physical dimensions when normal units are used.

It is a consequence of these transformations that if a signal or particle is moving with the speed c according to one observer, then it is moving with the same speed relative to the other. (As Fraser has expressed it, relativity replaces Newton's absolute rest by an absolute speed.) It turns out that light propagates with just such a speed, and so has the same speed according to all observers. This phenomenon, which runs quite contrary to Galilean thinking, was a consequence of Maxwell's description of electromagnetism, and it provided the motivation for the development of special relativity. The transformation laws can in fact be deduced from the constancy of the velocity of light, together with a few plausible symmetry assumptions. Experimentally this was checked to a high accuracy by the Michaelson-Morley and Kennedy-Thorndyke experiments.

3 General relativity

The single most important step in the development of general gelativity from special relativity is a change in the application of the idea of inertial observers, a change forced upon one by the distinctive properties of gravitation. As was already known to Galileo, a gravitational field accelerates all bodies in the same way, independently of their mass and internal constitution. A simple Newtonian argument clarifies this. The property of a body whereby it resists any attempt to change its state of motion is its *inertial mass* m_i. This enters the Newtonian equation of motion

$$\vec{F} = m_i \vec{w} \tag{3.1}$$

where $\vec{F}$ is the total force acting on the body and $\vec{w}$ is the acceleration. On the other hand the property which determines a body's response to a gravitational field is the *gravitational mass* m_g which enters the law for the gravitational force $\vec{F}_g$ acting on the body:

$$\vec{F}_g = m_g \vec{g} \tag{3.2}$$

where $\vec{g}$ is the gravitational field intensity, generated according to the inverse square law by a gravitating body. The constancy of $\vec{w}$ implies, from (3.1) and (3.2), that

$$m_i / m_g = \kappa \tag{3.3}$$

a constant, independent of the body chosen. By choosing the units of $\vec{g}$ suitably we can arrange that $\kappa = 1$, leading to $\vec{w} = \vec{g}$, independently of the body's mass.

Relation (3.3) has been experimentally verified to a very high accuracy (at best 10^{-12}) for a large range of masses, from a neutron (Witteborn and Fairbank, 1967), to an apple (Roll et al., 1964; Braginsky and Panov, 1971; etc.) to a planet (Williams et al., 1976).

The phenomenon is seen as even more remarkable when we realise that, according to special relativity, a (negative) contribution to the inertial mass of the body is its binding energy – something quite different in kind from the rest mass of the constituent particles. It might be understandable if the gravitational mass were proportional to the number of particles, so that each particle had a sort of unit gravitational charge, which would in turn be approximately equal to the inertial mass. But the accuracy of the experiments rules this out: it is the inertial mass, in all its various forms, and nothing else, that is related to the gravitational mass.

The gravitational field is not the only place in nature where an acceleration is independent of the nature of the body involved: the same thing happens with so-called inertial forces, the forces felt by an observer who is not inertial to which we have already referred in defining an inertial observer. In the case of uniform acceleration, for example, the form of the inertial forces can be derived immediately from Newton's laws of motion. If $\vec{F}$ denotes a "real" (i.e. not just inertial) force acting on an object, $\vec{r}$ denotes its position vector in an inertial frame and $\vec{r}'$ its position in a uniformly accelerating frame, then the two frames will be related by

$$\vec{r}' = \vec{r} - \frac{1}{2}\vec{w}t^2$$

and so, while Newton's law of motion in the inertial frame reads

$$m\ddot{\vec{r}} = \vec{F}$$

if we transform to the accelerating frame by substituting for $\vec{r}$ in terms of $\vec{r}'$ we have

$$m\ddot{\vec{r}}' = \vec{F} - m\vec{w}.$$

Thus in the accelerating frame every object is subjected to a "fictitious force" that is proportional to its mass (just as the gravitational force was proportional to the mass) in addition to any "real" force acting on it. A calculation would yield the same result in the case of the fictitious force due to a rotational acceleration (the combined Coriolis and centrifugal force). This proportionality to the mass, which arises automatically simply because we have created the force by transferring a term from the $m\ddot{\vec{r}}$ of Newton's equations from one side of the equation to the other, is a distinctive feature of the fictitious forces, distinguishing them from, say, the electrostatic force which depends on the charge rather than the mass.

The similarity between the gravitational force and fictitious forces makes it possible to annul the effect of the gravitational force by transforming to a coordinate system that is accelerating with precisely the gravitational acceleration. This is termed a *freely falling coordinate system* or *frame*. An example is the coordinate system defined by observers in a satellite in orbit, in free-fall under gravity, where objects float freely with no gravitational effects. So the question naturally arises, whether gravity itself is a fictitious force.

For the fictitious forces of acceleration and rotation it is possible to transform back to the original inertial frame in which the fictitious force is annulled everywhere. This is not the case with the gravitational force because the direction of the gravitational acceleration varies from place to place. A coordinate system that is freely falling for observers in Italy will be accelerating in the wrong direction for observers in Australia. So we could only regard gravity as a fictitious force if we were prepared to localise the idea of a coordinate system. We can seek to annul gravity, not everywhere but only sufficiently close to some particular freely falling observer. How close this is depends on the accuracy with which our measurements are to be carried out.

But to implement even this restricted idea of reducing gravity to a fictitious force, it is essential that the effect of gravity is precisely equivalent to the effect of an acceleration, so that by passing to a (local) freely falling frame, gravity can be cancelled exactly. The assertion that this is indeed the case is called the *principle of equivalence*. It entails two ideas: as far as mechanical effects are concerned, the gravitational force must be exactly proportional to the inertial mass (equation 3.3); and any non-mechanical effects of acceleration (for example, electromagnetic effects) must be exactly the same for gravity.

The electromagnetic effect of gravitation (manifested in the red-shift of electromagnetic radiation) has been verified to one per cent in laboratory experiments (Pound and Rebka, 1960). So, combining this with the overwhelmingly impressive verification of the mechanical equivalence, we shall here take the principle to be verified.

If we are prepared to work with a frame that is only local, then, by taking freely falling frames as our standard inertial frame, gravity can be abolished near any chosen observer, and the gravitational force becomes a fictitious force, attributable to the fact that one is using a non-inertial frame, accelerating relative to the (inertial) freely falling frame. This is the approach taken by general relativity. The alternative approach would be to insist that one uses a single global coordinate frame, in which gravity becomes a real force. But in this case there is no criterion for choosing the global frame (because of the arguments relating to the curvature of space-time given in section 1 above) and the very exact proportionality between the gravitational and inertial masses (equation 3.3) has to be attributed to an amazing coincidence.

4 Local, global and infinitesimal

To express precisely the idea of a "local inertial frame", which we have sketched rather vaguely in the previous section, we need some mathematical machinery which will be developed in the next two chapters. The key ideas will be this formalisation of local frames (to which we return shortly) and, corresponding to the locality of frames, the idea that there is no preferred global frame, so that we must be prepared to work with an arbitrary global frame; that is, in a general coordinate system, with no special relation between one possible global coordinate system and another. Thus it must be possible to transform our equations under general coordinate transformations, and not just under a special set such as the transformations between inertial frames of Galilean physics or special relativity. It was this idea that motivated Einstein, who called it the *principle of general covariance*. The name *general relativity* comes from this, expressing the passage from the restricted set of transformations in special relativity to the set of general coordinate transformations. The principle of general covariance demands that the laws of physics be expressed in a form that is valid in any coordinate system. In practice this is achieved by expressing them in terms of *tensors*, sets of numbers that change as the coordinate system is changed in a particular sort of way.

The idea of a local inertial frame is elusive because on examination it appears that what we really need is an infinitesimal inertial frame. For any specified accuracy of measurement we can find a freely falling coordinate frame defined in a region that is so small that gravitational effects cannot be detected within it, to the given accuracy. But as the required accuracy increases, so the size of the region has to shrink, and to define a "frame" in which all gravitational effects vanish we need somehow to take an ideal limit of finite coordinate systems as they shrink to zero.

This is formalised in the definition of the *tangent space* as the set of all tangent vectors to curves at a point. A tangent vector can be thought of as the limit of a finite displacement as it shrinks to zero, and so it has the right infinitesimal quality. In a way that will be given precise expression later, the tangent space approximates any finite region covered by a finite coordinate system, with the accuracy of the approximation increasing as the size shrinks. Thus the limit of approximate freely falling frames in a sequence of regions shrinking to zero will be realised as a frame on the tangent space.

Just as Newtonian physics is expressed by equations depending on inertial frames, whose validity for all observers depends on the transformation laws between inertial frames, so physics on curved manifolds is expressed by equations depending on the tangent spaces at various points. The simplest way to do this is to introduce a method of multiplying together tangent vectors and their duals, which produces the tensors referred to above, ensuring the general covariance of the equations.

From now on, when we speak of an inertial frame we shall denote this infinitesimal concept of a frame on the tangent space. But it is important to remember that, while this is a mathematical abstraction, it can be linked explicitly with actual finite coordinate systems constructed by particular physical processes such as radar measurements and theodolite readings, when they are carried out by a freely falling observer. We make this link in Chapter 9. The physical construction of coordinates is only possible, however, within a limited – though possibly very large – region. We have no reason to suppose that the measurements can be extended to give a coordinate system that covers the entire cosmos. And indeed, there is no reason to think that such a truly global coordinate system would exist, even mathematically.

So we have come to distinguish three levels of mathematical description. The infinitesimal, at which gravity is completely abolished and special relativity holds exactly with no gravitational forces; the local, where one can work with physically defined finite frames and where gravitational effects are seen as being expressed through the difference between the finite frame and the infinitesimal inertial frames; and the global level of the entire cosmos where a coordinate system, if it exists at all, is a purely mathematical construct with little or no physical significance.

Our description of the mathematical apparatus of curved manifolds will pass through these levels in the reverse order. First, we establish the idea of space-time as a whole in the form of a topological space whose only structure is a primitive idea of "closeness". Next, we shall introduce local coordinate systems (called *charts*) which make space-time into a *manifold*. Finally, we develop ideas connected with the tangent space and related infinitesimal objects.

1

THE BACKGROUND MANIFOLD STRUCTURE

1.1 Topological spaces

A *topological space* is a pair $(M, \boldsymbol{t})$ consisting of a set M and a family $\boldsymbol{t}$ of subsets of M such that:

i) the empty set $\emptyset$ and M itself belong to $\boldsymbol{t}$;
ii) the union of any number of elements of $\boldsymbol{t}$ is again an element of $\boldsymbol{t}$;
iii) the intersection of any finite number of elements of $\boldsymbol{t}$ is again an element of $\boldsymbol{t}$.

The family $\boldsymbol{t}$ is said to form a *topology* on M and its elements are called the *open sets* of M.

Let α range over a set of indices A, and denote a general element of $\boldsymbol{t}$ by U_α; if N is a subset of M, the family

$$\boldsymbol{t}' = \{U_\alpha \cap N\}_{\alpha \in A}$$

is a topology on N, called the *induced topology* on N.

A *neighbourhood* of a point (element) p of M is a set $U(p)$ containing an open set to which the point p belongs.

A topological space M is *connected* if it is not possible to write $M = A \cup B$ with $A,\ B\ \in \boldsymbol{t}$ and $A \cap B = \emptyset$.

A topological space is *separated* (or Hausdorff) if it is connected and any two distinct points have disjoint neighbourhoods (i.e. neighbourhoods with no points in common).

A family $\{U_\alpha\}$ of (open) sets of M forms a (open) *covering* of M if

$$\bigcup_{\alpha \in A} U_\alpha = M \quad .$$

A covering R is *finite* if $\{U_\alpha\}$ contains a finite number of elements; R is *locally finite* if any point p in M has a neighbourhood which intersects at most a finite number of elements of R.

A topological space M is *compact* if it is separated and any open covering R of M contains a finite covering R' of M; it is *paracompact* if for any open covering R there exists a locally finite covering R' such that every element of R' is contained in some element of R (Bourbaki, 1948).

1.2 Maps

Given two sets U and U', a *map*:

$$\Phi : U \longrightarrow U' \tag{1.2.1}$$

from U to U' associates to each point $p \in U$ a uniquely determined point $p' \in U'$:

$$p \longmapsto p' = \Phi(p) \quad . \tag{1.2.2}$$

The point p' is said to be the image of p under Φ and similarly the set $\Phi(U) := \{\Phi(p) : p \in U\}$ is the image in U' of U.

If for every $p' \in \Phi(U)$ there is only one point $p \in U$ such that $\Phi(p) = p'$, the map Φ is said to be one to one (*injective*).

If $\Phi(U) = U'$, the map is said to be *surjective* (mapping U onto U') and if it is both surjective and injective, then the map is said to be *bijective* (one–to–one and onto), Fig. 1-1.

If we are given a map $\Phi : U \to U'$ and a map $\Psi : U' \to U''$, then the map:

$$\Psi \circ \Phi : U \longrightarrow U''$$

defined by applying Φ and then Ψ is called the composition of Φ with Ψ.

If $\Phi : U \to U'$ is injective, then we can define a map $\Phi^{-1} : \Phi(U) \to U$ which satisfies:

$$\Phi \circ \Phi^{-1} = I_{\Phi(U)} \qquad \Phi^{-1} \circ \Phi = I_U \tag{1.2.3}$$

where I_U denotes the identity on U.

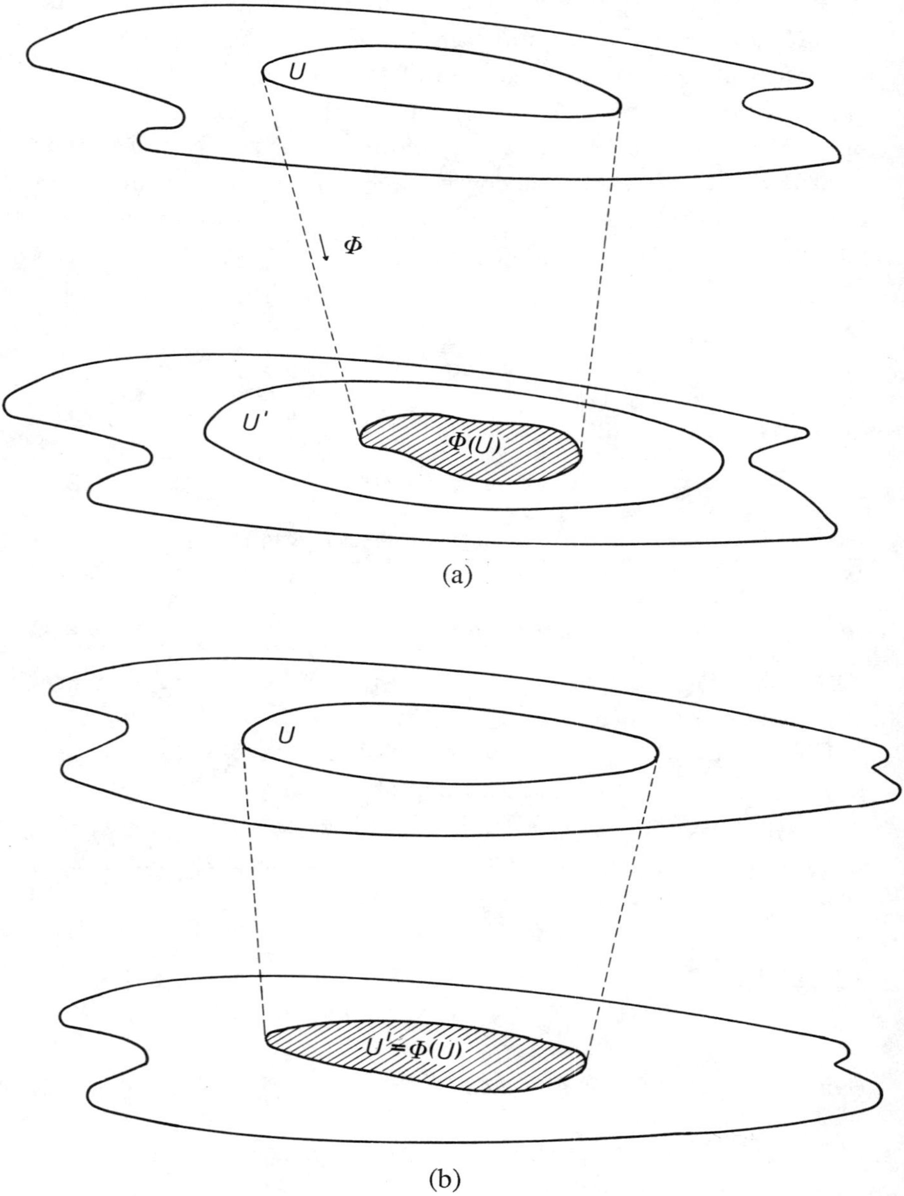

Fig. 1-1 Maps between manifolds; a)when Φ is one–to–one it is said to be *injective*; b)when $\Phi(U) = U'$ the map is said to be *surjective*.

Suppose now that $(U, \boldsymbol{t})$ and $(U', \boldsymbol{t}')$ are topological spaces. A map $\Phi: U \rightarrow U'$ is said to be *continuous* at a point $p \in U$ if $\Phi^{-1}(W')$ is a neighbourhood of p for every neighbourhood W' of $\Phi(p) \in U'$.

A map Φ is called *continuous* in U if it is continuous at every point p in U: this happens precisely when $\Phi^{-1}(W')$ is open for every open W' in U'.

If $\Phi : U \rightarrow U'$ is bijective, and if Φ and Φ^{-1} are continuous, then Φ is called a *homeomorphism* and U and U' are called homeomorphic.

More generally, if Φ is injective and Φ and Φ^{-1} are continuous with respect to the induced topology on $\Phi(U)$, then Φ is called a *homeomorphism into U'*.

If we are given more than two maps, the composition of maps satisfies the associative property:

$$\Phi \circ (\Psi \circ \Theta) = (\Phi \circ \Psi) \circ \Theta \quad . \tag{1.2.4}$$

1.3 Coordinate neighbourhoods

Given a topological space $(M, \boldsymbol{t})$ we define a *coordinate neighbourhood* (or chart) of M to be a pair $(U_\alpha, \varphi_\alpha)$, with U_α an element of $\boldsymbol{t}$ and φ_α a homeomorphism of U_α onto an open set of $\mathbb{R}^n$, the space of the n-tuples of real numbers $\{x^i\}_{i=1,2,\ldots,n}$. The numbers $x^1, x^2, \ldots$ constituting the image $\{x^i\}$ of a point $p \in U_\alpha$ under the map φ_α are called the local coordinates of p. A family of charts $\mathcal{A} = \{(U_\alpha, \varphi_\alpha)\}_{\alpha \in A}$ on M is said to form an *atlas* on M if

$$\bigcup_{\alpha \in A} U_\alpha = M \ .$$

Let $(U_\alpha, \varphi_\alpha)$ and (U_β, φ_β) be two charts on M with $U_\alpha \cap U_\beta \neq \emptyset$. A point $p \in U_\alpha \cap U_\beta$ can be expressed in terms of its two images in $\mathbb{R}^n$ as:

$$p = \varphi_\alpha^{-1}(x) = \varphi_\beta^{-1}(x') \tag{1.3.1}$$

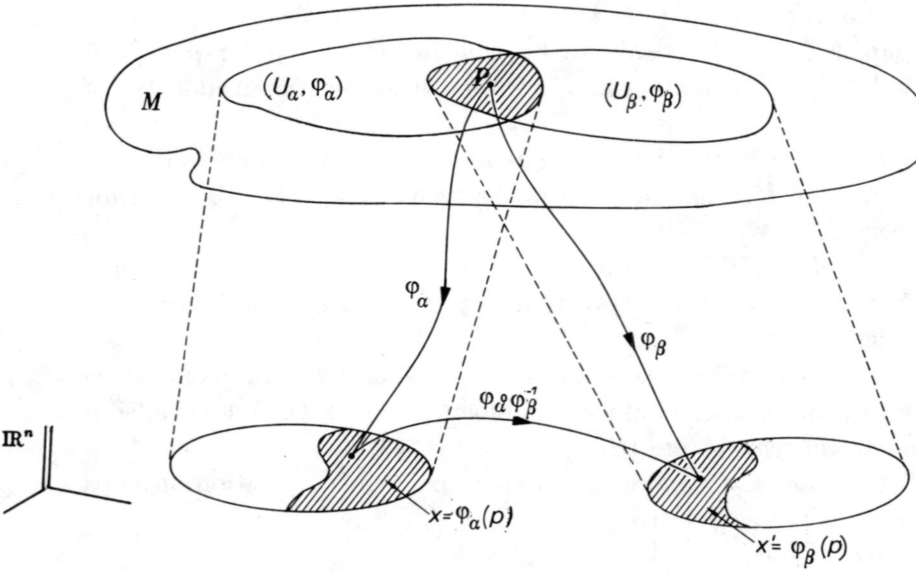

Fig. 1-2 Geometrical representation of a differentiable manifold

where we put $x \equiv \{x^i\}$, $x' \equiv \left\{x'^i\right\}$. It is easy to show that the composite map:

$$\varphi_\beta \circ \varphi_\alpha^{-1} : \varphi_\alpha(U_\alpha \cap U_\beta) \longrightarrow \varphi_\beta(U_\alpha \cap U_\beta) \tag{1.3.2}$$

is a homeomorphism between open sets of $\mathbb{R}^n$, Fig. 1-2. The inverse map is given by :

$$\left(\varphi_\beta \circ \varphi_\alpha^{-1}\right)^{-1} = \varphi_\alpha \circ \varphi_\beta^{-1} \ . \tag{1.3.3}$$

An atlas on M is a C^r-atlas if $\varphi_\beta \circ \varphi_\alpha^{-1}$ (and its inverse) for any pair (α, β) is an $\mathbb{R}^n$-valued C^r-function of n variables (i.e. has a continuous r-th derivative). The maps given by (1.3.2) and (1.3.3) are called coordinate transformations in M and can be written as

$$x' = \varphi_\beta \circ \varphi_\alpha^{-1}(x) \tag{1.3.4}$$

and its inverse:

$$x = \varphi_\alpha \circ \varphi_\beta^{-1}(x') \ . \tag{1.3.5}$$

1.4 Differentiable manifolds

A *differentiable manifold** of class C^r and dimension n is a Hausdorff topological space with a C^r-atlas. Any point in a C^r-manifold has a neighbourhood which is also a coordinate neighbourhood of M. Hereafter, a C^r-differentiable manifold will be simply referred to as a manifold unless stated otherwise.

A map $f : M \to \mathbb{R}$ is said to be a C^k-(real-valued) function at $p \in M$, if for any chart $(U_\alpha, \varphi_\alpha)$ containing p, there exists a neighbourhood U of p such that the composite map:

$$\tilde{f}_\alpha : \mathbb{R}^n \supset \varphi_\alpha(U) \to \mathbb{R}$$

defined by

$$\tilde{f}_\alpha(x^1, x^2, \ldots, x^n) \equiv f \circ \varphi_\alpha^{-1}(x) \tag{1.4.1}$$

is a C^k-differentiable function in $\varphi_\alpha(U)$.

A function f on M which is differentiable at every point of M, is said to be differentiable in M. The function $\tilde{f}_\alpha$ defined in (1.4.1) is called the coordinate representation of f at p relative to the chart $(U_\alpha, \varphi_\alpha)$. Denote by $\mathcal{F}$ the set of differentiable (real-valued) functions on M with the internal operations:

i) multiplication fg: $(fg)(p) = f(p)g(p)$ (product of numbers)
ii) addition $f+g$: $(f+g)(p) = f(p) + g(p)$ (sum of numbers).

These operations clearly satisfy:

$$\begin{aligned} fg &= gf \\ h(fg) &= (hf)g \\ f+g &= g+f \qquad\qquad \forall h, f, g \in \mathcal{F} \\ h(f+g) &= hf + hg \end{aligned} \tag{1.4.2}$$

with $+$ satisfying the axioms for a group. A set with such operations is termed an *Abelian ring* so $\mathcal{F}$ is termed the Abelian ring of (differentiable) functions on M.

A function $f \in \mathcal{F}$ is said to be *locally Lipschitz* of degree s ($s \in \mathbb{R}$ and $0 < s \leq 1$) if for any chart $(U_\alpha, \varphi_\alpha)$ on M and any

*See De Rham (1955); Choquet-Bruhat et al. (1977); Kobayashi and Nomizu (1969).

$x_0 \in U_\alpha$ there exist an open set W_α of U_α containing x_0 and a real number $k \geq 0$ such that: $\tilde{f}_\alpha$ is bounded in $\varphi_\alpha(W_\alpha) \subset \mathbf{R}^n$, and for each pair $p, p' \in W_\alpha$ the following condition holds:

$$\left|\tilde{f}_\alpha(x^1, x^2, \ldots, x^n) - \tilde{f}_\alpha(x'^1, x'^2, \ldots, x'^n)\right| \leq k|x - x'|^s \quad (1.4.3)$$

where $|x - x'| = \left[\sum_i \left(x^i - x'^i\right)^2\right]^{\frac{1}{2}}$. A locally Lipschitz function of degree 1 is simply termed Lipschitz.

1.5 Maps of manifolds

A manifold M is locally homeomorphic to an open set of $\mathbf{R}^n$, so every point $p \in M$ can be referred to by means of its local coordinates relative to some chart containing p. Similarly any relation among manifolds can conveniently be expressed in terms of local coordinates. Let M and N be two differentiable manifolds with dimension n and $\Psi : M \to N$ a map of M into N:

$$\Psi(p) = p' \qquad p \in M \quad p' \in N \ . \quad (1.5.1)$$

Let $(U, \varphi)_p$ and $(U', \varphi')_{p'}$ be two charts containing p and p', respectively such that:

$$\varphi(p) = \{x\} \qquad , \qquad \varphi'(p') = \{x'\} \ . \quad (1.5.2)$$

From the definitions we easily have:

$$p' = \varphi'^{-1}(x') = \Psi \circ \varphi^{-1}(x) \quad (1.5.3)$$

which leads to:

$$x' = \varphi' \circ \Psi \circ \varphi^{-1}(x) \ . \quad (1.5.4)$$

This is the same as relation (1.5.1) in coordinate form; the function

$$\tilde{\Psi}(x^1, x^2, \ldots, x^n) = \varphi' \circ \Psi \circ \varphi^{-1}(x)$$

is the coordinate representation of the map Ψ, Fig. 1-3.

In all our applications the map Ψ will be such that each coordinate representation $\tilde{\Psi}$ is differentiable; in this case we call Ψ itself differentiable (just as we did in the case of functions in $\mathcal{F}$). If $\Psi : M \to N$ is a homeomorphism with both Ψ and Ψ^{-1} differentiable, then we call Ψ a *diffeomorphism*.

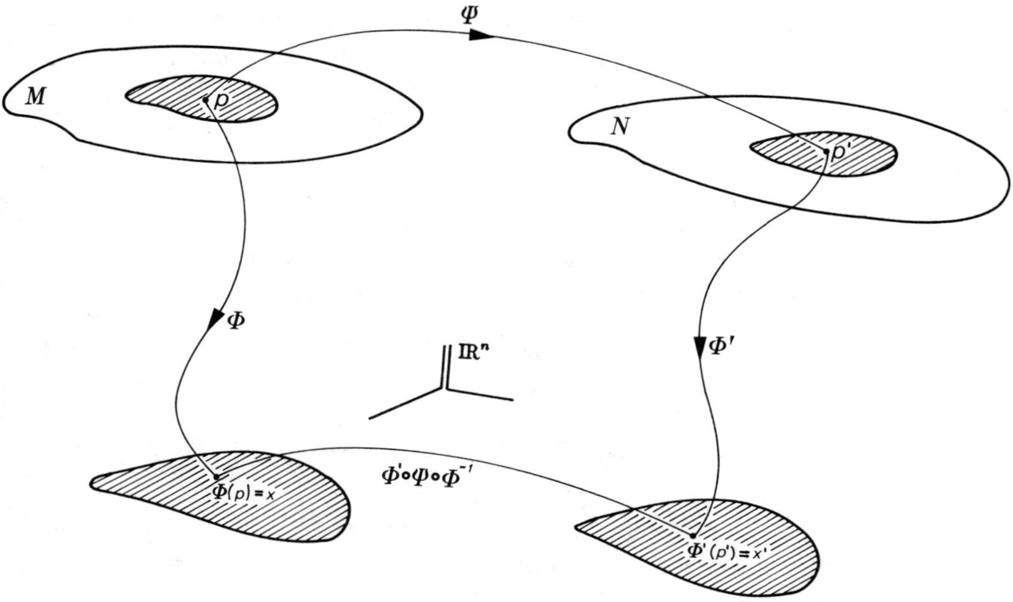

Fig. 1-3 Maps of manifolds and their coordinate representation.

If dim $(N) <$ dim (M), a C^r-map $(r \geq 0)$ $\Phi : N \rightarrow M$ is said to be an *immersion* if it is locally one–to–one; and, if for any $q \in N$, there exists a neighbourhood U of q, such that Φ^{-1}, restricted to $\Phi(U)$, is a C^r-map. The image $\Phi(N)$ is said to be a m-dimensional ($m =$ dim (N)) immersed submanifold of M. The set $\Phi(N)$ is said to be *imbedded* in M if Φ is a homeomorphism of N into its image in M, with the induced topology of M. An imbedded submanifold of M with dimension $m =$ dim $(M) - 1$, is termed a *hypersurface*.

1.6 The tangent space

Given a manifold M, a *curve* γ in M is a map $\gamma : \mathbb{R} \rightarrow M$,

$$t \longmapsto \gamma(t) \tag{1.6.1}$$

with $t \in (a, b)$ (or $t \in [a, b]$), $a, b \in \mathbb{R}$, $\gamma(t) \in M$. The variable t is called the parameter of the curve; if (a, b) is replaced by the closed interval $[a, b]$, the image of γ in M is called a segment.

Unless stated otherwise, γ will be assumed differentiable. The tangent to a curve γ at a point $p = \gamma(t)$ on it is a map:

$$\underset{p}{\dot{\gamma}} : \mathcal{F} \longrightarrow \mathbb{R} \tag{1.6.2}$$

defined as:

$$\underset{p}{\dot{\gamma}}(f) = \frac{d}{dt}\left[f \circ \gamma(t)\right]_{\gamma^{-1}(p)} \qquad f \in \mathcal{F} \tag{1.6.3}$$

where d/dt is the derivative with respect to t. From the properties of a derivative we readily deduce that the tangent is a linear map:

$$\underset{p}{\dot{\gamma}}\,(c_1 f_1 + c_2 f_2) = c_1 \underset{p}{\dot{\gamma}}(f_1) + c_2 \underset{p}{\dot{\gamma}}(f_2) \tag{1.6.4}$$

and satisfies the Leibniz rule:

$$\underset{p}{\dot{\gamma}}(f_1 f_2) = \underset{p}{\dot{\gamma}}(f_1) f_2(p) + f_1(p) \underset{p}{\dot{\gamma}}(f_2) \tag{1.6.5}$$

where $c_1,\ c_2 \in \mathbb{R}$; $f_1,\ f_2 \in \mathcal{F}$.

We define the tangent space at p, written $T_p(M)$, to be the set of all the maps $\mathcal{F} \to \mathbb{R}$ defined in this way for all differentiable curves through p. Of course, there will be many different curves having the same tangent vector: if we define the relation R_p of *having the same tangent vector at* p by:

$$\gamma R_p \gamma' \Longleftrightarrow \underset{p}{\dot{\gamma}}(f) = \underset{p}{\dot{\gamma}'}(f) \qquad \forall f \in \mathcal{F} \tag{1.6.6}$$

then R_p is an equivalence relation on the set of all curves through p and there is an equivalence class of curves corresponding to each tangent vector in p. When we do not want to specify which curve γ defines a tangent vector, we use letters u, v, etc., to indicate general tangent vectors.

The set of tangent vectors at p forms a vector space, in the formal sense, if we define an operation of addition for vectors $u, v \in T_p(M)$ by:

$$(u + v)(f) = u(f) + v(f) \qquad \forall f \in \mathcal{F} \tag{1.6.7}$$

and scalar multiplication of $u \in T_p(M)$ by $a \in \mathbb{R}$, by:

$$(au)(f) = a(u(f)) \ . \tag{1.6.8}$$

The neutral element, called the zero vector and denoted by o, is defined by $o(f) = 0$, satisfying:

$$u + o = u \quad , \quad ao = o \quad ; \quad u \in T_p(M) \ , \ a \in \mathbb{R} \ .$$

To each element $u \in T_p(M)$ corresponds an inverse under addition denoted by $-u$, defined by $(-u)(f) = -(u(f))$, such that:

$$u + (-u) = o \ .$$

The tangent space forms an Abelian group with respect to the operation $+$. We shall now deduce the coordinate representations of the above properties. Let $(U_\alpha, \varphi_\alpha)$ be a coordinate neighbourhood of p, partially containing a curve $\gamma(t)$ through p. We denote the coordinates of a general point on γ by x^i, thus:

$$x^i(t) = ((\varphi_\alpha \circ \gamma(t))^i \qquad i = 1, 2, \ldots, n \ . \tag{1.6.9}$$

These are the parametric equations of the curve $\gamma(t)$ relative to the given chart. From (1.6.9), relation (1.6.3) reads:

$$\begin{aligned} \dot{\underset{p}{\gamma}}(f) &= \frac{d}{dt}\left[f \circ \varphi_\alpha^{-1} \circ \varphi_\alpha \circ \gamma(t)\right]_{\gamma^{-1}(p)} \\ &= \frac{d}{dt}\left[\tilde{f}_\alpha\left(x^i(t), x^2(t), \ldots, x^n(t)\right)\right]_{\gamma^{-1}(p)} \\ &= \dot{\gamma}^i(p)\left(\partial_i \tilde{f}_\alpha\right)_{\varphi^{-1}(p)} \qquad i = 1, 2, \ldots, n \end{aligned} \tag{1.6.10}$$

where

$$\partial_i = \frac{\partial}{\partial x^i} \quad , \quad \dot{\gamma}^i(p) = \left(\frac{dx^i}{dt}\right)_{\gamma^{-1}(p)}$$

and

$$\dot{\gamma}^i \partial_i \equiv \sum_{i=1}^{n} \dot{\gamma}^i \partial_i \ .$$

The last relation is the Einstein summation convention which we shall adopt from now on; according to this, any pair of repeated indices appearing diagonally are meant to be summed over their entire range of values.

From (1.6.10) we see that the action of a tangent vector $\dot{\underset{p}{\gamma}}$ is specified in coordinates by the numbers $(\dot{\gamma}^i(p))_{i=1,2,\ldots,n}$, called the *components* of $\dot{\underset{p}{\gamma}}$ with respect to the given coordinates. Conversely, suppose a coordinate neighbourhood $(U_\alpha, \varphi_\alpha)$ is specified along with a set of numbers $(u^i)_{i=1,2,\ldots,n}$, and suppose p has coordinates p^i. If we define a curve γ by:

$$x^i(t) \equiv (\varphi_\alpha \circ \gamma(t))^i = u^i t + p^i$$

then we clearly have, on differentiating by t, that $\dot{\gamma}^i = dx^i/dt = u^i$. So the u^i are the components of a tangent vector $u = \dot{\gamma}$, whose action on a function f is given from (1.6.10) as:

$$u(f) = u^i \left(\partial_i \tilde{f}_\alpha\right) . \tag{1.6.11}$$

So coordinates set up a one–to–one correspondence between tangent vectors u and n-tuples of numbers u^i. In particular, if we take $u^i = \delta^i_j$, where δ^j_i is the Kronecker delta, so that $(u^i)_{i=1,2,\dots,n} = (0, 0, \dots, 1, \dots, 0)$ (a 1 in the jth position), then the corresponding vector acts on f according to:

$$f \longmapsto \delta^i_j \left(\partial_i \tilde{f}_\alpha\right) = \left(\partial_j \tilde{f}_\alpha\right) .$$

We denote this vector by ∂_j so that we have:

$$\partial_j f = \left(\partial_j \tilde{f}_\alpha\right) . \tag{1.6.12}$$

Hence, (1.6.11) can be written as:

$$u(f) = u^i \partial_i(f) \tag{1.6.13}$$

or

$$u = u^i \partial_i . \tag{1.6.14}$$

1.7 Bases in the tangent space

Recall that,* given n vectors $\{e_i\}$ of $T_p(M)$, they are called *linearly independent* if the relation

$$c^i e_i = o \tag{1.7.1}$$

for any n-tuple of real numbers $\{c^i\}$, implies that $c^i = 0$ for each i. If m (finite) is the greatest number of vectors for which linear independence holds, then m is called the dimension of the vector space. We shall shortly see that the dimension of $T_p(M)$ is equal to n, the dimension (in a different sense) of the manifold. Any collection of n linearly independent vectors $\{e_i\}$ is called a *basis* for $T_p(M)$. Any element u of $T_p(M)$ can be expressed uniquely as

*This argument applies to any vector space.

a linear combination of vectors of the basis; there always exist n numbers $u^i \in \mathbf{R}$, such that

$$u = u^i e_i \ ; \tag{1.7.2}$$

these numbers are termed the components of u at p with respect to the given basis. Conversely a set of vectors at p forms a basis for $T_p(M)$ if any vector in p can be expressed uniquely as a linear combination of them. As a consequence of this, we easily recognize from (1.6.14) that, corresponding to any chart containing p, the set of n vectors $\{\partial_i\}$ forms a basis for $T_p(M)$ called the *coordinate* (or *natural*) basis in p relative to the given chart. This shows that the dimension of $T_p(M)$ is in fact equal to n. With respect to this basis, the (coordinate) components of u are identical to the components defined after (1.6.10), namely:

$$u^i = \left(\frac{dx^i}{dt}\right)_{\gamma^{-1}(p)} \tag{1.7.3}$$

where t is the parameter on a curve which has the vector u as tangent at p.

1.8 Transformation properties of vector components

The tangent space at $p \in M$ admits more than one basis; in fact it admits infinitely many. We shall now discuss how to describe a change of basis. If $\{e_i\}$ and $\{e'_i\}$ are two bases in p, from the properties of a basis, we can write:

$$e'_i = A_i{}^j e_j \tag{1.8.1}$$

where $A_i{}^j$ form an $n \times n$ matrix of real numbers. Conversely the elements of the basis $\{e_i\}$ can be expressed in terms of the basis $\{e'_i\}$ as:

$$e_j = B_j{}^k e'_k \tag{1.8.2}$$

where $B_j{}^k$ form another $n \times n$ matrix of real numbers. Combining (1.8.1) and (1.8.2) we have:

$$e'_i = A_i{}^j B_j{}^k e'_k \tag{1.8.3}$$

which leads to:

$$\left(\delta_i^k - A_i{}^j B_j{}^k\right) e'_k = o \; . \tag{1.8.4}$$

From the properties of a basis and those of matrices, relation (1.8.4) implies that $A_i{}^j$ is a non-singular matrix, and:

$$B_j{}^k = \overset{-1}{A}_j{}^k . \tag{1.8.5}$$

All non-singular matrices induce a basis transformation in $T_p(M)$. Let us show in fact that given a basis $\{e_i\}$ and a non-singular matrix $C_i{}^j$ of real numbers, the set of vectors:

$$z_i = C_i{}^j e_j \tag{1.8.6}$$

is a basis for $T_p(M)$. Let $\{b^i\}$ be an n-tuple of real numbers and let us show that:

$$b^i z_i = o \Rightarrow b^i = 0 \qquad i = 1, 2, \ldots, n \; . \tag{1.8.7}$$

From (1.8.6) we have:

$$b^i C_i{}^j = 0 \; ;$$

multiplying both sides by $\overset{-1}{C}_j{}^k$ gives $b^i \delta_i^k = b^k = 0$, proving that $\{z_i\}$ is linearly independent and hence a basis for $T_p(M)$.

A change of the basis induces a change of the components of a vector. Let

$$e'_i = A_i{}^j e_j \tag{1.8.8}$$

be a general basis transformation at p, and let

$$u = u^i e_i = u'^i e'_i \tag{1.8.9}$$

define the components of a vector u with respect to the two bases. From (1.8.8) and (1.8.9) we easily have:

$$\left(u'^i A_i{}^k - u^k\right) e_k = o \tag{1.8.10}$$

which leads to:

$$u'^i = \overset{-1}{A}{}^i{}_j u^j \; . \tag{1.8.11}$$

This is expressed by saying that the components of a vector change *contravariantly* to the corresponding change of basis; they are also termed *contravariant*.

1.9 The tangent map

Let M and N be two manifolds and let $\Phi: M \to N$ be a map of M into N. The induced vectors in N are given by maps:

$$\Phi_*(u) : \mathcal{F}(N) \longrightarrow \mathbb{R} \qquad u \in T_p(M) \tag{1.9.1}$$

defined by:

$$(\Phi_*(u))(f) \equiv u(f \circ \Phi) \qquad f \in \mathcal{F}(N) . \tag{1.9.2}$$

From this and (1.6.3), we easily prove that

$$\Phi_* (\underset{p}{\dot{\gamma}}) = (\Phi \overset{\cdot}{\circ} \gamma)_{\Phi(p)} ; \tag{1.9.3}$$

in fact, for any $f \in \mathcal{F}(N)$, we have:

$$\begin{aligned}(\Phi_* (\underset{p}{\dot{\gamma}}))(f) &= \underset{p}{\dot{\gamma}}(f \circ \Phi) = \frac{d}{dt}\left[(f \circ \Phi) \circ \gamma(t)\right]_{\gamma^{-1}(p)} \\ &= \frac{d}{dt}\left[f \circ (\Phi \circ \gamma(t))\right]_{(\Phi\circ\gamma)^{-1}\circ\Phi(p)} = (\Phi \overset{\cdot}{\circ} \gamma)_{\Phi(p)}(f)\end{aligned} \tag{1.9.4}$$

which implies (1.9.3).

It is convenient to find the coordinate representation of (1.9.2). Let $(U_\alpha, \varphi_\alpha)$ and $(U'_\alpha, \varphi'_\alpha)$ be two coordinate neighbourhoods of $p \in M$ and $\Phi(p) \in N$ respectively and let

$$\varphi_\alpha(p) = \{x^i\} \quad , \quad \varphi'_\beta \circ \Phi(p) = \left\{x'^i\right\} \tag{1.9.5}$$

be the corresponding local coordinates. From (1.7.3) we then have, for any $u \in T_p(M)$:

$$\begin{aligned}(\Phi_*(u))^i_{\Phi(p)} &\equiv u'^i|_{\Phi(p)} = \left(\frac{dx'^i}{dt}\right)_{\Phi(p)} \\ &= \frac{d}{dt}\left[\left(\varphi'_\beta \circ \Phi\right) \circ \varphi_\alpha^{-1}(x)\right]^i_{\varphi_\alpha^{-1}(x)} \\ &= \left(\frac{\partial x'^i}{\partial x^k}\frac{dx^k}{dt}\right)_p = \left.\frac{\partial x'^i}{\partial x^k}u^k\right|_p .\end{aligned} \tag{1.9.6}$$

In the case where $M = N$ and $\Phi = \mathrm{Id}$, we can regard the transformations from one coordinate to another as a change of basis in

$T_p(M)$. Thus we can write:

$$\partial_i f \equiv \frac{\partial \tilde{f}_\alpha}{\partial x^i} = \frac{\partial}{\partial x^i}\left(f \circ \varphi_\beta^{-1} \circ \varphi_\beta \circ \varphi_\alpha^{-1}\right)$$
$$= \frac{\partial \tilde{f}_\beta}{\partial x'^j}\frac{\partial x'^j}{\partial x^i} \equiv \left(\partial'_j f\right)\left(\frac{\partial x'^j}{\partial x^i}\right) \qquad (1.9.7)$$

where $x'(x) = \varphi_\beta \circ \varphi_\alpha^{-1}$ from (1.3.4), so that:

$$\partial_i = \frac{\partial x'^j}{\partial x^i}\partial'_j \,. \qquad (1.9.8)$$

Hence from (1.8.8) we see that $\partial x'^j/\partial x^i$ are the components of the inverse of the matrix transformation from ∂_i to ∂'_j, and so (1.8.11) immediately gives (1.9.6).

In the case where Φ is a diffeomorphism (see 1.5), the map Φ^{-1} has the coordinate representation $\varphi_\alpha \circ \Phi^{-1} \circ \varphi_\beta = \tilde{\Phi}^{-1}$. Since Φ^{-1} is differentiable, so then is $\tilde{\Phi}^{-1}$ and (as we have just seen in the case $\Phi = Id$) its derivative $\partial x^j/\partial x'^i$ is the inverse matrix to the derivative $\partial x'^j/\partial x^i$ of $\tilde{\Phi}$. Hence this derivative is invertible, and so $\Phi_* : T_p(M) \to T_p(N)$ is invertible. We use this later in Sect. 2.1.

1.10 The cotangent space

Let f be a function of $\mathcal{F}(M)$; we define the differential of f at p to be the map:

$$\underset{p}{\boldsymbol{d}f} : T_p(M) \longrightarrow \mathbb{R} \qquad (1.10.1)$$

such that:

$$\underset{p}{\boldsymbol{d}f}(u) = u(f) \qquad \forall u \in T_p(M). \qquad (1.10.2)$$

This map is a linear map as one easily deduces from definition (1.6.7); in fact:

$$\underset{p}{\boldsymbol{d}f}(c_1 u + c_2 w) = c_1 \underset{p}{\boldsymbol{d}f}(u) + c_2 \underset{p}{\boldsymbol{d}f}(w) \qquad (1.10.3)$$

$c_1, c_2 \in \mathbb{R}$, $u, w \in T_p(M)$. The set of all linear maps from $T_p(M)$ into $\mathbb{R}$, of which $\underset{p}{\boldsymbol{d}f}$ is one, is called the *cotangent* space at p.

It is denoted by $\overset{*}{T}_p(M)$ and its general elements will be termed *covectors* and denoted by bold face letters as $\boldsymbol{u}$, $\boldsymbol{w}$, $\boldsymbol{\omega}$ .... Just as all vectors are tangent to curves, so we shall see below (Sect. 1.11) that all covectors are differentials of functions. As for the tangent space, we shall define an operation of addition defined by:

$$(\boldsymbol{w} + \boldsymbol{v})(u) = \boldsymbol{w}(u) + \boldsymbol{v}(u) \tag{1.10.4}$$

and a scalar multiplication of $\boldsymbol{w} \in \overset{*}{T}_p(M)$ by $a \in \mathbb{R}$ defined by:

$$(a\boldsymbol{w})(u) = a(\boldsymbol{w}(u)) \ . \tag{1.10.5}$$

The neutral element is in this case defined by $\boldsymbol{o}(u) = 0$ satisfying $\boldsymbol{w} + \boldsymbol{o} = \boldsymbol{w}$ and $a\boldsymbol{o} = \boldsymbol{o}$, $\forall u \in T_p(M)$ and $a \in \mathbb{R}$; the inverse element to a given $\boldsymbol{w}$ is defined as $(-\boldsymbol{w})(u) = -\boldsymbol{w}(u)$ and is such that $\boldsymbol{w} + (-\boldsymbol{w}) = \boldsymbol{o}$. With these operations the cotangent space is a linear vector space in the same sense as for the tangent space. (It is in fact the dual of the space $T_p(M)$.)

1.11 Bases in the cotangent space

A basis in the cotangent space is similarly defined as a set of n linearly independent covectors $\{\boldsymbol{e}^i\}$ which permit one to express any other covector in terms of components:

$$\boldsymbol{\omega} = \omega_i \boldsymbol{e}^i \qquad\qquad \omega_i \in \mathbb{R} \ , \ \boldsymbol{\omega} \in \overset{*}{T}_p(M) \ . \tag{1.11.1}$$

Although one can choose an arbitrary basis in $\overset{*}{T}_p(M)$, it is convenient to link its choice uniquely to that of a basis in the tangent space. Let $\{e_i\}$ be such a basis in $T_p(M)$; we shall show that the set of covectors $\{\boldsymbol{e}^i\}$ so defined that:

$$\boldsymbol{e}^i(e_j) = \delta^i_j \qquad , \tag{1.11.2}$$

is a basis for $\overset{*}{T}_p(M)$. From (1.10.2), (1.10.3) and (1.11.2) we have:

$$\boldsymbol{e}^i(u) = u^k \boldsymbol{e}^i(e_k) = u^i \qquad \forall u \in T_p(M). \tag{1.11.3}$$

If $\boldsymbol{\omega}$ is a general covector, we have from (1.10.2) and (1.11.3):

$$\boldsymbol{\omega}(u) = u^k \boldsymbol{\omega}(e_k) = \boldsymbol{\omega}(e_k)\boldsymbol{e}^k(u) \ ; \tag{1.11.4}$$

hence, from the arbitrariness of u we have the unique decomposition:

$$\boldsymbol{\omega} = \boldsymbol{\omega}(e_k)\boldsymbol{e}^k = \omega_k \boldsymbol{e}^k \tag{1.11.5}$$

where $\boldsymbol{\omega}(e_k) = \omega_k$. Since the expression $\boldsymbol{\omega} = \omega_k \boldsymbol{e}^k$ applies to any covector, the set $\{\boldsymbol{e}^i\}$ is a basis for $\overset{*}{T}_p(M)$. This basis is called the *dual* basis of $\{e_i\}$ in $T_p(M)$; in what follows we shall always refer to such a basis in $\overset{*}{T}_p(M)$. Suppose the basis in $T_p(M)$ is the natural basis $\{\partial_i\}$. We note that the differentials of the coordinate functions $\boldsymbol{d}x^i$ satisfy, from the definitions (1.6.12) and (1.10.2):

$$\boldsymbol{d}x^i(\partial_j) = \partial_j(x^i) = \frac{\partial x^i}{\partial x^j} = \delta^i_j \tag{1.11.6}$$

(clearly no distinction between x^i and $\tilde{x}^i_\alpha$ is here necessary if the latter is the ith coordinate in $\mathbb{R}^n$). In other words, the coordinate differentials constitute the dual basis to the natural basis. In the dual basis, the components of the differential of a function are given as follows:

$$(\boldsymbol{d}f)_i = \boldsymbol{d}f(\partial_i) = \partial_i(f) = \frac{\partial \tilde{f}_\alpha}{\partial x^i} \tag{1.11.7}$$

from (1.11.5). Since the $\boldsymbol{d}x^i$ form a basis, we can write any covector $\boldsymbol{\omega}$ as:

$$\boldsymbol{\omega} = \omega_i \boldsymbol{d}x^i \ . \tag{1.11.8}$$

Suppose $\boldsymbol{\omega}$ is given; then we can define a function f by $f(p) = \omega_i x^i(p)$ from which $\tilde{f}_\alpha(x) = \omega_i x^i$, and so from (1.11.7):

$$(\boldsymbol{d}f)_j = \frac{\partial(\omega^i x_i)}{\partial x^j} = \omega_j \ . \tag{1.11.9}$$

Hence, since $\boldsymbol{\omega}$ was arbitrary, any covector can be written as the differential of a function.

The link (duality) of $\{\boldsymbol{e}^i\}$ with $\{e_i\}$ permits us to deduce the transformation law of a dual basis as a consequence of a given basis change in $T_p(M)$. Let then

$$e'_i = A_i{}^j e_j \tag{1.11.10}$$

be a basis transformation, where A is a non-singular matrix. If $\left\{\boldsymbol{e}'^i\right\}$ is the dual of the new basis $\{e'_i\}$, we have from (1.11.2) and

(1.10.3):

$$\boldsymbol{e}'^j(e'_k) = A_k{}^i \boldsymbol{e}'^j(e_i) = \delta^j_k \ . \qquad (1.11.11)$$

Multiplying by $\overset{-1}{A}_l{}^k$:

$$\boldsymbol{e}'^j(e_l) = \overset{-1}{A}_l{}^j = \delta^k_l \overset{-1}{A}_k{}^j = \overset{-1}{A}_k{}^j \boldsymbol{e}^k(e_l) \ .$$

Thus $\boldsymbol{e}'^j$ has the same effect on a $\overset{-1}{A}_k{}^j \boldsymbol{e}^k$ as basis vector and so on all vectors (by linearity); in other words:

$$\boldsymbol{e}'^j = \overset{-1}{A}_k{}^j \boldsymbol{e}^k \ . \qquad (1.11.12)$$

Straightforwardly we deduce, for any covector $\boldsymbol{\omega}$, the transformation law:

$$\omega'_i = A_i{}^j \omega_j \ . \qquad (1.11.13)$$

The components of a covector transform *covariantly* to the law of the transformation of the tangent basis $\{e_i\}$.

1.12 The dual tangent map

We have learned in Sect.1.9 that a map $\Phi\colon M \to N$, where M and N are two manifolds, induces a map $\Phi_*\colon T_p(M) \to T_{\Phi(p)}(N)$, between the tangent spaces. We can now define a map:

$$\Phi^* : \overset{*}{T}_{\Phi(p)}(N) \longrightarrow \overset{*}{T}_p(M) \qquad (1.12.1)$$

by

$$(\Phi^*(\boldsymbol{\omega}))\,(u) = \boldsymbol{\omega}\,(\Phi_*(u)) \qquad \forall u \in T_p(M) \ . \qquad (1.12.2)$$

If we apply this to the differential $\boldsymbol{d}f$ we obtain:

$$\begin{aligned}(\Phi^*(\boldsymbol{d}f))\,(u) &= (\boldsymbol{d}f)\,(\Phi_*(u)) \\ &= (\Phi_*(u))\,(f) = u(f \circ \Phi) = \boldsymbol{d}(f \circ \Phi)(u) \qquad (1.12.3)\end{aligned}$$

which implies:

$$\Phi^*(\boldsymbol{d}f) = \boldsymbol{d}(f \circ \Phi) \ . \qquad (1.12.4)$$

If we choose charts $(U_\alpha, \varphi_\alpha)_p$ and $\left(U'_\beta, \psi_\beta\right)_{\Phi(p)}$ for neighbourhoods of p and $\Phi(p)$ respectively, then we can write (1.12.4) in component form, using (1.11.7) as:

$$\begin{aligned}(\Phi^*(\boldsymbol{d}f))_i &= \frac{\partial}{\partial x^i}(f \tilde{\circ} \Phi)_\alpha = \frac{\partial}{\partial x^i}(f \circ \Phi \circ \varphi_\alpha^{-1}) \\ &= \frac{\partial}{\partial x^i}(f \circ \psi_\beta^{-1} \circ \psi_\beta \circ \Phi \circ \varphi_\alpha^{-1}) = \frac{\partial}{\partial x^i}\left(\tilde{f}_\beta \circ \tilde{\Phi}\right) \quad (1.12.5)\end{aligned}$$

where $\tilde{\Phi} = \psi_\beta \circ \Phi \circ \varphi_\alpha^{-1}$. Writing the coordinates in U'_β as $\left\{x'^i\right\}$ and those in U_α as $\{x^k\}$, so that

$$\tilde{\Phi}^i = x'^i\left(\{x^k\}\right) \tag{1.12.6}$$

(1.12.5) becomes:

$$(\Phi^*(\boldsymbol{d}f))_i = \frac{\partial \tilde{f}_\beta}{\partial x'^j}\frac{\partial x'^j}{\partial x^i} = (\boldsymbol{d}f')_j \frac{\partial x'^j}{\partial x^i} \quad . \tag{1.12.7}$$

In the case where Φ = Id, this gives the relation between the coordinate components of $\boldsymbol{d}f$ in two different coordinate systems:

$$(\boldsymbol{d}f)_i = (\boldsymbol{d}f')_j \frac{\partial x'^j}{\partial x^i}. \tag{1.12.8}$$

Since any covector can be expressed as a differential, this equation holds for any covector. We could also, of course, derive this transformation law from the law (1.9.8) for the transformation of the coordinate basis, where the transformation matrix is the inverse of the matrix of the $(\partial x'^i/\partial x^j)$. Since

$$\frac{\partial x'^j}{\partial x^k}\frac{\partial x^k}{\partial x'^l} = \delta^j_l \quad ,$$

we see that the inverse matrix to $(\partial x'^i/\partial x^j)$ is just $(\partial x^j/\partial x'^l)$ and so (1.11.13) gives:

$$\omega'_i = \frac{\partial x^j}{\partial x'^i}\omega_j \ , \tag{1.12.9}$$

for the change in components of a covector under change of coordinates, agreeing with (1.12.8) above.

1.13 Tensors

An important vector space which can be constructed from the Cartesian product of a collection of vector spaces is the *tensor* *product.* Consider for instance the space $T_p \times T_p$ and the bilinear map which associates to any pair (u, v) of tangent vectors at p a new operator $u \otimes v$ which maps $\overset{*}{T}_p \times \overset{*}{T}_p$ into $\mathbb{R}$ as follows:*

$$(u \otimes v)(\boldsymbol{\omega}, \boldsymbol{\sigma}) = \boldsymbol{\omega}(u)\boldsymbol{\sigma}(v) \qquad \boldsymbol{\omega},\ \boldsymbol{\sigma} \in \overset{*}{T}_p(M)\ . \tag{1.13.1}$$

The space of bilinear maps $\boldsymbol{T}: \overset{*}{T}_p \times \overset{*}{T}_p \to \mathbb{R}$, endowed with an internal operation (addition +) and an external operation (multiplication by a number) defined by:

$$\begin{aligned}(\boldsymbol{T} + \boldsymbol{T}')\,(\boldsymbol{\omega}, \boldsymbol{\sigma}) &= \boldsymbol{T}(\boldsymbol{\omega}, \boldsymbol{\sigma}) + \boldsymbol{T}'(\boldsymbol{\omega}, \boldsymbol{\sigma}) \\ (a\boldsymbol{T})(\boldsymbol{\omega}, \boldsymbol{\sigma}) &= a\,(\boldsymbol{T}(\boldsymbol{\omega}, \boldsymbol{\sigma})) \qquad a \in \mathbb{R}\end{aligned} \tag{1.13.2}$$

is a linear vector space called the *tensor product* of T_p by itself and written as $\otimes^2 T_p(M)$; its elements are called twice-contravariant tensors. Now the bilinear maps of the type $u \otimes v$ as in (1.13.1) form a subset of $\otimes^2 T_p$ and have the properties:

$$\begin{aligned} u \otimes (v_1 + v_2) &= u \otimes v_1 + u \otimes v_2 \\ (u_1 + u_2) \otimes v &= u_1 \otimes v + u_2 \otimes v \\ u \otimes (av) &= a(u \otimes v) \\ (au) \otimes v &= a(u \otimes v)\ . \end{aligned} \tag{1.13.3}$$

They generate, however, the entire $\otimes^2 T_p(M)$ in the sense that every element of $\otimes^2 T_p(M)$ can be written as a sum of elements of the form $u \otimes v$. To see this explicitly let us have a basis $\{e_i\}$ in $T_p(M)$ and any element $\boldsymbol{T}$ of $\otimes^2 T_p(M)$. Then if $\{\boldsymbol{e}^i\}$ is the dual basis in $\overset{*}{T}_p(M)$ we have:

$$\boldsymbol{T}(\boldsymbol{\omega}, \boldsymbol{\sigma}) = \boldsymbol{T}(\omega_i \boldsymbol{e}^i, \sigma_j \boldsymbol{e}^j) = \omega_i \sigma_j \boldsymbol{T}(\boldsymbol{e}^i, \boldsymbol{e}^j) = \omega_i \sigma_j T^{ij} \tag{1.13.4}$$

using linearity and writing $T^{ij} = \boldsymbol{T}(\boldsymbol{e}^i, \boldsymbol{e}^j)$. From (1.11.5) and (1.13.1), this becomes:

$$\boldsymbol{T}(\boldsymbol{\omega}, \boldsymbol{\sigma}) = T^{ij} \boldsymbol{\omega}(e_i) \boldsymbol{\sigma}(e_j) = T^{ij} e_i \otimes e_j(\boldsymbol{\omega}, \boldsymbol{\sigma}) \tag{1.13.5}$$

---

*Here bilinear means that $(u \otimes v)(\omega, \sigma)$ is linear in both ω and σ, i.e.

$$\begin{aligned}(u \otimes v)(\omega' + \omega'', \sigma) &= (u \otimes v)(\omega', \sigma) + (u \otimes v)(\omega'', \sigma) \\ (u \otimes v)(a\omega, b\sigma) &= ab(u \otimes v)(\omega, \sigma)\ .\end{aligned}$$

hence:

$$T = T^{ij} e_i \otimes e_j \ . \qquad (1.13.6)$$

It is easy to show that the n^2 elements of $\otimes^2 T_p(M)$, that is $\left\{e_i \otimes e_j\right\}$, are linearly independent and therefore form a basis for $\otimes^2 T_p(M)$, proving that every tensor can be written as a linear combination of tensor products. The coefficients T^{ij} in (1.13.6) are the components of the tensor with respect to the given basis and possess a well-defined transformation property under a change of basis. Let:

$$e'_i = A_i{}^j e_j \qquad (1.13.7)$$

be such a basis transformation in $T_p(M)$; from (1.13.3) we easily deduce:

$$e'_i \otimes e'_j = A_i{}^k A_j{}^r e_k \otimes e_r \qquad k, r = 1, 2, \ldots n \qquad (1.13.8)$$

which leads, from (1.13.6), to:

$$T'^{ij} = \overset{-1}{A}{}^i{}_k \overset{-1}{A}{}^j{}_r T^{kr} \ . \qquad (1.13.9)$$

The above considerations are easily generalized to the tensor product of $T_p(M)$ by itself α-times:

$$\otimes^\alpha T_p(M) = T_p \underset{1}{\otimes} T_p \underset{2}{\otimes} \ldots \underset{\alpha}{\otimes} T_p \qquad (1.13.10)$$

defined as α-linear maps of $\overset{*}{T}_p \times \overset{*}{T}_p \times \ldots \times \overset{*}{T}_p$ (α-times) into $\mathbb{R}$. The general element of this space is a tensor T with components:

$$T = T^{i_1 i_2 \ldots i_\alpha} e_{i_1} \otimes e_{i_2} \otimes \ldots \otimes e_{i_\alpha} \qquad (1.13.11)$$

whose transformation properties are given by:

$$T'^{i_1 i_2 \ldots i_\alpha} = \overset{-1}{A}{}^{i_1}{}_{j_1} \overset{-1}{A}{}^{i_2}{}_{j_2} \cdots \overset{-1}{A}{}^{i_\alpha}{}_{j_\alpha} T^{j_1 j_2 \ldots j_\alpha} \ . \qquad (1.13.12)$$

The tensor T is said to be α-times contravariant.

What has been said before applies similarly to the cotangent space. Consider the bilinear map which associates to any pair of covectors $(\boldsymbol{\omega}, \boldsymbol{\sigma})$ a new operator, written as $\boldsymbol{\omega} \otimes \boldsymbol{\sigma}$ which maps $T_p \times T_p$ into $\mathbb{R}$ as follows:

$$(\boldsymbol{\omega} \otimes \boldsymbol{\sigma})(u, v) = \boldsymbol{\omega}(u) \boldsymbol{\sigma}(v) \qquad u, v \in T_p(M) \ . \qquad (1.13.13)$$

This map is bilinear in that:

$$
\begin{aligned}
(a\boldsymbol{\omega} \otimes \boldsymbol{\sigma}) &= a(\boldsymbol{\omega} \otimes \boldsymbol{\sigma}) \\
(\boldsymbol{\omega} \otimes a\boldsymbol{\sigma}) &= a(\boldsymbol{\omega} \otimes \boldsymbol{\sigma}) \\
(\boldsymbol{\omega} + \boldsymbol{\sigma}) \otimes \boldsymbol{\pi} &= \boldsymbol{\omega} \otimes \boldsymbol{\pi} + \boldsymbol{\sigma} \otimes \boldsymbol{\pi} \\
\boldsymbol{\omega} \otimes (\boldsymbol{\pi} + \boldsymbol{\rho}) &= \boldsymbol{\omega} \otimes \boldsymbol{\pi} + \boldsymbol{\omega} \otimes \boldsymbol{\rho} \,.
\end{aligned}
\tag{1.13.14}
$$

The space of such bilinear maps endowed with operations as in (1.13.2) is a vector space called the tensor product of $\overset{*}{T}_p$ by itself: $\otimes^2 \overset{*}{T}_p(M)$; its elements are called twice-covariant tensors. Also in this case the tensor product can be extended to any collection of cotangent spaces and in general we have:

$$\otimes^\beta \overset{*}{T}_p(M) = \overset{*}{T}_p \otimes \overset{*}{T}_p \otimes \ldots \otimes \overset{*}{T}_p \,. \tag{1.13.15}$$

As before we can prove that the general element of this space can be written as:

$$\boldsymbol{R} = R_{i_1 i_2 \ldots i_\beta} \boldsymbol{e}^{i_1} \otimes \boldsymbol{e}^{i_2} \otimes \ldots \otimes \boldsymbol{e}^{i_\beta} \tag{1.13.16}$$

where $\{\boldsymbol{e}^i\}$ is a basis in $\overset{*}{T}_p(M)$. From (1.13.7) and (1.11.12), the transformation properties of the components of $\boldsymbol{R}$ now read:

$$R'_{i_1 i_2 \ldots i_\beta} = A_{i_1}{}^{j_1} A_{i_2}{}^{j_2} \ldots A_{i_\beta}{}^{j_\beta} R_{j_1 j_2 \ldots j_\beta} \,. \tag{1.13.17}$$

The tensor $\boldsymbol{R} \in \otimes^\beta \overset{*}{T}_p(M)$ is said to be β-times covariant.

The spaces (1.13.10) and (1.13.15) can evidently be considered as particular examples of a more general type of tensor product space, given by:

$$\otimes^\alpha T_p(M) \otimes^\beta \overset{*}{T}_p(M) \tag{1.13.18}$$

consisting of $(\alpha + \beta)$-linear maps from

$$\left(\overset{*}{T}_p(M)\right)^\alpha \times \left(T_p(M)\right)^\beta$$

into $\mathbb{R}$, whose general element $\boldsymbol{S}$ is:

$$\boldsymbol{S} = S^{i_1 i_2 \ldots i_\alpha}{}_{j_1 j_2 \ldots j_\beta} e_{i_1} \otimes \ldots \otimes e_{i_\alpha} \otimes \boldsymbol{e}^{j_1} \otimes \ldots \otimes \boldsymbol{e}^{j_\beta}. \tag{1.13.19}$$

A change of basis in $T_p(M)$ induces as before a transformation of the tensor components as:

$$S'^{i_1 i_2 \dots i_\alpha}{}_{j_1 j_2 \dots j_\beta} = \overset{-1}{A}{}^{i_1}{}_{k_1} \dots \overset{-1}{A}{}^{i_\alpha}{}_{k_\alpha} A_{j_1}{}^{r_1} \dots A_{j_\beta}{}^{r_\beta} S^{k_1 \dots k_\alpha}{}_{r_1 \dots r_\beta} \,. \tag{1.13.20}$$

A tensor $\boldsymbol{S} \in \otimes^\alpha T_p \otimes^\beta \overset{*}{T}_p$ is called a mixed tensor of valence (α, β) (α-times contravariant and β-times covariant); the number $\alpha + \beta$ is called the *rank* of the tensor. This result has a most important converse: namely, if we are given a set of numbers $S^{i_1 \dots i_\alpha}{}_{j_1 \dots j_\beta}$ for each choice of basis, and if the sets are related by equation (1.13.20), then there exists a tensor $\boldsymbol{S}$ having these numbers as its components. To prove this one simply defines the action of $\boldsymbol{S}$ on any ordered set of α covectors and β vectors $(\boldsymbol{\omega}_1, \boldsymbol{\omega}_2, \dots, \boldsymbol{\omega}_\alpha, u_1 u_2 \dots, u_\beta)$ by:

$$\begin{aligned} &\boldsymbol{S}(\boldsymbol{\omega}_1, \boldsymbol{\omega}_2, \dots, \boldsymbol{\omega}_\alpha, \ u_1, u_2 \dots, u_\beta) \\ &\qquad = S^{i_1 i_2 \dots i_\alpha}{}_{j_1 j_2 \dots j_\beta} \underset{1}{\omega}{}_{i_1} \underset{2}{\omega}{}_{i_2} \dots \underset{\alpha}{\omega}{}_{i_\alpha} \underset{1}{u}{}^{j_1} \underset{2}{u}{}^{j_2} \dots \underset{\beta}{u}{}^{j_\beta} \,. \end{aligned} \tag{1.13.21}$$

The formula (1.13.20) then ensures that this is independent of the choice of basis.

The map $(u, v) \mapsto u \otimes v$ introduced earlier can be extended to define a product, called the tensor product, between any two tensors. Explicity, if $\boldsymbol{S}$ is a tensor of valence (r, s) and $\boldsymbol{S}'$ a tensor of valence (r', s'), we define a tensor $\boldsymbol{S} \otimes \boldsymbol{S}'$ of valence $(r + r', s + s')$ by the formula

$$\begin{aligned} &(\boldsymbol{S} \otimes \boldsymbol{S}') (\underset{1}{\boldsymbol{\omega}}, \dots, \underset{r+r'}{\boldsymbol{\omega}}, \underset{1}{u}, \dots, \underset{s+s'}{u}) \\ &\qquad = \boldsymbol{S} (\underset{1}{\boldsymbol{\omega}}, \dots, \underset{r}{\boldsymbol{\omega}}, \underset{1}{u}, \dots, \underset{s}{u}) \, \boldsymbol{S}' (\underset{r+1}{\boldsymbol{\omega}}, \dots, \underset{r+r'}{\boldsymbol{\omega}}, \underset{s+1}{u}, \dots, \underset{s+s'}{u}) \end{aligned} \tag{1.13.22}$$

or in components:

$$(\boldsymbol{S} \otimes \boldsymbol{S}')_{i_1 \dots i_{r+r'}}{}^{j_1 \dots j_{s+s'}} = S_{i_1 \dots i_r}{}^{j_1 \dots j_s} S'_{i_1 \dots i_{r+r'}}{}^{j_1 \dots j_{s+s'}} \,. \tag{1.13.23}$$

Tangent vectors can be regarded as tensors of valence (1,0) while the cotangents are tensors of valence (0,1).

We now introduce a process called *contraction* which transforms a tensor of valence (r, s) into one of valence $(r - 1, s - 1)$. Let $1 \leq p \leq r$, $1 \leq q \leq s$. Then if $\boldsymbol{S}$ is a tensor of valence (r, s), we define the contraction of $\boldsymbol{S}$ with respect to the pth contravariant and q'th covariant index as the tensor $\tilde{\boldsymbol{S}}$ of valence $(r - 1, s - 1)$

whose components are:

$$\tilde{S}^{i_1 \ldots i_{r-1}}{}_{j_1 \ldots j_{s-1}} = S^{i_1 \ldots i_{p-1} i i_{p+1} \ldots i_r}{}_{j_1 \ldots j_{q-1} i j_{q+1} \ldots j_s} \quad . \tag{1.13.24}$$

For this definition to make sense we need to know that there actually is a tensor whose components are given by this formula, and that this tensor does not depend on the choice of a basis (since the above formula uses components and so refers to a basis). Now the formula does define a tensor for it is easily seen that we can define $\tilde{S}$ equivalently by:

$$\tilde{S}\,(\underset{1}{\omega}, \ldots, \underset{r-1}{\omega}, \underset{1}{u}, \ldots, \underset{s-1}{u}) = \sum_i S\,(\underset{1}{\omega} \ldots e^i \ldots \underset{r-1}{\omega}, \underset{1}{u} \ldots e_i \ldots \underset{s-1}{u}) \;. \tag{1.13.25}$$

To see that it does not depend on the choice of basis, we replace e_i by $A_i{}^j e_j$ and e^i by $\overset{-1}{A}{}_k{}^i e^k$ and calculate:

$$\begin{aligned} \sum_k S\left(\ldots, \overset{-1}{A}{}_i{}^k e^i, \ldots, A_k{}^s e_s, \ldots\right) &= \overset{-1}{A}{}_i{}^k A_k{}^s S(\ldots, e^i, \ldots, e_s, \ldots) \\ &= S(\ldots, e^i, \ldots, e_i, \ldots) \;; \end{aligned} \tag{1.13.26}$$

so the expression is unchanged if we change the basis. We can apply contraction repeatedly to reduce the valence of a tensor.

1.14 Symmetry operations on tensors

The space of mixed tensors can be even more general than (1.13.18), for instance as :

$$\otimes^\alpha T_p \otimes^\beta \overset{*}{T}_p \otimes^\gamma T_p \otimes^\delta \overset{*}{T}_p \quad . \tag{1.14.1}$$

Elements of this space can be regarded as tensors of valence $(\alpha + \gamma,\ \beta + \delta)$ but their components should bear indices in the right order as shown by the sequence (1.14.1) itself:

$$S = S^{i_1 \ldots i_\alpha}{}_{j_1 \ldots j_\beta}{}^{l_1 \ldots l_\gamma}{}_{m_1 \ldots m_\delta} e_{i_1} \otimes \ldots . \otimes e_{i_\alpha} \otimes \ldots . \tag{1.14.2}$$

In what follows it is convenient to define properties of tensors by referring to the corresponding properties of the components. These properties will always be independent of what basis (natural or not) is used, and the reader should have no difficulty in translating them into coordinate independent form by reexpressing the equations as statements about the effects of tensors as multilinear maps.

A tensor of valence $(0, \beta)$ is *symmetric* with respect to a pair of indices (i, j) when :

$$R_{\ldots i \ldots j \ldots} = R_{\ldots j \ldots i \ldots} \tag{1.14.3}$$

it is *antisymmetric* with respect to them, if:

$$R_{\ldots i \ldots j \ldots} = -R_{\ldots j \ldots i \ldots} . \tag{1.14.4}$$

These symmetries are invariant (i.e. coordinate independent) properties of a tensor. Let $\boldsymbol{R}$ be a tensor of valence $(0, 2)$; we call the *symmetric part* of $\boldsymbol{R}$, the tensor with components:

$$R_{(ij)} = \frac{1}{2}(R_{ij} + R_{ji}) \tag{1.14.5}$$

and the *antisymmetric part* of $\boldsymbol{R}$, the tensor with components:

$$R_{[ij]} = \frac{1}{2}(R_{ij} - R_{ji}) \quad . \tag{1.14.6}$$

From (1.14.5) and (1.14.6) it follows identically that:

$$R_{ij} = R_{(ij)} + R_{[ij]} \quad . \tag{1.14.7}$$

The equation $R_{ij} = 0$ implies the independent relations:

$$R_{(ij)} = 0 \qquad R_{[ij]} = 0 \quad . \tag{1.14.8}$$

Conversely if $R_{(ij)} = 0$ the tensor R_{ij} is antisymmetric and if $R_{[ij]} = 0$ the tensor is symmetric. The above considerations from (1.14.3) to (1.14.7) and those which follow, apply similarly to contravariant tensors.

Definitions (1.14.5) and (1.14.6) extend to tensors of arbitrary valence $(0, \beta)$ or $(\alpha, 0)$. We define the *complete symmetrization* of $\boldsymbol{R}$ to be the tensor with components:

$$R_{(i_1 i_2 \ldots i_\beta)} = \frac{1}{\beta!} \sum_{\rho \in \sigma_\beta} R_{i_{\rho(1)} \ldots i_{\rho(\beta)}} \tag{1.14.9}$$

where σ_β is the set of all permutations of the integers $1, 2, \ldots, \beta$; we call the *complete antisymmetrization* of $\boldsymbol{R}$ the tensor with components:

$$R_{[i_1 \dots i_\beta]} = \frac{1}{\beta!} \left[\sum_{\rho \in \sigma_{\beta_+}} R_{i_{\rho(1)} \dots i_{\rho(\beta)}} - \sum_{\rho \in \sigma_{\beta_-}} R_{i_{\rho(1)} \dots i_{\rho(\beta)}} \right] \qquad (1.14.10)$$

where σ_{β_+} is the set of all even permutations of the integers $1, 2, \dots \beta$ and σ_{β_-} the set of all odd permutations.

Let us calculate (1.14.10) when $\beta = 3$ and $\beta = 4$. In the former case we have the following equivalent expressions:

$$\begin{aligned} R_{[ijk]} &= \frac{1}{6} \left(R_{ijk} + R_{kij} + R_{jki} - R_{ikj} - R_{kji} - R_{jik} \right) \\ &= \frac{1}{3} \left(R_{[ij]k} + R_{[ki]j} + R_{[jk]i} \right) \\ &= \frac{1}{3} \left(R_{i[jk]} + R_{k[ij]} + R_{j[ki]} \right) \ ; \end{aligned} \qquad (1.14.11)$$

in the latter case we have:

$$R_{[ijkl]} = \frac{1}{24} \left(R_{i[jkl]} - R_{j[ikl]} + R_{k[ijl]} - R_{l[ijk]} \right) \quad . \qquad (1.14.12)$$

Over a given group of indices, the symmetrization and antisymmetrization brackets can operate simultaneously; in this case they should be worked out independently as in the following examples:

$$R_{[ij](kl)} = \frac{1}{2} \left(R_{[ij]kl} + R_{[ij]lk} \right) = \text{then acting on } [ij] \qquad (1.14.13)$$

$$R_{i(j[kl])} = \frac{1}{2} \left(R_{i(jkl)} - R_{i(jlk)} \right) = \text{ then acting on the } (\dots) \qquad (1.14.14)$$

The brackets operations act on all the indices they contain; however a group of indices which is written between vertical bars as $|\dots|$, is not acted upon by the brackets operation; for example

$$R_{(i|jk|l)} = \frac{1}{2} \left(R_{ijkl} + R_{ljki} \right) \ . \qquad (1.14.15)$$

1.15 The metric tensor

A tensor of valence (0,2) at p, $\overset{*}{\boldsymbol{g}} \in \otimes^2 \overset{*}{T}_p(M)$, is called a *metric tensor* if it is symmetric:

$$\overset{*}{\boldsymbol{g}} = g_{ij} \boldsymbol{e}^i \otimes \boldsymbol{e}^j, \qquad g_{ij} = g_{(ij)} \qquad (1.15.1)$$

and the mapping $\overset{*}{g}: T_p \times T_p \to \mathbb{R}$ is non-degenerate, that is:

$$\overset{*}{g}(u, v) = 0 \qquad \forall v \in T_p(M) \tag{1.15.2}$$

implies $u = o$. The result of this mapping is called the *scalar product* of u and v and is usually denoted as $\overset{*}{g}(u, v)$ or simply $(u|v)$ if a metric $\overset{*}{g}$ is chosen and fixed. If $\{e_i\}$ is a basis in $T_p(M)$, then from (1.13.13) and (1.11.2) we have

$$(e_i|e_j) = g_{ij} \,. \tag{1.15.3}$$

In terms of components the scalar product of any two vectors may conveniently be written as:

$$(u|v) = g_{ij} u^i v^j \tag{1.15.4}$$

where $u = u^i e_i$ and $v = v^j e_j$.

We define the modulus of a vector by:

$$\|u\| = |(u|u)|^{\frac{1}{2}}. \tag{1.15.5}$$

The vector u is called *null* if $\|u\| = 0$. Any two vectors $u \neq o$ and $v \neq o$ are termed *orthogonal* to each other if:

$$(u|v) = 0 \,. \tag{1.15.6}$$

A collection of n non-null pairwise orthogonal vectors $\{u_\alpha\}$ forms a basis in $T_p(M)$. In fact writing

$$(u_a|u_b) = \varepsilon_{ab} \tag{1.15.7}$$

where

$$\begin{aligned} \varepsilon_{ab} &= 0 \qquad a \neq b \\ \varepsilon_{ab} &\neq 0 \qquad a = b \qquad ; \end{aligned}$$

any linear combination

$$c^a u_a = o \qquad c^a \in \mathbb{R} \tag{1.15.8}$$

implies $c^a = 0$, for taking the scalar product of (1.15.8) by u_b we have:

$$c^b (u_a|u_b) = c^b \varepsilon_{ab} = 0 \tag{1.15.9}$$

which implies $c^a = 0$. We note that the metric tensor $\overset{*}{g}$ is non-degenerate if and only if the set $\{g_{ij}\}$ forms a non-singular matrix. If $\overset{*}{g}$ is non-degenerate, the condition $(u|v) = 0$ for any $v \in T_p(M)$ implies by definition that the system of linear and homogeneous equations:

$$u^i g_{ij} = 0 \tag{1.15.10}$$

admits as its only solution the trivial $u^i = 0$, hence $\mathrm{Det}(g_{ij}) \equiv g \neq 0$. Conversely if the latter is true, the non-degeneracy of $\overset{*}{g}$ follows immediately.

Let us now show that the elements of the inverse matrix to g_{ij}, say g^{ij}, are components of a tensor of valence (2,0). To this purpose it suffices (by the remarks following (1.13.20)) to demonstrate that by a change of basis in $T_p(M)$, the quantities change contravariantly as the components of a tensor of valence (2,0). The matrix properties ensure that:

$$g^{ij} g_{jk} = \delta^i_k = g'^{ir} g'_{rk} \tag{1.15.11}$$

where g'_{ij} refer to the new basis in $\otimes^2 \overset{*}{T}_p(M)$ induced by a basis change in $T_p(M)$ and g'^{ij} are the elements of the inverse matrix to g'_{ij}. Recalling (1.13.17), the relation:

$$g'_{ij} = A_i{}^r A_j{}^s g_{rs} \tag{1.15.12}$$

can be written, in matrix form, as:

$$g' = A g A^{\mathrm{T}} \tag{1.15.13}$$

where A^{T} denotes the transpose of A. Inverting (1.15.13), this gives:

$$\overset{-1}{g'} = (\overset{-1}{A})^{\mathrm{T}} \overset{-1}{g} \overset{-1}{A} \tag{1.15.14}$$

which, written in components , becomes

$$g'^{ij} = A^i{}_r A^j{}_s g^{rs} \ . \tag{1.15.15}$$

This proves the tensorial character of the set g^{ij} which will be referred to as the contravariant metric tensor.

We are now in a position to define isomorphisms* of $\overset{*}{T}_p \to T_p$ and $T_p \to \overset{*}{T}_p$ by:

$$\boldsymbol{\omega} \longmapsto \omega \quad : \quad \omega^i = g^{ij}\omega_j \tag{1.15.16}$$

$$u \longmapsto \boldsymbol{u} \quad : \quad u_j = g_{ik}u^k . \tag{1.15.17}$$

Note that $g^{ij}\omega_j$ are the components of a vector because it is formed by first taking $g \otimes \boldsymbol{\omega}$, with components $g^{ij}\omega_k$, followed by contraction on the j and k indices. The same holds for $g_{ij}u^i$. When we are working with components we denote the components of ω by ω^i and those of $\boldsymbol{u}$ by u_i. The position of the indices distinguishes ω^i from the components of the cotangent vector ω_i, and similarly for u_i. When we are not using components, the pairs $\boldsymbol{\omega}, \omega$ and $u, \boldsymbol{u}$ are distinguished by the type face.

The operations in (1.15.17) are also called raising and lowering of indices. From (1.15.17) the scalar product (1.15.4) defined on $T_p(M)$ extends to $\overset{*}{T}_p(M)$ as:

$$(\boldsymbol{u}|\boldsymbol{v}) = (u|v) \; . \tag{1.15.18}$$

In terms of components the latter can be written:

$$g^{ij}u_i v_j = g_{ij}u^i v^j \quad . \tag{1.15.19}$$

Other convenient forms for the scalar product are:

$$(u|v) = u_i v^i = u^i v_i \quad . \tag{1.15.20}$$

Given a metric tensor $\overset{*}{\boldsymbol{g}}$ at p, since (g_{ij}) is a non-singular symmetric matrix, it is a well-known result of linear algebra that there exists a non-singular matrix $(A_i{}^j)$ such that:

$$\eta_{im} = A_i{}^j g_{jk} \overset{-1}{A}_m{}^k \tag{1.15.21}$$

*Given two groups G and G', a map $\Phi : G \mapsto G'$ is an isomorphism if it is bijective and preserves the group operation; that is, if for any two elements g and h of G, $\Phi(gh) = \Phi(g)\Phi(h)$. In general a map between groups (not necessarily a bijective map) which preserves the group operation is said to be homomorphism (Geroch 1985).

is a diagonal matrix of the form $\text{Diag}\left\{\underbrace{-1\ldots-1}_{s},\underbrace{+1\ldots+1}_{r}\right\}$ for some r and s (independent of the choice of A) with $r+s=n$. The number $\tau=r-s$ is the *signature* of the metric.

Let us now consider at $p\in M$ a set of n vectors $\{\lambda_a\}$; we call them orthonormal (with respect to $\overset{*}{g}$), if

$$(\lambda_a|\lambda_b)=\eta_{ab}\;. \tag{1.15.22}$$

Let $(U_\alpha,\varphi_\alpha)$ be a coordinate neighbourhood of p and write $\vartheta_a{}^i$ for the components of λ_a with respect to the corresponding natural basis in p:

$$\lambda_a=\vartheta_a{}^i\partial_i \qquad \vartheta_a{}^i\in\mathbf{R}\;. \tag{1.15.23}$$

Since the set $\{\lambda_a\}$ forms a basis in $T_p(M)$, the matrix $[\vartheta_a{}^i]$ is non-singular; under the coordinate transformation

$$\bar{x}^a=\bar{x}^a(x^i) \tag{1.15.24}$$

chosen so that:

$$\left(\frac{\partial\bar{x}^a}{\partial x^k}\right)_p=\overset{-1}{\vartheta}{}^a{}_k\quad, \tag{1.15.25}$$

the basis $\{\lambda_a\}_p$ becomes the natural basis at p relative to the new coordinates. In fact:

$$\lambda_a=\left(\frac{\partial}{\partial\bar{x}^a}\right)_p=\vartheta_a{}^i\partial_i\quad; \tag{1.15.26}$$

and as a consequence of this, the metric tensor components become at p:

$$\left(\frac{\partial}{\partial\bar{x}^a}\bigg|\frac{\partial}{\partial\bar{x}^b}\right)_p=\eta_{ab}\;. \tag{1.15.27}$$

A coordinate transformation of the form (1.15.24), which makes the corresponding natural basis orthonormal in p, always exists. We shall see later that the metric $\overset{*}{g}_p$ at p is usually given by a *metric field* , which assigns a metric $\overset{*}{g}_p$ to each different point p.

A manifold which has the above properties is called Riemannian.* Consequently if we carry out the coordinate transformations which make (1.15.27) hold at p, then at a different point p' we shall have in the $\bar{x}$-coordinates:

$$\begin{aligned}\bar{g}_{ab}(p') &= \left(\frac{\partial}{\partial \bar{x}^a}\middle|\frac{\partial}{\partial \bar{x}^b}\right)_{p'} \\ &= \left(\frac{\partial x^c}{\partial \bar{x}^a}\right)_{p'}\left(\frac{\partial x^d}{\partial \bar{x}^b}\right)_{p'}\left(\frac{\partial}{\partial x^c}\middle|\frac{\partial}{\partial x^d}\right)_{p'} \\ &= \left(\frac{\partial x^c}{\partial \bar{x}^a}\right)_{p'}\left(\frac{\partial x^d}{\partial \bar{x}^b}\right)_{p'} g_{cd}(p') \,. \end{aligned} \tag{1.15.28}$$

Since in general $(\partial x^i/\partial \bar{x}^a)_{p'} \neq \vartheta_a{}^i$ and $g_{ij}(p') \neq g_{ij}(p)$ we have:

$$\bar{g}_{ab}(p') \neq \eta_{ab} \quad . \tag{1.15.29}$$

The chart $(U_\alpha, \varphi_\alpha)$ for which the natural basis is orthonormal at $p \in U_\alpha$, but not necessarily at other points of U_α, is said to be *adapted to p*.

A Riemannian manifold always admits a chart adapted to any of its points. A Riemannian manifold which admits a single chart adapted to all of its points is said to be (globally) pseudo-Euclidean when $r < n$ and $s < n$, while if $r = n$ $(s = 0)$ or $s = n$ $(r = 0)$ the metric (and also the manifold) is said to be (properly) Euclidean.

There are infinitely many coordinate systems adapted to a given point p but any two such systems, $\left\{x'^i\right\}$ and $\{\bar{x}^i\}$ say, must have the metric equal to η_{ij} and so must satisfy:

$$\frac{\partial x'^i}{\partial \bar{x}^j}\frac{\partial x'^k}{\partial \bar{x}^l}\eta_{ik} = \eta_{jl} \quad . \tag{1.15.30}$$

We can regard this as a matrix equation of the form

$$AHA^T = H \,. \tag{1.15.31}$$

*Many authors restrict the term Riemannian to the case $s = 0$, calling the general case pseudo-Riemannian.

Taking determinants we see that

$$\operatorname{Det}\left[\frac{\partial x'^i}{\partial \bar{x}^j}\right] = \pm 1 \quad , \tag{1.15.32}$$

and more generally, for any (not necessarily adapted) coordinate system $\hat{x}^i$ we have that

$$\Delta \equiv \operatorname{Det}\left[\frac{\partial \hat{x}^i}{\partial \bar{x}^j}\right] \neq 0$$

and so either $\Delta > 0$ or $\Delta < 0$ for all points where it is defined. This means that the sets of all coordinates at p fall into two classes, R and L. Any two coordinate systems in R are related by the $\Delta > 0$ sign, as are any two coordinate systems in L; but a transformation from an R-coordinate to an L-coordinate, or viceversa, involves $\Delta < 0$. The designations R and L are interchangeable, we choose them arbitrarily. The coordinates in R are called *right-handed* and those in L are called *left-handed.* If there exists an atlas covering the whole manifold whose charts are related by a $\Delta > 0$ sign, then the manifold is termed *orientable.* In this case the classes R and L can be chosen globally. Otherwise they have to be chosen purely locally.

For the rest of this section, we shall assume that the signature of the metric is $n-2$; in this case, which is of importance in classical relativity, the metric is called Lorentzian.

Now let $V = T_p(M)$ for some $p \in M$, and let us consider some further properties of V with its Lorentzian metric g. A subspace W of the given Lorentzian vector space (V, g) is *space-like* if and only if the restriction of the metric to W, i.e., $g|_{\mathrm{W}}$ is positive-definite. This implies that all its vectors are space-like. A vector v is space-like if $(v|v) > 0$ while it is termed time-like if $(v|v) < 0$. The subspace W is said to be time-like if it contains a time-like vector, while it is *null* if it contains some null vector but no time-like ones.

We remark in passing that, given a metric at every point of M (a metric field), we say that a curve in M is time-like, space-like, or null if its tangent vector is time-like, space-like or null at all its points. Similarly a sub-space S of a Riemannian manifold (M, g)

is space-like if it does *not* contain a null or a time-like curve; it is time-like at p if it contains a time-like curve through p, it is null at p if it contains a null curve through p but no time-like ones. We now return to the vector space $V = T_p(M)$ and prove:

PROPOSITION 1

If v is orthogonal to a time-like vector u, then v is space-like.

Proof: Let $(v|u) = 0$ and $(u|u) < 0$; then choose an orthonormal basis so that:

$$v \equiv (v^0, \vec{a}) \quad , \quad u \equiv (u^0, \vec{b}) \; . \tag{1.15.33}$$

The assumed conditions can be written as:

$$a) \quad v^0 u^0 = \vec{a} \cdot \vec{b} \quad ; \quad b) \quad |u^0| > \|\vec{b}\| \tag{1.15.34}$$

where $\|\vec{b}\| = (\vec{b} \cdot \vec{b})^{1/2}$; these imply:

$$|v^0| < \frac{|\vec{a} \cdot \vec{b}|}{\|\vec{b}\|} < \|\vec{a}\| \; . \tag{1.15.35}$$

Hence v is space-like.

PROPOSITION 2

If k is a null vector and v is orthogonal to k, then v is either space-like or null and proportional to k.

Proof: As before let

$$k \equiv (k^0, \vec{c}) \quad , \quad v \equiv (v^0, \vec{a})$$

then, since from the hypothesis

$$|k^0| = \|\vec{c}\| \quad , \quad v^0 k^0 = \vec{c} \cdot \vec{a} \tag{1.15.36}$$

we have

$$|v^0| = \frac{|\vec{c} \cdot \vec{a}|}{\|\vec{c}\|} \leq \|\vec{a}\|$$

where the equality sign holds only if $\vec{c} = \alpha \vec{a}$ for some constant α. Hence v is space-like or is null and such that $v = \alpha^{-1} k$.

1.16 Raising and lowering of tensor indices

The isomorphism (1.15.17) between T_p and $\overset{*}{T}_p$ and vice versa can be extended to arbitrary tensor product spaces: given a tensor of valence (r, s) with components $S^{i_1 \dots i_r}{}_{j_1 \dots j_s}$, the linear combinations

$$g_{i_1 k} S^{i_1 \dots i_r}{}_{j_1 \dots j_s} \equiv S_k{}^{i_2 \dots i_r}{}_{j_1 \dots j_s} \tag{1.16.1}$$

are the components of a tensor of valence $(r-1, s+1)$. In (1.16.1) the index i_1 has been *lowered*; similarly the index j_1 can be *raised* by means of the tensor $(g)^{-1}$:

$$g^{j_1 l} S^{i_1 \dots i_r}{}_{j_1 \dots j_s} \equiv S^{i_1 \dots i_r l}{}_{j_2 \dots j_s} \; ; \tag{1.16.2}$$

so we obtain the components of a tensor of valence $(r+1, s-1)$. This operation can be extended to any of the indices i_α and j_α and repeatedly: for instance

$$\begin{aligned} R_{ijk\dots} &= g_{ir} g_{js} g_{kl} \dots R^{rsl\dots} \\ T^{ijk\dots} &= g^{ir} g^{js} g^{kl} \dots T_{rsl\dots} . \end{aligned} \tag{1.16.3}$$

The operations of lowering and raising the indices of a tensor do not change its rank.

Comparing (1.16.1) and (1.16.2) with (1.13.24), we recognize that the operations of lowering and raising lead to that of *contraction* over two (or more) indices of a tensor when both indices of the metric tensor sum with indices of the tensor components: for instance from (1.16.1)

$$g_{i_1 i_2} S^{i_1 i_2 \dots i_r}{}_{j_1 \dots j_s} = S_{i_2}{}^{i_2 i_3 \dots i_r}{}_{j_1 \dots j_s} . \tag{1.16.4}$$

The complete contraction of a tensor of even rank yields a scalar

$$g_{ik} g^{jl} S^i{}_j{}^k{}_l = S_{il}{}^{il} = S . \tag{1.16.5}$$

The complete contraction

$$g^{ij} R_{ij} = R^i{}_i \equiv \mathrm{Tr}(R^i{}_j) \tag{1.16.6}$$

of a symmetric tensor of valence $(0, 2)$ or $(2, 0)$ is called the *trace* of the tensor.

1.17 Alternating tensors

A very useful quantity is the alternating symbol defined by

$$\delta_{i_1 \ldots i_n} = \begin{cases} +1 & \text{when } i_1 \ldots i_n \text{ form} \\ & \quad \text{an even permutation of } 1\ 2 \ldots n \\ -1 & \text{when } i_1 \ldots i_n \text{ form} \\ & \quad \text{an odd permutation of } 1\ 2 \ldots n \\ 0 & \text{when any two indices are equal.} \end{cases} \tag{1.17.1}$$

Suppose we have a right-handed coordinate system $\{\bar{x}^i\}$ adapted to p. Let $\boldsymbol{\eta}$ be a tensor at p whose components at p in these coordinates are equal to $\delta_{i_1 \ldots i_n}$. Then in an arbitrary coordinate system $\{x^i\}$, this tensor has components:

$$\eta_{i_1 i_2 \ldots i_n} = \frac{\partial \bar{x}^{j_1}}{\partial x^{i_1}} \frac{\partial \bar{x}^{j_2}}{\partial x^{i_2}} \cdots \frac{\partial \bar{x}^{j_n}}{\partial x^{i_n}} \delta_{j_1 j_2 \ldots j_n} = J \delta_{i_1 i_2 \ldots i_n} \tag{1.17.2}$$

where J is the determinant of the matrix $[\partial \bar{x}^j / \partial x^i]$, termed the *Jacobian* of the transformation $x^i = x^i(\bar{x})$.

Since all right-handed systems adapted to p are related by a transformation with $J = 1$, this shows that if we take a different right-handed coordinate system adapted to p, then we still obtain the same tensor $\boldsymbol{\eta}$. Thus $\boldsymbol{\eta}$ depends only on J (which specifies which coordinates are used) and on the choice of which coordinates to call *right-handed* and which *left-handed.*

A very convenient way to express the Jacobian is obtained on recalling that the metric components in the coordinates $\{\bar{x}^i\}$ adapted to p, are η_{ij} (see (1.15.27)), so from the transformation law:

$$g_{ij} = \frac{\partial \bar{x}^r}{\partial x^i} \frac{\partial \bar{x}^s}{\partial x^j} \eta_{rs} \tag{1.17.3}$$

and the properties of the matrices, we have:

$$g = \epsilon J^2 \tag{1.17.4}$$

where $g = \operatorname{Det}[g_{ij}]$ and $\epsilon = \operatorname{sign} \operatorname{Det}[\eta_{ij}] = \pm 1$. From (1.17.4) we then have:

$$J = \pm (\epsilon g)^{\frac{1}{2}} \quad . \tag{1.17.5}$$

Following the terminology of (1.15) we call the coordinates right-handed in the case of the + sign and otherwise left-handed. Relation (1.17.2) becomes

$$\eta_{i_1 i_2 \dots i_n} = \pm(\epsilon g)^{\frac{1}{2}} \delta_{i_1 \dots i_n} \tag{1.17.6}$$

which implies:

$$\eta^{i_1 i_2 \dots i_n} = \pm\epsilon(\epsilon g)^{-\frac{1}{2}} \delta^{i_1 i_2 \dots i_n} . \tag{1.17.7}$$

Combining the alternating symbol with Kronecker deltas, a new alternating tensor is obtained which is called the generalized Kronecker delta:

$$\delta^{i_1 i_2 \dots i_n}_{j_1 j_2 \dots j_n} = \delta^{i_{a_1}}_{j_1} \delta^{i_{a_2}}_{j_2} \dots \delta^{i_{a_n}}_{j_n} \delta_{a_1 a_2 \dots a_n} . \tag{1.17.8}$$

This tensor is different than zero when all up (or down) indices differ from each other, and equals +1 (−1) when the up-indices form an even (odd) permutation of the down-indices. From (1.17.6), (1.17.7) and (1.17.8) the following numerical relations hold (see Veblen, 1962):

$$\eta^{i_1 i_2 \dots i_n} \eta_{j_1 j_2 \dots j_n} = \epsilon \delta^{i_1 i_2 \dots i_n}_{j_1 j_2 \dots j_n} \tag{1.17.9}$$

$$\delta^{i_1 i_2 \dots i_r i_{r+1} \dots i_k}_{j_1 j_2 \dots j_r i_{r+1} \dots i_k} = \frac{(n-r)!}{(n-k)!} \delta^{i_1 i_2 \dots i_r}_{j_1 j_2 \dots j_r} \tag{1.17.10}$$

$$\eta^{i_1 i_2 \dots i_r i_{r+1} \dots i_n} \delta^{a_{r+1} \dots a_n}_{i_{r+1} \dots i_n} = (n-r)!\ \eta^{i_1 \dots i_r a_{r+1} \dots a_n} \tag{1.17.11}$$

$$\delta^{i_1 \dots i_r i_{r+1} \dots i_k}_{j_1 \dots j_r j_{r+1} \dots j_k} \delta^{j_{r+1} \dots j_k}_{i_{r+1} \dots i_k} = \frac{(n-r)!}{(n-k)!} (k-r)! \delta^{i_1 \dots i_r}_{j_1 \dots j_r} \tag{1.17.12}$$

$$\eta^{i_1 i_2 \dots i_n} \eta_{i_1 i_2 \dots i_n} = \epsilon n! . \tag{1.17.13}$$

The generalized Kronecker delta has wide application in matrix algebra. Let $A_i^{\ j}$ be the general element of a $(n \times n)$-matrix; we have by definition:

i) determinant of a k-rowed minor:

$$A^{i_1 i_2 \dots i_k}_{j_1 j_2 \dots j_k} = \delta^{i_1 \dots i_k}_{a_1 \dots a_k} A_{j_1}^{\ a_1} \dots A_{j_k}^{\ a_k} \tag{1.17.14}$$

ii) determinant of the matrix:

$$A = \frac{\epsilon}{n!} \delta^{i_1 i_2 \dots i_n}_{j_1 j_2 \dots j_n} A_{i_1}^{\ j_1} A_{i_2}^{\ j_2} \dots . A_{i_n}^{\ j_n} = \frac{\epsilon}{n!} A^{i_1 \dots i_n}_{i_1 \dots i_n} \tag{1.17.15}$$

iii) determinant of the cofactor of a k-rowed minor:

$$C^{i_1 i_2 \dots i_k}_{j_1 j_2 \dots j_k} = \frac{1}{[(n-k)!]^2} \delta^{i_1 i_2 \dots i_n}_{j_1 j_2 \dots j_n} A^{j_{k+1} \dots j_n}_{i_{k+1} \dots i_n} \tag{1.17.16}$$

iv) the Laplace expansion of a matrix:

$$A^{i_1 \dots i_k}_{j_1 \dots j_k} C^{j_1 \dots j_k}_{a_1 \dots a_k} = k! A \delta^{i_1 \dots i_k}_{a_1 \dots a_k} \ . \tag{1.17.17}$$

Given an antisymmetric tensor $\boldsymbol{T}$, of valence $(k,0)$ $[(0,k)]$, $k < n$, we define its *dual* (or Hodge dual) to be the tensor:

$$\begin{aligned} \overset{*}{T}_{j_{k+1} \dots j_n} &= \frac{1}{k!} T^{i_1 \dots i_k} \eta_{i_1 \dots i_k j_{k+1} \dots j_n} \\ \overset{*}{T}^{j_{k+1} \dots j_n} &= \frac{1}{k!} T_{i_1 \dots i_k} \eta^{i_1 \dots i_k j_{k+1} \dots j_n} \ . \end{aligned} \tag{1.17.18}$$

From (1.17.9) and the further relation:

$$\delta^{i_1 \dots i_k i_{k+1} \dots i_n}_{j_1 \dots \dots \dots j_n} = (-1)^{k(n-k)} \delta^{i_{k+1} \dots i_n i_1 \dots i_k}_{j_1 \dots \dots \dots j_n} \tag{1.17.19}$$

we define the dual of the dual:

$$\begin{aligned} \overset{**}{T}{}^{a_1 a_2 \dots a_k} &= \frac{1}{(n-k)!} \overset{*}{T}_{j_{k+1} \dots j_n} \eta^{j_{k+1} \dots j_n a_1 \dots a_k} \\ &= \frac{\epsilon(-1)^{k(n-k)}}{k!} T^{i_1 i_2 \dots i_k} \delta^{a_1 \dots a_k}_{i_1 \dots i_k} = \epsilon(-1)^{k(n-k)} T^{a_1 \dots a_k} \ . \end{aligned} \tag{1.17.20}$$

A useful application of the above formulas is when dealing with tensors which are antisymmetric only with respect to selected pairs of indices. Consider for instance a tensor of valence $(4,0)$ say, which is antisymmetric only in the two pairs of indices:

$$T^{ijkl} = T^{[ij][kl]} \ . \tag{1.17.21}$$

From (1.17.18) we term *left-dual* the tensor:

$$T_{\overset{*}{ij}}{}^{kl} = \frac{1}{2} \eta_{ijrs} T^{rskl} \tag{1.17.22}$$

and *right-dual* the tensor:

$$T_{ij}{}^{\overset{*}{kl}} = \frac{1}{2} \eta^{klmn} T_{ijmn} \ . \tag{1.17.23}$$

We then call *double-dual* the tensor:

$$T^{\overset{*}{ij}\overset{*}{kl}} = \frac{1}{4} \eta^{ijrs} \eta^{klmn} T_{rsmn} \ . \tag{1.17.24}$$

A convenient way to express the double-dual is considering its mixed form; from (1.17.9) we have:

$$\overset{*}{T}{}^{ij}{}_{\overset{*}{kl}} = \frac{1}{4}\epsilon\delta^{ijrs}_{klmn}T_{rs}{}^{mn}. \tag{1.17.25}$$

From (1.17.8) we can, however, split the generalized Kronecker delta into the following terms:

$$\delta^{ijrs}_{klmn} = \delta^{ij}_{kl}\delta^{rs}_{mn} + \delta^{rs}_{kl}\delta^{ij}_{mn} + 2\delta^{r[i}_{kl}\delta^{j]s}_{mn} - 2\delta^{s[i}_{kl}\delta^{j]r}_{mn} \tag{1.17.26}$$

where:

$$\delta^{ab}_{cd} = \delta^a_c\delta^b_d - \delta^b_c\delta^a_d \quad . \tag{1.17.27}$$

Using (1.17.27) and (1.17.26) in (1.17.25) we finally have:

$$\overset{*}{T}{}^{ij}{}_{\overset{*}{kl}} = \epsilon\left[T_{kl}{}^{ij} + \frac{1}{2}\delta^{ij}_{kl}T + 2T^j_{[k}\delta^i_{l]} - 2T^i_{[k}\delta^j_{l]}\right] \tag{1.17.28}$$

where

$$T_r{}^s = T_{rm}{}^{sm} \quad ; \quad T = T_{mn}{}^{mn} \ . \tag{1.17.29}$$

It is easy to compute from (1.17.28) that

$$\overset{*}{\delta}{}^{ij}{}_{\overset{*}{kl}} = \epsilon\delta^{ij}{}_{kl} \ . \tag{1.17.30}$$

1.18 Exterior algebra

The completely skew-symmetric tensors of valence $(0, r)$ (respectively $(r, 0)$), where r is an integer ≥ 2, can be thought of as elements of a new linear vector space on M termed the *space of r-forms* (r-vectors), denoted $\Lambda^r\overset{*}{T}_p(M)$ ($\Lambda^r T_p(M)$; $p \in M$) and defined as the space of the r-linear maps,* (Cartan, 1945; Flanders, 1963):

$$\Lambda^r\overset{*}{T}_p(M) \ni \overset{(r)}{\alpha} : T_p(M) \times .... \times T_p(M) \to \mathbb{R} \tag{1.18.1}$$

*A similar definition holds for the r-tensors.

satisfying the following axioms:

a) $\overset{(r)}{\boldsymbol{\alpha}}(u_1, \ldots u_i, u_{i+1}, \ldots u_r) = -\overset{(r)}{\boldsymbol{\alpha}}(u_1, \ldots u_{i+1}, u_i, \ldots u_r)$

b) $\overset{(r)}{\boldsymbol{\alpha}}(au_1 + bw, u_2, \ldots, u_r)$

$$= a\,\overset{(r)}{\boldsymbol{\alpha}}(u_1, u_2, \ldots) + b\,\overset{(r)}{\boldsymbol{\alpha}}(w, u_2, \ldots) \quad (1.18.2)$$

These imply

c) $\overset{(r)}{\boldsymbol{\alpha}}(\ldots, u, \ldots, u, \ldots) = 0$

d) $\overset{(r)}{\boldsymbol{\alpha}}(au_1, \ldots, bu_i, \ldots, u_r) = ab\,\overset{(r)}{\boldsymbol{\alpha}}(u_1, \ldots, u_i, \ldots, u_r)$ (1.18.2′)

From properties a) and c) we have in general:

$$\overset{(r)}{\boldsymbol{\alpha}}(u_{i_1}, u_{i_2}, \ldots, u_{i_r}) = \delta_{i_1 i_2 \ldots i_r}\overset{(r)}{\boldsymbol{\alpha}}(u_1, u_2, \ldots, u_r) \quad (1.18.3)$$

and from $(1.18.2)'d$:

$$\overset{(r)}{\boldsymbol{\alpha}}(u_1, u_2, \ldots, u_r) = u_1^{[i_1} u_2^{i_2} \ldots u_r^{i_r]}\,\overset{(r)}{\boldsymbol{\alpha}}_{i_1 i_2 \ldots i_r} \quad (1.18.4)$$

where $u_k = u_k^j e_j$ and $\overset{(r)}{\boldsymbol{\alpha}}_{i_1 i_2 \ldots i_r} \equiv \overset{(r)}{\boldsymbol{\alpha}}(e_{i_1}, e_{i_2}, \ldots, e_{i_r})$. If $\{\boldsymbol{e}^i\}_p$ is the dual basis at $p \in M$, since $u^i = \boldsymbol{e}^i(u)$, (1.18.4) can be written as:

$$\overset{(r)}{\boldsymbol{\alpha}}(u_1, u_2, \ldots, u_r) = \boldsymbol{e}^{[i_1}(u_1)\boldsymbol{e}^{i_2}(u_2)\ldots\boldsymbol{e}^{i_r]}(u_r)\,\overset{(r)}{\boldsymbol{\alpha}}_{i_1 i_2 \ldots i_r}$$
$$\equiv \overset{(r)}{\boldsymbol{\alpha}}_{i_1 i_2 \ldots i_r}\boldsymbol{e}^{[i_1} \otimes \boldsymbol{e}^{i_2} \otimes \cdots \otimes \boldsymbol{e}^{i_r]}(u_1, \ldots u_r)\,. \quad (1.18.5)$$

The map

$$\boldsymbol{e}^{[i_1} \otimes \ldots \otimes \boldsymbol{e}^{ir]} \equiv \frac{1}{r!}\left[\boldsymbol{e}^{i_1} \wedge \boldsymbol{e}^{i_2} \wedge \ldots \wedge \boldsymbol{e}^{i_r}\right] \quad (1.18.6)$$

is an element of $\Lambda^r \overset{*}{T}_p$, since it satisfies the axioms (1.18.2), and, due to the arbitrariness of the u_k's, the set of all such maps, for all choices of $i_1 \ldots i_r$ with $i_1 < i_2 < \cdots < i_r$ constitutes a basis for $\Lambda^r \overset{*}{T}_p$. A general r-form can be written as:

$$\overset{(r)}{\boldsymbol{\alpha}} = \frac{1}{r!}\alpha_{i_1 i_2 \ldots i_r}\boldsymbol{e}^{i_1} \wedge \boldsymbol{e}^{i_2} \wedge \ldots \wedge \boldsymbol{e}^{i_r}\,. \quad (1.18.7)$$

The space $\Lambda^r \overset{*}{T}_p(M)$ has dimension $N = \binom{n}{r}$; thus an r-form, with $r > n$ yields identically zero.

The symbol $\wedge$ is termed *wedge* and the operation it symbolizes is called the *wedge* or *exterior product.* This can be thought of as a linear map:

$$\overset{*}{T}_p \times \ldots \times \overset{*}{T}_p \ni (\boldsymbol{\omega}_1, \boldsymbol{\omega}_2, \ldots, \boldsymbol{\omega}_r) \to \boldsymbol{\omega}_1 \wedge \boldsymbol{\omega}_2, \ldots \wedge \boldsymbol{\omega}_r \in \Lambda^r \overset{*}{T}_p \quad (1.18.8)$$

with the properties, which follow from (1.18.2) and (1.18.6):

$$(a\boldsymbol{\omega}_1 + b\boldsymbol{\omega}_2) \wedge \boldsymbol{\omega}_3 = a(\boldsymbol{\omega}_1 \wedge \boldsymbol{\omega}_3) + b(\boldsymbol{\omega}_2 \wedge \boldsymbol{\omega}_3)$$
$$\boldsymbol{\omega}_1 \wedge (a'\boldsymbol{\omega}_2 + b'\boldsymbol{\omega}_3) = a'(\boldsymbol{\omega}_1 \wedge \boldsymbol{\omega}_2) + b'(\boldsymbol{\omega}_1 \wedge \boldsymbol{\omega}_3)$$
$$\boldsymbol{\omega} \wedge \boldsymbol{\omega} = \boldsymbol{o} \qquad\qquad a, b, a', b' \in \mathbb{R}$$
$$\boldsymbol{\omega}_1 \wedge \boldsymbol{\omega}_2 = -\boldsymbol{\omega}_2 \wedge \boldsymbol{\omega}_1 \qquad \boldsymbol{\omega}_k \in \overset{*}{T}_p(M). \quad (1.18.9)$$

The exterior product applies to the spaces of r-forms (r-vectors) themselves; the exterior multiplication of $\Lambda^r\overset{*}{T}_p(M)$ with $\Lambda^s\overset{*}{T}_p(M)$ $(r + s \leq n)$ yields:

$$\left(\Lambda^r\overset{*}{T}_p(M)\right) \wedge \left(\Lambda^s\overset{*}{T}_p(M)\right) = \Lambda^{r+s}\overset{*}{T}_p(M) \qquad (1.18.10)$$

with the rule: $\overset{(r)}{\boldsymbol{\alpha}} \in \Lambda^r\overset{*}{T}_p(M)$, $\overset{(s)}{\boldsymbol{\beta}} \in \Lambda^s\overset{*}{T}_p(M)$:

$$\overset{(r)}{\boldsymbol{\alpha}} \wedge \overset{(s)}{\boldsymbol{\beta}} = (-1)^{rs}\, \overset{(r)}{\boldsymbol{\alpha}} \wedge \overset{(s)}{\boldsymbol{\beta}}\,. \qquad (1.18.11)$$

An r-form $\overset{(r)}{\boldsymbol{\alpha}}$ which can be expressed as the exterior product of covectors, is said to be *simple* or *decomposable.* A simple r-form provides a natural definition of the determinant of an $r \times r$ matrix:

$$\boldsymbol{\omega}_1 \wedge \boldsymbol{\omega}_2 \wedge \ldots \wedge \boldsymbol{\omega}_r(u_1, u_2, \ldots, u_r) = \mathrm{Det}[\boldsymbol{\omega}_a(u_b)] \qquad a, b = 1, 2 \ldots r\,. \qquad (1.18.12)$$

In fact, from (1.18.6), the left-hand side of this can be written as:

$$r!\boldsymbol{\omega}^{[1}(u_1)\boldsymbol{\omega}^2(u_2)\ldots\boldsymbol{\omega}^{r]}(u_r) = \delta_{i_1 i_2 \ldots i_r}\boldsymbol{\omega}^{i_1}(u_1)\boldsymbol{\omega}^{i_2}(u_2)\ldots\boldsymbol{\omega}^{i_r}(u_r)$$
$$= \mathrm{Det}\left[\boldsymbol{\omega}^i(u_j)\right] \qquad (1.18.13)$$

as stated. In particular, letting $r = n$, we have :

$$\boldsymbol{\omega}^1 \wedge \boldsymbol{\omega}^2 \wedge \ldots \wedge \boldsymbol{\omega}^n = \omega^1_{i_1}\omega^2_{i_2}\ldots\omega^n_{i_n}\boldsymbol{e}^{i_1} \wedge \boldsymbol{e}^{i_2} \wedge \ldots \wedge \boldsymbol{e}^{i_n}$$
$$= \omega^1_{i_1}\omega^2_{i_2}\ldots\omega^n_{i_n}\delta^{i_1 i_2 \ldots i_n}\boldsymbol{e}^1 \wedge \boldsymbol{e}^2 \wedge \ldots \wedge \boldsymbol{e}^n$$
$$= \mathrm{Det}[\omega^i_j]\ \boldsymbol{e}^1 \wedge \boldsymbol{e}^2 \wedge \cdots \wedge \boldsymbol{e}^n \qquad (1.18.14)$$

from (1.18.3). Of interest is the general relation:

$$\boldsymbol{e}^{i_1} \wedge \boldsymbol{e}^{i_2} \wedge \ldots \wedge \boldsymbol{e}^{i_r}(e_{j_1}, e_{j_2}, \ldots, e_{j_r}) = \delta^{i_1 i_2 \ldots i_r}_{j_1 j_2 \ldots j_r}\,, \qquad (1.18.15)$$

from (1.18.13) and (1.17.8).

Let M be a (at least) locally oriented manifold and consider a chart adapted to $p \in M$ (assume it to be an R-chart), then we term *natural volume element* the n-form:

$$\overset{(n)}{\boldsymbol{\eta}} = \boldsymbol{d\bar{x}}^1 \wedge \boldsymbol{d\bar{x}}^2 \wedge ... \wedge \boldsymbol{d\bar{x}}^n \tag{1.18.16}$$

where $\{\boldsymbol{d\bar{x}}^i\}$ is a dual coordinate basis which is orthonormal at p. Transforming to a general chart which agrees with the orientation of M, the volume element becomes:

$$\begin{aligned}\overset{(n)}{\boldsymbol{\eta}} &= \mathrm{Det}\left[\frac{\partial \bar{x}^i}{\partial x^j}\right] \boldsymbol{dx}^1 \wedge \boldsymbol{dx}^2 \wedge ... \wedge \boldsymbol{dx}^n \\ &= \sqrt{\epsilon g}\,\boldsymbol{dx}^1 \wedge \boldsymbol{dx}^2 \wedge ... \wedge \boldsymbol{dx}^n \\ &= \frac{1}{n!}\eta_{i_1 i_2 \ldots i_n} \boldsymbol{dx}^{i_1} \wedge \boldsymbol{dx}^{i_2} \wedge ... \wedge \boldsymbol{dx}^{i_n} \qquad \epsilon = \mathrm{sign}(g)\end{aligned} \tag{1.18.17}$$

from (1.17.5) and $\eta_{i_1 \ldots i_n} = (\epsilon g)^{\frac{1}{2}} \delta_{i_1 \ldots i_n}$.

We shall now introduce the concept of integration on a manifold; let $\overset{(n)}{\boldsymbol{\alpha}}$ be an arbitrary n-form and $\varphi : U \to \mathbb{R}^n$ a local chart on some $U \subset M$ with $U \ni p = \varphi^{-1}(x^i)$. We define the integral of $\overset{(n)}{\boldsymbol{\alpha}}$ on U to be the quantity:

$$\int_U \overset{(n)}{\boldsymbol{\alpha}} = \int_{\varphi(U)} \overset{(n)}{\boldsymbol{\alpha}}(\partial_1, \partial_2 \ldots \partial_n) dx^1 dx^2 \ldots dx^n \ . \tag{1.18.18}$$

On the right-hand side we have a standard Riemann integral on $\varphi(U) \subset \mathbb{R}^n$. If we specify $\overset{(n)}{\boldsymbol{\alpha}}$ to be the volume element (1.18.17), (1.18.18) yields the n-dimensional proper-volume of U as:

$$\int_U \overset{(n)}{\boldsymbol{\eta}} = \int_{\varphi(U)} (\epsilon g(x))^{\frac{1}{2}} dx^1 dx^2 \ldots dx^n \ . \tag{1.18.19}$$

The integral of a form on a set U of M, gives a value which is independent of the choice of the chart, provided the orientation is preserved. To prove this we need to recall the properties of integrals under change of variables; in that case we have

$$dx^1 dx^2 \ldots dx^n \to \left|\frac{\partial x}{\partial x'}\right| dx'^1 dx'^2 \ldots dx'^n \ ;$$

moreover from the properties of forms, we also have:

$$\overset{(n)}{\boldsymbol{\alpha}}(\partial_1 \ldots \partial_n) = \left|\frac{\partial x'}{\partial x}\right| \overset{(n)}{\boldsymbol{\alpha}}(\partial'_1 \ldots \partial'_n)$$

thus the value of the integral in (1.18.18) is independent of the coordinates in use.

Let us conclude this section by relating the dual (or Hodge) map of Section 17 to the exterior product notation. If the manifold is orientable and contains a metric, we define the dual map:

$$* : \Lambda^r \overset{*}{T} \to \Lambda^{n-r} \overset{*}{T} \tag{1.18.20}$$

such that for any set $(u_{r+1}, ..., u_n) \in T_p \underset{1}{\times} ... \underset{n-r}{\times} T_p$:

$$* \overset{(n-r)}{\alpha} (u_{r+1}, ..., u_n) \overset{(n)}{\boldsymbol{\eta}} = \overset{(r)}{\boldsymbol{\alpha}} \wedge \boldsymbol{u}_{r+1} \wedge ... \wedge \boldsymbol{u}_n \tag{1.18.21}$$

where $\boldsymbol{u}_k = \overset{*}{\boldsymbol{g}}(u_k, \cdot)$. From the definitions, (1.18.21) can be written explicitly as:

$$\begin{aligned} &* \overset{(n-r)}{\alpha}_{i_{r+1} \dots i_n} u_{r+1}^{[i_{r+1}} \dots u_n^{i_n]} \frac{1}{n!} \eta_{j_1 \dots j_n} \boldsymbol{e}^{j_1} \wedge ... \wedge \boldsymbol{e}^{j_n} \\ &= \frac{1}{r!} \overset{(r)}{\alpha}_{j_1 \dots j_r} u_{r+1\, j_{r+1}} \dots u_{n j_n} \boldsymbol{e}^{j_1} \wedge ... \wedge \boldsymbol{e}^{j_r} \wedge \boldsymbol{e}^{j_{r+1}} \wedge ... \wedge \boldsymbol{e}^{j_n} \; . \end{aligned} \tag{1.18.22}$$

From the arbitrariness of the set $\{u_k\}_{k=r+1 \dots n}$ we have:

$$\frac{1}{n!} * \overset{(n-r)}{\alpha}_{i_{r+1} \dots i_n} \eta_{j_1 \dots j_n} = \frac{1}{r!} \overset{(r)}{\alpha}_{j_1 \dots j_r} g_{j_{r+1} i_{r+1}} \dots g_{j_n i_n} \; , \tag{1.18.23}$$

contracting with $\eta^{j_1 \dots j_n}$ and recalling (1.17.13), we have :

$$* \overset{(n-r)}{\alpha}_{i_{r+1} \dots i_n} = \frac{\epsilon}{r!} \overset{(n)}{\alpha}_{j_1 \dots j_r} \eta^{j_1 \dots j_r j_{r+1} j_n} g_{j_{r+1} i_{r+1}} \cdots g_{j_n i_n} \tag{1.18.24}$$

as in (1.17.18).

1.19 Measure of lengths and the world-function

The proper volume defined in (1.18.19) is an invariant measure which plays a fundamental role in relativity. While (1.18.19) applies to all dimensions $n \geq 2$, the case with $n = 1$ requires some care since it leads to a natural definition of the measure of length on a smooth curve. Consider the curve, γ say, as a one-dimensional manifold with $t \in \mathbb{R}$. If $g^{(1)}$ is a metric on $\mathbb{R}$, we have that

$$\overset{(1)}{\boldsymbol{\eta}} = (\epsilon g^{(1)})^{\frac{1}{2}} \boldsymbol{dt} \; . \tag{1.19.1}$$

Following the arguments made in the previous section, we define as a measure of length on γ (one-dimensional proper volume) the quantity:

$$L = \int_\gamma \overset{(1)}{\boldsymbol{\eta}}(\dot{\gamma})\, dt = \int_\gamma \|\dot{\gamma}\| dt \tag{1.19.2}$$

where $\|\dot{\gamma}\|$ is the modulus of the tangent vector to the curve. If we consider the curve as being imbedded in a n-dimensional manifold then taking $g^{(1)}$ to be the induced metric $\gamma^* g$ on γ, (1.19.2) can be written, from (1.15.5) as:

$$L = \int_\gamma |(\dot{\gamma}|\dot{\gamma})|^{\frac{1}{2}} \, dt \ . \tag{1.19.3}$$

As expected, and as is easy to check, (1.19.3) does not depend on the parametrization on the curve. The quantity L is positive and is zero when the curve γ is null.

If the curve whose length we are measuring is entirely contained in a coordinate neighbourhood U of M, equation (1.19.3) can be rewritten as:

$$L = \int_\gamma \left| g_{ij} \frac{dx^i}{dt} \frac{dx^j}{dt} \right|^{\frac{1}{2}} dt \qquad \gamma \in U \ .$$

It is usual to write this integral as $L = \int_\gamma |ds^2|^{\frac{1}{2}}$ where

$$ds^2 = g_{ij} dx^i dx^j$$

is called the square of the "line element" in U.

Let us consider two points $p_1 = \gamma(t_1)$ and $p_2 = \gamma(t_2)$; there is an infinity of smooth curves joining them and so the function

$$L(p_1, p_2; \gamma) = \int_{t_1}^{t_2} |(\dot{\gamma}|\dot{\gamma})|^{\frac{1}{2}} dt \tag{1.19.4}$$

is a path-dependent function of the two points. If we keep the points fixed, we shall term *extremal* or *geodesic* a curve on which $L(p_1, p_2; \gamma)$ has an extremum. The length of a curve between two points plays a key role in the theory of measurements in curved manifolds. It turns out, however, more convenient to introduce a new quantity, termed the *world-function*, defined as:

$$\tilde{\Omega}(t_1, t_2; \gamma) = \frac{1}{2} L^2 \tag{1.19.5}$$

which, like L, depends on the parameters of two points and on the curve joining them. This was introduced by Ruse (1931) and Synge (1960) and we shall consider it further later on (see Sect. 3.7).

2

DIFFERENTIATION

2.1 Tensor fields and congruences

The algebraic structures which have been given to the manifold refer to each point separately. In order to proceed and do differential geometry on M, we need to introduce the concept of a field. A vector field is a map

$$X : M \to T(M) \equiv \bigcup_{p \in M} T_p(M) \tag{2.1.1}$$

such that $X(p) \in T_p(M)$. The space of the vector fields on M will be denoted by $\mathcal{T}(M)$. The above definition is naturally extended to covectors (in which case these will be called one-forms) and tensors of valence (r, s), the corresponding fields being elements of $\mathcal{T}^*(M)$ and $\mathcal{T}^{(r,s)}(M)$ respectively.

The tensor product and contraction defined in (1.13) can be extended to tensor fields in the obvious way: if T, S are tensor fields, then we define the field $T \otimes S$ by*:

$$(T \otimes S)(p) \equiv T(p) \otimes S(p); \tag{2.1.2}$$

*Readers interested in algebra may like to note that $T \otimes S$ can be regarded as a tensor product of T and S in the sense defined in Sec.1.13, provided that we regard $\mathcal{T}^{(r,s)}$ as a vector space generalized by having functions in $\mathcal{F}$ play the role of the scalars$\mathbb{R}$, with multiplication being defined by

$$(fT)(p) = f(p)T(p).$$

When the idea of a vector space is generalized in this way by allowing the scalars to lie in a ring it is called a *module*, in this case an $\mathcal{F}$-module. Then $\mathcal{T}$ is the dual of $\mathcal{T}$ as an $\mathcal{F}$-module, (see Choquet-Bruhat et al., 1977; Kobayashi and Nomizu, 1969).

while if C denotes the contraction operation defined by (1.13.24) we can define CT by:

$$(CT)(p) \equiv C(T(p)). \tag{2.1.3}$$

As a consequence of these operations we have the pairing:

$$\begin{gathered}(\boldsymbol{\omega}, X) \longrightarrow \boldsymbol{\omega}(X) \\ \mathcal{T}^* \times \mathcal{T} \longrightarrow \mathcal{F}\end{gathered} \tag{2.1.4}$$

taking a covector field and a vector field into a function, defined by:

$$(\boldsymbol{\omega}(X))(p) \equiv \boldsymbol{\omega}(p)(X(p)). \tag{2.1.5}$$

Note that, taking the case $r = 1$, $s = 1$ in (1.13.24), so that contraction maps tensors of rank $(1,1)$ into tensors of rank $(0,0)$, i.e. functions, we have that

$$\boldsymbol{\omega}(X) = C(\boldsymbol{\omega} \otimes X). \tag{2.1.6a}$$

We can define components of fields by taking a basis at each point. Thus vector fields $e_1, \ldots, e_n$ will be said to constitute a set of *basis fields* if $\{e_1(p), \ldots, e_n(p)\}$ is a basis for each p. They induce covector fields $\boldsymbol{e}^1, \ldots, \boldsymbol{e}^n$ and a general tensor field T of valence (r, s) can be written as:

$$T = T^{i_1 \ldots i_r}{}_{j_1 \ldots j_s} e_{i_1} \otimes \ldots \otimes e_{i_r} \otimes \boldsymbol{e}^{j_1} \otimes \ldots \otimes \boldsymbol{e}^{j_n} \ .$$

The only drawback is that there may not exist a set of $\{e_i\}$ which is a basis for the whole of M, in which case the above construction must be carried out only in some open neighbourhood U of the point under consideration. Finally, we note that a vector field X can act on a function $f \in \mathcal{F}$ to produce another function $X(f)$ according to:

$$(X(f))(p) \equiv X(p)f. \tag{2.1.6b}$$

We have seen (1.9, 1.12) that a smooth map $\Phi : M \to N$ from one manifold to another (or, as a special case, from one manifold to itself) induces maps:

$$\Phi_* : T_p(M) \longrightarrow T_{\Phi(p)}(N) \qquad \text{(see 1.9.2)} \tag{2.1.7}$$

$$\Phi^* : \overset{*}{T}_{\Phi(p)}(N) \longrightarrow \overset{*}{T}_p(M) \qquad \text{(see 1.12.2)} \tag{2.1.8}$$

We now see how far this extends to fields.

To take the covector case first, if we are given a covector field $\boldsymbol{\omega} \in \mathcal{T}^*(N)$ then we can define a field $\bar{\Phi}^*\boldsymbol{\omega} \in \mathcal{T}^*(M)$ by :

$$\begin{aligned} \left(\bar{\Phi}^*\boldsymbol{\omega}\right)(p) &= \Phi^*\left(\boldsymbol{\omega}(\Phi(p))\right) \\ \bar{\Phi}^* : \mathcal{T}^*(N) &\longrightarrow \mathcal{T}^*(M). \end{aligned} \tag{2.1.9}$$

In the case of vectors, the situation is more complicated, for if Φ is not injective – if for example there exist p and q in M such that $\Phi(p) = \Phi(q) = p' \in N$, say – then given a vector field X on M it is likely that $\Phi_*(X(p))$ and $\Phi_*(X(q))$ will be two different vectors in $T_{p'}(N)$; so we do not get a well-defined vector field on N. On the other hand if Φ is not surjective, the set of vectors $\Phi_*(X(p))$ as p ranges over M will not give vectors at all points of N and so will not be a vector field on the whole of N. Only if Φ is bijective is it possible, given a vector field X on M, to define a vector field $\bar{\Phi}_*X$ on N by:

$$(\bar{\Phi}_*X)(p') = \Phi_*(X(\Phi^{-1}(p'))) \qquad p' \in N\,. \tag{2.1.10}$$

In applications in differential geometry, when Φ is bijective, it is also usually a diffeomorphism in which case (cf. 1.9) $\Phi^* : \overset{*}{T}_{p'}(N) \to \overset{*}{T}_p(M)$ is invertible. This enables us to define a map:

$$\tilde{\Phi}_* : \mathcal{T}^*(M) \longrightarrow \mathcal{T}^*(N)$$

by

$$\left(\tilde{\Phi}_*\boldsymbol{\omega}\right)(p') = \Phi^{*-1}\left(\boldsymbol{\omega}(\Phi^{-1}(p'))\right). \tag{2.1.11}$$

It follows immediately from the definitions that $\tilde{\Phi}_*$ and $\bar{\Phi}^*$ are inverses of each other.

Since $\tilde{\Phi}_*$ and $\bar{\Phi}_*$ map in the same direction (M to N) it is natural to try to extend them to the whole algebra of tensor fields. It is not hard to see that for every valence (r, s) there is a unique $\mathbb{R}$-linear map:

$$\Phi_*^{(r,s)} : \mathcal{T}^{(r,s)}(M) \longrightarrow \mathcal{T}^{(r,s)}(N) \tag{2.1.12}$$

satisfying:

$$\Phi_*^{(r+r',s+s')}(T \otimes S) = \Phi_*^{(r,s)}(T) \otimes \Phi_*^{(r',s')}(S) \tag{2.1.13}$$

defined by

$$\Phi_*^{(r,s)}\left(T^{i_1\dots i_r}{}_{j_1\dots j_s} e_{i_1}\otimes\dots\otimes e_{i_r}\otimes\boldsymbol{\omega}^{j_1}\otimes\dots\otimes\boldsymbol{\omega}^{j_s}\right)$$
$$=\left(T^{i_1\dots i_r}{}_{j_1\dots j_s}\circ\varphi^{-1}\right)\left(\bar{\Phi}_* e_{i_1}\otimes\dots\otimes\tilde{\Phi}_*\boldsymbol{\omega}^{j_1}\otimes\dots\right) \quad (2.1.14)$$

for some (and hence, by linearity, any) collection of basis fields e_i etc.

The formalism just developed applies to functions (tensors of valence $(0,0)$) if we define:

$$\Phi_*^{(0,0)} f = f\circ\Phi^{-1}. \quad (2.1.15)$$

Finally we simplify the notation by using the single symbol $\bar{\Phi}_*$ to embrace all of $\bar{\Phi}_*$, $\tilde{\Phi}_*$, $\Phi_*^{(r,s)}$, it being clear, from what $\bar{\Phi}_*$ is operating on, which one is intended. We note that the map $\bar{\Phi}_*$ commutes with contraction in the sense that:

$$C\bar{\Phi}_* S = \bar{\Phi}_* CS. \quad (2.1.16)$$

Let us now introduce the concept of a congruence. Suppose that p is a fixed point in M and that we are given a collection of diffeomorphisms $\{\varphi_t : t\in(-\epsilon,\epsilon)\}$ with the following properties:

i) Each φ_t is a diffeomorphism $U\to M$ with U an open set;

ii) If t, u and $t+u$ are all in $(-\epsilon,\epsilon)$, then the composite

$$\varphi_t\circ\varphi_u : \varphi_u^{-1}(U)\cap U \longrightarrow M$$

agrees with $\varphi_{(t+u)}$ on $\varphi_u^{-1}(U)\cap U$.

iii) The map $\gamma_p : t\to\varphi_t(p)$ is differentiable, for each p.

If it were not for the fact that the parameter is restricted to $(-\epsilon, \epsilon)$ and the domains restricted to open sets, we should have what is called a *one-parameter group of diffeomorphisms*; in this case we refer to a family satisfying i) and ii) as a *local one-parameter group of local diffeomorphisms.* Putting $u=t=0$ in ii), shows that $\varphi_0\circ\varphi_0$ agrees with φ_0 in $\varphi_0^{-1}(U)\cap U = V$, say, with $p\in V$; so since φ_0 is one–to–one, φ_0 is the identity on V. From iii) the tangent to the curve γ_q at $t=0$ is a vector $X(q)$, say, and the assignment $q\to X(q)$ defines a vector field on U.

If U is a coordinate neighbourhood, we can write the coordinates of $\varphi_t(q)$ as a function of t and the coordinates of q, for example:

$$(\varphi_t(q))^i = \xi^i(q^1, ..., q^n; t). \tag{2.1.17}$$

Then differentiating gives:

$$X^i = \left.\frac{\partial \xi^i}{\partial t}\right|_{t=0}. \tag{2.1.18}$$

The local group property (ii), namely $\varphi_t \circ \varphi_u = \varphi_{t+u}$, gives, in terms of coordinates,

$$\xi^i(\xi^1(q^1...; u), \xi^2, ...; t) = \xi^i(q^1, ..., q^n; t+u). \tag{2.1.19}$$

Differentiating with respect to t and putting $t = 0$ gives

$$\left.\frac{\partial \xi^i}{\partial t}\right|_{(\xi^1(q^1,..;u),...,0)} = \left.\frac{\partial \xi^i}{\partial t}\right|_{(q^1,...,q^n;u)}$$

that is (from (2.1.18))

$$\frac{\partial \xi^i}{\partial t}(q^1, ..., q^n; u) = X^i(\xi^1(q^1, ..., q^n; u), ...). \tag{2.1.20}$$

It is possible to carry out the inverse process: suppose we are given a vector field X over some coordinate neighbourhood. Then equation (2.1.20) can be regarded as an ordinary differential equation for the function $\xi^i(u)$ (with $q^1, .., q^n$ as fixed parameters) subject to the initial conditions:

$$\xi^i(0) = q^i$$

(corresponding to $\varphi_0 = \text{Id}$). It can be shown that this equation has a unique solution, provided that the vector field is smooth* with t in some interval $[-\epsilon, \epsilon]$, and we restrict q to be in a neighbourhood of p with compact closure. In this way we construct $\xi^i(q, t)$ for each q, which in turn defines maps $\varphi_t(q)$ which can be shown to constitute a local one-parameter group of local diffeomorphisms. Note that the solution $\xi^i(q, t)$ is precisely the coordinates of the

*In fact provided it is Lipschitz (see Sect. 1.4).

curve γ_q through q, having tangent vector X: we refer to γ_q as an *integral curve* of the field X. If we reparametrize γ_q by defining

$$\gamma'_q(t) = \gamma_q(a+t)$$

for a fixed constant a, then γ'_q also has tangent vector X (γ'_q is the curve γ_r for $r = \gamma_q(a)$). So there are an infinite number of integral curves through q, corresponding to different choices of the parametrizations (different choices of a). Suppose that, in each case, we choose just one of these. Then we call the result a *congruence* and denote it by C_X; that is, a set of integral curves of a vector field such that there is one curve through every point of M.

2.2 The Lie derivative

Let X be a vector field on a set U in M such that a corresponding local one-parameter group of local transformations, $\{\varphi_t\}$, can be defined as in Sect. 2.1. If T is any tensor field, we have defined in the previous sections the field $\bar{\varphi}_{s*}T$ obtained by mapping T with the transformations φ_s. The *Lie derivative* of T with respect to s is defined to be (minus) the rate of change of T under this mapping:

$$\pounds_X T \equiv -\frac{d}{ds}\bar{\varphi}_{s*}T|_{s=0} = \lim_{s\to 0}\frac{1}{s}\left[T - \bar{\varphi}_{s*}T\right]. \tag{2.2.1}$$

We shall first examine this in detail for the cases where T is a function and then a vector field, later saying how it applies to general tensor fields. For T a function (a tensor field of rank $(0,0)$) the Lie derivative reduces to the ordinary directional derivative, because (2.2.1) becomes, from (2.1.15),

$$\pounds_X f = -\frac{d}{ds}(f\circ\varphi_s^{-1})$$

and so

$$(\pounds_X f)(p) = -\frac{d}{ds}f(\varphi_s^{-1}(p)) = -\frac{d}{ds}f(\varphi_{-s}(p))$$

because $\varphi_s\circ\varphi_{-s} = \mathrm{Id}$ locally. Hence the above expression becomes:

$$-\frac{d}{ds}f(\gamma_p(-s)) = \dot{\gamma}_p f = X(p)f = (X(f))(p)$$

i.e.

$$\pounds_X f = X(f). \tag{2.2.2}$$

(This result provides the motivation for the minus sign in the definition (2.2.1)). Later we shall also interpret this result as showing that $\pounds$ is a *convective derivative* in the sense of fluid mechanics. Now suppose T is a vector field Y. We can also associate a 1-parameter family $\{\psi_t\}$ with Y: let us write $\kappa_q \in C_Y$ for the integral curves of Y, with

$$\kappa_q(t) = \psi_t(q)$$

retaining γ_q for the integral curves of X in C_X (Fig. 2-1). The action of φ_s will move the curves κ_q into curves that no longer lie in C_Y; thus the curve κ_q through $q = \varphi_s^{-1}(p)$ is mapped to a curve $\varphi_s \circ \kappa_q$ that passes through $\varphi_s(q) = p$ but is different from κ_p. The induced vector field $\bar{\varphi}_{s*}Y$ at p is the vector $\varphi_{s*}(Y(q))$ (Cf. 2.1.10) tangent to $\varphi_s \circ \kappa_q$, which is different from $Y(p) = \dot{\kappa}_p$. The definition (2.2.1) then gives:

$$\begin{aligned}(\pounds_X Y)(p) &= -\frac{d}{ds}\varphi_{s*}\left(Y(\varphi_s^{-1}(p))\right)\Big|_{s=0} \\ &= \lim_{s\to 0}\frac{1}{s}\left[Y(p) - \varphi_{s*}(Y(\varphi_s^{-1}(p)))\right].\end{aligned}$$

When acting on a function f this gives

$$\begin{aligned}(\pounds_X Y)(p)f &= -\frac{d}{ds}\varphi_{s*}(Y(\varphi_s^{-1}(p)))f\Big|_{s=0} \\ &= -\frac{d}{ds}Y(\varphi_s^{-1}(p))(f\circ\varphi_s)\Big|_{s=0}.\end{aligned}$$

We write this in terms of an arbitrary coordinate system and use the fact that locally, $\varphi_s \circ \varphi_{-s} =$Id, i.e. $\varphi_s^{-1} = \varphi_{-s}$. Thus

$$\begin{aligned}(\pounds_X Y)(p)f &= -\frac{d}{ds}Y^j(\varphi_{-s}(p))(f\circ\varphi_s)_{,j}(q)\Big|_{s=0} \\ &= -\frac{d}{ds}\left[Y^j(\varphi_{-s}(p))\left(f_{,k}(p)\frac{\partial\varphi_s^k}{\partial x^j}(q)\right)\right]_{s=0}\end{aligned}$$

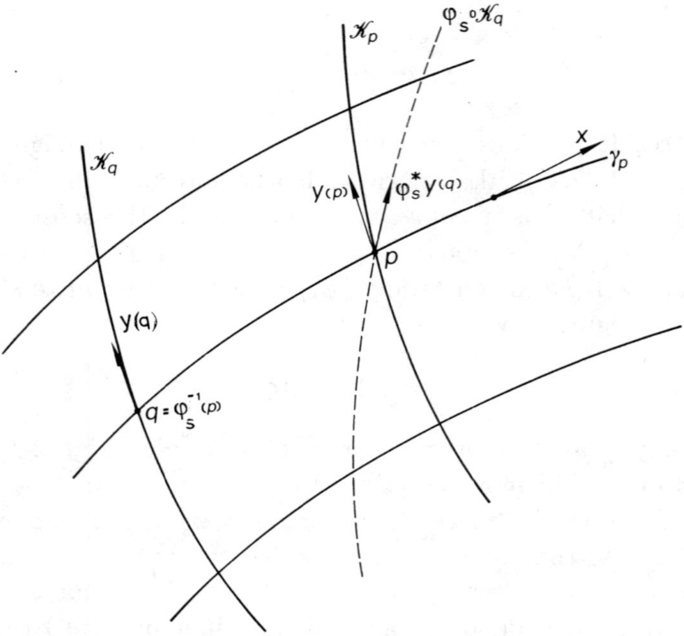

Fig. 2-1 Geometrical interpretation of the Lie derivative of a vector field along a given direction.

recalling that $q_s = \varphi_s^{-1}(p)$. We now do the differentiation, using the fact that

$$\lim_{s\to 0} \varphi_s = \mathrm{Id}, \quad \lim_{s\to 0} \frac{\partial \varphi_s^k}{\partial x^j} = \delta_j^k, \quad \frac{d}{ds}\varphi_{-s}^l(p) = \frac{d}{ds}\gamma_p(-s)^l = -X^l$$

and assuming that φ_s is sufficiently differentiable jointly in s and x^k to justify the above change in order of $\lim_{s\to 0}$ and $\partial/\partial x^j$ (e.g. φ_s twice differentiable suffices). Then we obtain:

$$\begin{aligned}(\pounds_X Y)(p)f &= X^l(p)Y^j{}_{,l}(p)f_{,j}(p) - Y^j(p)X^k_{,j}(p)f_{,k}(p) \\ &= X^l Y^j_{,l} f_{,j} + X^l Y^j f_{,jl} - X^l Y^j f_{,jl} \\ &\quad - Y^j X^k_{,j} f_{,k} \\ &= X^l(Y^j f_{,j})_{,l} - Y^j(X^k f_{,k})_{,j} \\ &= \{X(Y(f)) - Y(X(f))\}(p)\end{aligned} \tag{2.2.3}$$

where everything is evaluated at p and f is twice differentiable. This relation is usually written

$$\pounds_X Y = XY - YX \tag{2.2.4}$$

and the abbreviation

$$[X, Y] \equiv XY - YX$$

(called the *commutator* or *Lie bracket* of X and Y) is introduced, to write (2.2.4) as

$$\pounds_X Y = [X, Y]\,. \tag{2.2.5}$$

For general vector fields u, v, w in $U \in M$, the following properties hold:

$$[u, v] = -[v, u] \tag{2.2.6}$$

$$[fu, gv] = fg[u, v] + fu(g)v - gv(f)u \tag{2.2.7}$$

$$[u, [v, w]] + [w, [u, v]] + [v, [w, u]] = o \tag{2.2.8}$$

$$[w, u + v] = [w, u] + [w, v] \tag{2.2.9}$$

with f,$g \in \mathcal{F}$ and o denoting the zero vector. Here property (2.2.8) is known as the Jacobi identity. If we refer to a basis field in U, we have from (2.2.7):

$$\pounds_X u = [X, u] = X^i u^j [e_i, e_j] + X^i e_i(u^j) e_j - u^j e_j(X^i) e_i. \tag{2.2.10}$$

The Lie bracket of any two vectors of the basis can be written in terms of the same basis as:

$$[e_i, e_j] = C^k{}_{ij} e_k \tag{2.2.11}$$

where the components $C^k{}_{ij}$ are called the *structure coefficients*; they are antisymmetric

$$C^k{}_{ij} = C^k{}_{[ij]}. \tag{2.2.12}$$

A basis field in U is a coordinate basis (or holonomic) if and only if its structure coefficients vanish:

$$[\partial_i, \partial_j] = o. \tag{2.2.13}$$

(This will be discussed in more detail in Sect. 2.12).

Under a general basis transformation:

$$e'_i = A_i{}^j e_j \tag{2.2.14}$$

with $A_i{}^j$ at least C^1 in U, the structure coefficients transform as follows:

$$C'^m{}_{ij} = A_i{}^r A_j{}^s \overset{-1}{A}{}^m{}_k C^k{}_{rs} + 2A_i{}^r A_j{}^s e_{[s} \overset{-1}{A}{}^m{}_{r]} , \tag{2.2.15}$$

hence they are not the components of a tensor. From (2.2.7) we find the Lie bracket of any vector of the basis with respect to X:

$$\pounds_X e_i = [X, e_i] = X^k C^m{}_{ki} e_m - e_i(X^k) e_k. \tag{2.2.16}$$

In terms of a coordinate basis (2.2.10) and (2.2.16) read, from (2.2.13):

$$[X, Y]^i = X^k \partial_k Y^i - Y^k \partial_k X^i \tag{2.2.17}$$

$$[X, \partial_j]^i = -\partial_j X^i \tag{2.2.18}$$

(the first result agreeing with (2.2.3) above).

Before moving on to the Lie derivative of a one-form, we shall establish the Leibniz rule for tensor products, in the form

$$\pounds_X (T \otimes S) = (\pounds_X T) \otimes S + T \otimes (\pounds_X S). \tag{2.2.19}$$

To prove this, note that from (2.2.1)

$$\pounds_X (T \otimes S) = \lim_{s \to 0} \frac{1}{s} [T \otimes S - \bar{\varphi}_{s*} T \otimes \bar{\varphi}_{s*} S]$$

from (2.1.13) and hence

$$\begin{aligned} &\lim_{s \to 0} \frac{1}{s} [(T - \bar{\varphi}_{s*} T) \otimes S + \bar{\varphi}_{s*} T \otimes (S - \bar{\varphi}_{s*} S)] \\ &\qquad = (\pounds_X T) \otimes S + \lim_{s \to 0} (\bar{\varphi}_{s*} T) \otimes \lim_{s \to 0} \frac{1}{s} (S - \bar{\varphi}_{s*} S) \\ &\qquad = (\pounds_X T) \otimes S + T \otimes \pounds_X S \end{aligned} \tag{2.2.20}$$

as required (provided the limits exist). Next we note that Lie differentiation commutes with contraction: if CT denotes the contraction of the tensor T on some pair of indices, then

$$\pounds_X CT = \lim_{s \to 0} \frac{1}{s} [CT - \bar{\varphi}_{s*} CT] = \lim_{s \to 0} \frac{1}{s} [CT - C \bar{\varphi}_{s*} T] = C \pounds_X T \tag{2.2.21}$$

from (2.1.16).

Now if $\boldsymbol{\omega}$ is a one-form and Y a vector field, we have

$$\begin{aligned}(\pounds_X\boldsymbol{\omega})(Y) &= C(\pounds_X\boldsymbol{\omega})\otimes Y \\ &= C\left[\pounds_X(\boldsymbol{\omega}\otimes Y) - \boldsymbol{\omega}\otimes \pounds_X Y\right] && \text{(from 2.2.20)} \\ &= \pounds_X(C\boldsymbol{\omega}\otimes Y) - \boldsymbol{\omega}(\pounds_X Y) && \text{(from 2.2.21)} \\ &= X(\boldsymbol{\omega}(Y)) - \boldsymbol{\omega}(\pounds_X Y) && \text{(from 2.2.19)}\end{aligned} \tag{2.2.22}$$

If we take Y to be a member e_k of a basis one-form this becomes:

$$(\pounds_X\boldsymbol{\omega})_k = (\pounds_X\boldsymbol{\omega})(e_k) = X(\omega_k) - \boldsymbol{\omega}([X,e_k]) \tag{2.2.23}$$

from (2.2.5). From this, the Lie derivative of any basis form, $\boldsymbol{e}^i$ say, reads:

$$\begin{aligned}\pounds_X\boldsymbol{e}^i &= -\boldsymbol{e}^i([X,e_k])\boldsymbol{e}^k \\ &= -X^r C^m{}_{rk}\boldsymbol{e}^i(e_m)\boldsymbol{e}^k + e_k(X^r)\boldsymbol{e}^i(e_r)\boldsymbol{e}^k \\ &= \left[e_k(X^i) - X^r C^i{}_{rk}\right]\boldsymbol{e}^k.\end{aligned} \tag{2.2.24}$$

In terms of coordinates (2.2.23) and (2.2.24) become:

$$\pounds_X\boldsymbol{\omega} = \left[X^i\partial_i\omega_k + \omega_i\partial_k X^i\right]\boldsymbol{e}^k. \tag{2.2.25}$$

$$\pounds_X\boldsymbol{e}^i = (\partial_k X^i)\boldsymbol{e}^k. \tag{2.2.26}$$

Then if we specialize (2.2.19) to a tensor field of valence $(1,1)$, say, we have, in components:

$$\begin{aligned}\pounds_X T = \pounds_X(T^i{}_j e_i\otimes \boldsymbol{e}^j) &= X(T^i{}_j)e_i\otimes\boldsymbol{e}^j + T^i{}_j(\pounds_X e_i)\otimes\boldsymbol{e}^j \\ &\quad + T^i{}_j e_i\otimes(\pounds_X\boldsymbol{e}^j) \\ &= [X^r e_r(T^i{}_j) - T^s{}_j e_s(X^i) + T^i{}_s e_j(X^s) \\ &\quad + X^r(T^s{}_j C^i{}_{rs} - T^i{}_s C^s{}_{rj})]e_i\otimes\boldsymbol{e}^j\end{aligned} \tag{2.2.27}$$

(from (2.2.20) and (2.2.7)). In the case of a coordinate basis, this reads:

$$\pounds_X T = \left[X^r\partial_r T^i{}_j - T^s{}_j\partial_s(X^i) + T^i{}_s\partial_j(X^s)\right]\partial_i\otimes\boldsymbol{dx}^j. \tag{2.2.28}$$

The Lie derivative generalizes to an arbitrary manifold the concept of the convective derivative of a scalar field (i.e. a function) along a given direction. We have already seen in (2.2.2) that the Lie derivative of a function is

$$\pounds_X f = X(f). \tag{2.2.29}$$

If we are given a vector field Y (or a field of one-forms $\boldsymbol{\omega}$) in U, the Lie derivative reads from (2.2.10) and (2.2.16) (or (2.2.21) and (2.2.22)):

$$\pounds_X Y = X(Y^k)e_k + Y^k \pounds_X e_k \tag{2.2.30}$$

$$\pounds_X \boldsymbol{\omega} = X(\omega_k)e_k + \omega_k \pounds_X \mathbf{e}^k \tag{2.2.30$'$}$$

namely it is the convective derivative of the field components *corrected* by the spurious contributions to their change which arise from the change of the basis along C_X. For this reason the Lie derivative of a field along a given direction measures the *intrinsic* variation of the field along that direction and a field is said to be Lie constant (or Lie transported) along a given direction when its Lie derivative vanishes (see Chevalley, 1946).

2.3 The connector

The Lie derivative measures in an n-dimensional differentiable manifold the (intrinsic) variation of a field (scalar, vector, tensor) in a given direction.

The value of the Lie derivative at a point however depends on the behaviour of the field in the neighbourhood of that point. How is one to study the behaviour of a field on a single curve? Let $\gamma(t)$ be such a curve; given a scalar field $f \in \mathcal{F}$ on M, a differential operation which is meaningful in this case is the convective derivative:

$$\dot{\gamma}(f) = \underset{p}{e}_k(f)\dot{\gamma}^k_p = \frac{d}{dt}[f \circ \gamma(t)] \tag{2.3.1}$$

where $t \in (a, b)$ and $\{e_k\}_p$ $(p = \gamma(t))$ is a basis in $T_{\gamma(t)}(M)$. The result of the operation (2.3.1), being independent of the bases on $\gamma(t)$ (and hence also of the coordinates) is a scalar field on $\gamma(t)$. Operation (2.3.1) is no longer meaningful when it is applied to the components of a vector field u defined on $\gamma(t)$, because

$$\dot{\gamma}(u^i)|_{\gamma(t)} = e_k(u^i)\dot{\gamma}^k|_\gamma \qquad i = 1, 2, \ldots, n \tag{2.3.2}$$

is a set of n scalar quantities which are not the components of a vector. Changing the basis field in a neighbourhood of $p = \gamma(t)$

we have:

$$\begin{aligned} a) \quad & \underset{p}{e}'_i = A_i{}^k \underset{p}{e}_k \quad u'^i = \overset{-1}{A}{}^i{}_k u^k \\ b) \quad & \dot{\gamma}_p(u'^i) = \overset{-1}{A}{}^i{}_k \dot{\gamma}_p(u^k) + \dot{\gamma}_p(\overset{-1}{A}{}^i{}_k) u^k. \end{aligned} \tag{2.3.3}$$

The last term on the right hand side of (2.3.3)b destroys the vectorial character of the transformation unless the matrix $A^i{}_k$ is constant on $\gamma(t)$.

The directional derivative of the components of a vector field (or of a tensor field) is not a satisfactory operation. In order to introduce a differential operation which preserves the vectorial (tensorial) character of a field, we introduce on the manifold a new algebraic structure called a *connector*.

The underlying idea is that in order to differentiate a vector u along a curve we need to *move* the vector at $\gamma(t)$ along to $\gamma(t+\delta t)$ and compare it with the value of u there; if we can do this without introducing a particular coordinate system, then we will achieve a purely vectorial operation of differentiation. We stress that, although we make heavy use of an arbitrary coordinate system to investigate the properties of this operation (frequently subtracting the components of vectors at one point from those of a vector at another point) the operation of moving vectors along a curve is independent of any particular coordinate system, as are the final results that we shall obtain.

Suppose $\gamma : \mathbf{R} \to M$ is a path in the manifold M and that p, q are points on γ, say $\gamma(a) = p$, $\gamma(b) = q$. We aim to introduce a map, denoted by

$$\Gamma(a, b; \gamma) : T_p(M) \longrightarrow T_q(M) \tag{2.3.4}$$

which can be thought of as the effect of carrying vectors from p to q along γ. It is natural to impose the following restrictions on Γ.

i) *Linearity.* Γ respects the linear structure of the tangent spaces by being itself a linear map. Thus, given u and v at p, we have

$$\Gamma(a, b; \gamma)\,(\alpha u + \beta v) = \alpha\Gamma(a, b; \gamma)(u) + \beta\Gamma(a, b; \gamma)(v) \tag{2.3.5}$$

α and β being real numbers. In particular if $u = u^i \underset{p}{e}_i$ then

$$\Gamma(a,b;\gamma)(u) = u^i\Gamma(a,b;\gamma)(\underset{p}{e}_i) = u^i\Gamma(a,b;\gamma)_i{}^j \underset{q}{e}_j \tag{2.3.6}$$

where the components $\Gamma_i{}^j$ are defined as:

$$\Gamma(a,b;\gamma)\underset{p}{e}_i = \Gamma(a,b;\gamma)_i{}^j \underset{q}{e}_j.$$

Here and in what follows, we adopt the convention that the *first* index of Γ refers to a basis at the first point, the *second* one to that at the second point.

ii) *Consistency.* If we carry a vector from p to q, and then from q to r (where $r = \gamma(c)$, say), we require that this is the same as carrying the vector directly from p to r. This means that:

$$\Gamma(b,c;\gamma)\circ\Gamma(a,b;\gamma) = \Gamma(a,c;\gamma)$$

where the symbol "$\circ$" on the left-hand side denotes composition of maps.

In terms of the expressions (2.3.6), using bases at p, q and r, this is :

$$\Gamma(a,b;\gamma)_i{}^j\Gamma(b,c;\gamma)_j{}^k = \Gamma(a,c;\gamma)_i{}^k. \tag{2.3.7}$$

Furthermore we require that

$$\Gamma(a,a;\gamma)_i{}^j = \delta_i^j. \tag{2.3.8}$$

iii) *Parametrization independence.* Γ should not depend on the way the curve between $p = \gamma(a)$ and $q = \gamma(b)$ is parametrized, but only on the geometrical set of points defined by the curve. Thus if a new parameter s' is defined such that $s = \sigma(s')$, and the reparametrized curve is:

$$\gamma'(s') = \gamma(\sigma(s')) \tag{2.3.9a}$$

then

$$\Gamma(\sigma(a),\sigma(b);\gamma) = \Gamma(a,b;\gamma'). \tag{2.3.9b}$$

iv) *Differentiability.* We require that, as we vary p or q, or as we deform the whole path γ between p and q, so $\Gamma(a,b;\gamma)$ should change smoothly. Variation of a and b is easy to formulate; we simply demand that (for each i, j and γ):

$$\Gamma(a,b;\gamma)_i{}^j \qquad \text{is a smooth function of } a \text{ and } b. \tag{2.3.10}$$

Variation of the whole of γ needs more careful formulation, however, involving a few technical definitions from advanced calculus.

Recall that if f is an ordinary function from $\mathbb{R}$ to $\mathbb{R}$, we call f *continuous* if

$$|f(x+\delta x) - f(x)| \to 0 \qquad \text{as} \qquad \delta x \to 0$$

while f is *differentiable* if the quotient

$$\frac{[f(x+\delta x) - f(x) - f'(x)\delta x]}{\delta x}$$

tends to zero as δx tends to zero; i.e. if, for any small $\epsilon > 0$, we can find a number $\delta(\epsilon)$ such that for all $|\delta x| < \delta(\epsilon)$ we have:

$$|f(x+\delta x) - f(x) - f'(x)\delta x| < \epsilon|\delta x|. \tag{2.3.11}$$

Now, while f is a function only of x, the map $\Gamma(a,b;\gamma)$ is a function not only of real variables a, b but also of the whole curve γ. We demand that Γ should vary differentiably as the whole of γ is varied, in the sense of (2.3.11), with $f(\cdot)$ replaced by $\Gamma(a,b;\cdot)$ and x replaced by γ. For this to make sense, two modifications have to be made. First, $|\delta x|$ has to be turned into $\|\delta\gamma\|$, where this denotes a real number measuring the size of $\delta\gamma$ (to be fixed shortly). Second, $f'(x)\delta x$ has to be replaced by $\Gamma'(a,b;\gamma)_i{}^j(\delta\gamma)$ where, for each $(i,\,j)$, $\Gamma'(a,b;\gamma)_i{}^j$ is a linear function of the whole variation $\delta\gamma$. (Recall that $\delta\gamma$ is itself a function of the curve-parameter s, while $\Gamma'(a,b;\gamma)_i{}^j(\delta\gamma)$ is, for each pair $(i,\,j)$, a single number depending on the whole function $\delta\gamma$ in a linear way).

To complete the definition of Γ', we have to fix the measure $\|\delta\gamma\|$ of the size of $\delta\gamma$. We need to ensure that, if $\|\delta\gamma\|$ is small, then not only must $\delta\gamma^i(s)$ be small for each s, but also $\delta\dot{\gamma}^i(s)$ must be small. Otherwise we could have variations where each $\gamma^i(s)$ moved very little but where the total length of γ (measured by integrating $|\dot{\gamma}^i(s)|$) changed radically; and we would not want to regard this as a *small* variation from the point of view of carrying vectors along the curve. So we define $\|\ \|$ by setting, for any function $f : [h,d] \to \mathbb{R}$ with a continuous derivative:

$$\|f\| = |f(h)| + \sup_{s\in[h,d]} |\dot{f}(s)|. \tag{2.3.12}$$

We denote the space of functions f for which this is defined by C, and we take γ^i, for each i (and hence $\delta\gamma^i$), to belong to C. C is a vector space (functions can be added and multiplied by constants in the obvious way), and so the concept of a linear map on C makes sense. The required translation of (2.3.11) then reads as follows: for each γ with γ^i in C and each a, b in $[h, d]$ there exist linear functions of $\delta\gamma$, $\Gamma'(a,b;\gamma)_i{}^j$, such that for any $\epsilon > 0$ we can find a number $\delta(\epsilon) > 0$ so that, for all $\delta\gamma$ with $\delta\gamma^i \in C$ and $\|\delta\gamma^i\| < \delta(\epsilon)$ we have:

$$\left|\Gamma(a,b;\gamma+\delta\gamma)_i{}^j - \Gamma(a,b;\gamma)_i{}^j - \Gamma'(a,b;\gamma)_i{}^j(\delta\gamma)\right| < \epsilon \max_i \|\delta\gamma^i\|. \tag{2.3.13}$$

Finally we require that, in a given coordinate neighbourhood, the effect of Γ' should not be arbitrarily large; i.e. that there is a constant K such that:

$$\left|\Gamma'(a,b;\gamma)_i{}^j(\delta\gamma)\right| < K \max_i \|\delta\gamma^i\|. \tag{2.3.14}$$

2.4 Parallel propagation and geodesics

We now study the consequences of the requirements placed on Γ in the previous section. Suppose γ_1 and γ_0 are two curves with $\gamma_1(a) = \gamma_0(a)$. Define a one-parameter family interpolating between them by:

$$\gamma_t^i(s) = t\gamma_1^i(s) + (1-t)\gamma_0^i(s).$$

It is then simple to verify, from the definition of the derivative (2.3.13), that the chain rule holds in the sense that:

$$\frac{d}{dt}\Gamma(a,b;\gamma_t)_i{}^j = \Gamma'(a,b;\gamma_t)_i{}^j(\gamma'_t) \tag{2.4.1}$$

where $\gamma'^i_t(s) = d\gamma_t^i(s)/dt = \gamma_1^i(s) - \gamma_0^i(s)$. Integrating between 0 and 1 and taking moduli gives

$$\begin{aligned} &\left|\Gamma(a,b;\gamma_1)_i{}^j - \Gamma(a,b;\gamma_0)_i{}^j\right| \\ &\leq \int_0^1 \left|\Gamma'(a,b;\gamma_t)_i{}^j(\gamma'_t)\right| dt \\ &< K \max_i \|\gamma'^i_t\| = K \sup_{s\in[a,b]} \max_i |\dot\gamma_1^i(s) - \dot\gamma_0^i(s)| \end{aligned} \tag{2.4.2}$$

from (2.3.14) and (2.3.12). This gives a specific expression for the way in which Γ varies continuously as γ is varied.

Next we use the differentiability with respect to b to determine how Γ varies as we move along a fixed curve. Define:

$$\bar{\Gamma}(s_0;\gamma)_i{}^j = \frac{d}{ds}\Gamma(s_0,s;\gamma)_i{}^j\bigg|_{s=s_0}. \tag{2.4.3}$$

(This exists from (2.3.10)). Using (2.3.7) we find:

$$\begin{aligned}\frac{d}{ds}\Gamma(a,s;\gamma)_i{}^j\bigg|_{s=s_0} &= \frac{d}{ds}\left(\Gamma(a,s_0;\gamma)_i{}^k\Gamma(s_0,s;\gamma)_k{}^j\right)\bigg|_{s=s_0} \\ &= \Gamma(a,s_0;\gamma)_i{}^k\bar{\Gamma}(s_0;\gamma)_k{}^j. \end{aligned}\tag{2.4.4}$$

This is a differential equation for $\Gamma(a,s;\gamma)$ as a function of s, with the initial condition at $s=a$, provided by (2.3.8). Thus the map Γ transporting vectors is fixed once we fix $\bar{\Gamma}$.

We shall now compare γ with the straight line μ (in a given coordinate system) defined by:

$$\mu^i(t) = t\dot{\gamma}^i(0) + \gamma^i(0). \tag{2.4.5}$$

It will be helpful to reparametrize both μ and γ, defining curves μ_s and γ_s by:

$$\gamma_s(t) = \gamma(st) \qquad \mu_s(t) = \mu(st).$$

Then we have that:

$$\begin{aligned}\left|\Gamma(0,s;\gamma)_i{}^j - \Gamma(0,s;\mu)_i{}^j\right| &= \left|\Gamma(0,1;\gamma_s)_i{}^j - \Gamma(0,1;\mu_s)_i{}^j\right| \\ &\le K\sup_t\max_i|\dot{\gamma}^i_s(t) - \dot{\mu}^i_s(t)| \\ &= Ks\sup_t\max_i|\dot{\gamma}^i(st) - \dot{\gamma}^i(0)| \end{aligned}\tag{2.4.6}$$

from (2.4.2), where

$$\sup_t\max_i|\dot{\gamma}^i(st) - \dot{\gamma}^i(0)| = \epsilon_1(s) \to 0 \qquad (s\to 0)$$

since $\dot{\gamma}^i(st)\to\dot{\gamma}^i(0)$ as $s\to 0$. In other words, $\Gamma(0,s;\gamma)$ is closely approximated by $\Gamma(0,s;\mu)$ for small s. But now the differential equation (2.4.4) for Γ, gives in the case $a=0$ with (2.3.8)

$$\Gamma(0,s;\gamma)_i{}^j = \delta_i^j + s\bar{\Gamma}(0;\gamma)_i{}^j + s\epsilon_2(s)$$

($\epsilon_2(s) \to 0$ as $s \to 0$), and similarly for μ:

$$\Gamma(0, s; \mu)_i{}^j = \delta_i^j + s\bar{\Gamma}(0; \mu)_i{}^j + s\epsilon_3(s).$$

Subtracting, using (2.4.6) and letting $s \to 0$ gives

$$\bar{\Gamma}(0; \gamma)_i{}^j = \bar{\Gamma}(0; \mu)_i{}^j. \tag{2.4.7}$$

Thus from (2.4.5), $\bar{\Gamma}$ depends only on the initial tangent vector $\dot{\gamma}(0)$. We can repeat the argument, replacing $\gamma(0)$ by any $\gamma(a)$ to deduce that $\bar{\Gamma}(a; \gamma)_i{}^j$ depends only on $\dot{\gamma}(a)$.

Finally we shall show that this dependence is linear. We have:

$$\bar{\Gamma}(0; \mu)_i{}^j = \frac{d}{ds}\Gamma(0, s; \mu)_i{}^j\bigg|_{s=0} = \frac{d}{ds}\Gamma(0, 1; \mu_s)_i{}^j\bigg|_{s=0}$$
$$= \Gamma'(0, 1; \mu_s)_i{}^j(\mu'_s)\bigg|_{s=0}$$

(from (2.4.1) with $\mu'^i_s(t) = d\mu^i_s(t)/ds = t\dot{\mu}^i(st) = t\dot{\gamma}^i(0)$). Setting $s = 0$, μ_0 is the constant map $t \to \gamma(0)$, μ'_0 is the map $t \to t\dot{\gamma}(0)$, and so:

$$\bar{\Gamma}(0; \mu)_i{}^j = \Gamma'(0, 1; \mu_0)_i{}^j(\mu'_0)$$

depends linearly on μ'_0 (from the definition of Γ'), while μ'_0 depends linearly on $\dot{\gamma}(0)$. Thus $\bar{\Gamma}(0; \mu)$ depends linearly on $\dot{\gamma}$ and we can write:

$$\bar{\Gamma}(0; \mu)_i{}^j = -\Gamma^j{}_{ik}(\gamma(0))\dot{\gamma}^k(0), \tag{2.4.8}$$

the coefficients $\Gamma^j{}_{ik}$ depending only on $\gamma(0)$. More generally, using (2.4.7) for general a:

$$\bar{\Gamma}(a; \gamma)_i{}^j = -\Gamma^j{}_{ik}(\gamma(a))\dot{\gamma}^k(a). \tag{2.4.9}$$

Inserting this in (2.4.4) gives for the connector*

$$\frac{d}{ds}\Gamma(a, s; \gamma)_i{}^j|_s = -\Gamma(a, s; \gamma)_i{}^k\Gamma^j_{km}(\gamma(s))\dot{\gamma}^m(s) \tag{2.4.10}$$

*It may be convenient here to point out that, from (2.3.7) and (2.3.8), the derivative of the inverse map $\Gamma(s, a; \gamma)$ reads as follows

$$\frac{d}{ds}\Gamma(s, a; \gamma)_i{}^j\bigg|_s = \Gamma^k_{ir}\dot{\gamma}^r(s)\Gamma(s, a; \gamma)_k{}^j$$

where we recall the initial condition (2.3.8):

$$\Gamma(a,a;\gamma)_i{}^j = \delta_i^j. \tag{2.4.11}$$

The construction can be, and usually is, reversed: given a field of numbers $\Gamma^j_{km}(p)$, we define $\Gamma(a,s;\gamma)$ with (2.4.10) and (2.4.11) to produce a connector with the required properties.

What we have shown is that this gives the most general connector, subject to the conditions of the previous section. A vector $u(s)$ defined at each point $\gamma(s)$ of a curve is said to be *parallely propagated* if

$$u^i(s) = \Gamma(0,s;\gamma)_j{}^i u^j(0)\ . \tag{2.4.12}$$

A *geodesic* is a curve γ whose tangent vector is everywhere non-zero and is proportional to a parallely propagated vector; i.e.

$$\Gamma(0,s;\gamma)_i{}^j \dot\gamma^i(0) = \lambda(s)\dot\gamma^j(s) \tag{2.4.13}$$

for a differentiable function λ with $\lambda(s) > 0$. From (2.4.11) we can take $\lambda(0) = 1$. Differentiating (2.4.13) with respect to s and using (2.4.10) we obtain:

$$\left(\frac{d}{ds}\Gamma(0,s;\gamma)_i{}^j\right)_s \dot\gamma^i(0) = -\Gamma(0,s;\gamma)_i{}^k \Gamma^j_{km}(\gamma(s))\dot\gamma^m(s)\dot\gamma^i(0)$$
$$= -\Gamma^j_{km}(\gamma(s))\dot\gamma^k(s)\dot\gamma^m(s)\lambda(s) = \frac{d\lambda(s)}{ds}\dot\gamma^j(s) + \lambda(s)\frac{d\dot\gamma^j}{ds};$$

thus from (2.4.13):

$$\frac{d\dot\gamma^i}{ds} + \Gamma^i_{kj}\dot\gamma^k\dot\gamma^j = f(s)\dot\gamma^i \tag{2.4.14}$$

where $f(s) = d\ln(\lambda^{-1}(s))/ds$. This is the differential equation for a geodesic. Suppose we reparametrize γ by defining $s = \sigma(t)$:

$$\gamma'(t) = \gamma(\sigma(t)) \tag{2.4.15}$$

for a smooth, strictly monotonic function σ. Then:

$$\dot\gamma'(t) = \dot\gamma(\sigma(t))\frac{d\sigma}{dt} = \frac{1}{\lambda(\sigma(t))}\frac{d\sigma}{dt}\Gamma(0,\sigma;\gamma)\dot\gamma(0) \tag{2.4.16}$$

so from (2.4.15), (2.4.13) and (2.4.16):

$$\dot\gamma'(t) = \frac{d\sigma(t)}{dt}\frac{1}{\lambda(\sigma(t))(\mathrm{d}\sigma/\mathrm{d}t)_0}\Gamma(0,t;\gamma')\dot\gamma'(0) \tag{2.4.17}$$

since $\gamma(\sigma(t))$ is a reparametrization of $\gamma'(t)$ and $\lambda(\sigma(t)) \neq 0$. Now choose σ to satisfy

$$\frac{d\sigma}{dt} = \lambda(\sigma(t)) \tag{2.4.18}$$

a differential equation that always has a solution if λ is differentiable. Then (2.4.13) becomes

$$\dot{\gamma}'(t) = \Gamma(0, t; \gamma')\dot{\gamma}'(0) \tag{2.4.19}$$

(using $\lambda(0) = 1$ and hence $d\sigma/dt|_0 = 1$). We have thus proved the following:

THEOREM

Every geodesic can be parametrized so as to satisfy (2.4.19).

The parameter $t = \sigma^{-1}(s)$ that achieves this is called an *affine* parameter: in this case the tangent vector is parallely propagated and not just proportional to a parallely propagated vector. An affinely parametrized geodesic is termed affine and will be denoted by $\boldsymbol{\gamma}$. The affine parameter is not unique: if we make a further change by setting:

$$\boldsymbol{\gamma}'(u) = \boldsymbol{\gamma}'(\tau(u))$$

then repeating the analysis leading to (2.4.17), we see that if $d\tau(u)/du = (d\tau/du)_0$, then $\boldsymbol{\gamma}'$ will also satisfy (2.4.19) (with u instead of t). This will be the case if

$$\tau(u) = au + b \qquad a, b \in \mathbf{R}. \tag{2.4.20}$$

The differential equation for an affine geodesic now reads, from (2.4.19):

$$\left.\frac{d\dot{\gamma}^i}{dt}\right|_t + \Gamma^i_{jk}(\gamma(t))\dot{\gamma}^j(t)\dot{\gamma}^k(t) = 0. \tag{2.4.21}$$

2.5 Transformation properties of the connector

Since $\Gamma(a,b;\gamma)$ carries vectors at $p=\gamma(a)$ into vectors at $q=\gamma(b)$, let us denote by $\underset{q}{\breve{u}}$ the image at q of a vector $\underset{p}{u}$ at p under Γ:

$$\underset{q}{\breve{u}} = \Gamma(a,b;\gamma)_i{}^j \underset{p}{u}{}^i \underset{q}{e}_j \,. \tag{2.5.1}$$

The set of n^2 numbers $\Gamma_i{}^j$ forms a new mathematical entity termed *bitensor*, which behaves like a covector at p and a vector at q. In fact if we change the bases at p and q as follows:

$$\underset{p}{e}'_i = A_i{}^k(p)\underset{p}{e}_k \qquad \underset{q}{e}_i = A_i{}^k(q)\underset{q}{e}_k \tag{2.5.2}$$

from the linearity of Γ, we have:

$$\begin{aligned}\Gamma(a,b;\gamma)(\underset{p}{e}'_i) &= A_i{}^k(p)\Gamma(a,b;\gamma)(\underset{p}{e}_k) = A_i{}^k(p)\Gamma(a,b;\gamma)_k{}^j \underset{q}{e}_j \\ &= \Gamma'(a,b;\gamma)_i{}^j \underset{q}{e}'_j = \Gamma'(a,b;\gamma)_i{}^r A_r{}^j(q)\underset{q}{e}_j\end{aligned}$$

hence, equating the coefficients of $\underset{q}{e}_j$, we have:

$$\Gamma'(a,b;\gamma)_i{}^j = A_i{}^r(p)\Gamma(a,b;\gamma)_r{}^s \overset{-1}{A}_s{}^j(q). \tag{2.5.3}$$

Here the dashed Γ should not be confused with the derivative of Γ with respect to the curve γ, introduced in the previous section. As a consequence of (2.5.3), the coefficients Γ^j_{ik} do not transform as the components of a tensor. To show this, let us first differentiate (2.5.1) with respect to the parameter s on γ, to obtain, from (2.4.10):

$$\left.\frac{d\breve{u}^i}{ds}\right|_q = \left(\frac{d\Gamma(a,s;\gamma)_j{}^i}{ds}\right)_q \underset{p}{u}{}^j = -\underset{q}{\breve{u}}{}^k\Gamma^i_{km}(b)\dot{\gamma}^m(b). \tag{2.5.4}$$

Then let us transform the tangent basis at q so that

$$\begin{aligned}\frac{d\breve{u}'^i}{ds} = -\breve{u}'^k\Gamma'^i_{km}\dot{\gamma}'^m &= -\overset{-1}{A}_j{}^k\Gamma'^i_{km}\overset{-1}{A}_n{}^m\dot{\gamma}^n\breve{u}^j \\ &= \frac{d}{ds}\overset{-1}{A}_j{}^i\breve{u}^j + \overset{-1}{A}_j{}^i\frac{d\breve{u}^j}{ds}.\end{aligned} \tag{2.5.5}$$

Using (2.5.4) in (2.5.5), we obtain at q, from the arbitrariness of $\dot{\gamma}$ and u

$$-\overset{-1}{A}_j{}^k\overset{-1}{A}_n{}^m\Gamma'^i_{km} = \underset{q}{e}_n\left(\overset{-1}{A}_j{}^i\right) - \overset{-1}{A}_m{}^i\Gamma^m_{jn}$$

hence:

$$\Gamma'^{i}_{km} = A_k{}^r A_m{}^s \overset{-1}{A}_j{}^i \Gamma^j_{rs} - A_k{}^r A_m{}^s e_s \left(\overset{-1}{A}_r{}^i\right) . \tag{2.5.6}$$

The coefficients $\Gamma^i_{jk}(q)$ are the components of a new object called the *connection** at q.

2.6 The covariant derivative

Suppose we are given a curve $\gamma : [0,1] \to M$ with $p = \gamma(0)$ and $q = \gamma(1)$ and a vector at each point of γ:

$$X(s) \in T_{\gamma(s)}(M) \tag{2.6.1}$$

for each $s \in [0,1]$, whose components vary smoothly with s. At the point p we can consider the two vectors $X(0)$ and $\Gamma(s,0;\gamma)(X(s)) = \check{X}(s) \in T_p(M)$ with $\check{X}(0) = X(0)$; (Fig. 2-2). The difference between these will be again a vector at p, and hence so will be the derivative of $\check{X}(s)$ with respect to s. So we can define the *absolute derivative* of X at p along γ to be the quantity:

$$\begin{aligned} \frac{DX}{Ds} &= \lim_{s\to 0} \frac{1}{s}\left[\check{X}(s) - X(0)\right] = \left.\frac{d\check{X}(s)}{ds}\right|_{s=0} \\ &= \left(\frac{d}{ds}\Gamma(s,0;\gamma)X(s)\right)_{s=0} + \Gamma(s,0;\gamma)\left.\frac{dX(s)}{ds}\right|_{s=0} \\ &= \left[\left(\frac{dX^i(s)}{ds}\right)_{s=0} + \Gamma^i_{jk}(\gamma(0))X^j(0)\dot{\gamma}^k(0)\right] \underset{pi}{e}. \end{aligned} \tag{2.6.2}$$

Now suppose Y is a vector field and γ a given curve. We can define

$$X(s) \equiv \underset{\gamma(s)}{Y}$$

or, in components,

$$X(s)^i \equiv Y^i(\gamma(s))$$

* Γ^i_{jk} was originally called an affine connection but now is more properly called a linear connection because the term affine transformation has come to mean a transformation of the form $x \mapsto Ax + b$, as opposed to a linear transformation $x \mapsto Ax$.

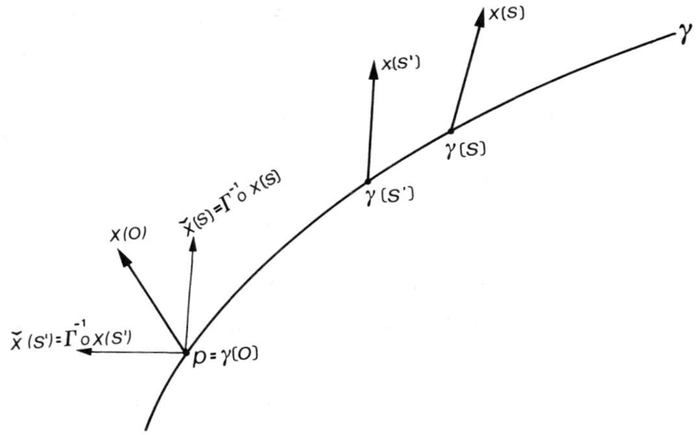

Fig. 2-2 The connector maps a vector field on a curve into a vector field at a point.

so that (2.6.2) becomes

$$\frac{DX(s)}{Ds} = \dot{\gamma}^k \left[\underset{p}{e}_k(Y^i) + \Gamma^i_{jk}(p)Y^j(p)\right] \underset{p}{e}_i \, . \tag{2.6.3}$$

The vectorial expression on the right-hand side is linear in $\dot{\gamma}^k$ and hence defines a tensor of valence (1, 1), whose components we write as

$$\nabla_k Y^i\big|_p \equiv \left[\underset{p}{e}_k(Y^i) + \Gamma^i_{jk}(p)Y^j(p)\right]. \tag{2.6.4}$$

We shall also introduce the following notation for the whole of the right-hand side of (2.6.3):

$$\nabla_{\dot{\gamma}} Y \equiv \dot{\gamma}^k \nabla_k Y^i\big|_p \underset{p}{e}_i.$$

The tensor $\nabla_k Y^i\big|_p$ is called the *covariant derivative* of Y at p, while $\nabla_{\dot{\gamma}} Y$ is called the *absolute derivative* of Y in the direction $\dot{\gamma}$.

We stress here that the notation "$\nabla_k Y^i$" is not to be interpreted as the directional derivative of the scalar field Y^i (the ith component of Y): to avoid ambiguity we shall always write the latter as $e_k(Y^i)$.

Next we extend the absolute derivative to one-forms. First, we define a connector $\tilde{\Gamma}(a, b; \gamma) : \overset{*}{T}_p(M) \to \overset{*}{T}_q(M)$ for covariant vectors (where $\gamma : [a,b] \to M$, $\gamma(a) = p$, $\gamma(b) = q$ as usual) by setting:

$$\left(\tilde{\Gamma}(a,b;\gamma)(\boldsymbol{\omega})\right)(u) = \boldsymbol{\omega}\left(\Gamma(b,a;\gamma)(u)\right) \tag{2.6.5}$$

for all $\boldsymbol{\omega} \in \overset{*}{T}_p(M)$, $u \in T_q(M)$. In terms of components, this reads:

$$\omega_j \tilde{\Gamma}(a,b;\gamma)^j{}_i u^i = \omega_j \Gamma(b,a;\gamma)_i{}^j u^i \tag{2.6.6}$$

so that the components of $\tilde{\Gamma}(\gamma)$ constitute the inverse matrix to $\Gamma(\gamma)$.

We can proceed exactly as for vectors. We reparametrize γ so that $p = \gamma(0)$ and $q = \gamma(s)$ and then given a one-form $\boldsymbol{\omega}$ in a neighbourhood of p, define

$$\breve{\boldsymbol{\omega}}_p(s) = \tilde{\Gamma}(s,0;\gamma)(\boldsymbol{\omega}_q(s)) \in \overset{*}{T}_p(M)$$

for each $q = \gamma(s)$. The absolute derivative is then given (as in (2.6.2)) by

$$\nabla_{\dot{\gamma}}\boldsymbol{\omega} = \lim_{s\to 0}\frac{1}{s}\left[\breve{\boldsymbol{\omega}}(s) - \boldsymbol{\omega}(0)\right] = \frac{d}{ds}\left(\tilde{\Gamma}(s,0;\gamma)(\boldsymbol{\omega}(s))\right)\bigg|_{s=0}. \tag{2.6.7}$$

To evaluate this in terms of components, we note that

$$\left(\tilde{\Gamma}(s,0;\gamma)(\boldsymbol{\omega}(s))\right)_i = \tilde{\Gamma}(s,0;\gamma)^j{}_i \omega_j(s) = \Gamma(0,s;\gamma)_i{}^j \omega_j(s)$$

from (2.6.6), taking the inverse, so that (2.6.7) becomes:

$$\begin{aligned}(\nabla_{\dot{\gamma}}\boldsymbol{\omega})_i &= \left(\frac{d}{ds}\Gamma(0,s;\gamma)_i{}^j\right)_{s=0}\omega_j(0) + \left(\frac{d\omega_i}{ds}\right)_{s=0} \\ &= \left(\frac{d\omega_i}{ds}\right)_{s=0} - \Gamma^j_{im}(0)\omega_j(0)\dot{\gamma}^m(0)\end{aligned}$$

from (2.4.10) and (2.3.8), i.e.

$$\nabla_{\dot{\gamma}}\boldsymbol{\omega} = \left[e_k(\omega_i) - \Gamma^j_{ik}\omega_j\right]_p \dot{\gamma}^k(0)\underset{p}{e}^i. \tag{2.6.8}$$

The quantity in brackets

$$\nabla_k \omega_i \equiv e_k(\omega_i) - \Gamma^j_{ik}\omega_j \tag{2.6.8$'$}$$

is the component of the covariant derivative at $p = \gamma(0)$ of the one-form $\boldsymbol{\omega}$ in the direction of $\underset{p}{e}_k$. Since this derivative is itself a one-form (as shown in (2.6.7)), the quantities $\nabla_k \omega_i$ are the components of a tensor of valence $(0, 2)$, written as $\nabla \boldsymbol{\omega}$.

We are now in the position to extend the absolute derivative to an arbitrary tensor field. The connector applied to $T \in T_p^{(r,s)}(M)$ can be defined by

$$(\Gamma^{(r,s)}(\gamma)T)(\boldsymbol{\omega}, \boldsymbol{\sigma}, ...u, v, ...) = T\left(\tilde{\Gamma}(\gamma)^{-1}\boldsymbol{\omega}, ..., \Gamma(\gamma)^{-1}u, ...\right).$$

It is then easily verified that this is consistent with tensor products, in the sense that

$$\Gamma^{(r+r', s+s')}(\gamma)(T \otimes S) = \Gamma^{(r,s)}(\gamma)T \otimes \Gamma^{(r',s')}(\gamma)S$$

and similar formulae for multiple tensor products. Thus if, in terms of a basis $\{e_i\}$, $\{\boldsymbol{e}^i\}$, we have

$$T = T^{ij...}{}_{pq...} e_i \otimes e_j \otimes ... \otimes \boldsymbol{e}^p \otimes \boldsymbol{e}^q \otimes ...$$

then

$$\Gamma^{(r,s)}(\gamma)T = T^{ij...}{}_{pq...}\Gamma(\gamma)e_i \otimes ... \otimes \tilde{\Gamma}(\gamma)\boldsymbol{e}^p \otimes ... \,. \tag{2.6.9}$$

Exactly as in (2.6.2), we define the absolute derivative of a tensor by:

$$\nabla_{\dot{\gamma}} T = \frac{d}{ds}\Gamma^{(r,s)}(\gamma_s)^{-1}(T(\gamma(s)))\bigg|_{s=0}. \tag{2.6.10}$$

To evaluate this in terms of components, we note that (2.6.9) implies that

$$\Gamma^{(r,s)}(\gamma_s)^{-1}(T(\gamma(s))) = T^{ij...}{}_{pq...}(\gamma(s))\Gamma(\gamma)^{-1}e_i \otimes ... \tag{2.6.11}$$

so that, repeating the calculations that lead to (2.6.8) gives

$$\nabla_{\dot{\gamma}} T = \nabla_k T^{ij...}{}_{nm...}\dot{\gamma}^k(p)\underset{p}{e}_i \otimes \underset{p}{e}_j \otimes ... \otimes \underset{p}{\boldsymbol{e}}^n \otimes \underset{p}{\boldsymbol{e}}^m \tag{2.6.12}$$

where

$$\begin{aligned}\nabla_k T^{ij...}{}_{nm...} &= \underset{p}{e}_k(T^{ij...}{}_{nm...}) + \Gamma^i_{rk}T^{rj...}{}_{nm...} \\ &+ \Gamma^j_{rk}T^{ir...}{}_{nm...} + \cdots - \Gamma^r_{nk}T^{ij..}{}_{rm..} - \Gamma^r_{mk}T^{ij...}{}_{nr...} - \cdots\end{aligned} \tag{2.6.13}$$

is the $)^{ij...}_{nm...}$ component of the covariant derivative of the tensor field T in the direction of e_k.

Let us now summarize the previous results. The absolute derivative along a curve $\gamma(s)$ of

i) a scalar field:

$$\nabla_{\dot\gamma} f = \dot\gamma(f) = e_k(f)\dot\gamma^k \tag{2.6.14}$$

ii) of a vector field: $Y = Y^i e_i$

$$\nabla_{\dot\gamma} Y = (\nabla_k Y^i)\dot\gamma^k e_i = \left[e_k(Y^i) + \Gamma^i_{jk} Y^j\right] \dot\gamma^k e_i \tag{2.6.15}$$

iii) of a one-form: $\boldsymbol{\omega} = \omega_i \boldsymbol{e}^i$

$$\nabla_{\dot\gamma} \boldsymbol{\omega} = (\nabla_k \omega_i)\dot\gamma^k \boldsymbol{e}^i = \left[e_k(\omega_i) - \Gamma^n_{ik}\omega_n\right] \dot\gamma^k \boldsymbol{e}^i \tag{2.6.16}$$

iv) of a tensor field of valence $(1,1)$: $T = T^i{}_j e_i \otimes \boldsymbol{e}^j$

$$\nabla_{\dot\gamma} T = (\nabla_k T^i{}_j)\dot\gamma^k e_i \otimes \boldsymbol{e}^j = \left[e_k(T^i{}_j) + \Gamma^i_{rk} T^r{}_j - \Gamma^s_{jk} T^i{}_s\right] \dot\gamma^k e_i \otimes \boldsymbol{e}^j. \tag{2.6.17}$$

From the definition of the absolute derivative, the following property holds:

$$\nabla_{\dot\gamma}(fY) = f\nabla_{\dot\gamma} Y + Y\dot\gamma(f) \qquad f \in \mathcal{F}, \quad Y \in T_p(M) \tag{2.6.18}$$

or, more generally

$$\nabla_{\dot\gamma}(T \otimes S) = (\nabla_{\dot\gamma} T) \otimes S + T \otimes (\nabla_{\dot\gamma} S); \tag{2.6.19}$$

while if C is a contraction

$$\nabla_{\dot\gamma}(CT) = C\nabla_{\dot\gamma} T. \tag{2.6.20}$$

2.7 Torsion and normal coordinates

Given a vector field in M, X say, and a congruence C_Y, two differential operations can be performed on X: the Lie derivative and the absolute derivative, both in the direction of Y. Let us now explore the relation between the two. From (2.6.2), (2.2.7) and (2.2.11) we have:

$$\begin{aligned}\nabla_Y X - \nabla_X Y &= [Y(X^i) - X(Y^i) + 2\Gamma^i_{[kj]} Y^j X^k] e_i \\ &= [Y, X] + (2\Gamma^i_{[kj]} + C^i_{kj}) Y^j X^k e_i.\end{aligned} \tag{2.7.1}$$

The quantity

$$T(Y, X) = \nabla_Y X - \nabla_X Y - [Y, X] = (2\Gamma^i_{[kj]} + C^i_{kj}) Y^j X^k e_i \tag{2.7.2}$$

defines a tensor field in U, of valence $(1, 2)$, called the *Torsion*:

$$T^i{}_{jk} = 2\Gamma^i_{[jk]} + C^i_{jk}. \tag{2.7.3}$$

In what follows we shall assume the torsion to be zero. This implies for example that:

$$\nabla_{[a}\nabla_{b]}f = 0 \tag{2.7.4}$$

for all functions f. A torsion-free connection is determined by the affine geodesics on M.

Since Γ^i_{jk} is not a tensor, as can be deduced from (2.5.6), one could envisage a change of basis such that, at p, $\Gamma'^i{}_{jk}|_p = 0$. We now show that such a possibility exists in a neighbourhood of each point of M. Call $U_{\mathcal{N}}(p) \ni p$ the set of all points which can be joined to p by a *unique* geodesic. It can be shown that, provided that the functions Γ^i_{jk} are sufficiently smooth, then $U_{\mathcal{N}}(p)$ is in fact a neighbourhood of p called a *normal neighbourhood.* Consider now a point $q \in U_{\mathcal{N}}(p)$ and call γ the affine geodesic joining p to q. If $\left\{\underset{p}{e}_i\right\}$ is a basis in $T_p(M)$, the set $\left\{\underset{q}{\breve{e}}_i = \Gamma(p,q;\gamma)_i{}^k \underset{p}{e}_k\right\}$ of the parallely propagated vectors of the basis, is still a basis in $T_q(M)$. With respect to $\left\{\underset{q}{\breve{e}}_i\right\}$, the vector $\dot{\gamma}$, tangent to the geodesic between p and q, has components at q:

$$\dot{\gamma}|_q = \dot{\gamma}^i(p)\underset{q}{\breve{e}}_i \tag{2.7.5}$$

from (2.5.1). If we parametrize the curve γ in such a way that $p = \gamma(0)$ and $q = \gamma(1)$, then we assign to each point q, the set of numbers:

$$y^i(s)|_q = \dot{\gamma}^i(0). \tag{2.7.6}$$

Since these are uniquely defined at $q \in U_{\mathcal{N}}(p)$, they form a coordinate system in $U_{\mathcal{N}}(p)$, termed *normal coordinates* based on p. From (2.7.6) we have at q:

$$\left.\frac{dy^i}{ds}\right|_q = \dot{\gamma}^i(0), \tag{2.7.7}$$

hence:

$$\left.\frac{d^2y^i}{ds^2}\right|_q = 0 \tag{2.7.8}$$

which implies, in the y-coordinates : ${\Gamma'}^i_{jk}\dot{\gamma}^j\dot{\gamma}^k|_q = 0$, from (2.4.21). This holds true at all points on γ, and hence also at p, where, since $\dot{\gamma}$ is arbitrary (all the geodesics stem from p), we have

$$ {\Gamma'}^i_{(jk)}|_p = 0. \tag{2.7.9} $$

If the torsion vanishes, then also ${\Gamma'}^i_{[jk]} = 0$ in $U_{\mathcal{N}}(p)$ (since we are using a coordinate basis and so ${C'}^i_{jk} = 0$), hence

$$ {\Gamma'}^i_{jk}|_p = 0. \tag{2.7.10} $$

The significance and importance of normal coordinates will be elucidated further in the following chapter.

2.8 Compatibility of the metric with the connection

We now introduce a further condition on the connection; that it preserves the inner product, in the sense that, for any $v, w \in T_p(M)$, and any γ with $\gamma(0) = p$:

$$ (\check{v}(t)|\check{w}(t)) = (v|w), \tag{2.8.1} $$

where we write as before $\check{v}(t) = \Gamma(0,t;\gamma)v$, $\check{w}(t) = \Gamma(0,t;\gamma)w$. From (2.6.14) we have :

$$ \nabla_{\dot{\gamma}}\left(\overset{*}{g}(\gamma(t))(\check{v},\check{w})_t\right) = 0. \tag{2.8.2} $$

Using the fact that $\nabla_{\dot{\gamma}}\check{v} = \nabla_{\dot{\gamma}}\check{w} = o$ from (2.5.4), we have:

$$ (\nabla_{\dot{\gamma}}\overset{*}{g}(\gamma(t)))(\check{v}(t),\check{w}(t)) = 0 $$

(from (2.6.20)). Since this holds for all curves γ and all v and w we must have:

$$ \nabla\overset{*}{g} = o \tag{2.8.3} $$

which in components reads:

$$ \nabla_k g_{ij} = 0. \tag{2.8.3$'$} $$

A metric on M which satisfies (2.8.3) is said to be compatible with the connection. From (2.6.17), (2.8.3) reads:

$$ e_k(g_{ij}) = \Gamma^n_{ik}g_{nj} + \Gamma^n_{jk}g_{in} \tag{2.8.4} $$

and the torsion being zero, we also have:

$$2\Gamma^n_{[ki]} = -C^n_{ki}. \qquad (2.8.5)$$

Combining (2.8.4) with (2.8.5), we have after some manipulation:

$$\Gamma^m_{ik} = \frac{1}{2}g^{mj}\left[e_k(g_{ij}) + e_i(g_{jk}) - e_j(g_{ik})\right] + \frac{1}{2}\left[C^m_{ki} + g^{jm}g_{ni}C^n_{jk} + g^{jm}g_{nk}C^n_{ji}\right]. \qquad (2.8.6)$$

In the case of a coordinate basis, $C^i_{jk} = 0$ holds and (2.8.6) reduces to:

$$\Gamma^m_{ik} = \frac{1}{2}g^{mj}\left[\partial_k g_{ij} + \partial_i g_{jk} - \partial_j g_{ik}\right]; \qquad (2.8.7)$$

the connection coefficients as in (2.8.7) are known as *Christoffel symbols*.

2.9 Parallelism

The linear connection on M generalizes to an n-dimensional manifold the concept of parallelism which is more familiar in three-dimensional Euclidean space.

For infinitesimal displacements (δs) along a curve γ, the map $\Gamma(\gamma)$ maps the basis $\{e_i\}_{\gamma(0)}$ into a basis $\{\check{e}_i\}_{\gamma(\delta s)}$ which is uniquely defined on γ provided we are in a Riemannian manifold and the connection is torsion-free and compatible with the metric. A general vector u at $\gamma(0)$ is mapped to a vector $\check{u}$ at $\gamma(\delta s)$, having there the same magnitude as u at $\gamma(0)$ and the same components with respect to $\{\check{e}_i\}_{\gamma(\delta s)}$ as u had with respect to $\{e_i\}_{\gamma(0)}$. The vector $\check{u}$ then, if referred to $\{\check{e}_i\}_{\gamma(\delta s)}$, makes no distinction between being at $\gamma(0)$ or at $\gamma(\delta s)$, therefore we call it the *parallel* vector to $u_{\gamma(0)}$ at $\gamma(\delta s)$. For finite displacements along γ, the parallel $\check{u}_{\gamma(s)}$ to $u_{\gamma(0)}$ is given by:

$$\check{u}^i_{\gamma(s)} = u^i_{\gamma(0)} - \int_0^s \Gamma^i_{jk}(\gamma(s'))\check{u}^j(s')\dot{\gamma}^k(s')ds'. \qquad (2.9.1)$$

Let v be a vector field defined on γ and assume that it is proportional to a parallely propagated vector on γ, i.e. (cf.2.4.12)

$$\check{v}_{\gamma(s)} = \lambda(s)v_{\gamma(s)}. \qquad (2.9.2)$$

From (2.9.2) and (2.6.2) the differential equation for parallel transport on γ reads at $\gamma(s)$:

$$\dot{\gamma}^k \nabla_k v = f(s) v, \tag{2.9.3}$$

where $f(s) = d(\ln \lambda(s))/ds$. From property (2.8.1) of the connection, the law of parallelism on γ, namely (2.9.3), preserves orthogonality; for example a parallely transported vector field, which is null ($\|v\| = 0$) somewhere on γ, will be null everywhere on it.

According to the definition (cf.(2.4.13)), the geodesic equation (2.4.14) is a particular case of (2.9.3), but while for a vector field tangent to a geodesic there always exists a reparametrization of the curve which brings to zero the function $f(s)$ in (2.9.3), this is not true for a general field parallely propagated along a curve γ. The geodesic equation then reads from (2.4.14) and (2.6.3):

$$\nabla_{\dot{\gamma}} \dot{\gamma}^k = f(s) \dot{\gamma}^k \,, \tag{2.9.4}$$

hence we prove the following:

PROPOSITION

A curve γ is a geodesic iff:

$$\nabla_{\dot{\gamma}} \dot{\gamma}^{[k} \dot{\gamma}^{i]} = 0. \tag{2.9.5}$$

Proof: If a curve is a geodesic, then (2.9.5) is trivial; conversely if the latter is true, then contraction with $g_{ij} \dot{\gamma}^j$ yields

$$\left(\nabla_{\dot{\gamma}} \dot{\gamma}^k \right) (\dot{\gamma} | \dot{\gamma}) = \frac{1}{2} \left[\frac{d}{ds} (\dot{\gamma} | \dot{\gamma}) \right] \dot{\gamma}^k \tag{2.9.6}$$

hence

$$\nabla_{\dot{\gamma}} \dot{\gamma}^k = \left[\frac{d}{ds} \ln |(\dot{\gamma} | \dot{\gamma})|^{\frac{1}{2}} \right] \dot{\gamma}^k \tag{2.9.7}$$

which implies that γ is a geodesic. ■

2.10 Applications of the covariant derivative

Expression (2.8.7) for the connection coefficients allows one to derive expressions which will be very useful in many applications.

Let U_α be a coordinate neighbourhood of M; contracting two indices in (2.8.7) leads to :

$$\Gamma^i_{ki} = \frac{1}{2} g^{in} \frac{\partial g_{in}}{\partial x^k}. \tag{2.10.1}$$

From the properties of determinants, we also have:

$$\delta^j_i g = g_{ik} M^{jk} \tag{2.10.2}$$

where M^{jk} is the adjoint of g_{jk}. If we partially differentiate (2.10.2) we get:

$$\frac{\partial g}{\partial x^l} = \frac{\partial g}{\partial g_{ik}} \frac{\partial g_{ik}}{\partial x^l} = M^{ik} \frac{\partial g_{ik}}{\partial x^l}; \tag{2.10.3}$$

but from the definition of M^{ik} the latter becomes:

$$\frac{\partial g}{\partial x^l} = g g^{ik} \frac{\partial g_{ik}}{\partial x^l} \tag{2.10.4}$$

hence (2.10.1) can be rewritten:

$$\Gamma^i_{ki} = \frac{\partial}{\partial x^k} \ln(|g|)^{\frac{1}{2}}. \tag{2.10.5}$$

Let us now contract the two lower indices of Γ^i_{jk} with the metric tensor; from (2.8.7) and (2.10.4) we get

$$g^{jk} \Gamma^i_{jk} = -\frac{1}{(|g|)^{\frac{1}{2}}} \frac{\partial}{\partial x^r} \left[(|g|)^{\frac{1}{2}} g^{ir} \right]. \tag{2.10.6}$$

Using (2.10.5) and (2.10.6) one obtains the following expressions for divergences:

$$\begin{aligned}
&a) \quad \nabla_i A^i = (|g|)^{-\frac{1}{2}} \frac{\partial}{\partial x^i} [(|g|)^{\frac{1}{2}} A^i] \\
&b) \quad \nabla_j A^{[ij]} = (|g|)^{-\frac{1}{2}} \frac{\partial}{\partial x^j} [(|g|)^{\frac{1}{2}} A^{[ij]}] \\
&c) \quad \nabla_j A^{(ij)} = g^{in} \left[(|g|)^{-\frac{1}{2}} \frac{\partial}{\partial x^j} [(|g|)^{\frac{1}{2}} A_n{}^j] - \frac{1}{2} \frac{\partial g_{kj}}{\partial x^u} A^{(kj)} \right].
\end{aligned} \tag{2.10.7}$$

A useful implication of the assumption that the manifold is torsion-free is

$$\nabla_{[i} A_{j]} = \partial_{[i} A_{j]}. \tag{2.10.8}$$

2.11 The exterior derivative

We shall now introduce a new operation, termed *exterior derivative* and denoted by $\boldsymbol{d}$, which maps r-forms into $(r+1)$-forms:

$$\boldsymbol{d} : \Lambda^r \overset{*}{T}(M) \longrightarrow \Lambda^{r+1} \overset{*}{T}(M), \tag{2.11.1}$$

and satisfies the following axioms:

$$\begin{aligned}
&a) \quad \boldsymbol{d}f = (\partial_i f)\boldsymbol{d}x^i \\
&b) \quad \boldsymbol{d}(\boldsymbol{\omega} + \boldsymbol{\sigma}) = \boldsymbol{d\omega} + \boldsymbol{d\sigma} \\
&c) \quad \boldsymbol{d}(\boldsymbol{\omega} \wedge \boldsymbol{\sigma}) = \boldsymbol{d\omega} \wedge \boldsymbol{\sigma} + (-1)^{\deg \omega} \boldsymbol{\omega} \wedge \boldsymbol{d\sigma} \\
&d) \quad \boldsymbol{d}(f\boldsymbol{\omega}) = \boldsymbol{d}f \wedge \boldsymbol{\omega} + f\boldsymbol{d\omega} \\
&e) \quad \boldsymbol{d}(\boldsymbol{d\omega}) = \boldsymbol{o}.
\end{aligned} \tag{2.11.2}$$

Here we refer in (2.11.2)a to some coordinate neighbourhood U of M, f is a differentiable function on U (a zero-form), $\boldsymbol{\omega}$ and $\boldsymbol{\sigma}$ are arbitrary forms, $\boldsymbol{o}$ denotes the zero p-form (with all components zero) for any p, and $\deg \boldsymbol{\omega} = r$ for $\boldsymbol{\omega} \in \Lambda^r \overset{*}{T}(M)$. We shall find such $\boldsymbol{d}$; suppose it is given, then from (2.11.2)e:

$$\boldsymbol{d}(\boldsymbol{d}x^i) = \boldsymbol{o} \tag{2.11.3}$$

hence, using this, (2.11.2)d and the linearity axiom b), we have for a general r-form:

$$\begin{aligned}
\boldsymbol{d}\omega &= \frac{1}{r!}\boldsymbol{d}\omega_{i_1 \dots i_r} \wedge \boldsymbol{d}x^{i_1} \wedge \dots \wedge \boldsymbol{d}x^{i_r} \\
&= \frac{1}{r!}\partial_j \omega_{i_1 \dots i_r} \boldsymbol{d}x^j \wedge \boldsymbol{d}x^{i_1} \wedge \dots \wedge \boldsymbol{d}x^{i_r} \\
&= (r+1)\partial_{[j}\omega_{i_1 \dots i_r]} \boldsymbol{d}x^j \otimes \boldsymbol{d}x^{i_1} \otimes \dots \otimes \boldsymbol{d}x^{i_r}
\end{aligned} \tag{2.11.4}$$

from (1.18.6) and (2.11.2)a. This operation clearly satisfies (2.11.2)c; in fact if $\boldsymbol{\omega} \in \Lambda^r \overset{*}{T}_p(M)$, $\boldsymbol{\sigma} \in \Lambda^s \overset{*}{T}_p(M)$, we have

$$\begin{aligned}
\boldsymbol{d}(\boldsymbol{\omega} \wedge \boldsymbol{\sigma}) &= \frac{1}{r!s!}\boldsymbol{d}(\omega_{i_1 \dots i_r}\sigma_{j_1 \dots j_s}) \wedge \boldsymbol{d}x^{i_1} \wedge \dots \wedge \boldsymbol{d}x^{i_r} \wedge \dots \wedge \boldsymbol{d}x^{j_s} \\
&= \frac{1}{r!s!}\left[\boldsymbol{d}(\omega_{i_1 \dots i_r})\sigma_{j_1 \dots j_s} + \omega_{i_1 \dots i_r}\boldsymbol{d}(\sigma_{j_1 \dots j_s})\right] \wedge \boldsymbol{d}x^{i_1} \dots \\
&= \boldsymbol{d\omega} \wedge \boldsymbol{\sigma} \\
&\quad + \frac{1}{r!s!}\omega_{i_1 \dots i_r}\boldsymbol{d}\sigma_{j_1 \dots j_s} \wedge \boldsymbol{d}x^{i_1} \wedge \dots \wedge \boldsymbol{d}x^{i_r} \wedge \dots \wedge \boldsymbol{d}x^{j_s} \\
&= \boldsymbol{d\omega} \wedge \boldsymbol{\sigma} + (-1)^r \boldsymbol{\omega} \wedge \boldsymbol{d\sigma},
\end{aligned} \tag{2.11.5}$$

from (1.18.11). Axiom e) is identically satisfied for any $\boldsymbol{\omega}$, in fact:

$$\begin{aligned} \boldsymbol{d}(\boldsymbol{d\omega}) &= \frac{1}{r!}\boldsymbol{d}\left(\partial_k \omega_{i_1 \dots i_r} \boldsymbol{dx}^k \wedge \boldsymbol{dx}^{i_1} \wedge \dots \wedge \boldsymbol{dx}^{i_r}\right) \\ &= \frac{1}{r!}\partial_j \partial_k \omega_{i_1 \dots i_r} \boldsymbol{dx}^j \wedge \boldsymbol{dx}^k \wedge \boldsymbol{dx}^{i_1} \wedge \dots \wedge \boldsymbol{dx}^{i_r} \qquad (2.11.6) \\ &= (r+2)(r+1)\partial_{[j}\partial_k \omega_{i_1 \dots i_r]} \boldsymbol{dx}^j \otimes \dots \otimes \boldsymbol{dx}^{i_r} \equiv \boldsymbol{o}. \end{aligned}$$

This expresses the equality of partial mixed derivatives on $\mathbb{R}^n$ and is known as the *Poincaré lemma.* The action of the $\boldsymbol{d}$-operation is independent of the coordinate system in which it is computed; this can be seen directly from (2.11.4), imposing a transformation $x \to x' = x'(x)$. However this property arises as a natural consequence of the general behaviour of $\boldsymbol{d}$ under mappings. Let us extend (1.12.4) to r-forms. If $\Phi : M \to N$ is a smooth map of M into N, M and N being differentiable manifolds, and $\overset{1}{\boldsymbol{\omega}} = f\boldsymbol{d}g$ is a one-form on N (f, $g \in \mathcal{F}(N)$), then

$$\Phi^* \overset{1}{\boldsymbol{\omega}} = (f \circ \Phi)\Phi^* \boldsymbol{d}g = (f \circ \Phi)\boldsymbol{d}(g \circ \Phi) \qquad (2.11.7)$$

from (1.12.1) and (1.12.4). We now define a mapping of r-forms on N into r-forms on M, calling it $\overset{r}{\Phi}{}^*$, by assuming the following basic properties:

$$\begin{aligned} \overset{r}{\Phi}{}^*(\overset{1}{\boldsymbol{\omega}}_1 \wedge \dots \wedge \overset{1}{\boldsymbol{\omega}}_r) &= (\Phi^* \overset{1}{\boldsymbol{\omega}}_1) \wedge \dots \wedge (\Phi^* \overset{1}{\boldsymbol{\omega}}_r) \\ \overset{r}{\Phi}{}^*(\overset{r}{\boldsymbol{\omega}} + \overset{r}{\boldsymbol{\sigma}}) &= (\overset{r}{\Phi}{}^* \overset{r}{\boldsymbol{\omega}}) + (\overset{r}{\Phi}{}^* \overset{r}{\boldsymbol{\sigma}}). \qquad (2.11.8) \end{aligned}$$

From (2.11.8) and (2.11.7) we have :

$$\begin{aligned} \overset{r}{\Phi}{}^* \overset{r}{\boldsymbol{\omega}} &= \overset{r}{\Phi}{}^* \left(\frac{1}{r!}\omega_{i_1 \dots i_r} \boldsymbol{dx}^{i_1} \wedge \dots \wedge \boldsymbol{dx}^{i_r}\right) \\ &= \frac{1}{r!}(\omega_{i_1 \dots i_r} \circ \Phi)\boldsymbol{d}(x^{i_1} \circ \Phi) \wedge \dots \wedge \boldsymbol{d}(x^{i_1} \circ \Phi) \quad (2.11.9) \end{aligned}$$

thus:

$$\begin{aligned} \boldsymbol{d}(\overset{r}{\Phi}{}^* \overset{r}{\omega}) &= \frac{1}{r!}\boldsymbol{d}(\omega_{i_1 \dots i_r} \circ \Phi) \wedge \boldsymbol{d}(x^{i_1} \circ \Phi) \wedge \dots \wedge \boldsymbol{d}(x^{i_r} \circ \Phi) \\ &= \frac{1}{r!}\Phi^*(\boldsymbol{d}\omega_{i_1 \dots i_r}) \wedge \Phi^* \boldsymbol{dx}^{i_1} \wedge \dots \wedge \Phi^* \boldsymbol{dx}^{i_r} \\ &= \overset{r+1}{\Phi}{}^* \boldsymbol{d}\overset{r}{\boldsymbol{\omega}} \qquad (2.11.10) \end{aligned}$$

from (2.11.8). The general relation (we drop the superscript for convenience)

$$\Phi^* d\omega = d\Phi^*\omega \tag{2.11.11}$$

essentially expresses the chain rule for partial derivatives.

2.12 Frobenius theorems

We have seen in (2.2.13) that if the basis is a coordinate basis, then the Lie bracket of any pair of basis vectors is zero. It is natural to ask the converse question: if all the Lie brackets are zero, is the basis a coordinate basis? The answer is, yes. In this section we shall discuss the theorem due to Frobenius, that allows us to conclude this.

In relativity theory, however, one frequently comes across bases where only some of the basis vectors commute (have zero Lie brackets), or where the basis vectors fall into two sets, the Lie brackets of vectors in one set only involving vectors in that set. In such cases it is not possible to represent the basis as a coordinate basis, but it is possible to choose coordinates that are very closely related to the basis, thus making calculations much simpler. For this reason we shall take a rather general version of the Frobenius theorem that also covers such cases.

We shall be concerned with coordinates $x^i, \ldots, x^n$ that can be divided into two sets, $\{x^1, \ldots, x^k\}$ and $\{x^{k+1}, \ldots, x^n\}$. To handle this we use capital latin letters A, B, .. running from 1 to k and Greek letters $\alpha, \beta, \ldots$ from $k+1$ to n. Suppose we have k vector fields $V_1, \ldots, V_k$ such that at each point p, the vectors $\{V_1(p), \ldots, V_k(p)\}$ are independent (expressed loosely: *independent vector fields*), and whose last components vanish; i.e.

$$V_{\mathrm{A}}^{\alpha} \equiv V_{\mathrm{A}}(x^{\alpha}) = 0 \qquad \mathrm{A} = 1, \ldots, k;\ \alpha = k+1, \ldots, n. \tag{2.12.1}$$

Here, when we wish to make explicit the coordinate system involved, we shall use the expression $V_{\mathrm{A}}(x^{\alpha})$ (V_{A} acting as a differential operator on x^{α}) for the αth component of V_{A}. This means that:

$$V_{\mathrm{A}} = V_{\mathrm{A}}^{\mathrm{B}} \partial_{\mathrm{B}}$$

and, since the matrix $\mathcal{V}$ with components $[V_{\mathrm{A}}^{\mathrm{B}}]$ is non-singular (the fields are independent)

$$\partial_{\mathrm{B}} = (\mathcal{V}^{-1})_{\mathrm{B}}^{\mathrm{A}} V_{\mathrm{A}}. \tag{2.12.2}$$

The Lie bracket satisfies

$$[V_{\mathrm{A}}, V_{\mathrm{B}}]^{\alpha} = 2\left(V_{[\mathrm{A}}^{\mathrm{C}} V_{\mathrm{B}],\mathrm{C}}^{\alpha} + V_{[\mathrm{A}}^{\beta} V_{\mathrm{B}],\beta}^{\alpha}\right) = 0$$

and so

$$[V_{\mathrm{A}}, V_{\mathrm{B}}] = k^{\mathrm{C}} \partial_{\mathrm{C}} = k^{\mathrm{C}} (\mathcal{V}^{-1})_{\mathrm{C}}^{\mathrm{D}} V_{\mathrm{D}}$$

from (2.12.2), i.e.

$$[V_{\mathrm{A}}, V_{\mathrm{B}}] = C_{\mathrm{AB}}^{\mathrm{D}} V_{\mathrm{D}} \tag{2.12.3}$$

for some functions $C_{\mathrm{AB}}^{\mathrm{D}}$ on M, called structure constants. In other words the structure constants are zero for indices outside the range $1, .., k$.

We shall state the Frobenius theorem in two parts, one giving the converse of this result, and the second giving the special case where the fields commute ($C_{\mathrm{AB}}^{\mathrm{D}} = 0$) in which case they can be related to a coordinate basis.

THEOREM 1

(Frobenius theorem for vectors). Let $\{V_{\mathrm{A}} : \mathrm{A} = 1, ..., k\}$ *be k vector fields independent at each point and satisfying* (2.12.3) *for functions* $C_{\mathrm{AB}}^{\mathrm{D}}$.

Case a. Without further restrictions, we can conclude that in a neighbourhood of any point there exist coordinates x^i *such that* (2.12.1) *holds.*

Case b. If in addition the fields commute ($C_{\mathrm{AB}}^{\mathrm{D}} = 0$), *then these coordinates can be chosen so that*

$$V_{\mathrm{A}}(x^i) = \delta_{\mathrm{A}}^i \qquad (i = 1, ..., n;\ \mathrm{A} = 1, ..., k).$$

Proof: This proceeds by induction on k. When $k = 1$ we have one vector field V_1, and so in fact both cases (a) and (b) are always satisfied in the trivial form of

$$[V_1, V_1] = o = 0.V_1 \ !$$

In this case start off with an arbitrary coordinate system $z^1, \dots, z^n$ near p. Choose vectors $\xi_2, \dots, \xi_n \in T_p(M)$ independent of $V_1(p)$ and then perform a fixed linear transformation of the coordinates ($z'^i = A^i{}_j z^j$) such that at p, $\xi^i_\alpha = \delta^i_\alpha$ ($\alpha = 2, \dots, n$). Let S be the surface $z'^1 = 0$. Then at p the vectors ξ_α are tangent to S and so V_1 is not tangent to S (since it is independent of the ξ_α). Thus there is some neighbourhood N of p such that V_1 is not tangent to S in N.

Let φ_s be the local one-parameter group corresponding to V_1: we choose N small enough to lie in the domain of φ_s. Define an "inverse coordinate patch" (a map θ from a subset of $\mathbb{R}^n$ to M) by

$$\theta(x^1, x^2, \dots, x^n) = \varphi_{x^1}\left(\chi(0, x^2, \dots, x^n)\right)$$

where χ is the map taking $(a^1, a^2, \dots, a^n) \in \mathbb{R}^n$ to the point with coordinates $z'^i = a^i$. To see that this really is the inverse of a coordinate map in a neighbourhood of $S \cap N$, it suffices to note that $\theta_*(\partial/\partial x^i)$ are independent vectors on S (because $\theta_*(\partial/\partial x^1) = V_1$ and $\theta_*(\partial/\partial x^\alpha) = \partial/\partial z'^\alpha$ for $\alpha = 2, \dots, n$ on S). Hence they are independent on a neighbourhood of $S \cap N$, from which it can be shown that θ is one–to–one in a small enough neighbourhood of $S \cap N$.

Since $\theta_*(\partial/\partial x^1) = V_1$ we have that, in the x-coordinates

$$V_1(x^i) = \delta^i_1 \qquad (2.12.3')$$

and so, in particular

$$V_1(x^\alpha) = 0 \qquad (\alpha = 2, \dots, n)$$

proving the theorem for $k = 1$. We now proceed by induction on k.

Suppose (2.12.1) is satisfied and assume that we have proved the result for $k - 1$ vector fields. To prove the result for k vector fields, start by choosing coordinates $\{y^i\}$, $i = 1, \ldots, n$, so that $V_1(y^i) = \delta^i_1$, as has just been done (Eq.(2.12.3′)) for the step $k = 1$. Define

$$W_1 = V_1 \qquad W_a = V_a - (V_a(y^1))V_1 \qquad (a = 2, \ldots, k). \quad (2.12.4)$$

Then $(W_1, \ldots, W_k)$ are still an independent set of fields but we have modified V_a for $a = 2, \ldots, k$ so that

$$W_a(y^1) = V_a(y^1) - (V_a(y^1))V_1(y^1) = 0. \quad (2.12.5)$$

We calculate, in the y-coordinates

$$[W_a, W_b]\,(y^1) = W_a(W_b(y^1)) - W_b(W_a(y^1)) = 0$$

and so $[W_a, W_b]$ is expressible in terms of the W_a (excluding $W_1 = V_1$); extending to W_1 we have:

$$[W_A, W_B] = D^C_{AB} W_C \quad (2.12.6)$$

and is zero in case b).

Let Σ be the surface $y^1 = 0$. Then the fields W_a are tangent to this surface, from (2.12.5), and we can use the inductive hypotheses with (2.12.6) to choose coordinates $x^2, \ldots, x^n$ in Σ so that $W_a(x^\alpha) = 0$ ($a = 2, \ldots, k$; $\alpha = k+1, \ldots, n$) in Σ (in case (b) $W_a(x^i) = \delta^i_a$). Thus $x^2, \ldots, x^n$ are defined in Σ as function of $y^2, \ldots, y^n$:

$$x^\xi = f^\xi(y^2, \ldots, y^n) \qquad (\xi = 2, \ldots, n).$$

This functional dependence allows us to define coordinates throughout a neighbourhood of p by

$$x^1 = y^1 \qquad x^\xi = f^\xi(y^2, \ldots, y^n) \qquad (\xi = 2, \ldots, n)\,. \quad (2.12.7)$$

Then for $\alpha = k+1, \ldots, n$ and $a = 2, \ldots, k$ we have from (2.12.6):

$$\frac{\partial}{\partial x^1} W^\alpha_a = W_1(W_a(x^\alpha)) = W_a(W_1(x^\alpha)) + D^b_{1a} W_b(x^\alpha)$$
$$= D^b_{1a} W_b(x^\alpha). \quad (2.12.8a)$$

In case (b) we have

$$\frac{\partial}{\partial x^1} W^i_a = W_a(\delta^i_1) = 0. \quad (2.12.8b)$$

In case (a) (2.12.8a) is a first-order homogeneous system of differential equations for the set $(W_2(x^\alpha), \ldots, W_k(x^\alpha))$ (considering $y^2, \ldots, y^n$ fixed) with initial conditions $W_a(x^\alpha) = 0$ on $y^1 = 0$ and so $W_a(x^\alpha) \equiv 0$. By construction $W_1(x^\alpha) = 0$ and so the result is proved for case (a). For case (b) (2.12.8)b has the initial conditions

$$W_a(x^\xi) = V_a(x^\xi) = \delta^\xi_a \qquad (a = 2, ..., k;\ \xi = 2, ..., n).$$

These conditions thus hold throughout the region, so that (2.12.4) gives the required result for the V_{A}. ■

There is another version of this theorem referring to one-forms rather then vectors. The geometrical idea is that the surfaces $S_c \equiv \{x^\alpha = c^\alpha\colon\ \alpha = k+1, ..., n\}$ for each set of constants c^α can be designated either by the set of vectors $\{V_{\mathrm{A}}\}$ which are tangent to these surfaces, or by specifying a set $\left\{\overset{\alpha}{\omega} : \alpha = k+1, ..., n\right\}$ of one-forms such that the set of tangent vectors to the surfaces comprises vectors V such that $\overset{\alpha}{\omega}(V) = 0$ for all $\alpha = k+1, ..., n$. The two versions of the theorem tell us when the vectors V_{A} are tangent to a set of surfaces, or when the forms $\overset{\alpha}{\omega}$ designate a set of surfaces in this way. We prove this second theorem by showing its equivalence to the first.

We use the following terminology: a set $\left\{\overset{\alpha}{\omega} :\ \alpha = k+1, ..., n\right\}$ of one-forms is called *integrable* if there exists, near any point, a coordinate system x^i such that

$$\overset{\alpha}{\omega}_{\mathrm{A}} \equiv \overset{\alpha}{\omega}(\partial_{\mathrm{A}}) = 0 \qquad (\alpha = k+1, ..., n;\ \mathrm{A} = 1, ..., k) \qquad (2.12.9)$$

(Since the vectors ∂_{A} form a set of basis fields for all tangent vectors to the coordinate surfaces $S_c \equiv \{x^\alpha = c^\alpha\}$, this is equivalent to the condition just described for designating these surfaces.) The theorem is then as follows:

THEOREM 2

(Frobenius theorem for 1-forms) let $\overset{k+1}{\omega}, ..., \overset{n}{\omega}$ *be* $(n-k)$ *1-forms, independent at each point and satisfying*

$$\overset{k+1}{\omega} \wedge \overset{k+2}{\omega} \wedge \ldots \wedge \overset{n}{\omega} \wedge \boldsymbol{d}\overset{\alpha}{\omega} = \boldsymbol{o} \qquad (2.12.10)$$

for each $\alpha = k+1, ..., n$. *Then* $\left\{\overset{k+1}{\omega}, ..., \overset{n}{\omega}\right\}$ *is integrable.*

Proof: We first choose additional one-forms $\overset{1}{\boldsymbol{\omega}}, \ldots, \overset{k}{\boldsymbol{\omega}}$ so that the combined set $\left\{\overset{1}{\boldsymbol{\omega}}, \ldots, \overset{n}{\boldsymbol{\omega}}\right\}$ is a basis at each point. Hence the set $\{\overset{i}{\boldsymbol{\omega}} \wedge \overset{j}{\boldsymbol{\omega}}\colon i, j = 1, \ldots, n\}$ is a basis for the set of all two-forms at each point, and we can therefore express $\boldsymbol{d}\overset{\alpha}{\boldsymbol{\omega}}$ in terms of them; separating out the two ranges of indices, we can write

$$\boldsymbol{d}\overset{\alpha}{\boldsymbol{\omega}} = A^{\alpha}_{\text{AB}}\overset{A}{\boldsymbol{\omega}} \wedge \overset{B}{\boldsymbol{\omega}} + B^{\alpha}{}_{\text{B}\beta}\overset{B}{\boldsymbol{\omega}} \wedge \overset{\beta}{\boldsymbol{\omega}} + C^{\alpha}_{\gamma\delta}\overset{\gamma}{\boldsymbol{\omega}} \wedge \overset{\delta}{\boldsymbol{\omega}} \tag{2.12.11}$$

(with $A^{\alpha}_{\text{AB}} = 0$ for $\text{A} \geq \text{B}$, $C^{\alpha}_{\gamma\delta} = 0$ for $\gamma \geq \delta$). Inserting this in (2.12.10) gives:

$$A^{\alpha}_{\text{AB}}\overset{k+1}{\boldsymbol{\omega}} \wedge \ldots \wedge \overset{n}{\boldsymbol{\omega}} \wedge \overset{A}{\boldsymbol{\omega}} \wedge \overset{B}{\boldsymbol{\omega}} = \boldsymbol{o}.$$

But the forms in this expression are independent, and hence $A^{\alpha}_{\text{AB}} = 0$. Hence (2.12.11) can be written

$$\boldsymbol{d}\overset{\alpha}{\boldsymbol{\omega}} = \overset{\alpha}{\boldsymbol{\theta}}_{\beta} \wedge \overset{\beta}{\boldsymbol{\omega}} \tag{2.12.12}$$

with $\overset{\alpha}{\boldsymbol{\theta}}_{\beta}$ a set of one-forms given by:

$$\overset{\alpha}{\boldsymbol{\theta}}_{\beta} = B^{\alpha}{}_{\text{B}\beta}\overset{B}{\boldsymbol{\omega}} + C^{\alpha}_{\gamma\beta}\overset{\gamma}{\boldsymbol{\omega}}.$$

We now relate the present situation, with 1-forms, to the previous formulation in terms of vectors. For each $p \in M$ define:

$$D_p \equiv \left\{V \in T_p(M) \mid \overset{\alpha}{\boldsymbol{\omega}}(V) = 0\ ;\ \alpha = k+1, \ldots, n\right\}.$$

The spaces D_p vary smoothly with p so that we can (locally) find vector fields $V_1, \ldots, V_k$ such that, for each p, the set $\{V_1(p), \ldots, V_k(p)\}$ is a basis for D_p. We will have proved the required result (2.12.9) if we can show that, for some coordinates, the V_{A} can be expressed in terms of the ∂_{A}, i.e. that $V^{\alpha}_{\text{A}} = 0$. But this is precisely the conclusion of Theorem 1, and so we need to show that the V_{A} satisfy the conditions of Theorem 1, part a), namely that $[V_{\text{A}}, V_{\text{B}}]$ is a linear combination of the V_{D}, i.e. that $[V_{\text{A}}, V_{\text{B}}](p) \in D_p$, or that $\overset{\alpha}{\boldsymbol{\omega}}([V_{\text{A}}, V_{\text{B}}]) = 0$. Now for any vectors and forms it can easily be shown by writing out in coordinates that

$$\boldsymbol{\omega}([V, W]) = V(\boldsymbol{\omega}(W)) - W(\boldsymbol{\omega}(V)) - \boldsymbol{d\omega}(V, W)\ . \tag{2.12.13}$$

Applying this to the present case

$$\overset{\alpha}{\boldsymbol{\omega}}([V_{\rm A}, V_{\rm B}]) = V_{\rm A}\left(\overset{\alpha}{\boldsymbol{\omega}}(V_{\rm B})\right) - V_{\rm B}\left(\overset{\alpha}{\boldsymbol{\omega}}(V_{\rm A})\right) - \boldsymbol{d}\overset{\alpha}{\boldsymbol{\omega}}(V_{\rm A}, V_{\rm B})$$
$$= \frac{1}{2}\left[-\overset{\alpha}{\boldsymbol{\theta}}_{\beta}(V_{\rm A})\overset{\beta}{\boldsymbol{\omega}}(V_{\rm B}) + \overset{\alpha}{\boldsymbol{\theta}}_{\beta}(V_{\rm B})\overset{\beta}{\boldsymbol{\omega}}(V_{\rm A})\right] = 0$$

(from (2.12.12)) as required. ■

For application in relativity theory we have a torsion-free connection so that, for a one-form $\boldsymbol{\omega}$, we can write

$$(\boldsymbol{d}\omega)_{ij} = \nabla_{[i}\omega_{j]}.$$

Thus the condition of the previous theorem can be written as

$$\overset{k+1}{\omega}_{[i}\overset{k+2}{\omega}_{j}\ldots\overset{n}{\omega}_{l}\nabla_{m}\overset{\alpha}{\omega}_{n]} = 0 \qquad (\alpha = k+1, \ldots, n)\,. \tag{2.12.14}$$

In the particular case $k = n - 1$, writing $\boldsymbol{\omega} \equiv \overset{n}{\boldsymbol{\omega}}$, this is

$$\omega_{[i}\nabla_{j}\omega_{k]} = 0.$$

If this is satisfied, then, from the theorem, there are coordinates such that $\boldsymbol{\omega}(\partial_{\rm A}) = 0$ $(A = 1, 2, ..., n-1)$; i.e. $\boldsymbol{\omega}$ has only an n-component; putting $x^n \equiv h$, a function on M, this means that

$$\boldsymbol{\omega} = f\boldsymbol{d}h. \tag{2.12.15}$$

Such a form is called *irrotational.*

In the case where $k = n - 2$ we can again apply the theorem: writing $\boldsymbol{\omega} \equiv \overset{n-1}{\boldsymbol{\omega}}$, $\boldsymbol{\nu} \equiv \overset{n}{\boldsymbol{\omega}}$ the conditions of the theorem are

$$\omega_{[i}\nu_{j}\nabla_{k}\omega_{l]} = 0 \qquad \omega_{[i}\nu_{j}\nabla_{k}\nu_{l]} = 0;$$

from these we conclude that $\boldsymbol{\omega}$ and $\boldsymbol{\nu}$ can be written as combinations of coordinate differentials $\boldsymbol{d}x^{n-1}$ and $\boldsymbol{d}x^{n}$, and so, setting $f \equiv x^{n-1}$, $g \equiv x^{n}$, we obtain

$$\boldsymbol{\omega} = \alpha\boldsymbol{d}f + \beta\boldsymbol{d}g$$
$$\boldsymbol{\nu} = \gamma\boldsymbol{d}f + \delta\boldsymbol{d}g \tag{2.12.16}$$

for real functions $\alpha, \beta, \gamma, \delta$.

If we are given a metric, then each vector field V defines a one-form $\boldsymbol{\omega}$ by

$$\boldsymbol{\omega} = \overset{*}{g}(V, \cdot)$$

and viceversa. So the above results on one-forms, can be applied to the 1-forms derived from vector fields. In this case it turns out that if $V_{k+1}, \dots V_n$ are vector fields, then the surfaces S_c defined by the corresponding forms $\overset{k+1}{\omega}, \dots, \overset{n}{\omega}$ are perpendicular to the vectors. This is easily seen, because, if $V \in T(S_c)$, then by definition $\overset{\alpha}{\omega}(V) = 0$. The equations (2.12.15) and (2.12.16) can be expressed as follows:

a) if a vector ℓ satisfies $\ell_{[i}\nabla_j \ell_{k]} = 0$ then $\ell_i = f g_{,i}$ and ℓ is perpendicular to the surfaces $g = \text{const}$.

b) if vectors ℓ, m satisfy:

$$\ell_{[i} m_j \nabla_k \ell_{l]} = \ell_{[i} m_j \nabla_k m_{l]} = 0$$

then $\ell_i = \alpha f_{,i} + \beta g_{,i}$; $m_i = \gamma f_{,i} + \delta g_{,i}$ and ℓ and m are both perpendicular to the two-surfaces $\{f = \text{const.},\ g = \text{const.}\}$.

A particularly interesting case occurs if vectors $V_{k+1}, \dots, V_n$ as well as generating forms that satisfy Theorem 2, themselves satisfy Theorem 1. It then turns out that we can find a single coordinate system that exhibits the conclusions of both theorems. One subsidiary condition is required. As in the proof of Theorem 2, let:

$$D_p = \left\{ V \in T_p(M) | (V_\alpha | V)_p = 0 \ ; \ \forall \alpha = k+1, \dots, n \right\}$$

and set

$$E_p = \left\{ V \in T_p(M) \mid (\exists\, c^\alpha)(V = c^\alpha \underset{p}{V}_\alpha) \right\}$$

to be the linear space spanned by the $\underset{p}{V}_\alpha$. If these vectors are linearly independent, then:

$$\dim\ D_p = k\ ,\ \dim\ E_p = n - k$$

We will say that D_p is *null* if the restriction of $\overset{*}{g}$ to D_p is singular (i.e. if there is a non-zero vector $V \in D_p$ such that $(V|W) = 0$ for all $W \in D_p$), and similarly for E_p. Clearly E_p is contained in the set of vectors perpendicular to D_p, and since that set has dimensions $n - k$ it must coincide with E_p; hence if D_p is null, the vector V above lies in E_p, and conversely if there exists a non zero V in $D_p \cap E_p$, then D_p is null and, by the same argument, so is E_p. Thus D_p and E_p are either both null or both non-null, and they are null if and only if $D_p \cap E_p \neq \{o\}$.

THEOREM 3

Let independent vector fields $V_{k+1}, ..., V_n$ be such that E_p is non-null for all p; (2.12.14) is satisfied with $\overset{\alpha}{\omega}_i \equiv (V_{k+1})_i$, and

$$\left[V_\alpha, V_\beta\right] = C^\gamma_{\alpha\beta} V_\gamma \qquad (\alpha, \beta, \gamma = k+1, ..., n).$$

Then there exists a coordinate system $x^1, ..., x^n$ in which

i) $\quad g_{\alpha\mathrm{B}} = 0 \qquad (B = 1, ..., k)$

ii) $\quad V_\alpha{}^{\mathrm{A}} = 0.$

Proof: Denote the coordinates of Theorem 1 by z^i, so that

$$V_\alpha(z^{\mathrm{A}}) = 0 \tag{2.12.17}$$

and the coordinates of Theorem 2 by w^i, so that the one-forms corresponding to V_β, viz:

$$\overset{\beta}{\boldsymbol{\omega}} \equiv \overset{*}{\boldsymbol{g}}(V_\beta, \cdot) \tag{2.12.18}$$

are linear combinations of the $\boldsymbol{d}w^\alpha$. Since both these sets are independent, we have conversely that:

$$\boldsymbol{d}w^\alpha = C^{\alpha\delta}\boldsymbol{\omega}_\delta. \tag{2.12.19}$$

In the w-coordinates, the fact that $\overset{\beta}{\boldsymbol{\omega}}$ can be expressed in terms of the $\boldsymbol{d}w^\alpha$ means that:

$$\overset{\beta}{\omega}_{\mathrm{A}} = 0. \tag{2.12.20}$$

We now simply take

$$x^{\mathrm{A}} = z^{\mathrm{A}}; \qquad x^\alpha = w^\alpha. \tag{2.12.21}$$

First we have to show that this really is a coordinate system, i.e. that the set $\{\boldsymbol{d}w^\alpha, \boldsymbol{d}z^{\mathrm{A}}\}$ is independent. So suppose that we had

$$e_\alpha \boldsymbol{d}w^\alpha = d_{\mathrm{A}}\boldsymbol{d}z^{\mathrm{A}} = e_\alpha C^\alpha_\delta \, \overset{\delta}{\boldsymbol{\omega}} \tag{2.12.22}$$

for some e_α, d_{A} (using (2.12.19)). Operating on V_γ:

$$e_\alpha C^\alpha{}_\delta \overset{\delta}{\boldsymbol{\omega}}(V_\gamma) = d_{\mathrm{A}}\boldsymbol{d}z^{\mathrm{A}}(V_\gamma) = d_{\mathrm{A}}V_\gamma(z^{\mathrm{A}}) = 0 \tag{2.12.23}$$

from (2.12.17), for all γ. Hence $e_\alpha C^\alpha_\delta V_\gamma \in E_p \cap D_p$. But since these are not null, we must have $e_\alpha C^{\alpha\delta} V_\delta = 0$; whence $e_\alpha = 0$ and $d_A = 0$ and linear independence follows. Condition ii) is then just equation (2.12.17), while (2.12.20) gives, in x-coordinates

$$g_{Ai} V^i_\beta = g_{A\alpha} V^\alpha_\beta = 0$$

which implies

$$g_{\alpha A} = 0. \tag{2.12.24}$$

■

2.13 Isometries on M

A diffeomorphism $\Phi : M \to M$ is an isometry on M if:

$$\Phi_*(\overset{*}{\boldsymbol{g}}) = \overset{*}{\boldsymbol{g}}. \tag{2.13.1}$$

Suppose we have a local one-parameter group of local diffeomorphisms which are isometries on M; then, calling C_ξ the congruence of the trajectories of the group and ξ the tangent field, from the definition of the Lie derivative we have:

$$\pounds_\xi \overset{*}{\boldsymbol{g}} = 0 \tag{2.13.2}$$

everywhere on M. Let U_α be a coordinate neighbourhood of M, containing C_ξ, then (2.13.2) reads:

$$\xi^k \partial_k g_{ij} + g_{kj} \partial_i \xi^k + g_{ik} \partial_j \xi^k = 0 \tag{2.13.3}$$

or, from (2.8.4):

$$\nabla_{(i} \xi_{j)} = 0. \tag{2.13.4}$$

A vector field which satisfies (2.13.2) is termed a Killing field, C_ξ a Killing congruence and (2.13.4) is termed the Killing equation. If γ is an affine geodesic in U_α, then $(\dot\gamma|\xi)$ is constant on γ; in fact:

$$\frac{d}{ds}(\dot\gamma|\xi) = \dot\gamma^k \dot\gamma_i \nabla_k \xi^i + \xi^i \dot\gamma^k \nabla_k \dot\gamma_i = 0 \tag{2.13.5}$$

from (2.13.4) and (2.4.21).

We now apply Theorem 3 of the preceding section to the case where V_α are Killing vectors. Let us then suppose we have E_p

non-null and:

$$
\begin{aligned}
&a) \qquad \pounds_{V_\alpha} \overset{*}{\boldsymbol{g}} = 0 \\
&b) \qquad \left[V_\alpha, V_\beta\right] = C^\gamma_{\alpha\beta} V_\gamma \qquad (2.13.6)\\
&c) \qquad V_{k[i} V_{|k+1|j} \cdots V_{|n|\ell} \nabla_m V_{|\alpha|n]} = 0
\end{aligned}
$$

with the V_α independent and using version (2.12.14) of the condition for Theorem 3. Then there is a coordinate system for which

$$g_{\alpha\text{A}} = 0 \qquad V_\alpha = V^\beta_\alpha \frac{\partial}{\partial x^\beta}. \qquad (2.13.7)$$

So (2.13.3) becomes:

$$V^\beta_\alpha \partial_\beta g_{ij} + g_{\beta j} \partial_i V^\beta_\alpha + g_{i\beta} \partial_j V^\beta_\alpha = 0. \qquad (2.13.8)$$

Put $i = \text{A}$, $j = \text{B}$ and use $g_{\beta\text{B}} = 0$ to give

$$V^\beta_\alpha \partial_\beta g_{\text{AB}} = 0$$

which implies

$$\partial_\alpha g_{\text{AB}} = 0 \qquad (2.13.9)$$

(i.e. g_{AB} depends only on $x^1, ..., x^k$). Now put $i = \text{A}$, $j = \gamma$ to give:

$$g_{\beta\gamma} \partial_{\text{A}} V^\beta_\alpha = 0$$

so that (from the non-nullity of E_p)

$$\partial_{\text{A}} V^\beta_\alpha = 0 \qquad (2.13.10)$$

and V^β_α depends only on $x^{k+1}, ..., x^n$.

To summarize: if a set of independent vector fields V_α spans a non-null subspace E_p of each T_p and satisfies (2.13.6)$a), b), c)$, then there are coordinates in which the metric takes the form:

$$
\begin{aligned}
ds^2 &= g_{\text{AB}}(x^1, ..., x^k) \boldsymbol{dx}^{\text{A}} \otimes \boldsymbol{dx}^{\text{B}} + g_{\alpha\beta}(x^i) \boldsymbol{dx}^\alpha \boldsymbol{dx}^\beta \\
V_\alpha &= V^\beta_\alpha \partial_\beta \qquad V^\beta_\alpha = V^\beta_\alpha(x^{k+1}, ..., x^n). \qquad (2.13.11)
\end{aligned}
$$

Note that this means that for fixed x^{A}, the metric $g_{\alpha\beta} \boldsymbol{dx}^\alpha \otimes \boldsymbol{dx}^\beta$ on the manifold with x^{A} constant and x^α as coordinates, has $(n-k)$ Killing vectors and so is a homogeneous space. A further simplification occurs if the Killing vectors commute:

$$\left[V_\alpha, V_\beta\right] = 0 \qquad (2.13.12)$$

Then from Theorem 1 we can choose the z-coordinates, which in Theorem 3 become $x^k, ..., x^n$, so that $V^\beta_\alpha = \delta^\beta_\alpha$. Inserting this in (2.13.8) with $i = \gamma$ and $j = \delta$ gives:

$$\partial_\alpha g_{\gamma\delta} = 0$$

so that $g_{\delta\gamma}$ depends only on $x^1, ..., x^k$.

3

THE CURVATURE

3.1 The Riemann tensor

The connection $\Gamma(\gamma)$ allows us to define the concept of parallelism on each curve γ in M. As mentioned in Sect. 2.9, this concept of parallelism is strictly path–dependent, so the parallel $\Gamma(\gamma_1)\circ u$ to a vector $u \in T_p(M)$ relative to a given curve γ_1 connecting $p = \gamma_1(0)$ to $q = \gamma_1(1)$ will differ from the parallel $\Gamma(\gamma_2)\circ u$ at q relative to a different curve γ_2 connecting p to q. Let us calculate this difference explicitly; consider the integral curves of two commuting vector fields Y and X respectively in some neighbourhood of p, so that from (2.2.7) and (2.2.16):

$$\begin{aligned} 0 = [X,Y] &= \pounds_Y(X^m e_m) = (\pounds_Y X^m)e_m + X^m(\pounds_Y e_m) \\ &= (\pounds_Y X^m)e_m + X^m(Y^k C^n_{km} e_n - e_m(Y^n)e_n) \end{aligned}$$

that is

$$Y(X^m) - X(Y^m) = X^r Y^k C^m_{rk}. \tag{3.1.1}$$

Choose coordinates as in Theorem 1 of (2.12) so that $X = \partial/\partial x^1, Y = \partial/\partial x^2$ and then set $s \equiv x^1, t \equiv x^2$. Consider the paths shown in Fig. 3-1 with coordinates $x^\alpha = \text{const.}$, for $\alpha \geq 3$. Then γ_0, γ_1 are integral curves of X, parametrized by s, while γ'_0, γ'_1 are integral curves of Y parametrized by t. An arbitrary vector u at p will be mapped into:

$$\check{u}_q = \Gamma(s_0, s_1; \gamma_1)\circ\Gamma(t_0, t_1; \gamma'_0)\circ u_p \tag{3.1.2}$$

moving along the path from p to p_1 and then from p_1 to q, from property (ii) of the connection (see Sect. 2.3) and to:

$$\check{u}'_q = \Gamma(t_0, t_1; \gamma'_1)\circ\Gamma(s_0, s_1; \gamma_0)\circ u_p \tag{3.1.3}$$

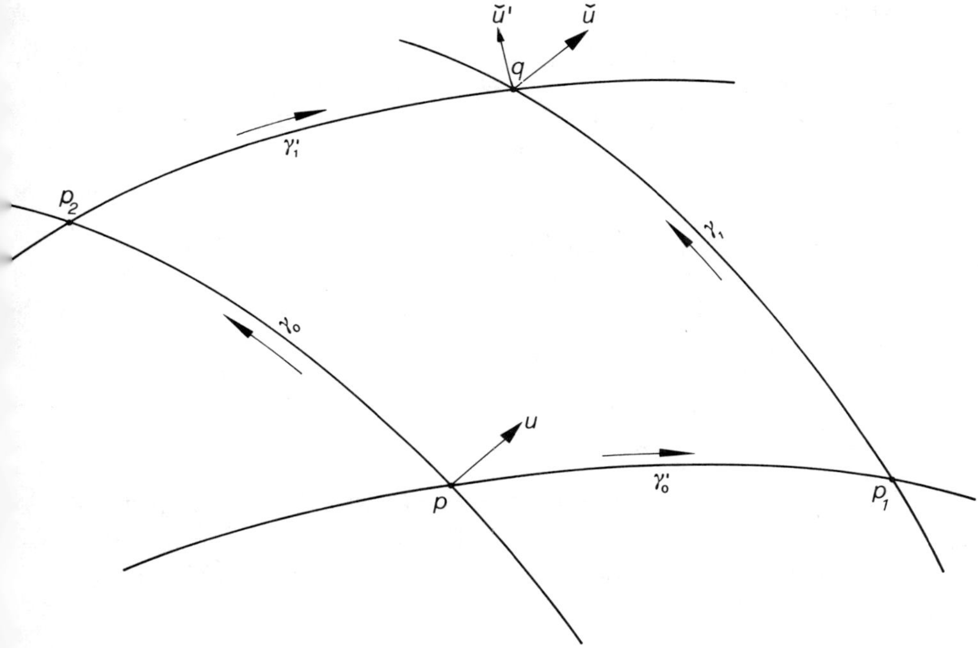

Fig. 3-1 An arbitrary vector at p will be mapped by the connector into a different vector in q according to the path considered. The difference between these two vectors in q depends on the curvature.

moving along the path from p to p_2 and then from p_2 to q. The difference now reads:

$$\breve{u}_q - \breve{u}'_q = [\Gamma(s_0, s_1; \gamma_1) \circ \Gamma(t_0, t_1; \gamma'_0) - \Gamma(t_0, t_1; \gamma'_1) \circ \Gamma(s_0, s_1; \gamma_0)] \circ u_p \,. \tag{3.1.4}$$

Assume that the points p_1 and p_2 are at small parameter distances from p, so the first term in (3.1.4) can be written in terms of components as:

$$\begin{aligned} &\Gamma(t_0, t_1; \gamma'_0)_i{}^n \Gamma(s_0, s_1; \gamma_1)_n{}^j \\ &= \left[\delta_i^n + \left(\frac{d}{dt}\Gamma_i{}^n\right)_p \delta t + \frac{1}{2}\left(\frac{d^2}{dt^2}\Gamma_i{}^n\right)_p \delta t^2 + O(\delta t^3)\right] \\ &\quad \times \left[\delta_n^j + \left(\frac{d}{ds}\Gamma_n{}^j\right)_{p_1} \delta s + \frac{1}{2}\left(\frac{d^2}{ds^2}\Gamma_n{}^j\right)_{p_1} \delta s^2 + O(\delta s^3)\right] \end{aligned} \tag{3.1.5}$$

where

$$\delta t = t_1 - t_0 \quad ; \quad \delta s = s_1 - s_0 \; .$$

Recalling (2.4.10), (2.4.11) and to the second order in the parameter distance from p, (3.1.5) becomes:

$$\begin{aligned}
&\Gamma(t_0, t_1; \gamma_0')_i{}^n \Gamma(s_0, s_1; \gamma_1)_n{}^j \\
&\quad = \delta_i^j - \Gamma^j_{im}(p)[X^m(p)\delta s + Y^m(p)\delta t] - e_r(\Gamma^j_{im}) Y^r X^m \delta t \delta s \\
&\qquad - \Gamma^j_{im}(p) \left(\frac{\partial X^m}{\partial t}\right)_p \delta t \delta s + \Gamma^n_{im}(p)\Gamma^j_{nr}(p) X^m Y^r \delta t \delta s \\
&\qquad + \frac{1}{2}\left[\left(\frac{d^2}{ds^2}\Gamma^j_i\right)_p \delta s^2 + \left(\frac{d^2}{dt^2}\Gamma^j_i\right)_p \delta t^2\right] \\
&\qquad + O(\delta t^3, \delta s^3).
\end{aligned} \tag{3.1.6}$$

The second term on the right-hand side of (3.1.4) will be just the same as (3.1.6) but for the interchange of s with t and Y with X. Performing straightforward calculations and using (3.1.1), we obtain:

$$\begin{aligned}
\check{u}_q - \check{u}'_q &= \left[e_r(\Gamma^j_{im}) - e_m(\Gamma^j_{ir}) + \Gamma^j_{nr}\Gamma^n_{im} - \Gamma^j_{nm}\Gamma^n_{ir} - \Gamma^j_{ik}C^k_{rm}\right] \\
&\quad \times X^m Y^r u^i e_j \delta t \delta s + O(\delta t^3, \delta s^3) \\
&\approx R^j{}_{irm} X^m Y^r u^i e_j \delta t \delta s
\end{aligned} \tag{3.1.7}$$

to second order, where:

$$R^j{}_{irm} = e_r(\Gamma^j_{im}) - e_m(\Gamma^j_{ir}) + \Gamma^j_{nr}\Gamma^n_{im} - \Gamma^j_{nm}\Gamma^n_{ir} - \Gamma^j_{ik}C^k_{rm}. \tag{3.1.8}$$

We shall also on occasion write the coefficient of $\delta t \delta s$ on the right-hand side of (3.1.7) as $R(u, Y, X)$. From its expression (3.1.7), the map R is multilinear, and by its construction it depends only on the connector (and not on the basis field e_i), and hence (if the connection is torsion-free), only on the metric $\overset{*}{g}$. It can thus be regarded as a tensor of valence (1, 3), known as the *Riemann tensor*, describing a basic structural property of $(M, \overset{*}{g})$ called the *curvature*.

As we shall later explain, the existence on M of a non-vanishing Riemann tensor expresses the impossibility of defining a coordinate system on any part of M that yields an orthonormal (coordinate) basis at every point of it; if this were possible, the manifold would be termed *flat* , otherwise it is termed *curved.* A natural consequence of the dependence of the connector on the path is the non-commutativity of covariant derivatives. Given a vector field X in some open set of M, a direct computation shows that:

$$\nabla_{[r}\nabla_{s]}X_i = \frac{1}{2}R^k{}_{isr}X_k + \frac{1}{2}T^k{}_{rs}\nabla_k X_i \qquad (3.1.9)$$

$$\nabla_{[r}\nabla_{s]}X^i = -\frac{1}{2}R^i{}_{ksr}X^k + \frac{1}{2}T^k{}_{rs}\nabla_k X^i \qquad (3.1.10)$$

from (2.6.2), (2.6.8′), (2.7.3) and (3.1.8).

In the absence of torsion, the terms in T become zero. From equation (3.1.10) we can easily calculate that if X, Y, Z are vector fields, then:

$$\nabla_X\nabla_Y Z - \nabla_Y\nabla_X Z = R(Z, X, Y) + \nabla_{[X,Y]}Z. \qquad (3.1.11)$$

Equation (3.1.11), or its equivalent (3.1.10), is called the Ricci identity. In deriving this, it was not in fact necessary that X and Y should be vector fields. Suppose, instead, that we have a map $\varphi : (s,t) \mapsto \varphi(s,t) \in M$ with:

$$X(s,t) = \varphi_*\left(\left.\frac{\partial}{\partial s}\right|_{(s,t)}\right), \qquad Y(s,t) = \varphi_*\left(\left.\frac{\partial}{\partial t}\right|_{(s,t)}\right)$$

i.e.

$$X^i = \frac{\partial}{\partial s}\varphi(s,t)^i, \qquad Y^i = \frac{\partial}{\partial t}\varphi(s,t)^i.$$

Thus X and Y differ from being vector fields in that they are not defined on a neighbourhood in M; and if $\varphi(s_1,t_1) = \varphi(s_2,t_2) = p$ but $X(s_1,t_1) \neq X(s_2,t_2)$, then X could take two (or more) different values at the same point p of M. In the notation of (2.6.2) we set, for any vector $Z(s,t)$, a function of s and t with $Z(s,t) \in T_{\varphi(s,t)}(M)$:

$$\left(\frac{DZ}{Ds}\right)^i \equiv \frac{\partial Z^i}{\partial s} + \Gamma^i_{jk}(\varphi(s,t))Z^j X^k \qquad (3.1.12)$$

$$\left(\frac{DZ}{Dt}\right)^i \equiv \frac{\partial Z^i}{\partial t} + \Gamma^i_{jk}(\varphi(s,t))Z^j Y^k \qquad (3.1.13)$$

and then a direct calculation gives:

$$\frac{D}{Ds}\frac{D}{Dt}Z - \frac{D}{Dt}\frac{D}{Ds}Z = R(Z, X, Y) \ . \qquad (3.1.14)$$

This becomes equivalent to (3.1.11) in the case where φ is injective, because then we can find vector fields $\bar{X}, \bar{Y}$ and $\bar{Z}$ such that $\bar{X} = X(s,t)$, and similarly for Y and Z for which, at $\varphi(s,t)$,

$$\frac{DZ}{Ds} = \nabla_X \bar{Z} = \nabla_{\bar{X}} \bar{Z} \quad etc.$$

$$[\bar{X}, \bar{Y}]\big|_{\varphi(s,t)} = 0.$$

3.2 Symmetry properties of the Riemann tensor and the Gaussian curvature

Since the Riemann tensor plays a key role in general relativity, it is essential to know how many independent components it has; to this end we need to know the symmetry properties of the Riemann tensor and the identities which follow from them.

A convenient way to do this is to refer the components (3.1.8) to a geodesic coordinate system $\{y^i\}$, adapted to a point p in M, assuming the manifold to be torsion-free. From the discussion of Sect. 2.7, from (2.7.10) and recalling that now $\Gamma'^i_{\ jk} = \Gamma'^i_{\ kj}$, we have at p

$$R'^i_{\ jkl} = 2\partial_{[k'}\Gamma'^i_{\ l]j} \tag{3.2.1}$$

where $\partial_{k'} = \partial/\partial y^k$.

An obvious symmetry property derives from the definition of R:

$$R'^i_{\ jkl} = -R'^i_{\ jlk} \ . \tag{3.2.2}$$

We can lower the contravariant index to obtain a tensor of valence (0, 4), i.e. $R_{ijkl} = g_{in}R^n_{\ jkl}$, so in geodesic coordinates, we readily have at p:

$$R'_{ijkl} = 2g'_{ir}\partial_{[k'}\Gamma'^r_{\ l]j} = \partial_{l'}\partial_{[i'}g'_{j]k} - \partial_{k'}\partial_{[i'}g'_{j]l} \tag{3.2.3}$$

from (2.8.7). Here we recognize two more symmetry properties of the Riemann tensor:

$$R'_{ijkl} = -R'_{jikl} \tag{3.2.4}$$

$$R'_{ijkl} = R'_{klij} \ . \tag{3.2.5}$$

From (3.2.3) and the symmetry of $\Gamma'^k_{\ ij}$ we have identically

$$R'_{i[jkl]} = 0. \tag{3.2.6}$$

The tensorial character of (3.2.2), (3.2.4), (3.2.5) and (3.2.6) ensures that those relations hold in a general coordinate system. Taking into account the symmetry properties (3.2.4) and (3.2.5), the identity (3.2.6) amounts, in an n-dimensional space, to $n(n-1)(n-2)(n-3)/24$ relations. Properties (3.2.2), (3.2.4) and (3.2.5) furthermore, allow one to express the Riemann tensor as a symmetric tensor of valence (0,2) over the bivector space $\mathcal{B}$; i.e. an element of the space $\overset{*}{\mathcal{B}}\otimes\overset{*}{\mathcal{B}}$, where $\overset{*}{\mathcal{B}}$ is the dual of the bivector space* $\mathcal{B}$ of dimensions $N = n(n-1)/2$. We can relabel the components $T^{[ij]}$ of any bivector by defining T^{A} $(\mathrm{A} = 1, 2, \ldots, N)$ according to $T^1 \equiv T^{10}$, $T^2 \equiv T^{20}$,...,$T^n \equiv T^{n0}$, $T^{n+1} \equiv T^{n1}, \ldots$; so that the index A labels the independent components. The Riemann tensor can then be written as a symmetric tensor R_{AB} in an N-dimensional space. Its independent components will then number $N(N+1)/2 = n(n-1)(n^2-n+2)/8$; but these are further reduced by the number of relations (3.2.6), so the total number of independent components of the Riemann tensor is:

$$\frac{n^2(n^2-1)}{12} \,. \tag{3.2.7}$$

In a four-dimensional space, this tensor has 20 independent components.

We can now better understand the significance and usefulness of the normal (geodesic) coordinates which were introduced in Sect. 2.7. If $U_{\mathcal{N}}(p)$ is a normal neighbourhood of $p \in M$, condition (2.7.10) implies the following property: the metric tensor, once expressed in terms of the normal coordinates y^i, is constant in $U_{\mathcal{N}}(p)$ to first order in the affine parameter distance s from p, measured on a geodesic γ through that point, and can be given the value η_{ij}. From (3.2.3) and (2.7.10), we have at $q = \gamma(s) \in U_{\mathcal{N}}(p)$:

$$g'_{ij}\big|_q = g'_{ij}\big|_p - \frac{1}{6}(R'_{irjt} + R'_{itjr})_p \dot{\gamma}^r(0)\dot{\gamma}^t(0)s^2 + O(s^3). \tag{3.2.8}$$

Let us evaluate $g'_{ij}\big|_p$. From (2.7.7), letting $\gamma(0) = p$, we have:

$$\dot{\gamma}(s) = \dot{\gamma}^i(0)\partial'_i$$

*We call bivector any antisymmetric tensor of valence (0, 2).

where $\{\partial'_i\}$ is the normal coordinate basis. Putting $s = 0$, we have $e_i = \partial'_i$ hence $g'_{ij}|_p = (\partial'_i|\partial'_j)_p = (e_i|e_j)_p$ and so, from the arbitrariness of e_i we can choose it so that $g'_{ij}|_p = \eta_{ij}$.

We shall now introduce the concept of Gaussian curvature of a two-surface embedded in an n–dimensional Riemannian manifold and so elucidate the physical meaning of the Riemann tensor. To this end we shall first give a few more definitions and theorems, the proofs of which we shall omit since they can be found in any specialized text on differential geometry (Helgason, 1962; Eisenhart, 1950).

On a differential manifold M with an affine connection, to any vector $u \neq o$ in $T_p(M)$ (any $p \in M$) there exists a unique geodesic $t \to \gamma(t)$ in M such that:

$$\gamma(0) = p \quad ; \quad \dot{\gamma}(0) = u \ .$$

It then follows that there always exists an open neighbourhood $S(o)$ of the origin ($u = o$) in $T_p(M)$ and an open neighbourhood U_p of p in M such that the mapping $u \to \gamma_u(1)$ is a diffeomorphism of $S(o)$ onto U_p. The mapping $u \to \gamma_u(1)$ is termed the *Exponential mapping* at p and is denoted by Exp_p. The open set $S(o)$ of $T_p(M)$ is termed *normal* if $\mathrm{Exp}_p\colon S(o) \to U_p$ is a diffeomorphism and for any $w \in S(o)$ and $0 \leq t \leq 1$, then $tw \in S(o)$. Thus the set $U_p = \mathrm{Exp}_p S(o)$ is termed a normal neighbourhhod of p in M. In a C^∞-manifold M with an affine connection, each point $p \in M$ has a normal neighbourhood which is a normal neighbourhood of any of its points. Let now $S_r(o)$, with $0 \leq r < \infty$, be a normal neighbourhood of the origin in $T_p(M)$ such that any $w \in S_r(o)$ has $|w| < r$ where $|w| = (\sum |w^i|^2)^{\frac{1}{2}}$ and components are with respect to some chosen basis; then the set $W_r(p) = \mathrm{Exp}_p S_r(o)$ is a normal neighbourhood of p in M. The sets $S_r(o)$ and $W_r(p)$ are termed *open balls* of radius r.

Let now ${}^2\Sigma$ be a two dimensional Riemannian manifold and $p \in {}^2\Sigma$; call ${}^2S_r(o)$ an open ball of radius r of $T_p({}^2\Sigma)$ and ${}^2W_r(p) = \mathrm{Exp}_p {}^2S_r(o)$ an open ball of radius r centred at p in ${}^2\Sigma$; we call the (*sectional*) *curvature* of ${}^2\Sigma$ at p the limit (which can be proved always to exist):

$$K = \lim_{r \to 0} \frac{\text{Area } {}^2S_r(o) - \text{Area } {}^2W_r(p)}{r^2 \text{ Area } {}^2S_r(o)} .$$

When ${}^2\Sigma$ is a two-surface then K is termed the *Gaussian curvature*. Given two unit tangent vectors u, v in $p \in M$ (Dim $M > 2$) and the two-surface ${}^2\Sigma$ spanned by the geodesics which stem from p with tangent vectors at p given by $\dot{\gamma} = \alpha u + \beta v$, $\alpha, \beta \in \mathbb{R}$; then one can show that the Gaussian curvature at p is given by:

$$K({}^2\Sigma) = \frac{R_{ijkl} u^i v^j u^k v^l}{(g_{ik} g_{jl} - g_{il} g_{jk}) u^i v^j u^k v^l} . \tag{3.2.9}$$

This justifies the name of curvature given customarily to the Riemann tensor.

3.3 Significance of a curvature tensor vanishing everywhere

Suppose that, at every point p in M,

$$R^i{}_{jkl} = 0. \tag{3.3.1}$$

Consider, then, two paths γ_0 and γ_1 from p to q:

$$\gamma_0(0) = \gamma_1(0) = p; \qquad \gamma_0(1) = \gamma_1(1) = q$$

say, and suppose that the paths are *homotopic*, meaning that we can find a smooth family of paths $\gamma_t (t \in [0, 1])$ with $\gamma_t(0) = p, \gamma_t(1) = q$ linking γ_0 and γ_1. The map $(s, t) \to \gamma_t(s) \equiv \varphi(s, t)$, say, allows us to apply verbatim the analysis of (3.1), using $X \equiv \varphi_*(\partial/\partial s)$, $Y \equiv \varphi_*(\partial/\partial t)$ as the commuting vector fields. If we construct a grid of the integral curves of X and Y, with spacing δt and δs respectively, then that analysis allows us to conclude that the difference between propagating a vector around $A \to B \to C$ and propagating it around $A \to D \to C$, for a cell $ABCD$ of the grid, is $O\left((|\delta t| + |\delta s|)^3\right)$. We can then use this, together with taking a limit as $\delta t, \delta s \to 0$, to show that propagating along γ_0 has the same effect as propagating along γ_1.

In fact this result can be obtained more smoothly by noting that if

$$Z^i(s,t) \equiv \Gamma(0,s;\gamma_t)_j{}^i u^j$$

where $u \in T_p(M)$ so that $Z^i(1,t)$ is the result of transporting u to q along γ_t, then by definition

$$\frac{DZ}{Ds} = 0$$

so, by (3.1.12) and (3.3.1):

$$\frac{D}{Ds}\left(\frac{DZ}{Dt}\right) = \frac{D}{Dt}\left(\frac{DZ}{Ds}\right) = 0$$

i.e.

$$\left(\frac{DZ}{Dt}\right)^i_{(s,t)} = \Gamma(0,s;\gamma_t)_j{}^i\left(\frac{DZ}{Dt}\right)^j_{(0,t)} = \Gamma(0,s;\gamma_t)_j{}^i\left(\frac{dZ}{dt}\right)^j_{(0,t)}$$

since $Z^i(0,t) = u^i$. Hence

$$\frac{d}{dt}Z(1,t) = \left.\frac{DZ}{Dt}\right|_{(1,t)} = 0$$

and $Z(1,t)$ is constant, independent of the path along which it is transported.

It can be shown that if we work in a neighbourhood U of M that is homeomorphic to $\mathbb{R}^n$, then all paths between p and q are homotopic. With this restriction the result of propagating a vector from p to q is unique, independent of the path chosen. Then choose at p, arbitrary in U, a basis of orthonormal vectors $\{u_{\hat{a}}\}_p$ so that

$$(u_{\hat{a}}|u_{\hat{b}})_p = \eta_{\hat{a}\hat{b}}. \tag{3.3.2}$$

From (3.3.1), a field of orthonormal bases $\{\check{u}_{\hat{a}}\}$ is then uniquely defined in U by parallel propagation; from the property (2.8.1) of the connection, we have everywhere:

$$(\check{u}_{\hat{a}}|\check{u}_{\hat{b}})_q = \eta_{\hat{a}\hat{b}} \qquad \forall q \in U. \tag{3.3.3}$$

The question now arises whether there exists a coordinate system $\{y^{\hat{a}}\}$ such that $\check{u}_{\hat{a}} = \partial/\partial y^{\hat{a}}$. In this case the system could be adapted to every point of U, which would then be pseudo-Euclidean (cf. Sect. 1.15). In view of Theorem 1 in Sect. 2.12, we need to calculate the Lie bracket of all pairs of basis vectors. Since the connection is assumed torsion-free we have, from (2.7.2)

$$[\breve{u}_{\hat{a}}, \breve{u}_{\hat{b}}] = \nabla_{\breve{u}_{\hat{a}}} \breve{u}_{\hat{b}} - \nabla_{\breve{u}_{\hat{b}}} \breve{u}_{\hat{a}}. \tag{3.3.4}$$

But since the $\breve{u}_{\hat{a}}$ are parallely propagated along any chosen curve in U we have $\nabla_{\dot{\gamma}} \breve{u}_{\hat{b}} = 0$ for all $\dot{\gamma}$, i.e. $\nabla_{\breve{u}_{\hat{a}}} \breve{u}_{\hat{b}} = 0$. Thus from (3.3.4) the $\breve{u}_{\hat{a}}$ commute and so from Theorem 1 of Sect. 2.12 there exists, in a neighbourhood of any point, a coordinate system with the required properties. Indeed, we can say more than this, because the coordinates satisfy:

$$y^{\hat{a}}(q) = y^{\hat{a}}(p) + \int_p^q dy^{\hat{a}} = y^{\hat{a}}(p) + \int_p^q \boldsymbol{\omega}^{\hat{a}} \tag{3.3.5}$$

where $\boldsymbol{\omega}^{\hat{a}} = \overset{*}{\boldsymbol{g}}(\cdot, u_{\hat{b}})\eta^{\hat{a}\hat{b}}$ and the relation $\boldsymbol{\omega}^{\hat{a}} = \boldsymbol{dy}^{\hat{a}}$ follows from $\boldsymbol{dy}^{\hat{a}}(\partial/\partial y^{\hat{c}}) = \delta^{\hat{a}}{}_{\hat{c}} = \eta^{\hat{a}\hat{b}}\overset{*}{\boldsymbol{g}}(u_{\hat{c}}, u_{\hat{b}})$ for all $\hat{c}$. The path independence of (3.3.5) is equivalent to the commuting of the $u_{\hat{a}}$ already proved; indeed by equation (2.12.13):

$$(\boldsymbol{d\omega}^{\hat{a}})(u_{\hat{c}}, u_{\hat{d}}) = u_{\hat{c}}(\overset{*}{\boldsymbol{g}}(u_{\hat{d}}, u^{\hat{a}})) - u_{\hat{d}}(\overset{*}{\boldsymbol{g}}(u_{\hat{c}}, u^{\hat{a}})) - \boldsymbol{\omega}^{\hat{a}}([u_{\hat{c}}, u_{\hat{d}}]) = 0$$

(where $u^{\hat{a}} = \eta^{\hat{a}\hat{b}}u_{\hat{b}}$) so that $\boldsymbol{d\omega}^{\hat{a}} = \boldsymbol{o}$ and the integral is path-independent for homotopic paths (from Stokes' theorem). Thus (3.3.5) defines $y^{\hat{a}}$ throughout *the whole of* U, and it does constitute a coordinate system because it implies that the $\boldsymbol{dy}^{\hat{a}}$ are linearly independent.

3.4 The Ricci tensor, the curvature scalar, the Weyl tensor

The curvature tensor allows one to define a lower rank tensor by a suitable contraction over two indices:

$$R^i{}_{jik} = R_j{}^i{}_{ki} =: R_{jk}. \tag{3.4.1}$$

This tensor is called the *Ricci tensor* ; it is symmetric with $n(n+1)/2$ independent components. Explicitly in terms of the connection coefficients, it reads:

$$R_{im} = e_r(\Gamma^r_{im}) - e_m(\Gamma^r_{ir}) + \Gamma^r_{nr}\Gamma^n_{im} - \Gamma^r_{nm}\Gamma^n_{ir} - \Gamma^r_{ik}C^k_{rm}. \tag{3.4.2}$$

Using a coordinate basis and recalling (2.10.1), (2.10.5) and (2.2.13), equation (3.4.2) can be written:

$$R_{im} = |g|^{-\frac{1}{2}}\partial_n(\Gamma^n_{im}|g|^{\frac{1}{2}}) - \partial_i\partial_m(\ln|g|^{\frac{1}{2}}) - \Gamma^r_{is}\Gamma^s_{mr}. \tag{3.4.3}$$

A further contraction of the two indices in the Ricci tensor yields a scalar called the curvature scalar:

$$R = R^i{}_i = R^{ij}{}_{ij}. \tag{3.4.4}$$

In a two-dimensional space, the Ricci tensor has three independent components while the curvature tensor has only one independent component. This means that in a two-dimensional space (surface) the curvature tensor can be expressed in terms of the curvature scalar alone:

$$R^{UV}{}_{ZT} = R\delta^{[U}_Z\delta^{V]}_T \qquad {\scriptstyle U,V,Z,T=1,2}\ . \tag{3.4.5}$$

In a three-dimensional space, the curvature tensor has as many independent components as the Ricci tensor, namely 6, therefore it can be expressed in terms of the latter and of its trace (the curvature scalar) alone:

$$R^{\alpha\beta}{}_{\gamma\delta} = 4\delta^{[\alpha}_{[\gamma}R^{\beta]}_{\delta]} - R\delta^{[\alpha}_{[\gamma}\delta^{\beta]}_{\delta]} \qquad \alpha,\beta,\gamma,\delta = 1,2,3. \tag{3.4.6}$$

In an $(n > 3)$-dimensional space the curvature tensor, having a number of independent components larger than that of the Ricci tensor, admits a decomposition in terms of a further tensorial quantity, called the Weyl tensor, which reads (Weyl, 1922):

$$C^i{}_{jkl} = R^i{}_{jkl} - \frac{2}{n-2}\left[\delta^i_{[k}R_{l]j} + g_{j[l}R^i_{k]}\right] + \frac{2R}{(n-1)(n-2)}\delta^i_{[k}g_{l]j}. \tag{3.4.7}$$

The Weyl tensor $C^i{}_{jkl}$ has the same symmetry properties as the curvature tensor, but, as can be deduced from (3.4.7), is trace-free:

$$C^i{}_{jik} = 0. \tag{3.4.8}$$

Equation (3.4.8) amounts to $n(n+1)/2$ relations which should be added to those which already constrain the curvature tensor, therefore the Weyl tensor has altogether

$$\frac{n(n+1)(n+2)(n-3)}{12}$$

independent components; it vanishes identically in any $n \leq 3$-dimensional manifold.

The Weyl tensor can be written in a more convenient form. Define:

$$E_{ijkl} = \frac{2}{(n-2)}(g_{i[k}R_{l]j} + g_{j[l}R_{k]i}) \tag{3.4.9}$$

$$g_{ijkl} = 2g_{i[k}g_{l]j}. \tag{3.4.10}$$

These tensors have the same symmetry properties of the curvature tensor; in terms of them, the Weyl tensor (3.4.7) can also be written as (Israel, 1970; Greenberg, 1972):

$$C_{ijkl} = R_{ijkl} - E_{ijkl} + \frac{R}{(n-1)(n-2)}g_{ijkl}. \tag{3.4.11}$$

Let us calculate the double-dual of the curvature tensor. From (1.17.28) we have:

$$\overset{*}{R}{}^{ij}{}_{kl}{}^{*} = \epsilon\left[R^{ij}{}_{kl} + \frac{1}{2}\delta^{ij}_{kl}R + 2R^{j}_{[k}\delta^{i}_{l]} - 2R^{i}_{[k}\delta^{j}_{l]}\right], \tag{3.4.12}$$

where $\epsilon = \text{sign}(g)$. Comparing this with (3.4.9) we can write:

$$\overset{*}{R}{}^{ij}{}_{kl}{}^{*} = \epsilon\left[R^{ij}{}_{kl} + \frac{1}{2}\delta^{ij}_{kl}R - (n-2)E^{ij}{}_{kl}\right], \tag{3.4.13}$$

hence:

$$E^{ij}{}_{kl} = \frac{1}{(n-2)}\left[R^{ij}{}_{kl} - \epsilon\overset{*}{R}{}^{ij}{}_{kl}{}^{*}\right] + \frac{R}{2(n-2)}\delta^{ij}_{kl}. \tag{3.4.14}$$

Using this in (3.4.11), we have:

$$C^{ij}{}_{kl} = \frac{1}{(n-2)}\left[(n-3)R^{ij}{}_{kl} + \epsilon\overset{*}{R}{}^{ij}{}_{kl}{}^{*}\right] - \frac{(n-3)R}{2(n-2)(n-1)}\delta^{ij}_{kl}. \tag{3.4.15}$$

The double-dual of the Weyl tensor is now easily calculated. From (1.17.20), (1.17.30) and (3.4.14) we obtain, after some algebra:

$$\overset{*}{C}{}^{ij}{}_{kl}{}^{*} = \epsilon C^{ij}{}_{kl} - \epsilon(n-4)\left[E^{ij}{}_{kl} - \frac{R}{2(n-2)}\delta^{ij}_{kl}\right]. \tag{3.4.16}$$

In a four-dimensional manifold with hyperbolic metric ($\epsilon = -1$), we have:

$$\overset{*}{C}{}^{ij}{}_{kl}{}^{*} = -C^{ij}{}_{kl}. \tag{3.4.17}$$

A very important feature of the Weyl tensor is connected with the transformation properties under conformal transformations of the metric. The latter is a change of the metric defined by:

$$\bar{g}_{ij} = \Omega g_{ij} \qquad ; \qquad \bar{g}^{ij} = \Omega^{-1} g^{ij} \tag{3.4.18}$$

where Ω is a non-zero real-valued function on M, called the conformal factor. Two metrics related as in (3.4.18) are termed conformally related. Under conformal transformations, the various geometrical quantities so far introduced transform as follows:

$$\bar{\Gamma}^i_{jk} = \Gamma^i_{jk} + \frac{1}{2}[\delta^i_j \varphi_k + \delta^i_k \varphi_j - g_{jk}\varphi^i] \, , \tag{3.4.19}$$

$$\begin{aligned} \bar{R}^i{}_{jkl} = R^i{}_{jkl} &+ \delta^i_{[l} \nabla_{k]} \varphi_j + g_{j[k} \nabla_{l]} \varphi^i \\ &+ \frac{1}{2} \delta^i_{[k} \varphi_{l]} \varphi_j - \frac{1}{2} g_{j[k} \varphi_{l]} \varphi^i - \frac{1}{2} \delta^i_{[k} g_{l]j} \varphi^r \varphi_r \, , \end{aligned} \tag{3.4.20}$$

$$\begin{aligned} \bar{R}_{jl} = R_{jl} - \frac{(n-2)}{4} & [2\nabla_l \varphi_j - \varphi_l \varphi_j + g_{jl} \varphi^r \varphi_r] \\ & - \frac{1}{2} g_{jl} \nabla_i \varphi^i \, , \end{aligned} \tag{3.4.21}$$

$$\bar{R} = \Omega^{-1} \left[R - (n-1) \nabla_r \varphi^r - \frac{(n-1)(n-2)}{4} \varphi_r \varphi^r \right] \tag{3.4.22}$$

where

$$\varphi_i = \partial_i (\ln \Omega) \, . \tag{3.4.23}$$

From (3.4.18) to (3.4.23), definition (3.4.7) written in terms of the metric $\bar{g}_{ij}$, yields:

$$\bar{C}^i{}_{jkl} = C^i{}_{jkl} \tag{3.4.24}$$

which shows the distinctive feature of the Weyl tensor of being *conformally invariant*, (Choquet-Bruhat et al., 1977).

3.5 The Bianchi identities

Expression (3.2.3) for the curvature tensor is usually understood as the definition of the curvature in terms of the second derivatives of the metric tensor. On the other hand, given a tensor field of valence (1, 3) on the manifold having the same symmetry properties as the curvature tensor (examples of these are in (3.4.9) and

(3.4.10)), one can ask the question whether a connection exists such that Eq. (3.2.3) holds. The answer to this question is in general, no! An indication of this is the fact that Eq. (3.2.1) implies in fact that a curvature tensor and the connection *must* satisfy a set of algebraic relations which are termed the *Bianchi identities.* (Unfortunately there is no simple relation involving the Riemann tensor alone that determines whether or not it belongs to some connection). Covariantly differentiating (3.1.8) and using geodesic coordinates in which ${C^i}_{jk} = 0, {\Gamma'^i}_{jk}\big|_p = 0$, we obtain at p:

$$\nabla_{m'} {R'^i}_{jkl} = \partial_{m'} \partial_{k'} {\Gamma'^i}_{jl} - \partial_{m'} \partial_{l'} {\Gamma'^i}_{jk}. \tag{3.5.1}$$

Antisymmetrising over the indices (k, l, m) we have identically (the Bianchi identities):

$$\nabla_{[m'} {R'^i}_{|j|kl]} = 0. \tag{3.5.2}$$

Of course the above relation holds in any coordinate system together with the following constraints which naturally stem from (3.2.6):

$$\nabla_{[m} {R^k}_{lij]} = 0. \tag{3.5.3}$$

If a metric is given in M, then similar relations hold for R_{ijkl}, i.e. in an arbitrary coordinate system:

$$\nabla_{[m} R_{kl]ij} = 0 \tag{3.5.2$'$}$$

$$\nabla_{[m} R_{kli]j} = 0. \tag{3.5.3$'$}$$

The number of independent relations implied by the latter form of the Bianchi identities is, from (3.5.2′) and taking into account (3.5.3′):

$$\frac{n^2(n-1)^2(n-2)}{12} - \frac{n^2(n-1)(n-2)(n-3)}{24} = \frac{n^2(n^2-1)(n-2)}{24}. \tag{3.5.4}$$

In a three-dimensional space, the Bianchi identities provide 3 independent relations, while when $n = 4$, they yield 20 independent relations. Contracting the Bianchi identities twice, we obtain after simple algebra:

$$\nabla_i (R^i{}_j - \frac{1}{2}\delta^i_j R) = 0. \tag{3.5.5}$$

In a four-dimensional space the tensor

$$G^i{}_j = R^i{}_j - \frac{1}{2}\delta^i_j R \tag{3.5.6}$$

is known as the Einstein tensor for the key role it plays in the theory of general relativity.

It is now convenient to write the Bianchi identities in terms of the Weyl tensor. From (3.4.7) we have:

$$R^{ij}{}_{kl} = C^{ij}{}_{kl} + \frac{2}{(n-2)}\left(\delta^i_{[k} S_{l]}{}^j - \delta^j_{[k} S_{l]}{}^i\right) \tag{3.5.7}$$

where we put

$$S_{kl} = R_{kl} - \frac{R}{2(n-1)} g_{kl}. \tag{3.5.8}$$

Then from (3.5.2) we have:

$$\nabla_{[m} C_{kl]}{}^{ij} = \frac{2}{(n-2)} (\delta^i_{[l} \nabla_m S_{k]}{}^j - \delta^j_{[l} \nabla_m S_{k]}{}^i). \tag{3.5.9}$$

Contraction over i and k yields, from (3.4.8) and (3.5.5):

$$\nabla_i C_{lm}{}^{ji} = 2\left(\frac{n-3}{n-2}\right) \nabla_{[m} S_{l]}{}^j. \tag{3.5.10}$$

Since the Weyl tensor is trace-free (see (3.4.8)), Eq. (3.5.10) amounts to $(n^2(n-1)/2) - n$ relations which, in a four-dimensional space, yield again 20 relations. In this case therefore ($n = 4$), Eq. (3.5.10) is equivalent to the full Bianchi identities, as in (3.5.2′).

With the use of the Bianchi identities we can now show that if the Gaussian curvature K at a point in M is independent of the two-surfaces through that point and that is true for all points in some region of M, then K is constant in that region. In fact (3.2.9) can be written as:

$$R_{ijkl} = 2K g_{i[k} g_{l]j} \,. \tag{3.5.11}$$

Covariant differentiation yields:

$$\nabla_m R_{ijkl} = 2\partial_m K g_{i[k} g_{l]j}$$

hence applying (3.5.2′) we obtain:

$$\partial_{[m}K g_{|i|k} g_{l]j} = 0.$$

Contraction with g^{ik} yields:

$$(n-2)\partial_{[m}K g_{l]j} = 0$$

which for $n > 2$ implies:

$$\partial_m K = 0\ . \tag{3.5.12}$$

3.6 The equation of geodesic deviation

Consider a one-parameter family of geodesics $\{\gamma_t | t \in [0,1]\}$. The interval between the points $\gamma_t(s)$ and $\gamma_{t+\delta t}(s)$ with the same value of s on neighbouring geodesics is expressed through the *connecting vector* Y, defined as:

$$Y^n = \left(\frac{d}{dt}\gamma_t(s)\right)^n \tag{3.6.1}$$

or equivalently:

$$Y = \varphi_*\left(\frac{\partial}{\partial t}\right) \tag{3.6.2}$$

where $\varphi(s,t) = \gamma_t(s)$, in the sense that

$$\left(\gamma_{t+\delta t}(s)\right)^n - \left(\gamma_t(s)\right)^n = \delta t Y^n + O(\delta t^2) \qquad (\delta t \to 0). \tag{3.6.3}$$

We shall derive the differential equation satisfied by Y. In the absence of torsion we have, from (3.1.13):

$$\frac{DY}{Ds} = \frac{DX}{Dt} \tag{3.6.4}$$

where $X \equiv \dot{\gamma}_t$, and hence:

$$\frac{D^2Y}{Ds^2} = \frac{D}{Ds}\frac{DX}{Dt} = \frac{D}{Dt}\frac{DX}{Ds} + R(X,X,Y) \tag{3.6.5}$$

from (3.1.14). But γ_t is a geodesic, so that $DX/Ds = 0$ (compare (3.1.12) and (2.4.21)). Hence (3.6.5) becomes:

$$\frac{D^2Y}{Ds^2} = R(X,X,Y) \tag{3.6.6}$$

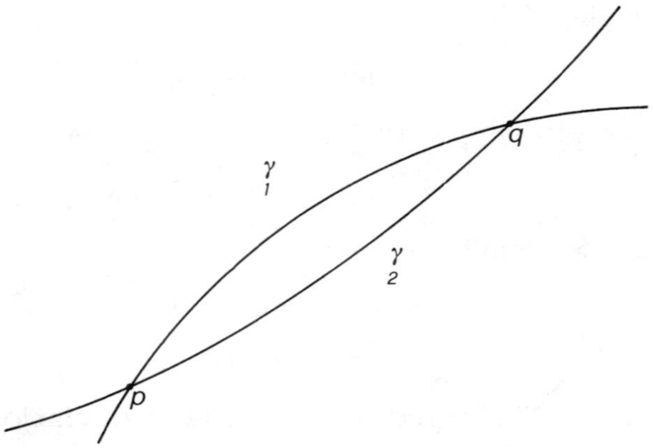

Fig. 3-2 When two neighbouring geodesics intersect twice, the points of intersection are termed *conjugate*.

or, in components:

$$\left(\frac{D^2Y}{Ds^2}\right)^i = R^i{}_{jkl}X^jX^kY^l. \tag{3.6.6'}$$

This is known as the *equation of geodesic deviation*. Neighbouring geodesics accelerate towards each other, with a rate directly measured by the curvature tensor, and eventually intersect. When two neighbouring geodesics intersect twice, the points of intersection are termed *conjugate*, (see Fig. 3-2).

If the curvature were zero everywhere in M, from the arguments of Sect. 3.3 it follows that

$$\frac{dY^k}{ds} = \text{const.} \tag{3.6.7}$$

so if this constant were zero, neighbouring geodesics would never intersect, manifesting the Euclidean concept of parallelism.

The deviation equation (3.6.6) is a set of four ordinary differential equations satisfied by the Y^i's along the geodesic curves γ; the vector field Y is called a *Jacobi field* on γ. We shall here outline a standard method of solution by iteration.

Let $\underset{p}{u}$ be an arbitrary vector at some point $p = \gamma(s_0)$ and expand it into the field $\check{u}$ of vectors parallel to $\underset{p}{u}$ on the curve, so as to have:

$$\check{u}^i(s) = \Gamma^i_{j_0} u^{j_0}; \qquad \frac{D\check{u}^i}{Ds} = 0 \tag{3.6.8}$$

(where for any geometrical object A, we set $A_{j_0} = A_j(s_0)$) then multiply (3.6.6) by $\check{u}_i$:

$$\frac{d^2(\check{u}_i Y^i)}{ds^2} = R^i{}_{jkl} X^j X^k Y^l \check{u}_i = \mathcal{K}^i{}_j Y^j \check{u}_i \tag{3.6.9}$$

where we put for convenience:

$$\mathcal{K}^i{}_j = R^i{}_{mnj} X^m X^n. \tag{3.6.10}$$

The particular case that will be needed in the next section is when our solution is defined in a closed interval $[s_0, s_1]$ on the curve, with no conjugate points on it and with given values at the ends of the range, i.e. $Y_0 = Y(s_0), Y_1 = Y(s_1)$. In this case let us introduce the symmetric Green's function for the operator d^2/ds^2,

$$G(s, s') = \begin{cases} \alpha(s - s_0)(s_1 - s') & s \leq s' \\ \alpha(s' - s_0)(s_1 - s) & s \geq s' \end{cases} \tag{3.6.11}$$

where $G(s_0, s') = G(s_1, s') = 0$ and :

$$\alpha = (s_1 - s_0)^{-1}; \tag{3.6.12}$$

and multiply (3.6.9) by Gds taking s' as a free parameter. Integration over the whole range yields

$$\int_{s_0}^{s_1} G \frac{d^2(\check{u}_i Y^i)}{ds^2} ds = \int_{s_0}^{s_1} G\mathcal{K}^i{}_j Y^j \check{u}_i ds = -\int_{s_0}^{s_1} \frac{dG}{ds} \frac{d}{ds}(\check{u}_i Y^i) ds \tag{3.6.13}$$

after integration by parts. From (3.6.11) we have

$$\frac{dG}{ds} = \begin{cases} \alpha(s_1 - s') & s \leq s' \\ -\alpha(s' - s_0) & s \geq s' \end{cases} \tag{3.6.14}$$

so that (3.6.13) can be written as:

$$\begin{aligned} -\alpha \int_{s_0}^{s'} (s_1 - s') \frac{d}{ds}(\check{u}_i Y^i) ds + \alpha \int_{s'}^{s_1} (s' - s_0) \frac{d}{ds}(\check{u}_i Y^i) ds \\ = -\alpha(s_1 - s')[\check{u}_{i'} Y^{i'} - \check{u}_{i_0} Y^{i_0}] + \alpha(s' - s_0)[\check{u}_{i_1} Y^{i_1} - \check{u}_{i'} Y^{i'}] \\ = \int_{s_0}^{s_1} G\mathcal{K}^i{}_j Y^j \check{u}_i ds. \end{aligned} \tag{3.6.15}$$

From the arbitrariness of $\check{u}$, (3.6.15) leads finally to:

$$-Y^{j'} + \alpha(s_1 - s')\Gamma_{i_0}{}^{j'} Y^{i_0} + \alpha(s' - s_0)\Gamma_{i_1}{}^{j'} Y^{i_1} = \int_{s_0}^{s_1} G\mathcal{K}^i{}_r Y^r \Gamma_i{}^{j'} ds. \tag{3.6.16}$$

Since s' is arbitrary and $Y^{i_0} = (Y_0)^i, Y^{i_1} = (Y_1)^i$ are given, (3.6.16) provides an integral equation for Y which can be solved by standard iterative methods.

3.7 The covariant derivative of the world-function

The equation of geodesic deviation is essential for determining the derivatives of the world-function introduced in Sect. 1.19. These derivatives provide natural examples of two-point tensors which will play an important role in the definition of physical measurements on curved manifolds, (cf. Chap. 9), (Synge, 1960).

For this purpose we need to consider only world-functions which refer to pairs of points lying on a common geodesic. In a normal neighbourhood $U_{\mathcal{N}}$ of M we can introduce a map $\Omega: U_{\mathcal{N}} \times U_{\mathcal{N}} \to \mathbf{R}$ defined as (see 1.19.5):

$$\Omega(p_0, p_1) = \tilde{\Omega}(s_0, s_1; \gamma) = \frac{1}{2}(s_1 - s_0)^2 (X \mid X) \tag{3.7.1}$$

where $X = \dot{\gamma}_{p_0 \to p_1}$ denotes the tangent vector to the (unique) geodesic connecting p_0 to p_1, parametrised so that

$$\gamma_{p_0 \to p_1}(s_0) = p_0 \qquad \gamma_{p_0 \to p_1}(s_1) = p_1 .$$

We want to investigate the derivatives of Ω, with respect to independent variations of its two end points, and so we consider varying p_0 and p_1 along curves κ_0 and κ_1, respectively, both parametrised by t. If $\gamma_t = \gamma_{\kappa_0(t) \to \kappa_1(t)}$ denotes the corresponding varied geodesic from $\kappa_0(t)$ to $\kappa_1(t)$, parametrised by s, then we have a one–parameter family of geodesics with connecting vector Y, say, as described in the previous section. By definition we have

$$\frac{DX}{Ds} = 0 \, , \tag{3.7.2}$$

$$\frac{DX}{Dt} = \frac{DY}{Ds} \, . \tag{3.7.2'}$$

To perform the required differentiation we then write

$$\frac{d}{dt}\Omega(\gamma_t(s_0), \gamma_t(s_1)) = \frac{\partial\Omega}{\partial x^{i_0}} Y^{i_0} + \frac{\partial\Omega}{\partial x^{i_1}} Y^{i_1} = (s_1 - s_0)^2 \left(\frac{DX}{Dt}|X\right) \tag{3.7.3}$$

where $\{x^{i_0} = x^i(s_0)\}$ are local coordinates for $\gamma(s_0)$. From (3.7.2′) we have:

$$\left(\frac{DX}{Dt}\middle|X\right) = \left(\frac{DY}{Ds}\middle|X\right) \tag{3.7.4}$$

but, since from (3.6.6)

$$\left(\frac{D^2Y}{Ds^2}\middle|X\right) = (R(X, X, Y) \mid X) \equiv 0$$

the latter implies, from (3.7.2):

$$\left(X\middle|\frac{DY}{Ds}\right) = \frac{d}{ds}(X|Y) = a = \text{const.} \tag{3.7.5}$$

If we integrate along $\gamma(s)$ from s_0 to s_1, we get:

$$a = (s_1 - s_0)^{-1} \left[(X|Y)\right]_{s_0}^{s_1} \tag{3.7.6}$$

hence (3.7.3) becomes:

$$\frac{d\Omega}{dt} = \Omega_{i_0} Y^{i_0} + \Omega_{i_1} Y^{i_1} = (s_1 - s_0)\left[X_{i_1} Y^{i_1} - X_{i_0} Y^{i_0}\right] \tag{3.7.7}$$

thus

$$\Omega_{i_0} = -(s_1 - s_0)X_{i_0} \qquad \Omega_{i_1} = +(s_1 - s_0)X_{i_1}. \tag{3.7.8}$$

From the latter we have:

$$\Omega_{i_0}\Omega^{i_0} = \Omega_{i_1}\Omega^{i_1} = (s_1 - s_0)^2(X|X) = 2\Omega. \tag{3.7.9}$$

Both derivatives in (3.7.8) are two-point functions, therefore differentiating one of them, Ω_{i_0} say, again with respect to t, yields:

$$\frac{D\Omega_{i_0}}{Dt} = \Omega_{i_0 j_0} Y^{j_0} + \Omega_{i_0 j_1} Y^{j_1} = -(s_1 - s_0)\left(\frac{DX_i}{Dt}\right)_{s=s_0} \tag{3.7.10}$$

where $\Omega_{i_0 j_0} = \nabla_j \Omega_{i_0}|_{x(s_0)}$ etc.

With the same procedure we put, from (3.7.2′), $(DX_i/Dt)_{s=0} = (DY_i/Ds)_{s=0}$, but now we need a solution for the derivative of the connecting vector field. This can be obtained from (3.6.16); at a general point $\gamma(s')$, we have:

$$Y^{j'} = \alpha(s_1 - s')\Gamma_{i_0}{}^{j'} Y^{i_0} + \alpha(s' - s_0)\Gamma_{i_1}{}^{j'} Y^{i_1} - \int_{s_0}^{s_1} G\mathcal{K}^i{}_k Y^k \Gamma_i{}^{j'} ds. \tag{3.7.11}$$

Contracting as in Sect. 3.6, with an arbitrary vector field $\check{u}$, parallely propagated along the curve γ of the integration, we obtain from (3.7.11):

$$\begin{aligned} \check{u}_{j'} Y^{j'} &= \alpha(s_1 - s')\check{u}_{i_0} Y^{i_0} + \alpha(s' - s_0)\check{u}_{i_1} Y^{i_1} - \int_{s_0}^{s_1} G\mathcal{K}^i{}_k Y^k \check{u}_i ds, \\ \alpha &= (s_1 - s_0)^{-1}. \end{aligned} \tag{3.7.12}$$

We differentiate this with respect to s' and obtain

$$\check{u}_{j'} \frac{DY^{j'}}{Ds'} = -\alpha\check{u}_{i_0} Y^{i_0} + \alpha\check{u}_{i_1} Y^{i_1} - \int_{s_0}^{s_1} \frac{dG}{ds'} \mathcal{K}^i{}_k Y^k \check{u}_i ds. \tag{3.7.13}$$

From (3.6.11) we have

$$\frac{dG}{ds'} = \begin{cases} -\alpha(s - s_0) & s \leq s' \\ \alpha(s_1 - s) & s \geq s' \end{cases} \tag{3.7.14}$$

hence (3.7.13) can be written as:

$$\begin{aligned} \check{u}_{j'} \frac{DY^{j'}}{Ds'} = &- \alpha\check{u}_{j'}(\Gamma_{i_0}{}^{j'} Y^{i_0} - \Gamma_{i_1}{}^{j'} Y^{i_1}) \\ &+ \alpha\check{u}_{j'} \int_{s_0}^{s'} (s - s_0)\mathcal{K}^i{}_k Y^k \Gamma_i{}^{j'} ds \\ &- \alpha\check{u}_{j'} \int_{s'}^{s_1} (s_1 - s)\mathcal{K}^i{}_k Y^k \Gamma_i{}^{j'} ds\ . \end{aligned}$$

From the arbitrariness of $\check{u}$ we finally have, for a general s' on γ:

$$\begin{aligned} \frac{DY^{j'}}{Ds'} = &- \alpha(\Gamma_{i_0}{}^{j'} Y^{i_0} - \Gamma_{i_1}{}^{j'} Y^{i_1}) \\ &+ \alpha\left[\int_{s_0}^{s'} (s - s_0)\mathcal{K}^i{}_k Y^k \Gamma_i{}^{j'} ds \right. \\ &\left. - \int_{s'}^{s_1} (s_1 - s)\mathcal{K}^i{}_k Y^k \Gamma_i{}^{j'} ds\right]\ . \end{aligned} \tag{3.7.15}$$

When $s' = s_0$, (3.7.15) gives:

$$\left.\frac{DY^i}{Ds'}\right|_{s'=s_0} = -\alpha Y^{i_0} + \alpha \Gamma_{i_1}{}^{i_0} Y^{i_1} - \alpha \int_{s_0}^{s_1} (s_1 - s) \mathcal{K}^j{}_k Y^k \Gamma_j{}^{i_0} ds. \tag{3.7.16}$$

To zero order in the curvature, that is neglecting the term containing $\mathcal{K}^i{}_j$ in (3.7.11), we have for a general $s' \equiv s$:

$$Y^j|_s = \alpha(s_1 - s)\Gamma_{i_0}{}^j Y^{i_0} + \alpha(s - s_0)\Gamma_{i_1}{}^j Y^{i_1} + O(\text{Riem}). \tag{3.7.17}$$

Using this value in the integrand on the left-hand side of (3.7.16) we obtain to first order in the curvature:

$$\begin{aligned}\left.\frac{DY^i}{Ds'}\right|_{s'=s_0} = &- \alpha(Y^{i_0} - \Gamma_{j_1}{}^{i_0} Y^{j_1}) \\ &- \alpha^2 Y^{r_0} \int_{s_0}^{s_1} (s_1 - s)^2 \mathcal{K}^j{}_k \Gamma_{r_0}{}^k \Gamma_j{}^{i_0} ds \\ &- \alpha^2 Y^{r_1} \int_{s_0}^{s_1} (s_1 - s)(s - s_0) \mathcal{K}^j{}_k \Gamma_{r_1}{}^k \Gamma_j{}^{i_0} ds \\ &+ O(|\text{Riem}|^2).\end{aligned} \tag{3.7.18}$$

Finally, recalling (3.7.2) and (3.7.10), we have:

$$\begin{aligned}\Omega_{i_0 j_0} &= g_{i_0 j_0} + \alpha g_{i_0 l_0} \int_{s_0}^{s_1} (s_1 - s)^2 \mathcal{K}^r{}_k \Gamma_r{}^{l_0} \Gamma_{j_0}{}^k ds \\ &\quad + O(|\text{Riem}|^2) \\ \Omega_{i_0 j_1} &= -g_{i_0 k_0} \Gamma_{j_1}{}^{k_0} + \alpha g_{i_0 l_0} \int_{s_0}^{s_1} (s_1 - s)(s - s_0) \mathcal{K}^r{}_k \Gamma_{j_1}{}^k \Gamma_r{}^{l_0} ds \\ &\quad + O(|\text{Riem}|^2).\end{aligned} \tag{3.7.19}$$

With a very similar procedure one calculates all the other higher derivatives of the world-function to first order in the curvature and its gradients. For the purpose of future applications, let us write the expressions of the third and fourth covariant derivatives at $s = s_0$:

$$\begin{aligned}&\Omega_{i_0 j_0 k_0} \\ &= -3\alpha^2 \int_{s_0}^{s_1} (s_1 - s)^2 S_{abcr} X^r \Gamma_{i_0}{}^a \Gamma_{j_0}{}^b \Gamma_{k_0}{}^c ds \\ &\quad + \frac{3}{2}\alpha^2 \int_{s_0}^{s_1} (s_1 - s)^3 X^p X^q \nabla_c S_{abpq} \Gamma_{i_0}{}^a \Gamma_{j_0}{}^b \Gamma_{k_0}{}^c ds\end{aligned}$$

$$+ O(|\mathrm{Riem}|^2) \tag{3.7.20}$$

$$\begin{aligned}
&\Omega_{i_0 j_0 k_0 l_0} \\
&= 3\alpha^3 \int_{s_0}^{s_1} (s_1 - s)^2 \Gamma_{i_0}{}^a \Gamma_{j_0}{}^b \Gamma_{k_0}{}^c \Gamma_{l_0}{}^d S_{abcd} ds \\
&\quad - 3\alpha^3 \int_{s_0}^{s_1} (s_1 - s)^3 X^q (\nabla_d S_{abcq} + \nabla_c S_{abdq}) \Gamma_{i_0}{}^a \Gamma_{j_0}{}^b \Gamma_{k_0}{}^c \Gamma_{l_0}{}^d ds \\
&\quad + \frac{3}{2}\alpha^3 \int_{s_0}^{s_1} (s_1 - s)^4 X^p X^q \nabla_d \nabla_c S_{abpq} \Gamma_{i_0}{}^a \Gamma_{j_0}{}^b \Gamma_{k_0}{}^c \Gamma_{l_0}{}^d ds \\
&\quad + O(|\mathrm{Riem}|^2) ,
\end{aligned} \tag{3.7.21}$$

where we have introduced for convenience the tensor:

$$S_{ijkm} = -\frac{1}{3}\left(R_{ikjm} + R_{imjk}\right) . \tag{3.7.22}$$

The calculation of the derivatives of the world-function higher than the fourth is somewhat cumbersome and the resulting expressions are sufficiently complicated to discourage any further discussion about them at least for the purposes of the applications we have in mind (cf. Chap. 9).

It must be stressed here that in all the calculations which lead to the derivatives of Ω, it was assumed, albeit implicitly, that the world-function was C^r-differentiable with r as large as needed, and that the geodesics along which the integrations were performed were smooth and without conjugate points. It is of interest now to deduce the limiting values of the world-function and of its derivatives when the points they depend on are made to coincide. After some simple algebra we obtain:

$$\begin{aligned}
\lim_{s_1 \to s_0} \Omega = \lim_{s_1 \to s_0} \Omega_i &= 0 && \text{(from 3.7.1 and 3.7.8)} \\
\lim_{s_1 \to s_0} \Omega_{ij} &= g_{ij}(s_0) && \text{(from 3.7.19)} \\
\lim_{s_1 \to s_0} \Omega_{ijk} &= 0 && \text{(from 3.7.20)} \\
\lim_{s_1 \to s_0} \Omega_{ijkl} &= S_{ijkl}(s_0) && \text{(from 3.7.21)}
\end{aligned} \tag{3.7.23}$$

It is worth pointing out here that these limiting values do not depend on the path along which the two points have been made to coincide.

3.8 Maximally symmetric spaces

A main property of a manifold with a metric is the number of isometries it admits. For each one-parameter family of isometries, a Killing one-form exists which satisfies Eq. (2.13.4). This can be more conveniently written in a coordinate neighbourhood U of M as:

$$\partial_{(j}\xi_{i)} = \Gamma^n_{ij}\xi_n. \tag{3.8.1}$$

In an n-dimensional space, (3.8.1) is a system of $n(n+1)/2$ relations which overdetermine the n unknowns ξ_i. But if a solution exists, then we can show that the Killing vector field thereby defined is uniquely determined in some U if the set $\left\{\xi_i, \partial_{[j}\xi_{i]}\right\}$ is known at a point $p \in U$. From (3.1.9) we have:

$$\nabla_{[k}\nabla_{j]}\xi_i = \frac{1}{2}R^m{}_{ijk}\xi_m \tag{3.8.2}$$

(we are in a torsion-free case); antisymmetrising over the indices (ijk) and recalling (3.2.6) we obtain:

$$\nabla_{[k}\nabla_j\xi_{i]} = 0. \tag{3.8.3}$$

The Killing equation (2.13.4) and the rule (1.14.11) imply:

$$\nabla_{[k}\nabla_{j]}\xi_i = -\frac{1}{2}\nabla_i\nabla_k\xi_j, \tag{3.8.4}$$

so combining this with (3.8.2), yields:

$$\nabla_i\nabla_k\xi_j = R^m{}_{ikj}\xi_m. \tag{3.8.5}$$

Now let q be any point such that there is a geodesic $\gamma(s)$ from p to q with $p = \gamma(0)$. For ease of notation, choose a vector field X in a neighbourhood of the geodesic such that $X_{\gamma(s)} = \dot{\gamma}(s)$. Then we have:

$$\begin{aligned}\left(\frac{D^2}{Ds^2}\boldsymbol{\xi}(\gamma(s))\right)_i &= X^k\nabla_k(X^j\nabla_j\xi_i)\\ &= \left(\delta^j_i\frac{d}{ds} - \Gamma^j_{im}X^m\right)\left(\delta^k_j\frac{d}{ds} - \Gamma^k_{jl}X^l\right)\xi_k(\gamma(s)) \end{aligned} \tag{3.8.6}$$

(modifying the definition (3.1.12) to covectors). Expanding these two alternative expressions for $D^2\boldsymbol{\xi}/Ds^2$, using the geodesic equation $X^k X^j{}_{;k} = 0$ and (3.8.5) in the first, gives:

$$X^k X^j R^m{}_{jki}\xi_m = \frac{d^2}{ds^2}\xi_i(\gamma(s)) - 2\Gamma^k_{im}X^m\frac{d}{ds}\xi_k(\gamma(s)) + \left(-\frac{d}{ds}(\Gamma^k_{il}X^l) + \Gamma^j_{im}\Gamma^k_{jl}X^m X^l\right)\xi_k(\gamma(s)). \quad (3.8.7)$$

This is a system of second-order ordinary differential equations in the independent variable s (all the functions being evaluated at $\gamma(s)$ and so regarded as functions of s) for the dependent variables $\xi_i(\gamma(s))$. So $\xi_i(\gamma(s))$ is uniquely determined at each s, and hence $\xi_i(q)$ is determined, once we know $\xi_i(p)$ and $d\xi_i(\gamma(s))/ds|_{s=0}$. But the latter is determined by knowing $\nabla_j\xi_i$ and ξ_i at p, and so $\xi_i(q)$ is uniquely determined at each point q connectable to p by a geodesic. To determine ξ_i at an arbitrary point p' we can now construct a chain of neighbourhoods $U_1, U_2, \ldots, U_n$ with $p \in U_1$ and $p' \in U_n$ such that each point in U_i can be joined by a geodesic to any point in $U_i \cup U_{i-1}$. Then we can successively determine ξ_i in $U_1, U_2, \ldots, U_n$ and hence at p'.

The linearity of the Killing equations (3.8.1) implies that if $\boldsymbol{\xi}$ and $\boldsymbol{\eta}$ are Killing forms, so is $\alpha\boldsymbol{\xi} + \beta\boldsymbol{\eta}$ for any constants α and β. Thus the set of Killing forms constitute a real linear vector space. We now ask for the dimensions of this space, i.e. the maximum number of linearly independent Killing forms that can exist in a given n–dimensional Riemannian manifold. We have seen that the value of the Killing form at any point can be derived from the values of the form ξ_i and its derivative $\nabla_{[j}\xi_{i]} = \nabla_j\xi_i$ at a given point p in terms of the linear equations (3.8.7); which implies that if ξ_i and $\nabla_{[j}\xi_{i]}$ vanish at one point p, then they vanish everywhere. Hence the maximum number of independent Killing forms is equal to the maximum number of independent values of the pair $\left(\xi_i(p), \nabla_{[j}\xi_{i]}(p)\right)$; since, given more than this number, we could find a non-trivial linear combination which vanishes at p along with its derivative and so vanishes everywhere. Thus there are at most $n{+}n(n-1)/2 = n(n+1)/2$ independent Killing forms on the manifold.

A manifold which admits all the Killing forms which are allowed, is termed *maximally symmetric*. A manifold does not however admit in general all the allowed Killing forms, because the quantities ξ_i and $\nabla_{[j}\xi_{i]}$ cannot necessarily be prescribed independently. They are in fact connected by the integrability conditions of equation (3.8.5). The latter read, extending (3.1.9) to tensors:

$$2\nabla_{[m}\nabla_{i]}\nabla_k\xi_j = R^n{}_{jim}\nabla_k\xi_n + R^n{}_{kim}\nabla_n\xi_j. \tag{3.8.8}$$

Differentiating (3.8.5) again gives

$$\nabla_r\nabla_k\nabla_j\xi_i = \nabla_r R_{ijk}{}^m\xi_m + R_{ijk}{}^m\nabla_{[r}\xi_{m]}, \tag{3.8.9}$$

and antisymmetrising over the last two differentiation indices, and comparing (3.8.8) gives

$$\begin{aligned}&\left(\nabla_i R^n{}_{mjk} - \nabla_m R^n{}_{ijk}\right)\xi_n \\ &\quad = \left(R^n{}_{jim}\delta^s_k - R^n{}_{kim}\delta^s_j + R^n{}_{ijk}\delta^s_m - R^n{}_{mjk}\delta^s_i\right)\nabla_{[s}\xi_{n]}\,. \end{aligned}\tag{3.8.10}$$

In order for the sets $\{\xi_i\}$ and $\left\{\nabla_{[i}\xi_{j]}\right\}$ to be independently specifiable, it must be the case that

$$\nabla_i R^l{}_{mjk} - \nabla_m R^l{}_{ijk} = 0 \tag{3.8.11}$$

$$R^{[l}{}_{jim}\delta^{s]}_k - R^{[l}{}_{kim}\delta^{s]}_j + R^{[l}{}_{ijk}\delta^{s]}_m - R^{[l}{}_{mjk}\delta^{s]}_i = 0. \tag{3.8.12}$$

These solutions give the constraints on the curvature tensors for the space to be maximally symmetric. Contracting (3.8.12) over k and s and recalling (3.2.6), we obtain:

$$(n-1)R^l{}_{jim} + R_{ij}\delta^l_m - R_{mj}\delta^l_i = 0. \tag{3.8.13}$$

A further contraction over j and m yields:

$$nR^l_i - R\delta^l_i = 0. \tag{3.8.14}$$

Combining (3.8.13) with (3.8.14), we deduce the expression of the curvature tensor of a maximally symmetric manifold:

$$R_{mikr} = \frac{R}{n(n-1)}(g_{ir}g_{mk} - g_{ik}g_{mr}). \tag{3.8.15}$$

Contracting now equation (3.8.11) over the indices i and k, we obtain

$$R_{,i} = 0 \tag{3.8.16}$$

which shows that maximally symmetric manifolds have a constant curvature scalar. Comparison with (3.5.11) shows that

$$\frac{R}{n(n-1)} \equiv K \tag{3.8.17}$$

is the Gaussian curvature of any two-surface imbedded in the maximally symmetric space. In terms of K, the Ricci tensor (3.8.14) becomes:

$$R_{ij} = (n-1)Kg_{ij}. \tag{3.8.18}$$

4

SPACE-TIME AND TETRAD FORMALISM

4.1 The space-time manifold and the physical observer

The properties of a manifold have been studied so far without assigning to it any role in the context of a physical theory and therefore without prescribing specific properties and dimensions. In the following however we shall be interested in the description of a Universe which is dominated by gravity, a phenomenon which manifests itself as effects due to the curvature of the underlying manifold.

A class of manifolds will then be selected as being suitable to describe a physical world. We shall call them *space-times* and require them to be four-dimensional, paracompact, connected, Hausdorff and without boundaries. Furthermore, since quantum field theories should be described on them, we demand that they admit a spinor structure (Geroch, 1986). The requirement that a space-time manifold is paracompact ensures that it always admits a metric with Lorentzian signature, while the other requirements stem naturally from physical considerations. In fact it would be meaningless (within the context of our present theories, which are cast in terms of interactions between neighbouring points) to talk of a Universe made of disconnected regions, or in which it would not be possible to find disjoint neighbourhoods of two events. And, for the same reason, we must assume that the manifold is open (without boundaries) since our model of physical interactions requires that every point has a neighbourhood similar to the Minkowski space of special relativity – which is the basis of the principle of equivalence. The property of admitting a spinor structure is closely related to the assumed paracompactness together with the possibility of defining in a continuous way the concepts of physical observer and of physical measurement.

We now proceed to the development of a theoretical scheme which allows us to make these concepts precise. A measurement is meaningful only when it is referred to an observer who can interpret it in terms of coordinate independent quantities. These measurements will be observer dependent, so a criterion should also be given for comparing the measurements made by different observers. A physical observer is a collection of measuring devices such as a clock, a theodolite and a light gun. This instrumentation however needs to be confined within a sufficiently small space (the observer's laboratory), and the measurements carried out over a sufficiently small interval of time so that the curvature effects can be neglected and an unambiguous (local) splitting of the space-time into space and time is possible according to the principle of equivalence.

A physical observer therefore is to be defined by a narrow congruence of time-like world-lines representing the points at fixed positions in his laboratory; call X the tangent vector field to this congruence and give a parametrization (τ say) such that $(X|X) = -1$ on the congruence. A space-like section Σ, which we can arrange to be a set of points on the congruence having the same value of the parameter τ, will represent the observer's three-dimensional space-laboratory at some given time. In view of the assumption that Σ is very small compared with the typical range of the background curvature, it is convenient in the following discussion to consider the congruence of lines as a single curve γ and the section Σ as a point on it (Fig. 4-1). We shall then call any time-like world-line γ an *observer* and refer to it, for brevity, by means of the vector field X tangent to γ (speaking of "the observer X").

Choose a fixed point $p \in \gamma$, define $X|_p = u$ and let h, π be the linear operators from $T_p(M)$ to itself defined by

$$\pi(w) = -(u|w)u \tag{4.1.1}$$

$$h(w) = w - \pi(w)\ . \tag{4.1.2}$$

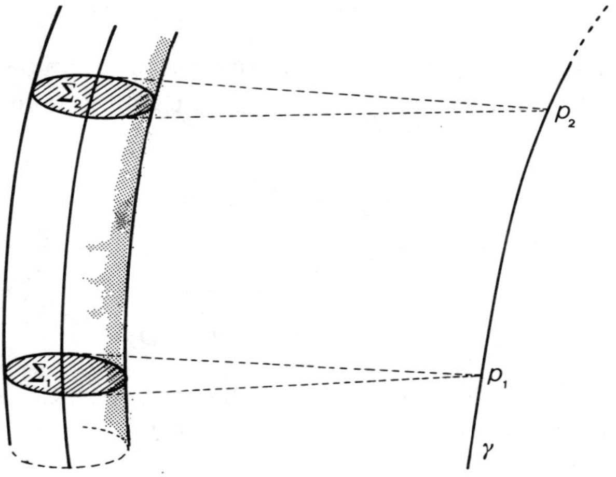

Fig. 4-1 The congruence of lines representing a physical observer is approximated to a single line and its space-like sections to points.

In terms of components:

$$\pi(w)^k = \pi_i^k w^i \qquad \pi_i^k = -u_i u^k \tag{4.1.3}$$

$$h(w)^k = h_i^k w^i \qquad h_i^k = \delta_i^k + u_i u^k \ . \tag{4.1.4}$$

If we define subspaces $T_{\perp p}(M), T_{\parallel p}(M)$ by:

$$T_{\perp p}(M) = \Big\{ v \in T_p(M) \mid (v|u) = 0 \Big\} \tag{4.1.5}$$

$$T_{\parallel p}(M) = \Big\{ v \in T_p(M) \mid \exists\ \lambda \in R,\ v = \lambda u \Big\} \tag{4.1.6}$$

then we see that

$$\pi : T_p(M) \longrightarrow T_{\parallel p}(M)$$
$$h : T_p(M) \longrightarrow T_{\perp p}(M)$$

and that π and h are the identity maps when restricted to $T_{\parallel p}(M)$ and $T_{\perp p}(M)$, respectively, i.e.

$$\pi(w) = w \qquad w \in T_{\parallel p}(M)$$
$$h(u) = u \qquad u \in T_{\perp p}(M) \ . \tag{4.1.7}$$

By the definition of h we have:

$$w = \pi(w) + h(w) \qquad \forall w \in T_p(M) \tag{4.1.8}$$

so that $T_p(M) = T_{\parallel p}(M) + T_{\perp p}(M)$ (any vector w can be written as the sum of a vector $\pi(w)$ in $T_{\parallel p}(M)$ and a vector $h(w)$ in $T_{\perp p}(M)$). Moreover, if any v satisfied

$$v \in T_{\parallel p}(M) \cap T_{\perp p}(M),$$

then we should have $v = \lambda X$ and $(v|X) = 0$; so that $v = o$. In such circumstances it is customary to write:

$$T_p(M) = T_{\parallel p}(M) \oplus T_{\perp p}(M). \tag{4.1.9}$$

It is easily shown that in this case the representation of w as a sum of vectors, one of $T_{\parallel p}(M)$ and one of $T_{\perp p}(M)$ is unique and that the correspondence

$$w \longleftrightarrow (\pi(w), h(w))$$

between $T_p(M)$ and the space $T_{\parallel p} \times T_{\perp p}(M)$ of pairs of vectors, one from each subspace, is an isomorphism. We say that $T_p(M)$ is the *direct sum* of the subspaces (Trautman, 1965). The maps h and π are termed respectively the parallel and transverse projection operators of u at p.

The space-time distance between the point p and a point p' infinitesimally close to p can then be written, from (4.1.8) and (4.1.9), as:

$$\delta s^2 = g_{ij}\delta x^i \delta x^j = -c^2 \delta T_X^2 + \delta L_X^2 \tag{4.1.10}$$

where:

$$\delta T_X^2 = -\frac{1}{c^2}\pi_{ij}\delta x^i \delta x^j \tag{4.1.11}$$

$$\delta L_X^2 = h_{ij}\delta x^i \delta x^j \ . \tag{4.1.12}$$

The special relativistic form of the space-time distance in (4.1.10) clearly suggests that (4.1.11) and (4.1.12) are respectively the infinitesimal time interval and space distance between the events p and p', as measured by X at p. In Chapter 9 we will show that this is the correct interpretation; now instead we shall further develop the concept of the reference frame.

4.2 Construction of a tetrad

A physical observer finds it natural and convenient to use a local system of Cartesian coordinate axes, which provide him or her with an instantaneous inertial frame. The result of a measurement, referred to these Cartesian axes, should be compared with the invariant projection on these axes of the tensor which characterises the phenomenon under consideration. Using the setting of the previous section, let us choose in the three-dimensional vector space $T_{\perp p}(M), (p = \gamma(\tau))$, three mutually orthogonal unit space-like vectors $\lambda_{\hat{\alpha}}$, with $\hat{\alpha} = 1, 2, 3$:

$$(\lambda_{\hat{\alpha}}|\lambda_{\hat{\beta}})_p = \delta_{\hat{\alpha}\hat{\beta}}. \tag{4.2.1}$$

Write $X \equiv \lambda_{\hat{0}}$, so the set $\{\lambda_{\hat{a}}\}_p$, $\hat{a} = \hat{0}, \hat{1}, \hat{2}, \hat{3}$, forms an orthonormal basis, termed a *tetrad* (Pirani, 1956b); hence:

$$(\lambda_{\hat{a}}|\lambda_{\hat{b}})_p = \eta_{\hat{a}\hat{b}}. \tag{4.2.2}$$

If ∂_i is a given coordinate basis on M, then from the properties of bases, there exist functions $\lambda_{\hat{a}}{}^i$ and $\bar{\lambda}^{\hat{b}}{}_i$ such that:

$$\lambda_{\hat{a}} = \lambda_{\hat{a}}{}^i \partial_i \qquad \partial_i = \bar{\lambda}^{\hat{b}}{}_i \lambda_{\hat{b}} \ . \tag{4.2.3}$$

From these it follows that:

$$\lambda_{\hat{a}}{}^i \bar{\lambda}^{\hat{b}}{}_i = \delta^{\hat{b}}_{\hat{a}} \tag{4.2.4}$$

$$(\partial_i|\lambda_{\hat{b}}) = \bar{\lambda}^{\hat{a}}{}_i \eta_{\hat{a}\hat{b}} = \lambda_{\hat{b}i} \tag{4.2.5}$$

$$(\partial_i|\partial_j) = \bar{\lambda}^{b}{}_i \lambda_{\hat{b}j} = g_{ij}. \tag{4.2.6}$$

Note that if we define one-forms $\boldsymbol{\lambda}^{\hat{a}}$ by $\boldsymbol{\lambda}^{\hat{a}} = \bar{\lambda}^a{}_i \boldsymbol{dx}^i$ then (4.2.4) becomes $\boldsymbol{\lambda}^{\hat{a}}(\lambda_{\hat{b}}) = \delta^{\hat{a}}_{\hat{b}}$; so the $\boldsymbol{\lambda}^{\hat{a}}$ are one-forms dual to $\lambda_{\hat{a}}$.

In what follows we shall omit the bar over the tetrad components of the coordinate basis, it being understood that tetrad indices are raised and lowered by the Minkowski metric $\eta_{\hat{a}\hat{b}}$. From (4.2.6) the following relations hold:

$$\lambda^{\hat{\alpha}}{}_i \lambda_{j\hat{\alpha}} = h_{ij} \qquad \lambda^{\hat{0}}{}_i \lambda_{j\hat{0}} = \pi_{ij} \tag{4.2.7}$$

$$\sqrt{-g} = \mathrm{Det}|\lambda_{i\hat{a}}| \ . \tag{4.2.8}$$

Given a tensor at p, its tetrad components are coordinate invariants:

$$T^{\hat{a}\hat{b}\dots} = T^{ij\dots}\lambda^{\hat{a}}{}_{i}\lambda^{\hat{b}}{}_{j}\dots \tag{4.2.9a}$$

The inverse relation holds:

$$T^{ij\dots} = T^{\hat{a}\hat{b}\dots}\lambda_{\hat{a}}{}^{i}\lambda_{\hat{b}}{}^{j}\cdots \tag{4.2.9b}$$

and more generally, for a mixed tensor:

$$T^{\hat{a}\hat{b}\dots}_{\hat{c}\hat{d}\dots} = T^{ij\dots}_{rs\dots}\lambda^{\hat{a}}{}_{i}\lambda^{\hat{b}}{}_{j}\dots\lambda_{\hat{c}}{}^{r}\lambda_{\hat{d}}{}^{s}\dots \tag{4.2.10}$$

From (4.2.7) the following correspondence between projected quantities and tetrad components is easily established:

$$\begin{aligned}
v^i_{\perp} &= h^i_j v^j = \lambda_{\hat{\alpha}}{}^{i}v^{\hat{\alpha}}\\
v^i_{\parallel} &= \pi^i_j v^j = \lambda_{\hat{0}}{}^{i}v^{\hat{0}}\\
h_{ij}v^i v^j &= \delta_{\hat{\alpha}\hat{\beta}}v^{\hat{\alpha}}v^{\hat{\beta}}\\
\pi_{ij}v^i v^j &= -(v^{\hat{0}})^2\,.
\end{aligned} \tag{4.2.11}$$

The above relations hold at p.

Suppose we are given a smooth tetrad field along a time-like congruence in M, each line being an observer, so that the components $\lambda_{i\hat{a}}$ are differentiable on M. The tetrad field is not in general a basis that could come from coordinates, since the equations needed for this,

$$\lambda_{\hat{a}} = \frac{\partial}{\partial\hat{x}^a} \qquad a = 0, 1, 2, 3$$

cannot in general be integrated. The degree of non-integrability is described by the numbers $C^{\hat{a}}{}_{\hat{b}\hat{c}}$, where (cf. (2.2.11)):

$$[\lambda_{\hat{b}}, \lambda_{\hat{c}}] = C^{\hat{a}}{}_{\hat{b}\hat{c}}\lambda_{\hat{a}}\,. \tag{4.2.12}$$

We saw in Section 2.12 (Theorem 1) that, if these were zero, then there would exist coordinates $\hat{x}_a$ such that: $\lambda_{\hat{a}} = \partial_{\hat{a}}$. The equivalent condition in terms of one-forms (see Sect. 2.12; Theorem 2) is that there exist coordinates $\hat{x}^a$ such that $\boldsymbol{\lambda}^{\hat{a}} = \boldsymbol{d}\hat{x}^a$, if and only if $\boldsymbol{d\lambda}^{\hat{a}} = 0$, i.e. if $\partial_{[j}\lambda^{\hat{a}}_{i]} = 0$.

4.3 Relations among tetrads and the Lorentz group

Let us consider two time-like observers u and u' whose world-lines γ and γ' intersect at p. They carry two distinct tetrad frames which we denote by $\{\lambda_{\hat{a}}\}$ and $\{\lambda'_{\hat{b}}\}$; we shall now consider the general relation among the measurements made in these two frames. If w is a vector at p, then with respect to the two frames we can write

$$w = w^{\hat{a}}\lambda_{\hat{a}} = w'^{\hat{b}}\lambda'_{\hat{b}}; \tag{4.3.1}$$

scalar multiplication with $\lambda^{\hat{c}}$ yields, from (4.2.2):

$$w^{\hat{c}} = w'^{\hat{b}}(\lambda'_{\hat{b}}|\lambda^{\hat{c}}) = L_{\hat{b}}{}^{\hat{c}}w'^{\hat{b}} \tag{4.3.2}$$

say. The coefficients $L_{\hat{b}}{}^{\hat{c}}$ identify a non-singular matrix since, from their definition, we have

$$\lambda'_{\hat{a}} = L_{\hat{a}}{}^{\hat{b}}\lambda_{\hat{b}}. \tag{4.3.3}$$

However, for arbitrary w, we have:

$$\eta_{\hat{c}\hat{d}}w^{\hat{c}}w^{\hat{d}} = \eta_{\hat{c}\hat{d}}L_{\hat{a}}{}^{\hat{c}}L_{\hat{b}}{}^{\hat{d}}w'^{\hat{a}}w'^{\hat{b}} = \eta_{\hat{a}\hat{b}}w'^{\hat{a}}w'^{\hat{b}} \tag{4.3.4}$$

hence

$$\eta_{\hat{a}\hat{b}} = L_{\hat{a}}{}^{\hat{c}}L_{\hat{b}}{}^{\hat{d}}\eta_{\hat{c}\hat{d}}\ . \tag{4.3.5}$$

The set $\mathbf{L}$ of matrices which satisfy (4.3.5) forms a group known as the Lorentz group. We shall in addition impose the conditions

$$L_{\hat{0}}{}^{\hat{0}} > 0, \tag{4.3.6}$$

which implies that future-pointing time-like vectors remain future-pointing, and

$$\mathrm{Det}(\mathbf{L}) = 1 \tag{4.3.7}$$

which means that right-handed frames are not mapped into left-handed ones. The Lorentz transformations which satisfy conditions (4.3.6) and (4.3.7) are termed *orthochronous* and *proper*, respectively. We denote the Lorentz group with these conditions by $\mathcal{L}$*.

---

* $\mathcal{L}$ comprises precisely the elements of the general Lorentz group $\mathbf{L}$ that can be connected to the identity.

Let $\lambda_{\hat{0}}$ be the time-like vector of a tetrad, then consider the subgroup $\tilde{\mathcal{L}} \subset \mathcal{L}$ of matrices which leave $\lambda_{\hat{0}}$ unchanged; i.e.

$$\lambda'_{\hat{0}} = \tilde{L}_{\hat{0}}{}^{\hat{a}}\lambda_{\hat{a}} = \lambda_{\hat{0}} \tag{4.3.8}$$

which implies

$$\tilde{L}_{\hat{0}}{}^{\hat{a}} = \delta^{\hat{a}}_{\hat{0}} \; . \tag{4.3.9}$$

This together with (4.3.5), shows that such matrices have the form:

$$\tilde{L} \equiv \begin{pmatrix} 1 & 0 \\ 0 & \tilde{L}^{\hat{\beta}}_{\hat{\alpha}} \end{pmatrix} \qquad \hat{\alpha}, \hat{\beta} = 1, 2, 3 \tag{4.3.10}$$

where $\left\{\tilde{L}_{\hat{\alpha}}{}^{\hat{\beta}}\right\}$, a 3×3 matrix, satisfies $\sum_{\hat{\alpha}} \tilde{L}_{\hat{\alpha}}{}^{\hat{\beta}}\tilde{L}_{\hat{\alpha}}{}^{\hat{\gamma}} = \delta^{\hat{\beta}\hat{\gamma}}$ and $\mathrm{Det}(\tilde{L}_{\hat{\alpha}}{}^{\hat{\beta}}) = 1$. In other words $\left\{\tilde{L}_{\hat{\alpha}}{}^{\hat{\beta}}\right\}$ is a rotation matrix and the group $\tilde{\mathcal{L}}$ is termed the *internal rotation group*. We can now see how to decompose any element of $\mathcal{L}$ into a product of a movement of $\lambda_{\hat{0}}$ and such an $\tilde{L}$. For a general L note that

$$\eta_{\hat{b}\hat{c}}L_{\hat{0}}{}^{\hat{b}}L_{\hat{0}}{}^{\hat{c}} = -(L_{\hat{0}}{}^{\hat{0}})^2 + \sum_{\alpha=1}^{3}(L_{\hat{0}}{}^{\hat{\alpha}})^2 = \eta_{\hat{0}\hat{0}} = -1 \tag{4.3.11}$$

and so $L_{\hat{0}}{}^{\hat{0}} \geq 1$ and we can define $\chi > 0$ by $L_{\hat{0}}{}^{\hat{0}} = \cosh\chi$. Then (4.3.11) gives:

$$\sum_{\alpha=1}^{3}(L_{\hat{0}}{}^{\hat{\alpha}})^2 = \sinh^2\chi$$

so that we can write

$$L_{\hat{0}}{}^{\hat{\alpha}} = \sinh\chi \; n^{\hat{\alpha}}$$

for a three-component object $n^{\hat{\alpha}}$ satisfying

$$\sum_{\hat{\alpha}=1}^{3}(n^{\hat{\alpha}})^2 = 1 \; .$$

Given $L_{\hat{0}}{}^{\hat{b}}$ (i.e. given χ and $n^{\hat{\alpha}}$) we can construct a "canonical" Lorentz transformation $\bar{L}_{\hat{a}}{}^{\hat{b}}$ agreeing with $L_{\hat{a}}{}^{\hat{b}}$ in the $({}_{\hat{0}}{}^{\hat{b}})$-

component; i.e. $\bar{L}_{\hat{0}}{}^{\hat{b}} = L_{\hat{0}}{}^{\hat{b}}$; simply set:

$$\bar{L}_{\hat{0}}{}^{\hat{0}} = L_{\hat{0}}{}^{\hat{0}} = \cosh\chi; \qquad \bar{L}_{\hat{0}}{}^{\hat{\alpha}} = L_{\hat{0}}{}^{\hat{\alpha}} = \sinh\chi n^{\hat{\alpha}}$$
$$\bar{L}_{\hat{\beta}}{}^{\hat{0}} = n^{\hat{\beta}}\sinh\chi; \qquad \bar{L}_{\hat{\beta}}{}^{\hat{\alpha}} = n^{\hat{\alpha}}n_{\hat{\beta}}(\cosh\chi - 1) + \delta^{\hat{\alpha}}_{\hat{\beta}}.$$

The difference between $L_{\hat{a}}{}^{\hat{b}}$ and $\bar{L}_{\hat{a}}{}^{\hat{b}}$ can be measured by multiplying $\bar{L}_{\hat{a}}{}^{\hat{b}}$ by $\overset{-1}{L}_{\hat{a}}{}^{\hat{b}}$ whose components are $L^{\hat{b}}{}_{\hat{a}}$, forming

$$M_{\hat{a}}{}^{\hat{b}} = \bar{L}_{\hat{a}}{}^{\hat{c}}L^{\hat{b}}{}_{\hat{c}}. \tag{4.3.12}$$

M is then a Lorentz transformation, and the agreement of L and $\bar{L}$ in their first column ensures that M is an internal rotation: explicitly, if $\eta'_{\hat{a}} = M_{\hat{a}}{}^{\hat{b}}\eta_{\hat{b}}$, then

$$\eta'_{\hat{0}} = M_{\hat{0}}{}^{\hat{b}}\eta_{\hat{b}} = \bar{L}_{\hat{0}}{}^{\hat{c}}L^{\hat{b}}{}_{\hat{c}}\eta_{\hat{b}} = L_{\hat{0}}{}^{\hat{c}}L^{\hat{b}}{}_{\hat{c}}\eta_{\hat{b}} = \delta^{\hat{b}}_{\hat{0}}\eta_{\hat{b}} = \eta_{\hat{0}}.$$

Thus writing $M_{\hat{a}}{}^{\hat{b}} = \tilde{L}^{\hat{b}}{}_{\hat{a}}$ we can invert (4.3.12) to form:

$$L^{\hat{b}}{}_{\hat{d}} = \tilde{L}^{\hat{b}}{}_{\hat{a}}\bar{L}^{\hat{a}}{}_{\hat{d}}\ .$$

Thus any Lorentz matrix can be written as a product of $\bar{L}$, depending on the numbers

$$L_{\hat{0}}{}^{\hat{\alpha}} = \sinh\chi n^{\hat{\alpha}} \in \mathbb{R}^3$$

and $\tilde{L}$, an internal rotation. This means that the Lorentz group $\mathcal{L}$ is topologically the direct product of $\mathbb{R}^3$ with the rotation group, *Rot*.

It is well known that the rotation group is doubly connected: if one considers the closed curve

$$\gamma : \theta \longmapsto \begin{pmatrix} \cos\theta & \sin\theta & 0 \\ -\sin\theta & \cos\theta & 0 \\ 0 & 0 & 1 \end{pmatrix} \tag{4.3.13}$$
$$[0, 2\pi] \longrightarrow Rot$$

then γ cannot be deformed to the constant curve $\theta \mapsto I$ while keeping its endpoints fixed; but if we traverse γ twice, defining

$$\gamma^{(2)} : [0, 2\pi] \in \theta \longmapsto \begin{cases} \gamma(2\theta) & 0 \le \theta < \pi \\ \gamma(2\theta - \pi) & \pi \le \theta \le 2\pi \end{cases}$$

then $\gamma^{(2)}$ can be deformed to the constant curve. Because of the direct product structure of $\mathcal{L}$, the same is true of it.

4.4 The propagation laws for tetrads

The changes in the tetrad components of a given quantity due to the motion of the tetrad itself along the observer world-lines is much less simple than the case previously considered. The absolute derivative of any of the $\lambda_{\hat{a}}$'s along γ, which we shall denote with a dot, is a new vector that can be expressed in terms of the tetrad itself:

$$\dot{\lambda}_{\hat{a}} \equiv \dot{\gamma}^i \nabla_i \lambda_{\hat{a}} = \Lambda_{\hat{a}\hat{b}} \lambda^{\hat{b}} \tag{4.4.1}$$

where $\Lambda_{\hat{a}\hat{b}}$ are some coefficients. A reasonable requirement is that under the dot operation, the tetrad preserves its orthonormality; this is assured if $\Lambda_{\hat{a}\hat{b}}$'s constitute an antisymmetric matrix. The physical interpretation of the coefficients follows from (4.4.1) on separating the time ($\hat{a} = \hat{0}$) from the space ($\hat{a} = \hat{\alpha}$) components:

$$\begin{aligned} &a) \qquad \dot{\lambda}_{\hat{0}} = \Lambda_{\hat{0}\hat{\beta}} \lambda^{\hat{\beta}} = \dot{X} \\ &b) \qquad \dot{\lambda}_{\hat{\alpha}} = \Lambda_{\hat{\alpha}\hat{\beta}} \lambda^{\hat{\beta}} - \Lambda_{\hat{\alpha}\hat{0}} X \,. \end{aligned} \tag{4.4.2}$$

From (4.4.2)a it follows that

$$\Lambda_{\hat{0}\hat{\alpha}} = \dot{X}_{\hat{\alpha}} \tag{4.4.3}$$

i.e. $\Lambda_{\hat{0}\hat{\alpha}}$ is the $\hat{\alpha}$-component of the observer's four-acceleration; from (4.4.2)b we then have:

$$\Lambda_{\hat{\alpha}\hat{\beta}} = (\dot{\lambda}_{\hat{\alpha}} | \lambda_{\hat{\beta}}) \tag{4.4.4}$$

which identifies in the *space* triad $\{\lambda_{\hat{\alpha}}\}$ an instantaneous axis of rotation about which the vectors of the triad rotate rigidly. In terms of coordinate components, (4.4.4) can be written as:

$$\Lambda_{[ij]} = \Lambda_{\hat{\alpha}\hat{\beta}} \lambda^{\hat{\alpha}}_i \lambda^{\hat{\beta}}_j = \lambda_{\hat{\alpha}r;s} \lambda^{\hat{\alpha}}_i X^s h^r_j. \tag{4.4.5}$$

The coefficients $\Lambda_{\hat{\alpha}\hat{\beta}}$ are termed *Fermi rotation coefficients* and if they vanish all along γ, then the tetrad is said to be Fermi transported. For such a frame we have, from (4.4.2)b and (4.4.3):

$$\dot{\lambda}_{\hat{\alpha}} = \dot{X}_{\hat{\alpha}} X. \tag{4.4.6}$$

Suppose a vector w has constant components in a Fermi transported frame. Then:

$$\dot{w} = w^{\hat{\alpha}} \dot{\lambda}_{\hat{\alpha}} + w^{\hat{0}} \dot{X} = (w|\dot{X})X - (w|X)\dot{X} \tag{4.4.7}$$

from (4.4.6), i.e.

$$\dot{w} = w^j(\dot{X}_j X - X_j \dot{X}) \,. \tag{4.4.8}$$

A vector satisfying this law is said to be Fermi-Walker transported.

Fermi-Walker transport preserves scalar product and orthogonality to the world-line and coincides with parallel transport when the line γ is a geodesic ($\dot{X} = 0$). The tangent to a world-line γ is naturally Fermi-Walker transported along itself because it trivially satisfies (4.4.7); the vectors of a spatial triad defined on γ may or may not be Fermi transported along it, but certainly they must not be parallely propagated (for $\dot{X} \neq 0$) if we want to preserve the tetrad structure on γ. It seems therefore that Fermi transport is more important than parallel transport when dealing with reference frames in general (Fermi, 1922; Walker, 1932).

4.5 The Ricci rotation coefficients

We recall from (2.6.8′) (now putting a $\hat{}$ on the indices to distinguish tetrad components from unhatted coordinate components) that the covariant derivative of a one-form $\boldsymbol{\varphi}$ in tetrad components is given by:

$$(\nabla_{\hat{a}} \boldsymbol{\varphi})_{\hat{b}} = \lambda_{\hat{a}}(\varphi_{\hat{b}}) - \Gamma^{\hat{c}}{}_{\hat{b}\hat{a}} \varphi_{\hat{c}}$$

where $\nabla_{\hat{a}} \equiv \nabla_{\lambda_{\hat{a}}}$. If we take $\boldsymbol{\varphi}$ to be the dual basis form $\boldsymbol{\lambda}^{\hat{d}}$ so that $\varphi_{\hat{b}} = \boldsymbol{\lambda}^{\hat{d}}(\lambda_{\hat{b}}) = \delta^{\hat{d}}_{\hat{b}}$ then this gives:

$$(\nabla_{\hat{a}} \boldsymbol{\lambda}^{\hat{d}})_{\hat{b}} = -\Gamma^{\hat{d}}{}_{\hat{b}\hat{a}}. \tag{4.5.1}$$

Lowering the index, we define:

$$\Gamma_{\hat{c}\hat{b}\hat{a}} = \eta_{\hat{c}\hat{d}} \Gamma^{\hat{d}}{}_{\hat{b}\hat{a}} = -(\nabla_{\hat{a}} \lambda_{\hat{c}})_{\hat{b}} \tag{4.5.2}$$

i.e.

$$\Gamma_{\hat{c}\hat{b}\hat{a}} = -\lambda_{\hat{c}k;i} \lambda_{\hat{b}}{}^k \lambda^i_{\hat{a}}. \tag{4.5.3}$$

In the context of the tetrad formalism, these components of the connection are known as the *Ricci rotation coefficients*.

The requirement that the orthonormality of the tetrad is preserved under displacement along any of the tetrad directions implies that:

$$\Gamma_{\hat{a}\hat{b}\hat{c}} = \Gamma_{[\hat{a}\hat{b}]\hat{c}}. \tag{4.5.4}$$

For a torsion-free connection, from (2.7.3) we have

$$C^{\hat{a}}{}_{\hat{b}\hat{c}} = 2\Gamma^{\hat{a}}{}_{[\hat{c}\hat{b}]}$$

while from (2.8.6) using the fact that the tetrad components of the metric are the constants $\eta_{\hat{a}\hat{b}}$,we have (lowering the index):

$$\Gamma_{\hat{a}\hat{b}\hat{c}} = \frac{1}{2}(C_{\hat{a}\hat{c}\hat{b}} + C_{\hat{b}\hat{a}\hat{c}} + C_{\hat{c}\hat{a}\hat{b}}). \tag{4.5.5}$$

This relation shows that the Ricci rotation coefficients can be determined without reference to the metric or connection except insofar as the metric is implicit in the orthogonality of the tetrad.

From (4.5.3) and (4.5.1) we recognise that:

$$\Lambda_{\hat{a}\hat{b}} = -\Gamma_{\hat{a}\hat{b}\hat{0}} \; . \tag{4.5.6}$$

Other properties of the congruence can be defined in terms of the Ricci rotation coefficients; these are the *expansion* $\Theta_{\hat{\alpha}\hat{\beta}}$ and the *vorticity* $\omega_{\hat{\alpha}\hat{\beta}}$:

$$\Theta_{\hat{\alpha}\hat{\beta}} \equiv -\Gamma_{\hat{0}(\hat{\alpha}\hat{\beta})} = -\lambda_{i(\hat{\alpha}} \pounds_X \lambda^i{}_{\hat{\beta})} \tag{4.5.7}$$

$$\omega_{\hat{\alpha}\hat{\beta}} \equiv -\Gamma_{\hat{0}[\hat{\alpha}\hat{\beta}]} = -\Lambda_{\hat{\alpha}\hat{\beta}} - \lambda_{i[\hat{\alpha}} \pounds_X \lambda^i{}_{\hat{\beta}]}. \tag{4.5.8}$$

In terms of coordinate components, these can be written from (4.5.3) as:

$$\Theta_{\hat{\alpha}\hat{\beta}} = \Theta_{ij}\lambda_{\hat{\alpha}}{}^i\lambda^j_{\hat{\beta}} \qquad \Theta_{ij} = X_{(r;s)}h^r_i h^s_j \tag{4.5.7$'$}$$

$$\omega_{\hat{\alpha}\hat{\beta}} = \omega_{ij}\lambda_{\hat{\alpha}}{}^i\lambda^j_{\hat{\beta}} \qquad \omega_{ij} = X_{[r;s]}h^r_i h^s_j \; . \tag{4.5.8$'$}$$

We shall defer to forthcoming sections a thorough discussion on these quantities. However it is worth mentioning here a more convenient way of expressing vorticity. From (4.5.1), relation (4.5.8) can also be written:

$$\omega_{\hat{\alpha}\hat{\beta}} = -\frac{1}{2}X_i(\pounds_{\lambda_{\hat{\beta}}} \lambda_{\hat{\alpha}})^i \; . \tag{4.5.9}$$

This explicitly shows that when $\omega_{\hat{\alpha}\hat{\beta}} = 0$ (the congruence is vorticity-free), then $\pounds_{\lambda_{\hat{\beta}}} \lambda_{\hat{\alpha}}$ is orthogonal to X and so the vectors of the triad $\{\lambda_{\hat{\alpha}}\}$ satisfy $\pounds_{\lambda_{\hat{\beta}}} \lambda_{\hat{\alpha}} = C^{\hat{\gamma}}{}_{\hat{\beta}\hat{\alpha}}\lambda_{\hat{\gamma}}$; thus the conditions of the Frobenius theorem (Theorem 1 of Sect. 2.12) imply the existence of a family of three-dimensional hypersurfaces orthogonal to the congruence. The reverse is also true, as is easy to show.

Let us now derive the transformation properties of the quantities we have introduced. Under a general tetrad transformation, the structure coefficients $C_{\hat{a}\hat{b}\hat{c}}$ as in (4.2.12) do *not* transform as Lorentz tensors; in fact we have from (4.5.1):

$$C'_{\hat{a}\hat{b}\hat{c}} = L_{\hat{c}}{}^{\hat{r}} L_{\hat{a}}{}^{\hat{s}} L_{\hat{b}}{}^{\hat{t}} C_{\hat{s}\hat{t}\hat{r}} - 2\eta_{\hat{r}\hat{s}} L_{[\hat{c}}{}^{\hat{r}} L_{\hat{b}]}{}^{\hat{t}} \partial_{\hat{t}} L_{\hat{a}}{}^{\hat{s}} \tag{4.5.10}$$

where we write:

$$\partial_{\hat{t}} = \lambda_{\hat{t}}{}^{i} \frac{\partial}{\partial x^i} \,. \tag{4.5.11}$$

The Ricci rotation coefficients have the same behaviour:

$$\Gamma'_{\hat{a}\hat{b}\hat{c}} = L_{\hat{a}}{}^{\hat{r}} L_{\hat{b}}{}^{\hat{s}} L_{\hat{c}}{}^{\hat{t}} \Gamma_{\hat{r}\hat{s}\hat{t}} - \eta_{\hat{r}\hat{s}} L_{\hat{b}}{}^{\hat{s}} L_{\hat{c}}{}^{\hat{t}} \partial_{\hat{t}} L_{\hat{a}}{}^{\hat{r}}), \tag{4.5.12}$$

whereas the vorticity and the expansion behave as Lorentz tensors and the acceleration as a Lorentz vector with respect to the internal rotation group. The Fermi rotation coefficients do not behave as Lorentz tensors even with respect to this transformation. In fact we have

$$\Lambda'_{\hat{\alpha}\hat{\beta}} = \tilde{L}_{\hat{\alpha}}^{\hat{\rho}} \tilde{L}_{\hat{\beta}}^{\hat{\sigma}} \Lambda_{\hat{\rho}\hat{\sigma}} + \delta_{\hat{\rho}\hat{\sigma}} \dot{\tilde{L}}_{\hat{\alpha}}{}^{\hat{\rho}} \tilde{L}_{\hat{\beta}}^{\hat{\sigma}} \tag{4.5.13}$$

where

$$\dot{\tilde{L}}_{\hat{\alpha}}{}^{\hat{\rho}} = X^i \frac{\partial}{\partial x^i} \tilde{L}_{\hat{\alpha}}{}^{\hat{\rho}} \,.$$

This relation ensures that a triad transformation along the congruence always exists which makes the $\Lambda_{\hat{\alpha}\hat{\beta}}$'s vanish. Finally the remaining coefficients $\Gamma_{\hat{\alpha}\hat{\beta}\hat{\gamma}}$ behave as Lorentz tensors with respect to the internal rotation group, if this satisfies the condition:

$$\lambda_{\hat{\gamma}}{}^{i} \frac{\partial}{\partial x^i} \tilde{L}_{\hat{\beta}}^{\hat{\alpha}} = 0 \tag{4.5.14}$$

identically for $\hat{\gamma} = 1, 2, 3$.

4.6 Differential operators related to a tetrad frame

Given a congruence C_X (with X a unit time-like vector) the use of the transverse projecting operator $h = I + X \otimes X$, or $h_i^{\ j} = \lambda_{\hat{\alpha}i} \lambda^{\hat{\alpha}j}$, where $\lambda_{\hat{\alpha}}$ are smooth fields of triads orthogonal to X, allows one to reduce the tangent space $T(M) = \cup_p T_p(M)$ to the smaller space $T_\perp = \cup_p T_{\perp p}(M)$ of vectors orthogonal to X. We denote the set of vector fields taking values in $T_\perp(M)$ by $\mathcal{T}_\perp(M)$.

There is a natural connection on this space, defined by the covariant derivative

$$^\perp\nabla_Z Y = h(\nabla_Z Y) \qquad Y \in \mathcal{T}_\perp(M) \quad Z \in \mathcal{T}(M) \tag{4.6.1}$$

satisfying the same structural relations as the usual covariant derivative ∇_Z, namely

$$^\perp\nabla_Z(fY) = Z(f)h(Y) + f\,^\perp\nabla_Z Y = Z(f)Y + f\,^\perp\nabla_Z Y$$
$$^\perp\nabla_{gZ} Y = g\,^\perp\nabla_Z Y$$

for $f, g \in \mathcal{F}(M)$, $Z \in \mathcal{T}(M)$, $Y \in \mathcal{T}_\perp(M)$. In components, (4.6.1) reads

$$(^\perp\nabla_Z Y)^i = (\nabla_Z Y)^i + X^i X_j (\nabla_Z Y)^j = (\nabla_Z Y)^i - X^i Y^j (\nabla_Z X)_j \tag{4.6.2}$$

(using $X_i Y^i = 0$). Writing this in tetrad components, with $\lambda_{\hat{0}} = X$, gives

$$(^\perp\nabla_Z Y)^{\hat{b}} = Z(Y^{\hat{b}}) + Z^{\hat{a}} Y^{\hat{c}} (\Gamma^{\hat{b}}{}_{\hat{c}\hat{a}} - \Gamma_{\hat{c}\hat{0}\hat{a}} \delta^{\hat{b}}_{\hat{0}}) = Z(Y^{\hat{b}}) + Z^{\hat{a}} Y^{\hat{c}}\,^\perp\Gamma^{\hat{b}}{}_{\hat{c}\hat{a}} \tag{4.6.3}$$

where we write $^\perp\Gamma^{\hat{b}}{}_{\hat{c}\hat{a}} = \Gamma^{\hat{b}}{}_{\hat{c}\hat{a}} - \Gamma_{\hat{c}\hat{0}\hat{a}} \delta^{\hat{b}}_{\hat{0}}$. Coordinate components yield a similar result from (4.6.2), namely

$$\begin{aligned} (^\perp\nabla_Z Y)^i &= Z(Y^i) + Z^n Y^j (\Gamma^i{}_{jn} - \lambda^{\hat{a}}_n \lambda_{\hat{b}j} \Gamma^{\hat{b}}{}_{\hat{0}\hat{a}} \lambda_{\hat{0}}{}^i) \\ &= Z(Y^i) + Z^n Y^j\,^\perp\Gamma^i{}_{jn} \\ ^\perp\Gamma^i{}_{jn} &= \Gamma^i{}_{jn} - \lambda^{\hat{a}}_n \lambda_{\hat{b}j} \lambda_{\hat{0}}{}^i \Gamma^{\hat{b}}{}_{\hat{0}\hat{a}}. \end{aligned} \tag{4.6.4}$$

The projector h can be regarded as a metric on $T_\perp(M)$, if we define

$$^\perp(Y|W) = h_{ij} Y^i W^j$$

(though for $Y, W \in T_{\perp p}$ this is identical to $(Y|W)$) and we then find that $^\perp\nabla$ is compatible with h in the sense that

$$Z(^\perp(Y|W)) = ^\perp(^\perp\nabla_Z Y|W) + ^\perp(Y|^\perp\nabla_Z W). \tag{4.6.5}$$

We can construct a *curvature* from ${}^{\perp}\nabla$ by defining

$${}^{\perp}R(Z,V)Y = {}^{\perp}\nabla_Z{}^{\perp}\nabla_V Y - {}^{\perp}\nabla_V{}^{\perp}\nabla_Z Y - {}^{\perp}\nabla_{[Z,V]}Y.$$

Substituting directly from (4.6.2) gives:

$${}^{\perp}R(Z,V)Y = h(R(Z,V)Y) + (Y|\nabla_Z X)\nabla_V X - (Y|\nabla_V X)\nabla_Z X. \tag{4.6.6}$$

In the particular case where all the vector fields involved are orthogonal to X, (4.6.3) becomes

$$({}^{\perp}\nabla_Z Y)^{\hat\beta} = Z(Y^{\hat\beta}) + Z^{\hat\alpha}Y^{\hat\gamma}\,{}^{\perp}\Gamma^{\hat\beta}{}_{\hat\gamma\hat\alpha} \tag{4.6.7}$$

while (4.6.6) becomes, on specializing to the case where Z, V, Y are the triad basis field $\lambda_{\hat\alpha}, \lambda_{\hat\delta}, \lambda_{\hat\beta}$

$${}^{\perp}R_{\hat\alpha\hat\delta\hat\beta\hat\gamma} = R_{\hat\alpha\hat\delta\hat\beta\hat\gamma} - \Gamma_{\hat\beta\hat0\hat\alpha}\Gamma_{\hat\gamma\hat0\hat\delta} + \Gamma_{\hat\beta\hat0\hat\delta}\Gamma_{\hat\gamma\hat0\hat\alpha}\,. \tag{4.6.8}$$

If X is hypersurface-orthogonal, then ${}^{\perp}R$ is the Riemannian curvature of the surfaces orthogonal to X and this equation is thus known as Gauss' equation, with the coefficients $\Gamma_{\hat\beta\hat0\hat\alpha}$ called the second fundamental form of the surface. The more general case will be considered in Sect. 8.1.

The curvature ${}^{\perp}R$ defined in (4.6.6) can be interpreted as follows: consider a closed path γ through a point p on a given curve ζ, say, of C_X and a vector $Y \in T_{\perp p}(M)$; then parallely propagate Y along γ (with the full connection) and simultaneously project it to $T_{\perp\gamma(s)}$, s being a parameter on γ (see Fig. 4-2). The vector $Y' \in T_{\perp p}(M)$ which will result at p, after moving one loop on γ, differs from the original vector Y by an amount measured by ${}^{\perp}R$. In this construction, the vector field $\dot\gamma$ is in general not orthogonal to $X_{\gamma(s)}$; if we restrict ourselves to a path γ', say, whose tangent $\dot\gamma'$ is everywhere orthogonal to X, then it would not be possible to close γ' on the same point $p = \zeta(\tau)$ unless X is hypersurface-orthogonal; hence, upon leaving p, the curve γ' can only cross ζ again at a different point $p' = \zeta(\tau')$, say (see Fig. 4-3). If we now perform the same construction as before and then parallely propagate the resulting vector $Y'_{(p')}$ from p' to p along ζ and project it to $T_{\perp p}(M)$, we find that the two resulting vectors Y and $h \circ \Gamma(p',p;\zeta)Y'_{(p')}$ in p differ by an amount measured by a new tensor

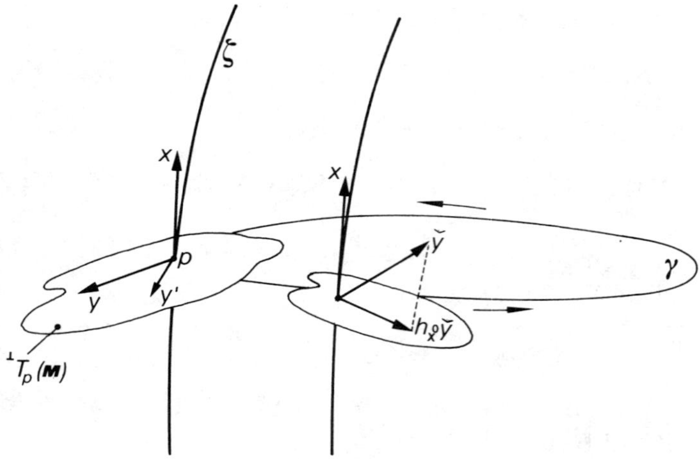

Fig. 4-2 Geometrical interpretation of the curvature tensor ${}^{\perp}R$.

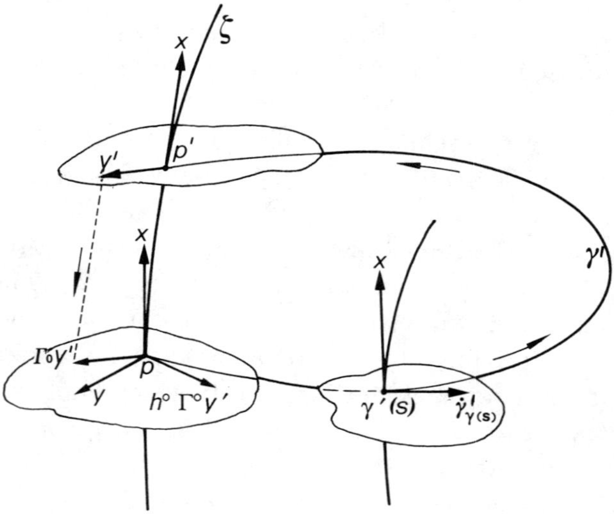

Fig. 4-3 Geometrical interpretation of the curvature tensor P.

P, defined in general as

$$P(Z,V)(Y) = 2^{\perp}\nabla^{\perp}_{[Z}\nabla_{V]}Y -^{\perp}\nabla_{h[Z,V]}Y \tag{4.6.9}$$

for any $V \in T_{\perp p}(M)$. From (4.6.1) this becomes:

$$\begin{aligned} P(Z,V)(Y) &= h(\nabla_Z[h(\nabla_V Y)]) - h(\nabla_V[h(\nabla_Z Y)]) - h(\nabla_{h[Z,V]}Y) \\ &= {}^{\perp}R(Z,V)(Y) - h(\nabla_X Y)(X|[Z,V]) \ . \end{aligned} \tag{4.6.10}$$

If we specify as before $Z = \lambda_{\hat{\alpha}}, V = \lambda_{\hat{\delta}}, Y = \lambda_{\hat{\beta}}$, the latter becomes:

$$P_{\hat{\alpha}\hat{\delta}\hat{\beta}\hat{\gamma}} = {}^{\perp}R_{\hat{\alpha}\hat{\delta}\hat{\beta}\hat{\gamma}} - 2\Gamma_{\hat{\beta}\hat{\gamma}\hat{0}}\Gamma_{\hat{0}[\hat{\alpha}\hat{\delta}]}. \tag{4.6.11}$$

If X is hypersurface-orthogonal, the vorticity coefficients $\Gamma_{\hat{0}[\hat{\alpha}\hat{\delta}]}$ (cf. 4.5.2) vanish, the points p and p' on ζ coincide, and so also do the tensors P and ${}^{\perp}R$. Nonetheless even if X is not hypersurface orthogonal and the points p and p' remain distinct on ζ, the tensors P and ${}^{\perp}R$ will coincide if the tetrad in p is Fermi transported along ζ since, in this case, the Fermi rotation coefficients $\Gamma_{\hat{\beta}\hat{\gamma}\hat{0}}$ (cf. 4.5.6) vanish.

5

SPINORS AND THE CLASSIFICATION OF THE WEYL TENSOR

5.1 Outline

In this chapter we give an introduction to topics whose full development would take us well beyond the scope of this book. Our main aim is to describe the scheme of classification for the algebraic structure of the Weyl tensor at any given point of space-time, due to Petrov (Petrov, 1969).

While this can be done using tensor calculus, much of the work in this area becomes clearer if one uses spinor calculus and so we shall include an outline of this.

Sections 5.2 to 5.4 describe various algebraic techniques connected with the group $\mathrm{SL}(2,\mathbb{C})$. In Section 5.5 these are used to define spinors, objects analogous to vectors with two complex valued components depending on the choice of tetrads. Sections 5.6 and 5.7 describe Riemannian geometry (connection and curvature) using spinors, including the spinor equivalent of the Weyl tensor. In Section 5.8 the case of non-vanishing torsion is considered and then, after a section (the 5.9th) on conformal properties, we describe, in Section 5.10, the classification of the Weyl tensor, first in spinor form and then in the tensor translation.

5.2 The group $\mathrm{SL}(2,\mathbb{C})$

$\mathrm{SL}(2,\mathbb{C})$ denotes the group of all complex ("$\mathbb{C}$"), 2×2 ("2") matrices ("L" = linear) with determinant equal to one ("S" = special). It is, as we shall see, intimately related to the Lorentz group. $\mathrm{SL}(2,\mathbb{C})$ acts on pairs (η^1, η^2) of complex numbers. We denote a general member of such a pair by $\eta^{\mathrm{A}}|_{A=1,2}$. Components will also be indexed by primed numbers $(1', 2')$, a general component being $\lambda^{\mathrm{A}'}|_{A'=1',2'}$. The reason for the distinction, as well as the significance of upper or lower indices, will emerge when these pairs are given geometrical interpretations in Sect. 5.5.

Since the elements of SL(2, $\mathbb{C}$) have unit determinant, we need a technique for analysing determinants. To this end, introduce the totally antisymmetric matrices

$$[\epsilon_{\mathrm{X}'\mathrm{Y}'}] = [\epsilon_{\mathrm{AB}}] = \begin{bmatrix} 0 & 1 \\ -1 & 0 \end{bmatrix} = [\epsilon^{\mathrm{AB}}] = [\epsilon^{\mathrm{X}'\mathrm{Y}'}]$$

where $\mathrm{X}', \mathrm{Y}' = 1', 2'$; $\mathrm{A}, \mathrm{B} = 1, 2$. Thus if $A \in \mathrm{SL}(2, \mathbb{C})$, we have

$$A^{\mathrm{A}}{}_{\mathrm{C}} A^{\mathrm{B}}{}_{\mathrm{D}} \epsilon_{\mathrm{AB}} = \epsilon_{\mathrm{CD}}. \tag{5.2.1}$$

We shall use ϵ_{AB} and ϵ^{AB} for *lowering and raising the index* just as g_{ij} and g^{ij} are used. Thus if a pair $(\eta_{\mathrm{A}})_{\mathrm{A}=1,2}$ has been defined, it is understood that η^{A} are defined by

$$\eta^{\mathrm{A}} = \epsilon^{\mathrm{AB}} \eta_{\mathrm{B}} \tag{5.2.2}$$

(Note that the order of indices is vital: $\epsilon^{\mathrm{AB}}\eta_{\mathrm{B}} \neq \eta_{\mathrm{B}}\epsilon^{\mathrm{BA}}$, the adjacent summed index-pair "$^{\mathrm{B}}{}_{\mathrm{B}}$" slopes down as ↘ in our conventions.) For consistency, if we define λ^{A}, then λ_{A} are understood to be given by:

$$\lambda_{\mathrm{A}} = \lambda^{\mathrm{B}} \epsilon_{\mathrm{BA}} \tag{5.2.3}$$

(again note the ↘ orientation of the pair "$^{\mathrm{B}}{}_{\mathrm{B}}$"). To link SL(2, $\mathbb{C}$) with the Lorentz group $\mathcal{L}$, we first link $\mathbb{R}^4$ with the set Herm (2, $\mathbb{C}$) of the 2×2 complex Hermitian matrices; we shall then act with $\mathcal{L}$ on $\mathbb{R}^4$ and with SL(2, $\mathbb{C}$) on the set Herm(2, $\mathbb{C}$) of such matrices. To make this link, introduce four 2×2 Hermitian matrices $\sigma_0, \ldots, \sigma_3$ by:

$$\sigma_0 = \frac{1}{\sqrt{2}}\begin{pmatrix} 1 & 0 \\ 0 & 1 \end{pmatrix} \qquad \sigma_1 = \frac{1}{\sqrt{2}}\begin{pmatrix} 0 & 1 \\ 1 & 0 \end{pmatrix}$$
$$\sigma_2 = \frac{1}{\sqrt{2}}\begin{pmatrix} 0 & i \\ -i & 0 \end{pmatrix} \qquad \sigma_3 = \frac{1}{\sqrt{2}}\begin{pmatrix} 1 & 0 \\ 0 & -1 \end{pmatrix} . \tag{5.2.4}$$

These define a map* $S: X \to \sigma_i X^i$ from $\mathbb{R}^4$ to Herm$(2, \mathbb{C})$. Explicitly:

$$S(X) = \frac{1}{\sqrt{2}} \begin{bmatrix} X^0 + X^3 & X^1 + iX^2 \\ X^1 - iX^2 & X^0 - X^3 \end{bmatrix} \tag{5.2.5}$$

from which it is clear that S is a one–to–one, onto, real linear map. A calculation shows that:

$$\mathrm{Det} S(X) = -\frac{1}{2}\eta_{ij} X^i X^j \tag{5.2.6}$$

and so Herm$(2, \mathbb{C})$ can be identified, via S, with $\mathbb{R}^4$, the determinant corresponding to $(-1/2)$ times the Lorentz metric. Index the components of σ_i as $\sigma_i{}^{AX'}$, thus

$$\sigma_i = \begin{bmatrix} \sigma_i{}^{11'} & \sigma_i{}^{12'} \\ \sigma_i{}^{21'} & \sigma_i{}^{22'} \end{bmatrix}$$

so that

$$S(X)^{AX'} = X^i \sigma_i{}^{AX'}. \tag{5.2.7}$$

Since the σ_i are Hermitian matrices, we have:

$$\overline{\sigma_i{}^{AX'}} = \sigma_i{}^{XA'}. \tag{5.2.8}$$

Using ϵ_{AB} we can write (5.2.6) as

$$S(X)^{AX'} S(X)^{BY'} \epsilon_{AB} = -\frac{1}{2}\eta_{ij} X^i X^j \epsilon^{X'Y'}$$

so

$$\sigma_i{}^{AX'} \sigma_j{}^{BY'} \epsilon_{AB} X^i X^j = -\frac{1}{2}\eta_{ij} X^i X^j \epsilon^{X'Y'}$$

or

$$\sigma_{(i}{}^{AX'} \sigma_{j)A}{}^{Y'} = +\frac{1}{2}\eta_{ij}\epsilon^{X'Y'}. \tag{5.2.9}$$

Multiplying by $\epsilon_{X'Y'}$ then gives

$$\sigma_i{}^{AX'} \sigma_{jAX'} = -\eta_{ij} \tag{5.2.10}$$

(noting that the left-hand side is automatically symmetric in i and j). Consequently:

$$-S(X)^{AX'} \sigma^i{}_{AX'} = -\sigma_j{}^{AX'} \sigma^i{}_{AX'} X^j = X^i$$

---

*Here by X we mean a general set $\{X^i\} \in \mathbb{R}^4$ of real numbers.

from (5.2.10) i.e. for any $H^{AX'} \in \text{Herm}(2, \mathbb{C})$,

$$\overset{-1}{S}\left(H^{AX'}\right)^i = -\sigma^i{}_{AX'}H^{AX'}; \tag{5.2.11}$$

thus the map

$$\left[H^{AX'}\right]_{A=1,2\ X'=1',2'} \longmapsto \left[-H^{AX'}\sigma^i{}_{AX'}\right]_{i=0,\ldots,3}$$

is the inverse map to S, restoring X. Since $\sigma_i^{AX'}$ and $-\sigma^i_{AX'}$ are the components of inverse maps, we have:

$$\sigma_i{}^{AX'}\sigma^i{}_{BY'} = -\delta^A{}_B\delta^{X'}{}_{Y'}. \tag{5.2.12}$$

We shall now show that, under the correspondence between Herm(2, ℂ) and $\mathbb{R}^4$ set up by S, transformations by SL(2, ℂ) in Herm (2, ℂ) correspond to transformations by the Lorentz group in $\mathbb{R}^4$. If $A \in \text{SL}(2, \mathbb{C})$, the transformation we shall study is the map P_A from Herm(2, ℂ) to itself defined by:

$$P_A: H \to AHA^\dagger \tag{5.2.13}$$

(where $A^\dagger$ denotes the transpose of the complex conjugate of A). This is a real-linear map with an inverse $\overset{-1}{P}_A$.

Define a linear map $L_A: \mathbb{R}^4 \to \mathbb{R}''$ by:

$$L_A = \overset{-1}{S} \circ P_A \circ S \tag{5.2.14}$$

i.e.

$$S(L_A X) = A(S(X))A^\dagger \tag{5.2.15}$$

for all $X \in \mathbb{R}^4$. Taking determinants and using (5.2.6) we have:

$$\begin{aligned}\text{Det}\,(S(L_A X)) &= -\frac{1}{2}\eta_{ij}\,(L_A X)^i\,(L_A X)^j \\ &= |\text{Det}A|^2\text{Det}(S(X)) = -\frac{1}{2}\eta_{ij}X^iX^j\end{aligned}$$

i.e. $L_A X$ has the same Lorentz length as X. Moreover the transformation L_A can be continuously changed to the identity by changing A to the identity; so L_A is a Lorentz transformation.

The map $\boldsymbol{L}: \text{SL}(2, \mathbb{C}) \to \mathcal{L}$ by $A \mapsto L_A$ satisfies

$$L_A \circ L_B = S^{-1} \circ P_A \circ P_B \circ S = S^{-1} \circ P_{AB} \circ S = L_{AB} \tag{5.2.16}$$

(a property expressed by saying that $\boldsymbol{L}$ is a homomorphism (see Sect. 1.15)). It can in fact be shown that $\boldsymbol{L}$ is onto (i.e. all Lorentz transformations have the form L_A for some A). The idea of the proof is to note that matrices close to the identity can be written uniquely as

$$\begin{pmatrix} 1+a & b \\ c & (1+bc)/(1+a) \end{pmatrix} \tag{5.2.17}$$

with a, b, c complex. Thus a neighbourhood of the identity is 6-dimensional, being parametrized by the real and imaginary parts of a, b and c. But the Lorentz group, as we have shown, is also six-dimensional and one can show that any homomorphism of one connected n-dimensional group into another must be onto.

The question then arises, is $\boldsymbol{L}$ one–to–one? The answer is no! For we need only consider the kernel of $\boldsymbol{L}$ consisting of those matrices mapped to the identity, i.e. those A for which $P_A = \mathit{identity}$, or

$$AHA^\dagger = H \qquad \forall H \in \mathrm{Herm}(2, \mathbb{C}).$$

By considering special choices of H, one can easily show that this is true only if $A = \pm I$, so that the kernel of $\boldsymbol{L}$ consists of $\{+I, -I\}$, and $\boldsymbol{L}$ is precisely two-to-one: $\overset{-1}{\boldsymbol{L}}(L)$ consists of two matrices differing in sign, for each $L \in \mathcal{L}$. If we ask, whether $\mathrm{SL}(2, \mathbb{C})$ is simply connected, or whether it is, like the Lorentz group, doubly connected, we find that the two-to-one nature of $\boldsymbol{L}$ makes it simply connected. Explicitly, we can define the path $\bar{\gamma}$ by

$$\mathbb{R} \ni \theta \overset{\bar{\gamma}}{\longmapsto} \begin{pmatrix} e^{i\theta/2} & 0 \\ 0 & e^{-i\theta/2} \end{pmatrix} \in \mathrm{SL}(2, \mathbb{C}).$$

Then $\boldsymbol{L}(\bar{\gamma}(\theta))$ is given by calculating

$$\bar{\gamma}(\theta) S(X) \bar{\gamma}(\theta)^\dagger = \begin{pmatrix} z+t & e^{i\theta}(x+iy) \\ e^{-i\theta}(x+iy) & z-t \end{pmatrix} = S(\gamma(\theta)(X)) \tag{5.2.18}$$

i.e. $\boldsymbol{L}(\bar{\gamma}(\theta)) = \gamma(\theta)$, where γ is the closed path in $\mathcal{L}$ given by (4.3.13) which is responsible for the $\mathcal{L}$'s being multiply connected. But, clearly $\bar{\gamma}(2\pi) = -I$, while $\bar{\gamma}(0) = I$: the curve does not close.

So the closed curves in SL(2, $\mathbb{C}$) are deformable to the constant map.

5.3 Lie algebras

A group such as SL(2, $\mathbb{C}$) can be regarded as a manifold: (5.2.17), for example, defines six coordinates in a neighbourhood of the origin, and the full manifold structure is fixed by requiring that, in any other chart, the transformations defined by multiplying by elements of the group are diffeomorphisms. In these circumstances, the group is called a Lie group. If G is any Lie group and $g \in G$, define $R(g)$ to be the transformation obtained by multiplying by g on the right:

$$R(g)x = xg \qquad \forall x \in G. \tag{5.3.1}$$

Let $\underline{g}$ $(= T_e(G))$ denote the tangent space at the identity e to G. Given any vector X in $\underline{g}$, let X^R be the vector field on G defined by mapping X to different points by the transformation $R(g)$; thus

$$\underset{g}{X}^R = R(g)_* X. \tag{5.3.2}$$

Given two vectors $X, Y \in \underline{g}$, define:

$$[X, Y] \stackrel{\text{def}}{=} [X^R, Y^R]_e \in \underline{g}. \tag{5.3.3}$$

(The square brackets on the right denoting the commutator defined in the Sect. 2.2.) This defines a multiplication operation on $\underline{g}$, taking two vectors X and Y into a third $[X, Y]$. A vector space with a multiplication (satisfying the axioms one expects for such a structure) is called an Algebra, and $\underline{g}$ equipped with this multiplication, is called the Lie Algebra of G. More generally, one calls a Lie Algebra any algebra that satisfies the axioms just alluded to, together with two identities characteristic of $\underline{g}$, namely:

$$[X, Y] = -[Y, X] \tag{5.3.4}$$

$$[X, [Y, Z]] + [Y, [Z, X]] + [Z, [X, Y]] = 0. \tag{5.3.5}$$

To represent this product more explicitly, let $\gamma: [0, 1] \to G$ be a curve with $\dot{\gamma}(0) = X$. Then for a smooth function f we have from

(1.6.3) and (5.3.2):

$$R(g)_*X(f) = \left(X^R(f)\right)(g)$$
$$= \frac{d}{ds}f\left(R(g)\gamma(s)\right)\Big|_{s=0} = \frac{d}{ds}f(\gamma(s)g)\Big|_{s=0}. \quad (5.3.6)$$

Similarly, if $Y = \dot{\chi}(0)$, we have

$$\left(Y^R(f)\right)(g) = \frac{d}{dt}f(\chi(t)g)\Big|_{t=0}. \quad (5.3.7)$$

Combining these:

$$\left(X^RY^R(f)\right)(g) = \frac{d}{ds}\left(Y^R(f)(\gamma(s)g)\right)\Big|_{s=0}$$
$$= \frac{d}{ds}\left\{\frac{d}{dt}f(\chi(t)\gamma(s)g)\Big|_{t=0}\right\}\Big|_{s=0}.$$

So at the origin:

$$\left(X^RY^R(f)\right)e = \frac{\partial^2}{\partial s\partial t}f(\chi(t)\gamma(s))\Big|_{s=t=0} \quad (5.3.8)$$

and hence:

$$[X,Y](f) = \frac{\partial^2}{\partial s\partial t}\left[f(\chi(t)\gamma(s)) - f(\gamma(s)\chi(t))\right]\Big|_{s=t=0}. \quad (5.3.9)$$

We see that if the group is Abelian ($hg = gh$, for all g, h) then the arguments of the two occurrences of f on the right-hand side of (5.3.9) are equal and in this case $[X,Y] = 0$. Both $\mathrm{SL}(2,\mathbb{C})$ and $\mathcal{L}$ are groups of matrices: $\mathrm{SL}(2,\mathbb{C})$ is a subset of the vector space $M_2^{\mathbb{C}}$ of all 2×2 complex matrices and $\mathcal{L}$ is a subset of the space $M_4^{\mathbb{R}}$ of all 4×4 real matrices. More generally, let G be a subset of $M_n^{\mathbb{K}}(\mathbb{K} = \mathbb{R}$ or $\mathbb{C})$ with the group product given by matrix multiplication. If γ is a curve in G, $\gamma(s)$ is an s-dependent matrix and $d\gamma(s)/ds$ is a matrix. Thus tangent vectors can be regarded as matrices in $M_n^{\mathbb{K}}$. In this case denoting the components of a general matrix Z in $M_n^{\mathbb{K}}$ by $[Z^a{}_b]_{a,b=1,\ldots,n}$, if $Z \mapsto \bar{f}(Z)$ is a function on $M_n^{\mathbb{K}}$, we have:

$$\frac{\partial^2}{\partial s\partial t}\bar{f}(\chi(t)\gamma(s)) = \frac{\partial}{\partial s}\left[\frac{\partial\bar{f}}{\partial Z^a{}_b}(\chi(t)\gamma(s))\dot{\chi}(t)^a{}_c\gamma(s)^c{}_b\right]$$
$$= \frac{\partial^2\bar{f}}{\partial Z^a{}_b\partial Z^p{}_q}\dot{\chi}(t)^a{}_c\gamma(s)^c{}_b\chi(t)^p{}_d\dot{\gamma}(s)^d{}_q$$

$$+\frac{\partial \bar{f}}{\partial Z^a{}_b}\dot{\chi}(t)^a{}_c\dot{\gamma}(s)^c{}_b\,. \qquad (5.3.10)$$

Substituting in (5.3.9), we obtain, on taking $\bar{f}$ to be any extension of f:

$$[X,Y](f)=\frac{\partial \bar{f}}{\partial Z^a_b}(Y^a_c X^c_b - X^a_c Y^c_b)$$

or

$$[X,Y]=YX-XY \qquad (5.3.11)$$

in matrix notation with matrix multiplication. In the case of SL$(2,\mathbb{C})$, G is the sub-manifold of matrices g satisfying

$$\mathrm{Det}g=1$$

and so for any path in $G, \gamma\colon \mathbf{R}\to G$, we have $(d/dt)(\mathrm{Det}g)=0$; hence $\underline{g}$ consists of vectors $X^{\mathrm{A}}{}_{\mathrm{B}}$ $(=\frac{d}{dt}\gamma(t)^{\mathrm{A}}{}_{\mathrm{B}})$ such that:

$$X^{\mathrm{A}}{}_{\mathrm{B}}\frac{\partial}{\partial g^{\mathrm{A}}_{\mathrm{B}}}(\mathrm{Det}g)\Big|_e=0$$

i.e.

$$0=X^{\mathrm{A}}{}_{\mathrm{B}}\left((\bar{g}^1)^{\mathrm{B}}{}_{\mathrm{A}}(\mathrm{Det}g)\right)_e=X^{\mathrm{A}}{}_{\mathrm{A}}.$$

Writing $s\ell(2,\mathbb{C})$ for the Lie Algebra of SL$(2,\mathbb{C})$, we have:

$$s\ell(2,\mathbb{C})=\{X\colon X^{\mathrm{A}}{}_{\mathrm{A}}=0\} \qquad (5.3.12)$$

the set of trace-free matrices.

For the Lorentz group, the matrices are restricted by (4.3.5) and the defining conditions for the corresponding Lie Algebra $\underline{\ell}$ is

$$\ell^a{}_b\frac{\partial}{\partial L^a{}_b}\left(\eta_{pq}L^p{}_r L^q{}_s-\eta_{rs}\right)\Big|_e=0$$

i.e.

$$\ell^a{}_b(\eta_{as}\delta^b_r+\eta_{ra}\delta^b_s)=0$$

so that

$$\underline{\ell}=\{\ell\colon \eta_{as}\ell^s{}_b=-\eta_{bs}\ell^s{}_a\} \qquad (5.3.13)$$

the set of 4×4 matrices which become antisymmetric on "lowering the index" with η.

Now we have a map $\boldsymbol{L}$ taking $\mathrm{SL}(2,\mathbb{C})$ to $\mathcal{L}$, which is a diffeomorphism when restricted to a neighbourhood of the identity. Thus $\boldsymbol{L}_*: s\ell(2,\mathbb{C}) \to \underline{\ell}$ (the induced map of tangent spaces acting on $T_e(\mathrm{SL}(2,\mathbb{C}))$) is a linear isomorphism. In fact, it is easy to see that, since $\boldsymbol{L}$ is a group homomorphism, $\boldsymbol{L}_*$ is a Lie Algebra isomorphism, setting up an exact correspondence between $s\ell(2,\mathbb{C})$ and $\underline{\ell}$. If $\gamma: [0,1] \to \mathrm{SL}(2,\mathbb{C}), \gamma(0) = e$, defines a vector $X = \dot{\gamma}(0) \in s\ell(2,\mathbb{C})$, then $\boldsymbol{L}_*X$ can be calculated from (5.2.14) by

$$(\boldsymbol{L}_*X)^a{}_b = \frac{d}{dt}(L_{\gamma(t)})^a{}_b\Big|_{t=0} = S^{-1} \circ \frac{d}{dt}P_{\gamma(t)}\Big|_{t=0} \circ S \,. \tag{5.3.14}$$

Now from (5.2.13) we have for any $H \in \mathrm{Herm}(2,\mathbb{C})$:

$$\frac{d}{dt}P_{\gamma(t)}H\Big|_{t=0} = \frac{d}{dt}\left(\gamma(t)H\gamma(t)^\dagger\right)\Big|_{t=0}$$

so

$$\left(\frac{d}{dt}P_{\gamma(t)}H\Big|_{t=0}\right)^{\mathrm{AC}'} = X^{\mathrm{A}}{}_{\mathrm{B}}H^{\mathrm{BC}'} + H^{\mathrm{AB}'}\overline{X}^{\mathrm{C}'}{}_{\mathrm{B}'}$$

or

$$\left(\frac{d}{dt}P_{\gamma(t)}\Big|_{t=0}\right)^{\mathrm{AC}'}{}_{\mathrm{BE}'} = X^{\mathrm{A}}{}_{\mathrm{B}}\delta^{\mathrm{C}'}{}_{\mathrm{E}'} + \overline{X}^{\mathrm{C}'}{}_{\mathrm{E}'}\delta^{\mathrm{A}}{}_{\mathrm{B}} \,.$$

Thus substituting in (5.3.14), using (5.2.7) and (5.2.11) and antisymmetrising over a and b :

$$\begin{aligned}(\boldsymbol{L}_*X)^{ab} &= -\,\sigma^{[a}{}_{\mathrm{AC}'}\left(X^{\mathrm{A}}{}_{\mathrm{B}}\delta^{\mathrm{C}'}{}_{\mathrm{E}'} + \overline{X}^{\mathrm{C}'}{}_{\mathrm{E}'}\delta^{\mathrm{A}}{}_{\mathrm{B}}\right)\sigma^{b]\mathrm{BE}'} \\ &= 2\left(S^{ab}{}_{\mathrm{A}}{}^{\mathrm{B}}X^{\mathrm{A}}{}_{\mathrm{B}} + \overline{S}^{ab}{}_{\mathrm{C}'}{}^{\mathrm{E}'}\overline{X}^{\mathrm{C}'}{}_{\mathrm{E}'}\right)\end{aligned} \tag{5.3.15}$$

with

$$S^{ab}{}_{\mathrm{A}}{}^{\mathrm{B}} = -\frac{1}{2}\sigma^{[a}{}_{\mathrm{AC}'}\sigma^{b]\mathrm{BC}'} \tag{5.3.16a}$$

$$\overline{S}^{ab}{}_{\mathrm{C}'}{}^{\mathrm{E}'} = -\frac{1}{2}\sigma^{[a}{}_{\mathrm{BC}'}\sigma^{b]\mathrm{BE}'} . \tag{5.3.16b}$$

We note that, from (5.2.12) and (5.3.16a):

$$S^a{}_{b\mathrm{A}}{}^{\mathrm{B}}S^b{}_{a\mathrm{P}}{}^{\mathrm{Q}} = \frac{1}{4}\left(\delta^{\mathrm{Q}}{}_{\mathrm{A}}\delta^{\mathrm{B}}{}_{\mathrm{P}} + \epsilon_{\mathrm{PA}}\epsilon^{\mathrm{BQ}}\right)$$

i.e.

$$S^a{}_{b\mathrm{AB}} S^b{}_{a\mathrm{PQ}} = \frac{1}{2}\epsilon_{\mathrm{A}(\mathrm{Q}}\epsilon_{\mathrm{P})\mathrm{B}} \tag{5.3.17}$$

while:

$$\overline{S}^a{}_{b\mathrm{C}'}{}^{\mathrm{E}'} S^b{}_{a\mathrm{P}}{}^{\mathrm{Q}} = 0. \tag{5.3.18}$$

Hence from (5.3.15) and (5.3.12)

$$(\boldsymbol{L}_* X)^a{}_b S^b{}_{a\mathrm{P}}{}^{\mathrm{Q}} = X^{\mathrm{Q}}{}_{\mathrm{P}}$$

or

$$\left(\overset{-1}{\boldsymbol{L}}{}_*\chi\right)^{\mathrm{Q}}{}_{\mathrm{P}} = \chi^a{}_b S^b{}_{a\mathrm{P}}{}^{\mathrm{Q}}\ . \tag{5.3.19}$$

Thus we now have maps $\boldsymbol{L}_*$, $\overset{-1}{\boldsymbol{L}}{}_*$ given by (5.3.15), (5.3.19) transforming between $\underline{\ell}$ (cf. (5.3.13)) and $s\underline{\ell}(2, \mathbb{C})$ (cf. (5.3.12)).

5.4 Bivector algebra

A bivector at a point p is an antisymmetric tensor of valence $(0, 2)$ at p. Here we consider the purely algebraic object consisting of an antisymmetric array of numbers $[F_{ij}]_{i,j=0,\ldots,3}$. We have already seen that when the F_{ij} are real numbers, they can be regarded as constituting an element of the Lie Algebra $\underline{\ell}$ of the Lorentz group (with an index lowered). The map $\boldsymbol{L}_*$ (cf. (5.3.15) and (5.3.19)) then sets up a one-to-one correspondence between such objects and arrays $X^{\mathrm{A}}{}_{\mathrm{B}}$ which, on lowering an index to form:

$$X_{\mathrm{CB}} = X^{\mathrm{A}}{}_{\mathrm{B}}\epsilon_{\mathrm{AC}}, \tag{5.4.1}$$

are symmetric

$$X_{\mathrm{CB}} = X_{\mathrm{BC}}.$$

We now generalize the construction to the case where the F_{ij} may be complex. The equation (5.3.15) for $\boldsymbol{L}_* X$ will automatically produce a real array, and so we no longer have a map setting up a one-to-one correspondence between the (complex) F_{ij} and X_{AB}. Instead we generalize (5.3.19) as follows: given F_{ij}, define:

$$F_{\mathbf{1}}{}^{\mathrm{A}}{}_{\mathrm{B}} = -F_{ij} S^{ij}{}_{\mathrm{B}}{}^{\mathrm{A}} \tag{5.4.2a}$$

$$F_{\mathbf{2}}{}^{\mathrm{C}'}{}_{\mathrm{E}'} = -F_{ij} \overline{S}^{ij}{}_{\mathrm{E}'}{}^{\mathrm{C}'}. \tag{5.4.2b}$$

(If F is real, we have ${F_2}^{C'}{}_{E'} = \overline{{F_1}^{C}{}_{E}}$; otherwise these are two independent quantities.) We shall see shortly that F_{ij} can be reconstructed from ${F_1}^{A}{}_{B}$ and ${F_2}^{C'}{}_{E'}$. To interpret these, define:

$$F_{1ij} = 2S_{ijA}{}^{B}{F_1}^{A}{}_{B}. \tag{5.4.3a}$$

From (5.4.2a) this is

$$F_{1ij} = \gamma_{ij}{}^{kl}F_{kl} \tag{5.4.3b}$$

where:

$$\gamma_{ij}{}^{kl} = -2S_{ijA}{}^{B}S^{kl}{}_{B}{}^{A}. \tag{5.4.4}$$

To calculate γ, we use (5.3.15) and (5.3.19) to write out the equation:

$$\boldsymbol{\chi} = \boldsymbol{L}_* \circ \overset{-1}{\boldsymbol{L}}_*(\boldsymbol{\chi})$$

for $\boldsymbol{\chi} \in \underline{\ell}$ giving from (5.4.4)

$$\chi_{ij} = \gamma_{ij}{}^{kl}\chi_{kl} + \overline{\gamma}_{ij}{}^{kl}\chi_{kl}$$

so that

$$\left(\gamma_{ij}{}^{kl} + \overline{\gamma}_{ij}{}^{kl}\right) = \delta_i^{[k}\delta_j^{l]} \tag{5.4.5}$$

fixing the real part of γ. To calculate the imaginary part, we use (5.4.4), (5.3.16) and (5.2.8) to write:

$$\begin{aligned}\gamma_{ijkl} - \bar{\gamma}_{ijkl} &= \frac{1}{2}\left\{\sigma_{[i|AC'|}\sigma_{j]}{}^{BC'}\sigma_{[k|BE'}\sigma_{l]}{}^{AE'}\right.\\ &\quad \left. - \sigma_{[i|CA'|}\sigma_{j]}{}^{CB'}\sigma_{[k|EB'|}\sigma_{l]}{}^{EA'}\right\}\\ &= \frac{1}{2}\left\{\sigma_{[i|AC'|}\sigma_{j]}{}^{BC'}\sigma_{[k|BE'|}\sigma_{l]}{}^{AE'}\right.\\ &\quad \left. - \sigma_{[i|AC'|}\sigma_{[l}{}^{BC'}\sigma_{k]|BE'|}\sigma_{j]}{}^{AE'}\right\}.\end{aligned} \tag{5.4.6}$$

This is antisymmetric on all indices, and so is equal to i$\alpha\delta_{ijkl}$ with α real. Thus we have, from (5.4.5) and (5.4.6):

$$\gamma_{ij}{}^{kl} = \frac{1}{2}\left(\delta_i^{[k}\delta_j^{l]} + i\alpha\delta_{ij}{}^{kl}\right). \tag{5.4.7}$$

Now

$$\begin{aligned}\gamma_{ij}{}^{kl}\gamma_{kl}{}^{pq} &= 4S_{ijA}{}^{B}S^{kl}{}_{B}{}^{A}S_{kl}{}^{D}{}_{C}S^{pqC}{}_{D}\\ &= 2S_{ij}{}^{AB}\epsilon_{BD}\epsilon_{CA}S^{pqDC} = \gamma_{ij}{}^{pq}\end{aligned} \tag{5.4.8}$$

where the second step is from (5.3.17); from the latter then, γ can be thought of as a projection operator. From (5.4.7), however, and recalling (1.17.9):

$$\gamma_{ij}{}^{kl}\gamma_{kl}{}^{pq} = \frac{1}{4}\left\{\delta_i^{[p}\delta_j^{q]}(1+4\alpha^2) + 2\mathrm{i}\alpha\delta_{ij}{}^{pq}\right\}$$

so that $\alpha = \pm\frac{1}{2}$. A direct evaluation using (5.2.4) gives $\gamma_{0123} = +\frac{\mathrm{i}}{4}$, so $\alpha = +\frac{1}{2}$ and

$$\gamma_{ij}{}^{kl} = \frac{1}{2}\left(\delta_i^{[k}\delta_j^{l]} + \frac{1}{2}\mathrm{i}\delta_{ij}{}^{kl}\right); \qquad (5.4.9a)$$

while conjugating

$$\overline{\gamma}_{ij}{}^{kl} = \frac{1}{2}\left(\delta_i^{[k}\delta_j^{l]} + \frac{1}{2}\mathrm{i}\delta_{ij}{}^{kl}\right). \qquad (5.4.9b)$$

For an antisymmetric array F_{ij}, we define the dual $\overset{*}{F}_{ij}$ by (cf. (1.17.18)):

$$\overset{*}{F}_{ij} = \frac{1}{2}\delta_{ij}{}^{kl}F_{kl}$$

clearly $\overset{**}{F} = -F$. Then the action of γ on F can be written from (5.4.9) and (5.4.3b) as:

$$F_{\mathbf{1}ij} \equiv \gamma_{ij}{}^{kl}F_{kl} = \frac{1}{2}(F_{ij} + \mathrm{i}\overset{*}{F}_{ij}). \qquad (5.4.10)$$

Similarly, if we define

$$F_{\mathbf{2}ij} = 2\overline{S}_{ij\mathrm{C}'}{}^{\mathrm{E}'}F_{\mathbf{2}}{}^{\mathrm{C}'}{}_{\mathrm{E}'} \qquad (5.4.11a)$$

then we find that :

$$F_{\mathbf{2}ij} = \overline{\gamma}_{ij}{}^{kl}F_{kl} = \frac{1}{2}(F_{ij} - \mathrm{i}\overset{*}{F}_{ij}). \qquad (5.4.11b)$$

A tensor P of the form $P_{ij} = F_{ij} + \mathrm{i}\overset{*}{F}_{ij}$ is called self-dual. The term (suggesting that P is equal to its own dual – an impossibility since $\overset{**}{P} = -P$) is misleading; in fact:

$$\overset{*}{P} = -\mathrm{i}P.$$

Similarly a tensor of the form $P_{ij} = F_{ij} - \mathrm{i}\overset{*}{F}_{ij}$ satisfies $\overset{*}{P} = \mathrm{i}P$ and is called anti-self-dual. Since

$$F_{ij} = F_{\mathbf{1}ij} + F_{\mathbf{2}ij},$$

we refer to F_1 and F_2 as the self-dual and anti-self-dual parts of F. This together with (5.4.3a) and (5.4.11), allows us to give the promised reconstruction of F, from $F_1{}^{\mathrm{A}}{}_{\mathrm{B}}$ and $F_2{}^{\mathrm{E}'}{}_{\mathrm{C}'}$, determining respectively the self-dual and anti-self-dual parts, as

$$F_{ij} = 2\left(S_{ij\mathrm{A}}{}^{\mathrm{B}} F_1{}^{\mathrm{A}}{}_{\mathrm{B}} + \overline{S}_{ij}{}^{\mathrm{E}'}{}_{\mathrm{C}'} F_2{}^{\mathrm{C}'}{}_{\mathrm{E}'}\right). \tag{5.4.12}$$

From (5.3.18) applied to this, we note, from (5.4.2a, b) that:

$$F_1{}^{\mathrm{A}}{}_{\mathrm{B}} \equiv -F_{ij} S^{ij}{}_{\mathrm{B}}{}^{\mathrm{A}} = -F_{1ij} S^{ij}{}_{\mathrm{B}}{}^{\mathrm{A}} \tag{5.4.13a}$$

$$F_2{}^{\mathrm{C}'}{}_{\mathrm{E}'} \equiv -F_{ij} \overline{S}^{ij}{}_{\mathrm{C}'}{}^{\mathrm{E}'} = -F_{2ij} \overline{S}^{ij}{}_{\mathrm{C}'}{}^{\mathrm{E}'}. \tag{5.4.13b}$$

Thus the map

$$[F_{ij}] \longmapsto \left[-F_{ij} S^{ij}{}_{\mathrm{B}}{}^{\mathrm{A}}\right]$$

depends only on the self-dual part, mapping the anti-self-dual part to zero, and has a partial inverse:

$$[X^{\mathrm{A}}{}_{\mathrm{B}}] \longmapsto \left[2S_{ij\mathrm{A}}{}^{\mathrm{B}} X^{\mathrm{A}}{}_{\mathrm{B}}\right]$$

restoring the self-dual part; with corresponding maps with $\overline{S}_{ij\mathrm{E}'}{}^{\mathrm{C}'}$ and the anti-self-dual part.

If we use the σ-matrices to convert F directly to spinor form we can define:

$$F_{\mathrm{AA}'\mathrm{BB}'} = \sigma^i{}_{\mathrm{AA}'} \sigma^j{}_{\mathrm{BB}'} F_{ij}. \tag{5.4.14}$$

From (5.3.16) and (5.2.12) we have

$$\sigma^i_{\mathrm{AA}'} \sigma^j_{\mathrm{BB}'} S_{ij\mathrm{PQ}} = \frac{1}{2} \epsilon_{\mathrm{A}'\mathrm{B}'} \epsilon_{\mathrm{A}(\mathrm{P}} \epsilon_{\mathrm{Q})\mathrm{B}},$$

so using this and (5.4.12) in (5.4.14), gives:

$$F_{\mathrm{AA}'\mathrm{BB}'} = \epsilon_{\mathrm{A}'\mathrm{B}'} F_{1\mathrm{AB}} + \epsilon_{\mathrm{AB}} F_{2\mathrm{A}'\mathrm{B}'}. \tag{5.4.15}$$

We may also apply this procedure to $\overset{*}{F}_{ij}$, using:

$$\overset{*}{F}_{ij} = -\mathrm{i}\left(F_{1ij} - F_{2ij}\right) = -2\mathrm{i}\left(S_{ij\mathrm{A}}{}^{\mathrm{B}} F_1{}^{\mathrm{A}}{}_{\mathrm{B}} - \overline{S}_{ij}{}^{\mathrm{E}'}{}_{\mathrm{C}'} F_2{}^{\mathrm{C}'}{}_{\mathrm{E}'}\right) \tag{5.4.16}$$

yielding

$$\overset{*}{F}_{\mathrm{AA}'\mathrm{BB}'} = \mathrm{i}\left(-\epsilon_{\mathrm{A}'\mathrm{B}'} F_{1\mathrm{AB}} + \epsilon_{\mathrm{AB}} F_{2\mathrm{A}'\mathrm{B}'}\right). \tag{5.4.17}$$

The bivector algebra just developed is frequently used to decompose an arbitrary array $\{T_{AB}\}_{A,B=1,2}$ into symmetric and antisymmetric parts, as follows. We write:

$$T_{AB} = T_{[AB]} + T_{(AB)}. \tag{5.4.18}$$

As a two-by-two antisymmetric matrix, $T_{[AB]}$ satisfies

$$T_{[AB]} = \alpha\epsilon_{AB}$$

whence contracting

$$T_A{}^A = T_{[AB]}\epsilon^{AB} = 2\alpha$$

so that (5.4.18) becomes:

$$T_{AB} = \frac{1}{2}T_C{}^C\epsilon_{AB} + T_{(AB)}. \tag{5.4.19}$$

In terms of tensors, $T_{(AB)}$ is equivalent to the self-dual bivector $S_{ij}{}^{AB}T_{(AB)}$, and so we see that the array $\{T_{AB}\}$ is equivalent to a scalar $T_C{}^C$ and a self-dual bivector.

5.5 Spinors

In describing space-time we have successively introduced a manifold structure, a connection and a metric (with the connection being determined by the metric if one requires the torsion to vanish). To define spinors, one needs one further element of structure, called a spin structure. The metric defines at every point p the set L_p of all tetrads at that point. A spin structure is a manifold $\hat{L}_p$ defined at each point p, together with a continuous map $\pi: \hat{L}_p \mapsto L_p$ such that

(i) π is a two-to-one covering map: two points of $\hat{L}_p$ correspond to each one of L_p, and π, restricted to a neighbourhood of one of the points, is a diffeomorphism.

(ii) $\hat{L}_p$ is simply connected*.

The construction of a spin structure can be thought of as the provision of a new two-valued physical aspect of a tetrad,

*Note that L_p is doubly connected like the Lorentz group.

sometimes called *spin-entanglement*. Specifying a frame together with its spin-entanglement gives a spin-entangled frame, or more briefly, a *spin-frame*; rotating the frame through 2π produces a frame with the opposite spin-entanglement, in a way we shall describe shortly.

First we shall show that each A in SL(2, $\mathbb{C}$) defines a transformation on $\hat{L}_p$. To see this, choose a path $\gamma\colon [0,1] \to \mathrm{SL}(2,\mathbb{C})$ with $\gamma(0) = I, \gamma(1) = A$. Take any $x \in \hat{L}_p$ and let $\lambda = \pi(x)$ be the corresponding ordinary tetrad. Define a path χ in L_p starting at λ by*:

$$\chi(t)_j = \boldsymbol{L}(\gamma(t))^i_j \lambda_i$$

or in matrix notation $\chi(t) = \lambda \boldsymbol{L}(\gamma(t))$. To each $\chi(t)$ there correspond two points in $\hat{L}_p$: we choose one, $\lambda(t)$, in such a way that $\lambda(0) = x$ and $\lambda(t)$ varies continuously with t (this can be done by property (i) of π). Then we define the result of acting with A on x to be $\lambda(1)$, written $\lambda(1) = xA$. From the fact that $\hat{L}_p$ is simply connected, it can be shown that xA does not depend on the choice of γ. Note that from (5.3.1):

$$\pi(xA) = \pi(x)\boldsymbol{L}(A).$$

A particular case occurs when $A = -I$ (I is the 2×2 unit matrix). Since $\boldsymbol{L}(-I) = 1_{\mathcal{L}}$ (the identity in the Lorentz group) (5.3.1) shows that:

$$\pi(x(-I)) = \pi(x).$$

So either $x(-I) = x$, or $x(-I)$ is the other spin-frame associated with the same ordinary frame as x. To see which is the case, we use the construction just described, with:

$$\gamma(t) = \begin{pmatrix} e^{i\pi t} & 0 \\ 0 & e^{-i\pi t} \end{pmatrix}.$$

Then $\boldsymbol{L}(\gamma(t))$ is a rotation through $2\pi t$. If we had $x(-I) = x$, then $\lambda(t)$ would return to its starting point and so, since $\hat{L}_p$ is simply connected, λ could be deformed to a constant curve. But such

*Hereafter in this section we shall omit the subscript ˆ over the tetrad indices to ease notation; all the indices in this chapter are therefore tetrad indices.

a transformation would allow us to deform $\boldsymbol{L}(\gamma(\cdot))$ to a constant curve, which is known to be impossible (cf. (5.2.18) and (4.3.13)). Thus $x(-I) \neq x$. We have deduced that rotation through 2π of the frame in L_p corresponds to a reversal of the spin entanglement.

A vector can be defined by giving its components in a frame: though we had chosen not to do so in Chapter 1, we could have defined a vector as something that associated a set of numbers $(X^i)_{i=0,\ldots,3}$ with each frame in such a way that when one changed frame, the numbers obeyed the correct transformation law. Another way of saying the same thing is to regard a vector at p as a collection of pairs of the form (X, λ) $(X \in \mathbb{R}^4, \lambda \in L_p)$ with two pairs (X, λ), (X', λ') being associated with the same vector if the frames are related by:

$$\lambda' = \lambda \overset{-1}{L} \qquad (\text{i.e. } \lambda'_i = (\overset{-1}{L})^j{}_i \lambda_j)$$

for a Lorentz transformation L, and the components are related by

$$X' = LX.$$

Such pairs are called equivalent, and a vector can be regarded as a set of equivalent pairs.

To define a spinor, we repeat this idea using spin-entangled frames instead of frames and $\mathrm{SL}(2, \mathbb{C})$ instead of the Lorentz group. This will specify a spinor by giving its components in each spin frame, the components being two complex numbers (i.e. an element of $\mathbb{C}^2$), corresponding to the action of $\mathrm{SL}(2, \mathbb{C})$ on such elements. So we define pairs: (η, f) (η', f') (where f, f' are spin-frames in $\hat{L}_p$ and η, η' are in $\mathbb{C}^2$) as *equivalent* , and write

$$(\eta, f) \approx (\eta', f'),$$

if there is an A in $\mathrm{SL}(2, \mathbb{C})$ such that

$$f' = f\overset{-1}{A}\ , \ \eta' = A\eta. \tag{5.5.1}$$

The set of all pairs (η', f') equivalent to (η, f) is written $[\eta, f]$. Any such set, with f a spin-frame at p and η in $\mathbb{C}^2$, is called a spinor at p. We write $S_p(M)$ for the set of all spinors at p, and $SM \equiv \cup_{p\in M} S_p(M)$ for the set of all spinors everywhere. We think of (f, η) and (f', η') as two different representations of a single object, a spinor; $\eta = (\eta^1, \eta^2)$ is composed of the components of this object in the spin-frame f, while η' is composed of the components in the spin-frame f'. A spinor is defined formally as an equivalence class under $\approx$. We see that a spinor is analogous to a vector, in that it has components for each frame. But unlike a vector, it does not have anything else! We cannot (on this approach) first give a geometrical picture of a spinor and then define its components; rather it is defined through its components.

Like $T_p(M)$, $S_p(M)$ is a vector space, with addition and multiplication by scalars being defined in the same way as for vectors. Similarly we can form the dual space $\overset{*}{S}_p(M)$ of cospinors and tensor spaces by taking tensor products.

We can now give an interpretation to the matrix ϵ_{AB} introduced earlier. Suppose that $[f, \eta]$, $[f, \chi]$ are two spinors at x. Then the number $\epsilon_{AB}\eta^A\chi^B$ does not depend on the choice of the spin-frame f; because changing to a different f', $f' = f\overset{-1}{A}$, changes η^A to $\eta'^A = A^A{}_B\eta^B$, and similarly for χ, so that

$$\epsilon_{AB}\eta'^A\chi'^B = \epsilon_{AB}A^A{}_C A^B{}_D \eta^C\chi^D = \epsilon_{CD}\eta^C\chi^D$$

from (5.2.1). So we can define an *inner product*, which we also denote by ϵ, between spinors according to:

$$\epsilon\left([f, \eta], [f, \chi]\right) \equiv \epsilon_{AB}\eta^A\chi^B.$$

This function works in exactly the same way as the metric for vectors. Given a spinor $\eta \in S_p(M)$ we define a cospinor $\boldsymbol{\eta}^\flat$ by

$$\boldsymbol{\eta}^\flat(\theta) = \epsilon(\eta, \theta) \qquad \forall \theta \in S_p(M).$$

The map $\eta \mapsto \boldsymbol{\eta}^\flat$ is an isomorphism, so we have an inverse $\boldsymbol{\theta} \mapsto \theta^\sharp : \overset{*}{S}_p(M) \to S_p(M)$. Hence we also have an inner product on $\overset{*}{S}_p(M)$, also called ϵ, defined by

$$\epsilon(\theta, \varphi) = \epsilon(\theta^\sharp, \varphi^\sharp).$$

The components ϵ^{AB} of the inner product constitute exactly the same matrix $\begin{bmatrix} 0 & 1 \\ -1 & 0 \end{bmatrix}$. This then, provides the geometrical basis for the conventions of lowering and raising the index with ϵ, introduced earlier; it is just the coordinate representation of this map between spinors and cospinors.

Spinors, being complex, admit of an operation having no analogue for vectors: complex conjugation. But the complex conjugate of the components of a spinor do not constitute another spinor of the type we have defined so far. For, suppose a spinor η has components η^{A} in a spin-frame f and has components η'^{A} in a spin-frame f', i.e.

$$\eta = [\eta^{A}, f] = [\eta'^{A}, f']$$

so that $f' = f\overset{-1}{A}$, $A\eta = \eta'$ for $A \in \mathrm{SL}(2, \mathbb{C})$. Then, taking the complex conjugate, we obtain:

$$\bar{A}\bar{\eta} = \bar{\eta}',$$

so $\bar{\eta}$ are a set of components which transform by the application of $\bar{A}$ instead of A. We call this a *dotted spinor* or *primed spinor* (so called because originally the indices were distinguished with diacritical dots, but now with primes). The definition mirrors that of ordinary spinors: on $\hat{L}M \times \mathbb{C}$ we define an equivalence relation $\approx$ by

$$(f, \eta) \approx (f', \eta')$$

if there is an $A \in \mathrm{SL}(2, \mathbb{C})$ such that $f' = f\bar{A}'$, $\bar{A}\eta = \eta'$. A dotted spinor is an equivalence class under $\approx$ and we denote the components relative to a spin frame f by $(\eta^{1'}, \eta^{2'})$, and the space of all such at p is $\bar{S}_p(M)$.

As we have just seen, there is a map of complex conjugation from spinors to dotted spinors:

$$S_p(M) \longrightarrow \bar{S}_p(M)$$

and taking duals

$$\overset{*}{\bar{S}}_p(M) \longrightarrow \overset{*}{S}_p(M).$$

Both maps are invertible, the inverse being formed by again complex-conjugating the components, and both maps and their inverses are all denoted by a bar, thus $\eta \to \bar{\eta}$.

Just as one can build up tensors of various valences, by taking tensor products of vectors and covectors, so one can build up spinor-tensors by taking tensor products of spinors of the above four kinds. The components of these objects are indexed by indices that can be upper, lower, primed or unprimed and which transform in the corresponding ways. For example, an element of

$$S_p(M) \otimes S_p(M) \otimes \overset{*}{\bar{S}}_p(M)$$

has components denoted by $\eta^{AB}{}_{C'}$. We shall refer to all kinds of spin-tensors simply as *spinors*, not reserving the word merely for elements of $S_p(M)$.

It is now straightforward to verify that the notation already adopted for the components of the maps S and its inverse is consistent with this: namely, that the numbers $\sigma_i{}^{AX'}$ introduced in (5.2.1) as components of the matrices $\sigma_0, \ldots, \sigma_3$ are in fact the components of a member of

$$\overset{*}{T}_p(M) \otimes S_p(M) \otimes \bar{S}_p(M)$$

as indicated by the indices. This is because if we apply to these numbers a transformation appropriate to change of spin frame by A, then the components are transformed to

$$L_{Ai}{}^j A^C{}_B \overline{A}^{X'}{}_{Y'} \sigma_j{}^{BY'}$$

which constitute the 2×2 matrices

$$\begin{aligned} L_{Ai}{}^j A \sigma_j A^\dagger = L_{Ai}{}^j A^\dagger(S(\lambda_j))A = L_{Ai}{}^j S(L_A \lambda_j) \\ = L_{Ai}{}^j \sigma_k L_{Aj}{}^k = \sigma_i. \end{aligned}$$

In other words, these components are unchanged by a change of frame, and so define a unique tensor, independent of what frame is used. Similarily, the quantities $S^i{}_{jA}{}^B$ and $\overline{S}^i{}_{jC'}{}^{E'}$, defined in terms of the σ's in the last section, are components of the tensors indicated by the indices. If we use ϵ to lower the index on a traceless matrix $F^A{}_B$, we can define (from (5.5.2))

$$F_{AB} = F^C{}_B \epsilon_{CA}.$$

So if $F^A{}_A = 0$, we have:

$$0 = F^A{}_A = \epsilon^{AB} F_{AB} = F_{21} - F_{12}.$$

Thus F_{AB} is symmetric : $F_{[AB]} = 0$, and conversely, if F_{AB} is symmetric, $F^{A}{}_{A} = 0$. This means that $S^{i}{}_{jA}{}^{B}$ (which maps the Lie Algebras as described in the last section) gives, on adjusting the indices, a quantity $S^{ij}{}_{AB}$ mapping skew-symmetric tensors (the Lie algebra of the Lorentz group with one index lowered) isomorphically to symmetric spinors, by

$$F_{ij} \longmapsto F_{ij} S^{ij}{}_{AB}.$$

We have said that a spinor is much less "concrete" than a vector, being specified only through its components. But we can to some extent give a geometric interpretation of a spinor with components η_A by considering the associated covector:

$$k_a = \sigma_a{}^{AX'} \eta_A \overline{\eta}_{X'}.$$

We have that, from (5.2.12):

$$k_a k^a = -\epsilon^{AB} \epsilon^{X'Y'} \eta_A \eta_B \bar{\eta}_{X'} \bar{\eta}_{Y'} = 0$$

so that k^a is null and, from (5.2.4), past directed. Clearly $\boldsymbol{k}$ does not contain all the informations in η, since if we alter η to $e^{i\omega}\eta$ (ω real) $\boldsymbol{k}$ is unaltered. To extract the remaining information, consider the bivector

$$T^{ab} = 2\left(S^{ab}{}_{AB}\eta^A\eta^B + \overline{S}^{ab}{}_{X'Y'}\bar{\eta}^{X'}\bar{\eta}^{Y'}\right).$$

From (5.3.16), (5.2.12)

$$T^{ab} k_b = 0. \tag{5.5.2}$$

Thus the rank of the matrix $[T^{ab}]$ is less than 4, and so, since the rank of an antisymmetric matrix must be even, is at most 2. Thus we can write:

$$T^{ab} = u^{[a} v^{b]}. \tag{5.5.3}$$

From (5.4.16) we also have:

$$\overset{*}{T}{}^{ab} k_b = 0$$

which implies that k^a is a linear combination of u^a and v^a. Thus we can write

$$T^{ab} = u^{[a} k^{b]}.$$

(5.5.5) then gives $u^a k_a = 0$. Thus u^a (which is undetermined up to the addition of a multiple of k^a) generates through linear combinations with k, a unique two-plane orthogonal to and containing k.

Thus a spinor determines (and, it can be shown, is determined by, up to a sign) a past-directed null vector and a two-plane orthogonal to and containing it. This configuration of a vector with an orthogonal two-plane containing it is called the Penrose-flag representation of the spinor (Penrose, 1960; Penrose and Rindler, 1984; Geroch, 1968).

5.6 The spinor connection

Given a spinor field η (i.e. an assignment of a spinor η_p to each point p) and a vector $\underset{p}{v}$ at p, we want to give a meaning to $\nabla_v \eta$, the rate of change of η in the direction of v. Recall that, in the case of a vector field, X, if the tetrad field λ_a is covariantly constant in the direction of v ($\nabla_v \lambda_a = o$), then $\nabla_v X = \nabla_v (X^a \lambda_a) = \nabla_v (X^a) \lambda_a$ so that $(\nabla_v X)^a = \nabla_v (X^a)$. We adopt this as the definition of the covariant derivative of a spinor field: if the spin-frame field f has a corresponding ordinary tetrad field λ_a such that

$$\nabla_v \lambda_a = o \tag{5.6.1}$$

then we require

$$(\nabla_v \eta)^{\mathrm{A}} \stackrel{*}{=} \nabla_v (\eta^{\mathrm{A}}). \tag{5.6.2}$$

(The $*$ over $=$ signifies that this is true only in the special spin-frame satisfying (5.6.1).)

We now need to write this in a general basis. Suppose a general basis field $\tilde{f}$ is related to f by $\tilde{f} = f \overset{-1}{A}$. Then the components of η are

$$\tilde{\eta}^{\mathrm{A}} = A^{\mathrm{A}}{}_{\mathrm{B}} \eta^{\mathrm{B}}$$

and so the components of the covariant derivative become

$$\begin{aligned} (\widetilde{\nabla_v \eta})^{\mathrm{A}} &= A^{\mathrm{A}}{}_{\mathrm{B}} (\nabla_v \eta)^{\mathrm{B}} = A^{\mathrm{A}}{}_{\mathrm{B}} \nabla_v (\eta^{\mathrm{B}}) \\ &= \nabla_v (A^{\mathrm{A}}{}_{\mathrm{B}} \eta^{\mathrm{B}}) - (\nabla_v A^{\mathrm{A}}{}_{\mathrm{B}}) \eta^{\mathrm{B}} \end{aligned} \tag{5.6.3}$$

i.e

$$(\tilde{\nabla}_v \eta)^{\text{A}} = \nabla_v (\tilde{\eta}^{\text{B}}) + v^m \Gamma^{\text{A}}{}_{\text{C}m} \tilde{\eta}^{\text{C}}$$

where

$$\Gamma^{\text{A}}{}_{\text{C}m} = -A^{\text{A}}{}_{\text{B},m} \overset{-1}{A}{}^{\text{B}}{}_{\text{C}}. \tag{5.6.4}$$

If we perform an analogous calculation for a vector field X with components in the corresponding ordinary frame related by $\tilde{X}^a = L_A{}^a{}_b X^b$, then we obtain

$$(\tilde{\nabla}_v X)^a = \nabla_v (\tilde{X}^a) + v^m \tilde{\Gamma}^c{}_{dm} \tilde{X}^c$$

where

$$\tilde{\Gamma}^c{}_{dm} = -L_A{}^c{}_{b,m} (\overset{-1}{L}{}_A)^b{}_d. \tag{5.6.5}$$

Since L_A and A are related as described in section (5.3), we can use these expressions to relate the two forms of Γ. Choose a fixed point p and write A_0 for the fixed matrix equal to the value $A(p)$ of A at p. Then

$$\begin{aligned}
\frac{\partial}{\partial x^m} \boldsymbol{L} \left(A \overset{-1}{A}{}_0 \right)^c{}_d \Big|_p &= \frac{\partial}{\partial x^m} \left(L_A{}^c{}_b \overset{-1}{L}{}_{A_0}{}^b{}_d \right) \Big|_p \\
&= -\tilde{\Gamma}^c{}_{dm} \qquad \text{(from (5.6.5))} \\
&= \boldsymbol{L}_*{}^c{}_{d\text{A}}{}^{\text{B}} \left(\frac{\partial}{\partial x^m} A^{\text{A}}{}_{\text{B}} \overset{-1}{A}_0{}^{\text{B}}{}_{\text{C}} \right) \\
&= \boldsymbol{L}_*{}^c{}_{d\text{A}}{}^{\text{B}} (-\Gamma^{\text{A}}{}_{\text{B}m}) \quad \text{(from (5.6.4))} .
\end{aligned}$$

Using (5.3.19) to invert this we obtain

$$\Gamma^{\text{A}}{}_{\text{C}m} = \Gamma^b{}_{am} S^a{}_{b\text{C}}{}^{\text{A}} \tag{5.6.6}$$

and we revert to our usual practice of denoting the general frame by λ_a, dropping the $\tilde{}$, from now on. Substituting in (5.6.3) then gives:

$$(\nabla_v \eta)^{\text{A}} = \nabla_v (\eta^{\text{A}}) + \Gamma^{\text{A}}{}_{\text{C}m} \eta^{\text{C}} v^m . \tag{5.6.7}$$

Complex conjugation gives, for a primed spinor α

$$(\nabla_v \alpha)^{\text{PA}'} = \nabla_v (\alpha^{\text{A}'}) + \bar{\Gamma}^{\text{A}'}{}_{\text{C}'m} \alpha^{\text{C}'} v^m \tag{5.6.8}$$

where

$$\bar{\Gamma}^{\text{A}'}{}_{\text{C}'m} = \Gamma^a{}_{bm} \overline{S}^b{}_{a\text{C}'}{}^{\text{A}'} \tag{5.6.9}$$

while the derivatives of cospinors can be obtained from the requirements that:

$$\nabla_v(\lambda_A \eta^A) = (\nabla_v \lambda)_A \eta^A + \lambda_A (\nabla_A \eta)^A$$

which results in

$$(\nabla_v \lambda)_A = \nabla_v(\lambda_A) - \Gamma^C{}_{Am} \lambda_C v^m \tag{5.6.10}$$

and for a primed co-spinor, similarly:

$$(\nabla_v \beta)_{A'} = (\nabla_v \beta_{A'}) - \bar{\Gamma}^{C'}{}_{A'm} \beta_{C'} v^m. \tag{5.6.11}$$

Finally, we note that we can use the fact that (5.3.15) is inverse to (5.3.19) to invert (5.6.6) as:

$$\Gamma^a{}_{bm} = 2\left(S^a{}_{bA}{}^B \Gamma^A{}_{Bm} + \overline{S}^a{}_{bA'}{}^{B'} \overline{\Gamma}^{A'}{}_{B'm}\right). \tag{5.6.12}$$

5.7 The spinor curvature

For this section only, indices m, n will denote coordinate components, while $a, b, c, \ldots$ (still without hats) will denote tetrad components.

If X and Y are vector fields and η a spinor field, we can compare $\nabla_X \nabla_Y \eta$ with $\nabla_Y \nabla_X \eta$ as we did for vectors. A calculation identical to (3.1.11) gives:

$$(\nabla_X \nabla_Y \eta - \nabla_Y \nabla_X \eta)^A - \left(\nabla_{[X,Y]}\eta\right)^A = \Re^A{}_{Bmn} X^n Y^m \eta^B \tag{5.7.1}$$

where

$$\Re^A{}_{Bnm} \equiv \Gamma^A{}_{Bm,n} - \Gamma^A{}_{Bn,m} + \Gamma^A{}_{Cn}\Gamma^C{}_{Bm} - \Gamma^A{}_{Cm}\Gamma^C{}_{Bn}$$

are the components of the *spinor curvature*. We see that this is the same expression as for the Riemann curvature, but with the first two (tetrad) indices replaced by spinor indices. We can make the similarity more explicit by defining matrices:

$$\overset{s}{\Gamma}_m = [\Gamma^A{}_{Bm}]_{A,B=1,2} \in s\ell(2, \mathbb{C})$$
$$\Re_{nm} = [\Re^A{}_{Bnm}]_{A,B=1,2} \in s\ell(2, \mathbb{C})$$

and Γ_m and R_{nm} for the corresponding matrices for the Riemannian connection taking values in the Lorentz Lie algebra $\underline{\ell}$. Thus

$$R_{nm} = \Gamma_{m,n} - \Gamma_{n,m} + [\Gamma_n, \Gamma_m]$$

where $[A, B]$ is the matrix commutator $AB - BA$.

We have seen in (5.3.11) that this commutator is the product in the Lie algebra, and so is preserved by $\overset{-1}{\boldsymbol{L}}_*$ mapping the Lie algebra $\underline{\ell}$ to $s\underline{\ell}(2, \mathbb{C})$. Hence using (5.3.19):

$$\overset{-1}{\boldsymbol{L}}_* R_{nm} = \overset{s}{\Gamma}_{m,n} - \overset{s}{\Gamma}_{n,m} + [\overset{s}{\Gamma}_n, \overset{s}{\Gamma}_m] = \mathfrak{R}_{nm}$$

or, in components

$$\mathfrak{R}^{\mathrm{A}}{}_{\mathrm{B}mn} = R^a{}_{bmn} S^b{}_{a\mathrm{B}}{}^{\mathrm{A}}. \tag{5.7.2}$$

Converting to tetrad indices we obtain:

$$\mathfrak{R}^{\mathrm{A}}{}_{\mathrm{B}cd} \equiv \lambda_c{}^m \lambda_d{}^m \mathfrak{R}^{\mathrm{A}}{}_{\mathrm{B}nm} = R^a{}_{bcd} S^b{}_{a\mathrm{B}}{}^{\mathrm{A}}. \tag{5.7.3}$$

The pair (a, b) is antisymmetric for R, and so for fixed A, B the elements $[\mathfrak{R}^{\mathrm{A}}{}_{\mathrm{B}}{}^a{}_b]_{a,b=0,\ldots,3}$ constitute an element of $\underline{\ell}^{\mathbb{C}}$ (tensors with complex components that are antisymmetric with one index lowered).

We saw in (5.4.1) and (5.4.2) that such an element is completely specified by the two quantities:

$$\boldsymbol{r_1}{}^{\mathrm{A}}{}_{\mathrm{BCD}} = -\mathfrak{R}^{\mathrm{A}}{}_{\mathrm{B}}{}^{ij} S_{ij\mathrm{CD}} = R^{klij} S_{kl}{}^{\mathrm{A}}{}_{\mathrm{B}} S_{ij\mathrm{CD}} \tag{5.7.4a}$$

and

$$\boldsymbol{r_2}{}^{\mathrm{A}}{}_{\mathrm{BC'D'}} = -\mathfrak{R}^{\mathrm{A}}{}_{\mathrm{B}}{}^{ij} \bar{S}_{ij\mathrm{C'D'}} = R^{klij} S_{kl}{}^{\mathrm{A}}{}_{\mathrm{B}} \bar{S}_{ij\mathrm{C'D'}} \tag{5.7.4b}$$

through:

$$\mathfrak{R}^{\mathrm{A}}{}_{\mathrm{B}ij} = 2\left(S_{ij\mathrm{P}}{}^{\mathrm{Q}} \boldsymbol{r_1}{}^{\mathrm{A}}{}_{\mathrm{B}}{}^{\mathrm{P}}{}_{\mathrm{Q}} + \overline{S}_{ij\mathrm{P'}}{}^{\mathrm{Q'}} \boldsymbol{r_2}{}^{\mathrm{A}}{}_{\mathrm{B}}{}^{\mathrm{P'}}{}_{\mathrm{Q'}}\right). \tag{5.7.5}$$

Define

$$R_{\mathrm{AA'BB'CC'DD'}} = \sigma^i{}_{\mathrm{AA'}} \sigma^j{}_{\mathrm{BB'}} \sigma^k{}_{\mathrm{CC'}} \sigma^l{}_{\mathrm{DD'}} R_{ijkl}. \tag{5.7.6}$$

From (5.4.15) applied to the first two indices,

$$\sigma^i{}_{\mathrm{AA'}} \sigma^j{}_{\mathrm{BB'}} R_{ijkl} = \epsilon_{\mathrm{A'B'}} \mathfrak{R}_{\mathrm{AB}kl} + \epsilon_{\mathrm{AB}} \bar{\mathfrak{R}}_{\mathrm{A'B'}kl}$$

so substituting this with (5.7.5) in (5.7.6) and using (5.4.14), gives

$$R_{AA'BB'CC'DD'} = \epsilon_{A'B'}\epsilon_{C'D'}\boldsymbol{r}_{1ABCD} + \epsilon_{A'B'}\epsilon_{CD}\boldsymbol{r}_{2ABC'D'} + \epsilon_{AB}\epsilon_{CD}\overline{\boldsymbol{r}}_{1A'B'C'D'} + \epsilon_{AB}\epsilon_{C'D'}\overline{\boldsymbol{r}}_{2A'B'CD}. \tag{5.7.7}$$

The spinor equivalent of the Ricci tensor is now given by contracting over AA′, CC′; from (5.2.10) and (5.7.6):

$$\begin{aligned} R^{AA'}{}_{BB'AA'DD'} &= \sigma^{i}{}_{BB'}\sigma^{l}{}_{DD'}R_{jl} \\ &= -\epsilon_{B'D'}\boldsymbol{r}_{1}{}^{A}{}_{BAD} - \boldsymbol{r}_{2DBB'D'} - \epsilon_{BD}\overline{\boldsymbol{r}}_{1}{}^{A'}{}_{B'A'D'} - \overline{\boldsymbol{r}}_{2D'B'BD} \end{aligned} \tag{5.7.8}$$

from (5.7.7), while a further contraction gives the Ricci scalar as:

$$R = 2\left(\boldsymbol{r}_{1}{}^{AD}{}_{AD} + \overline{\boldsymbol{r}}_{1}{}^{A'D'}{}_{A'D'}\right). \tag{5.7.9}$$

Equation (5.7.8) can be simplified on noting that, from (5.7.4*b*)

$$\overline{\boldsymbol{r}}_{2D'B'BD} = \boldsymbol{r}_{2DBB'D'}. \tag{5.7.10}$$

We can now show from the cyclic identity for R that a similar reality condition holds in (5.7.9). We have:

$$0 = \delta^{ijkl}R^{n}{}_{jkl} = 2\overset{*}{R^{n}{}_{j}{}^{ij}} = 2\overset{*}{R^{ijn}}{}_{j}. \tag{5.7.11}$$

Repeating the analysis leading to (5.7.8) with R^{nkij} and using (5.4.17) gives

$$\sigma_{iAA'}\sigma_{jBB'}\sigma_{kCC'}\sigma_{lDD'}\overset{*}{R^{ijkl}} = \mathrm{i}(\epsilon_{A'B'}\epsilon_{C'D'}\boldsymbol{r}_{1ABCD} - \epsilon_{AB}\epsilon_{CD}\overline{\boldsymbol{r}}_{1A'B'C'D'} + \epsilon_{A'B'}\epsilon_{CD}\boldsymbol{r}_{2ABC'D'} - \epsilon_{AB}\epsilon_{C'D'}\overline{\boldsymbol{r}}_{2A'B'CD}). \tag{5.7.12}$$

Contracting on the B, D indices, and then on the A, C indices, and using (5.7.10), gives

$$\boldsymbol{r}_{1}{}^{AD}{}_{AD} = \overline{\boldsymbol{r}}_{1}{}^{A'D'}{}_{A'D'} \tag{5.7.13}$$

from (5.7.9).

Now $\boldsymbol{r}_{1}{}^{A}{}_{BAD} = \epsilon^{AC}\boldsymbol{r}_{1CBAD}$ is antisymmetric in B, D and so from (5.4.19):

$$\boldsymbol{r}_{1}{}^{A}{}_{BAD} = \frac{1}{2}\epsilon_{BD}\boldsymbol{r}_{1}{}^{A}{}_{CA}{}^{C} = -\frac{R}{8}\epsilon_{BD} \tag{5.7.14}$$

from (5.7.13). Inserting this in (5.7.8), with (5.7.10), gives

$$\begin{aligned} \boldsymbol{r}_{2DBB'D'} &= \frac{1}{2}\left[R^{AA'}{}_{BB'AA'DD'} - \frac{R}{4}\epsilon_{BD}\epsilon_{B'D'}\right] \\ &= -\frac{1}{2}\left[R_{ij} - \frac{R}{4}g_{ij}\right]\sigma^{i}{}_{BB'}\sigma^{j}{}_{DD'} \end{aligned} \tag{5.7.15}$$

i.e. $\boldsymbol{r}_1$ corresponds to the trace-free Ricci tensor.

In equation (5.4.19) we decomposed a two-index spinor into its symmetric and antisymmetric parts; now we generalise this to handle the 4-index spinor $\boldsymbol{r}_{\mathbf{1},\mathbf{2}\,ABCD}$. The difference between this and its totally symmetric part is, following (5.4.19):

$$\boldsymbol{r}_{\mathbf{1}\,ABCD} - \boldsymbol{r}_{\mathbf{1}\,(ABCD)} = \frac{1}{3}\left[(\boldsymbol{r}_{\mathbf{1}\,ABCD} - \boldsymbol{r}_{\mathbf{1}\,ACDB}) + (\boldsymbol{r}_{\mathbf{1}\,ABCD} - \boldsymbol{r}_{\mathbf{1}\,ADBC})\right]$$
$$= \frac{1}{6}\left[\boldsymbol{r}_{\mathbf{1}\,AE}{}^{E}{}_{D}\epsilon_{BC} + \boldsymbol{r}_{\mathbf{1}\,AE}{}^{E}{}_{C}\epsilon_{BD}\right] = \frac{R}{48}\left[\epsilon_{AD}\epsilon_{BC} + \epsilon_{AC}\epsilon_{BD}\right].$$

Putting $\Psi_{ABCD} = \boldsymbol{r}_{\mathbf{1}\,(ABCD)}$ thus gives:

$$\boldsymbol{r}_{\mathbf{1}\,ABCD} = \Psi_{ABCD} + \frac{R}{24}\epsilon_{A(D}\epsilon_{|B|C)}. \qquad (5.7.16)$$

Since $\boldsymbol{r}_2$ and R contain all the information in the Ricci tensor, we might expect that Ψ corresponded to the rest of the Riemann tensor, i.e. to the Weyl tensor. To verify that this is so, we can repeat the analysis, replacing R_{ijkl} by K_{ijkl}, where

$$K^{ijkl} = \frac{1}{2}\left[R^{ijkl} - \overset{*}{R}{}^{ij\overset{*}{kl}}\right]$$

(cf. (3.4.15)). This can be decomposed into self-dual and anti-self-dual parts as

$$K^{ijkl} = \left[R^{\underset{\mathbf{2}}{ij}\underset{\mathbf{2}}{kl}} + R^{\underset{\mathbf{1}}{ij}\underset{\mathbf{1}}{kl}}\right]$$

where a **1** (resp.**2**) under a pair of indices denotes taking the self-dual (resp. anti-self-dual) part. Thus $K^{ijkl}S_{ij}{}^{AB}S_{ij}{}^{CD} = \boldsymbol{r}_{\mathbf{1}}{}^{ABCD}$. Passing to Ψ we have, from (5.7.16), (5.3.17) and (5.4.15):

$$\Psi_{ABCD} = K^{ijkl}S_{ijAB}S_{klCD} + \frac{R}{12}S^{i}{}_{jAB}S^{j}{}_{iCD} = C^{ijkl}S_{ijAB}S_{klCD}. \qquad (5.7.17)$$

Thus Ψ depends only on the Weyl tensor and (since R_{ijkl} can be reconstructed from R, Ψ and $\boldsymbol{r}_2$) it contains all the information in the Weyl tensor. Ψ is called the Weyl spinor. Proceeding as in equations (5.7.6), (5.7.7) we obtain

$$C_{AA'BB'CC'DD'} \equiv \sigma^{i}{}_{AA'}\sigma^{j}{}_{BB'}\sigma^{k}{}_{CC'}\sigma^{l}{}_{DD'}C_{ijkl}$$
$$= \epsilon_{A'B'}\epsilon_{C'D'}\Psi_{ABCD} + \epsilon_{AB}\epsilon_{CD}\bar{\Psi}_{A'B'C'D'}$$

which inverts to

$$C_{ijkl} = \sigma_i{}^{AA'}\sigma_j{}^{BB'}\sigma_k{}^{CC'}\sigma_l{}^{DD'}\left(\epsilon_{A'B'}\epsilon_{C'D'}\Psi_{ABCD} + \epsilon_{AB}\epsilon_{CD}\bar{\Psi}_{A'B'C'D'}\right). \tag{5.7.18}$$

The Bianchi identities take the form:

$$\nabla^i R^*_{ijkl} = 0$$

which from (5.7.12) gives (with $\nabla^{AA'} = \sigma^{iAA'}\nabla_i$):

$$\left(\nabla^{A}{}_{B'}\boldsymbol{r}_{\mathbf{1}ABCD} - \nabla_{B}{}^{A'}\bar{\boldsymbol{r}}_{\mathbf{2}A'B'C'D'}\right)\epsilon_{C'D'} + \left(-\nabla_{B}{}^{A'}\bar{\boldsymbol{r}}_{\mathbf{1}A'B'C'D'} + \nabla^{A}{}_{B'}\boldsymbol{r}_{\mathbf{2}ABC'D'}\right)\epsilon_{CD} = 0.$$

Contracting on C′, D′, this gives:

$$\nabla^{A}{}_{B'}\boldsymbol{r}_{\mathbf{1}ABCD} = \nabla_{B}{}^{A'}\bar{\boldsymbol{r}}_{\mathbf{2}A'B'CD}. \tag{5.7.19}$$

For a vacuum space-time (when $R = 0$ and $\boldsymbol{r}_{\mathbf{2}} = 0$) this gives

$$\nabla^{A}{}_{B'}\Psi_{ABCD} = 0. \tag{5.7.20}$$

5.8 The torsion case

If torsion is included, section 5.6 and equations (5.7.1), (5.7.10) are unchanged. But now

$$R^i{}_{[jkl]} = 2\left[T^i{}_{[jl;k]} + 2T^m{}_{[jk}T^i{}_{|m|l]}\right]. \tag{5.8.1}$$

Also

$$\begin{aligned} R_{ijkl} - R_{klij} &= \frac{3}{2}\left[R_{i[jkl]} + R_{l[jkl]} + R_{j[ilk]} + R_{k[jil]}\right] \\ &= K_{ilj;k} + K_{kil;j} + K_{jki;l} + K_{ljk;i} + 8T^m{}_{[k|[i}S_{j]m|l]} \\ &\quad + 2\left[T^m{}_{kl}A_{imj} + T^m{}_{ij}A_{lmk}\right] \\ &\equiv V_{ijkl} \end{aligned} \tag{5.8.2}$$

say, where:

$$\begin{aligned} K_{ilj} &= -T_{ilj} + T_{lji} - T_{jil} \\ A_{ilj} &= T_{ilj} - T_{jli} \\ S_{ilj} &= T_{ilj} + T_{jli}\ . \end{aligned}$$

So the symmetric part of R is given by:

$$\frac{1}{2}\left(R_{ijkl}+R_{klij}\right)=R_{ijkl}-V_{ijkl}\equiv M_{ijkl} \tag{5.8.3}$$

say.

We can now define spinor parts of M, analogous to R in the last section:

$$\begin{aligned}\boldsymbol{m}_{\mathbf{1}\mathrm{ABCD}}&=M_{ijkl}S^{ij}{}_{\mathrm{AB}}S^{kl}{}_{\mathrm{CD}}=\boldsymbol{r}_{\mathbf{1}\mathrm{ABCD}}-V_{ijkl}S^{ij}{}_{\mathrm{AB}}S^{kl}{}_{\mathrm{CD}}\\ \boldsymbol{m}_{\mathbf{2}\mathrm{ABX'Y'}}&=M_{ijkl}S^{ij}{}_{\mathrm{AB}}\overline{S}^{kl}{}_{\mathrm{X'Y'}}=\boldsymbol{r}_{\mathbf{2}\mathrm{ABX'Y'}}-V_{ijkl}S^{ij}{}_{\mathrm{AB}}\overline{S}^{kl}{}_{\mathrm{X'Y'}}.\end{aligned}$$

The contractions of M are:

$$M^{i}{}_{jil}=R_{jl}-K^{i}{}_{lj;i}-K_{lj}{}^{i}{}_{;i}+2T_{[l;j]}+2T^{mi}{}_{[j}T_{|im|l]}+T^{m}{}_{lj}T_{m}\equiv M_{jl} \tag{5.8.4}$$

where $T_m=T^{i}{}_{im}$, and

$$M^{ij}{}_{ij}=R. \tag{5.8.5}$$

On the other hand M does not satisfy the cyclic identity; we have from (5.8.1) and (5.8.2):

$$\begin{aligned}M_{i[jkl]}=\frac{1}{2}T_{i[jl;k]}+T^{m}{}_{[jk}T_{|im|l]}-\frac{1}{2}T_{[jkl];i}+\frac{1}{2}A_{[j|i|l;k]}\\ +T^{m}{}_{[jk}T_{l]mi}+T^{m}{}_{i[k}A_{j|m|l]}\equiv U_{ijkl}\end{aligned} \tag{5.8.6}$$

say. Thus repeating the analysis of the last section, we have

$$R=2\left(\boldsymbol{m}_{\mathbf{1}}{}^{\mathrm{AD}}{}_{\mathrm{AD}}+\overline{\boldsymbol{m}}_{\mathbf{1}}{}^{\mathrm{A'D'}}{}_{\mathrm{A'D'}}\right) \tag{5.8.7}$$

(cf. (5.7.9));

$$\overline{\boldsymbol{m}}_{\mathbf{2}\mathrm{D'B'DD}}=\boldsymbol{m}_{\mathbf{2}\mathrm{DBB'D'}} \tag{5.8.8}$$

(cf. (5.7.10)), and

$$\begin{aligned}\sigma_{i\mathrm{AA'}}\sigma_{j\mathrm{BB'}}\sigma_{k\mathrm{CC'}}\sigma_{l\mathrm{DD'}}M^{ij\overset{*}{kl}}=-\mathrm{i}(\epsilon_{\mathrm{A'B'}}\epsilon_{\mathrm{C'D'}}\boldsymbol{m}_{\mathbf{1}\mathrm{ABCD}}-\epsilon_{\mathrm{AB}}\epsilon_{\mathrm{CD}}\overline{\boldsymbol{m}}_{\mathbf{1}\mathrm{A'B'C'D'}}\\ -\epsilon_{\mathrm{A'B'}}\epsilon_{\mathrm{CD}}\boldsymbol{m}_{\mathbf{2}\mathrm{ABC'D'}}+\epsilon_{\mathrm{AB}}\epsilon_{\mathrm{C'D'}}\overline{\boldsymbol{m}}_{\mathbf{2}\mathrm{A'B'CD}}).\end{aligned} \tag{5.8.9}$$

(cf. (5.7.12), the sign altering because we are dualizing on the second pair of indices). Contracting and using (5.7.8) we obtain:

$$M_{ij}{}^{\overset{*}{ij}}=\frac{1}{2}\delta_{ijkl}M^{ijkl}=\frac{1}{2}\delta_{ijkl}U^{ijkl}=2\mathrm{i}\left(\boldsymbol{m}_{\mathbf{1}\mathrm{AB}}{}^{\mathrm{AB}}-\overline{\boldsymbol{m}}_{\mathbf{1}\mathrm{A'B'}}{}^{\mathrm{A'B'}}\right). \tag{5.8.10}$$

Giving, from (5.8.7),

$$\boldsymbol{m}_{\mathbf{1}\,\mathrm{AB}}{}^{\mathrm{AB}} = \frac{R}{4} + \frac{1}{8\mathrm{i}}\delta_{ijkl}U^{ijkl} = \frac{M}{4}$$

where $M = R - \frac{1}{2}\mathrm{i}\delta_{ijkl}U^{ijkl}$. We deduce:

$$\boldsymbol{m}_{\mathbf{1}}{}^{\mathrm{A}}{}_{\mathrm{BAD}} = -\frac{M}{8}\epsilon_{\mathrm{BD}} \tag{5.8.11}$$

(cf. (5.7.14)). So repeating the analysis of (5.7.8)

$$\boldsymbol{m}_{\mathbf{2}\,\mathrm{DBB'D'}} = -\frac{1}{2}\left(M_{ij} - \frac{R}{4}g_{ij}\right)\sigma^{i}{}_{\mathrm{BB'}}\sigma^{j}{}_{\mathrm{DD'}} \tag{5.8.12}$$

(the trace-free part of M_{ij}), and

$$\boldsymbol{m}_{\mathbf{1}\,\mathrm{ABCD}} = \chi_{\mathrm{ABCD}} + \frac{M}{24}\epsilon_{\mathrm{A(D}}\epsilon_{|\mathrm{B}|\mathrm{C)}} \tag{5.8.13}$$

where $\chi_{\mathrm{ABCD}} = \boldsymbol{m}_{\mathbf{1}\,(\mathrm{ABCD})}$.

Thus the totally symmetric spinor definable in the presence of torsion corresponds not to the Weyl tensor, but to the analogous complex-valued tensor constructed using M_{ijkl}.

5.9 Conformal spinors

We have noted that $C^{i}{}_{jkl}$ is unchanged if g_{ij} is replaced by

$$\bar{g}_{ij} = \Omega g_{ij};$$

(cf. (3.4.24)). Such a replacement implies that a frame $\{\lambda_i\}_{i=0,\ldots,3}$ should be replaced by $\{\bar{\lambda}_i\}_{i=0,\ldots,3}$ where

$$\bar{\lambda}_i = \Omega^{-\frac{1}{2}}\lambda_i.$$

The problem is that our definition of spinors (cf. (5.5.1)) requires us to know how spinors transform under this conformal rescaling of the tetrad, since hitherto we have only considered Lorentz transformations of the tetrad. The choice of the transformation involves a certain amount of freedom, and we obtain different sorts of spinors (referred to as spinors of different conformal weight) depending on how the choice is made. Let α be a real number. Then we modify (5.5.1) by enlarging the set of tetrads $\hat{L}_p$ to include all such rescaled tetrads. And we require that pairs (η, f), $(\bar{\eta}, \bar{f})$

should be equivalent with conformal weight α if there is an A in $\mathrm{SL}(2,\mathbb{C})$ and a conformal factor Ω such that:

$$\bar{f} = \Omega^{-\frac{1}{2}} f \overset{-1}{A} \quad , \quad \bar{\eta} = \Omega^{\frac{\alpha}{2}} A\eta. \tag{5.9.1}$$

The set of all pairs (η', f') equivalent with weight α to (η, f) is written $[\eta, f]_\alpha$ and defines a spinor with conformal weight α. The alternating spinor ϵ_{AB} is defined to have conformal weight zero, so that it has the same numerical value in all frames; while for $\sigma_i{}^{\mathrm{AX}'}$ to keep its numerical value, we require it to have conformal weight $\alpha = \frac{1}{2}$ on each of its spinor indices, so that the conformal weight factor of $\Omega^{\frac{1}{2}}$ under conformal rescaling for the spinor indices will cancel the $\Omega^{-\frac{1}{2}}$ factor introduced by the tetrad index. If we also require $\sigma^i{}_{\mathrm{AX}'}$ to be numerically invariant (as we shall do), then this implies that the "i" is not a vector index; if it were it would acquire a factor of $\Omega^{\frac{1}{2}}$ under the change of tetrads, while "lowering the index" with ϵ_{AB} shows that the spinor indices also have a conformal weight of $\frac{1}{2}$, so that the conformal weights would not cancel. Thus, the "i" index has the character of a conformally weighted vector. (This shows that we could equally well have chosen a different system, under which this "i" is a vector index and the spinor indices have weight $-\frac{1}{2}$ each to cancel it.) Thus if we use $\sigma_i{}^{\mathrm{AX}'}$ to produce the spinor-equivalent of a vector, defining for example

$$F_{\mathrm{AA'BB'}} = \sigma_{i\mathrm{AA'}}\sigma_{j\mathrm{BB'}}F^{ij}$$

then we have a conformal weight of $\frac{1}{2}$ per index. If F is antisymmetric and we write

$$F_{\mathrm{AA'BB'}} = \varphi_{\mathrm{AB}}\epsilon_{\mathrm{A'B'}} + \bar{\varphi}_{\mathrm{A'B'}}\epsilon_{\mathrm{AB}}$$

then φ_{AB} will need a conformal weight of $+1$ per index to balance the conformal weight of $F_{\mathrm{AA'BB'}}$.

Particular care has to be exercised in the case of geometrical objects that depend on the metric, such as the Riemann tensor. If we change the metric by a conformal transformation, then the tensor changes intrinsically as a geometrical object; but its tetrad components change as well by virtue of the fact that the tetrad is changing. Applying this to the Weyl spinor, we note first that the Weyl tensor changes according to

$$C^{ijkl}[\bar{g}] = \bar{g}^{jp}\bar{g}^{kq}\bar{g}^{lr}C^{i}{}_{pqr}[\bar{g}] = \bar{g}^{jp}\bar{g}^{kq}\bar{g}^{lr}C^{i}{}_{pqr}[g] = \Omega^{-3}C^{ijkl}[g]$$

from (3.4.24). On the other hand, the spinor components of a spinor Ψ_{ABCD} derived from a *fixed* tensor would change, as we have noted, according to a conformal weight of +1 per index:

$$\bar{\Psi}_{\text{ABCD}} = \Omega^2\Psi_{\text{ABCD}}$$

(cf. (5.9.1)). Combining the change in C^{ijkl} with the change in the components of its spinor representative, we obtain:

$$\bar{\Psi}_{\text{ABCD}}[\bar{g}] = \Omega^{-1}\Psi_{\text{ABCD}}[g]. \tag{5.9.2}$$

An equation frequently encountered in spinor calculus is:

$$\nabla^{\text{AX}'}\chi_{\text{AB}\ldots} = 0 \tag{5.9.3}$$

where

$$\nabla^{\text{AX}'} \equiv \sigma^{i\text{AX}'}\nabla_i \tag{5.9.4}$$

and $\chi_{\text{AB}\ldots}$ is a totally symmetric spinor with an arbitrary number m of indices. (5.9.3) is called *the zero mass spin $m/2$ wave equation.* Suppose that χ has conformal weight α, and consider a conformally rescaled metric in which (5.9.3) becomes:

$$\bar{\nabla}^{\text{AX}'}\bar{\chi}_{\text{AB}\ldots} = 0 \tag{5.9.5}$$

where $\bar{\nabla}$ denotes covariant differentiation using the Levi-Civita connection of $\bar{g}$ and expressed in $\bar{g}$-tetrads. From (3.4.19) and (5.6.6) we have:

$$\bar{\Gamma}^{\text{B}}_{\text{C}i} = \Gamma^{\text{B}}_{\text{C}i} + S^{m}{}_{i\text{C}}{}^{\text{B}}\varphi_m$$

where $\varphi_m = \partial_m(\ln\Omega)$, and so from (5.3.16) and (5.2.12):

$$\sigma^{i\text{AX}'}\bar{\Gamma}^{\text{B}}{}_{\text{C}i} = \sigma^{i\text{AX}'}\Gamma^{\text{B}}{}_{\text{C}i} + \frac{1}{4}\varphi_m\left[\sigma^{m}{}_{\text{C}}{}^{\text{X}'}\epsilon^{\text{AB}} - \sigma^{m\text{BX}'}\delta^{\text{A}}{}_{\text{C}}\right]. \tag{5.9.6}$$

Hence

$$\begin{aligned}\bar{\nabla}^{\text{AX}'}\bar{\chi}_{\text{AB}\ldots} &= \sigma^{i\text{AX}'}\left(\partial_i\bar{\chi}_{\text{AB}\ldots} - \bar{\Gamma}^{\text{C}}{}_{\text{A}i}\bar{\chi}_{\text{CB}\ldots} - \cdots\right)\\ &= \Omega^{\frac{\alpha}{2}}\left\{\nabla^{\text{AX}'}\chi_{\text{AB}\ldots} + \frac{\alpha}{2}\varphi_i\sigma^{i\text{AX}'}\chi_{\text{AB}\ldots} + \frac{m+2}{4}\varphi_i\sigma^{i\text{AX}'}\chi_{\text{AB}\ldots}\right\}\end{aligned}$$

from (5.9.6), using the symmetry of χ. Thus if

$$\alpha = -\left(1 + \frac{m}{2}\right) \tag{5.9.7}$$

we have that

$$\bar{\nabla}^{AX'}\bar{\chi}_{AB\ldots} = \Omega^{\frac{\alpha}{2}}\nabla^{AX'}\chi_{AB\ldots}. \tag{5.9.8}$$

Under these circumstances we see that equations (5.9.3) and (5.9.5) are equivalent and the wave equation is said to be conformally invariant.

We have already seen in (5.7.20) that the Weyl spinor satisfies the wave equation in a vacuum space-time. On conformal rescaling to a new space-time, the Ricci tensor will become non-zero and so the Weyl tensor will no longer satisfy this equation. However, if we define a new spinor χ_{ABCD} of conformal weight -3 (cf. (5.9.7) with $m = 4$), and such that

$$\chi_{ABCD} \stackrel{*}{=} \Psi_{ABCD} \tag{5.9.9}$$

for frames orthonormal with respect to the vacuum metric, the χ_{ABCD} will satisfy the wave equation in all conformally related metrics. The components of χ in a conformally related frame given by $\bar{g}_{ij} = \Omega g_{ij}$ are

$$\bar{\chi}_{ABCD} = \Psi_{ABCD}[g]\Omega^{-\frac{3}{2}} = \Omega^{-\frac{1}{2}}\bar{\Psi}_{ABCD}[\bar{g}]$$

from (5.9.2). The spinor χ plays a vital role in the conformal treatment of infinity.

5.10 The Weyl spinor and the Petrov classification

We now turn to our main goal: the analysis of the symmetric spinor Ψ_{ABCD} derived from the Weyl tensor. We first show that any symmetric spinor χ with k indices can be written as:

$$\chi_{A_1\ldots A_k} = \overset{(1)}{\alpha}_{(A_1}\overset{(2)}{\alpha}_{A_2}\cdots\overset{(k)}{\alpha}_{A_k)} \tag{5.10.1}$$

for one-index spinors $\overset{(1)}{\alpha}\cdots\overset{(k)}{\alpha}$. To see this, we note that χ is determined by the numbers:

$$\chi_0 = \chi_{11\ldots1}, \quad \chi_1 = \chi_{21\ldots1}, \quad \ldots, \chi_k = \chi_{22\ldots2}\,.$$

Let ℓ be the largest integer for which $\chi_\ell \neq 0$. Define spinors $\overset{(i)}{\alpha}$ by:

$$\overset{(i)}{\alpha}{}_1 = 1, \overset{(i)}{\alpha}{}_2 = 0 \qquad (i = \ell+1, \ldots, k)$$
$$\overset{(j)}{\alpha}{}_1 = x_j, \overset{(j)}{\alpha}{}_2 = 1 \qquad (j = 1, \ldots, \ell)$$

where the x_j are to be determined so that

$$\chi_{A_1 \cdots A_k} = \beta \overset{(1)}{\alpha}{}_{(A_1} \cdots \overset{(\ell)}{\alpha}{}_{A_\ell} \overset{(\ell+1)}{\alpha}{}_{A_{\ell+1}} \cdots \overset{(k)}{\alpha}{}_{A_k)} \tag{5.10.2}$$

for some β. Note that this form automatically ensures that $\chi_r = 0$ for $r > \ell$, as required, while for $r \leq \ell$ we need

$$\chi_r = \beta \frac{a_r}{\binom{k}{r}} \tag{5.10.3}$$

where

$$a_r = \sum x_{i_1} x_{i_2} \cdots x_{i_{\ell-r}}$$

the sum being over all sets of indices $(i_1, \cdots, i_{\ell-r})$ with $1 \leq i_1 < i_2 < \cdots < i_{\ell-r} \leq \ell$. But a_r is the coefficient of $x^{\ell-r}$ in the polynomial

$$P(x) \equiv (x + x_1) \cdots (x + x_\ell)$$

and so (5.10.3) is equivalent to

$$P(x) = \frac{1}{\beta} \sum_{r=0}^{\ell} \chi_r \binom{k}{r} x^{\ell-r}. \tag{5.10.4}$$

Clearly, given any set of numbers (χ_r), if we set $\beta = \chi_0$, then the right-hand side of (5.10.4) is a polynomial with leading coefficient 1 (monic) which can be factorized to define $x_1 \cdots x_\ell$ as (minus) its roots. Thus (5.10.2) can be satisfied using these x's, and (5.10.1) follows on redefining $\overset{(1)}{\alpha}$ as $\beta \overset{(1)}{\alpha}$.

Applied to the Weyl spinor Ψ, this defines four spinors $\overset{(1)}{\eta} \cdots \overset{(4)}{\eta}$ such that:

$$\Psi_{ABCD} = \overset{(1)}{\eta}{}_{(A} \overset{(2)}{\eta}{}_{B} \overset{(3)}{\eta}{}_{C} \overset{(4)}{\eta}{}_{D)}. \tag{5.10.5}$$

They are called the principal spinors of the Weyl tensor. The definition is not unique, in that they can be multiplied by any four numbers whose product is 1; but (as can be seen from the foregoing proof) this is the only freedom available apart from relabeling. As we saw in Sect 5.5, a spinor defines a real null vector, and multiplying the spinor by a factor multiplies the null vector by the squared modulus of the factor. Thus the four principal spinors define four unique null directions, called the principal null directions of the Weyl tensor. Since $\eta_{\mathrm{A}}\eta^{\mathrm{A}} = 0$ for any spinor, if η is a principal spinor, it satisfies:

$$\Psi_{\mathrm{ABCD}}\eta^{\mathrm{A}}\eta^{\mathrm{B}}\eta^{\mathrm{C}}\eta^{D} = 0. \tag{5.10.6}$$

The converse is also true: we can perform a Lorentz transformation to ensure that $\ell = k$, after which if we set:

$$\eta^1 = -x \qquad \eta^2 = 1$$

then

$$\Psi_{\mathrm{ABCD}}\eta^{\mathrm{A}}\eta^{\mathrm{B}}\eta^{\mathrm{C}}\eta^{D} = \sum_{r=0}^{4} \Psi_r \binom{4}{r} (-x)^{4-r}$$

and so comparison with (5.10.4) shows that the roots of (5.10.6) give precisely the principal spinors. To translate (5.10.6) into vector terms, we substitute from (5.7.17) to write it as

$$T^{ij}T^{kl}C_{ijkl} = 0 \tag{5.10.7}$$

where $T^{ij} = S^{ij}{}_{\mathrm{AB}}\eta^{\mathrm{A}}\eta^{\mathrm{B}}$. Set

$$k^i = \sigma^i{}_{\mathrm{AX}'}\eta^{\mathrm{A}}\bar{\eta}^{\mathrm{X}'}$$

and let ℓ^i be any complex vector not proportional to k, satisfying

$$k^i\ell_i = 0 \qquad \ell_i\ell^i = 0. \tag{5.10.8}$$

Translating this into spinor terms, shows that

$$\ell_i = \sigma_i{}^{\mathrm{AX}'}\eta_{\mathrm{A}}\theta_{\mathrm{X}'}$$

for some spinor $\theta_{\mathrm{X}'}$ not proportional to $\bar{\eta}_{\mathrm{X}'}$. Then

$$\epsilon^{\mathrm{X}'\mathrm{Y}'} = 2\frac{\bar{\eta}^{[\mathrm{X}'}\theta^{\mathrm{Y}']}}{\bar{\eta}_{\mathrm{Z}'}\theta^{\mathrm{Z}'}}.$$

Substituting for $\epsilon^{\mathrm{X}'\mathrm{Y}'}$ in the equations

$$S^{ij}{}_{\mathrm{AB}} = -\frac{1}{2}\sigma^{[i}{}_{\mathrm{AX}'}\sigma^{j]}{}_{\mathrm{BY}'}\epsilon^{\mathrm{X}'\mathrm{Y}'}$$

(cf. (5.3.16)) gives:

$$T^{ij} \propto k^{[i}\ell^{j]} \, . \tag{5.10.9}$$

Thus (5.10.7) becomes

$$k^i k^k C_{ijkl} \ell^j \ell^l = 0$$

for any ℓ satisfying (5.10.8). But this implies, from the symmetry on j and l of $k^i k^k C_{ijkl}$, that

$$k^i k^k C_{ijkl} \ell^j m^l = 0$$

for m and ℓ orthogonal to k and hence that

$$k^i k^k k_{[p} C_{i]jk[l} k_{q]} = 0. \tag{5.10.10}$$

All the steps of the argument can be reversed, so that (5.10.10) is the vector equivalent of (5.10.6), and so a condition for k to be a principal null direction.

It can occur that some of the $\eta^{(i)}$ are proportional to each other (in which case they can be made equal by multiplying by a factor). In this situation the Weyl tensor is called *algebraically special*. The various possibilities, with their terminology, are as follows:

Type	Conditions
I	–
II	$\eta^{(1)} = \eta^{(2)}$
III	$\eta^{(1)} = \eta^{(2)} = \eta^{(3)}$
D	$\eta^{(1)} = \eta^{(2)}, \eta^{(3)} = \eta^{(4)}$
N	$\eta^{(1)} = \eta^{(2)} = \eta^{(3)} = \eta^{(4)}$
O	$\Psi = 0$

In each case it is understood that there exists a labeling and scaling under which the conditions hold, with the η's otherwise pair-wise linearly independent.

The spinor version of the classification, just given, is by far the easiest to understand. But historically it was first found in terms of the eigen-bivectors of the Weyl tensor; that is bivectors F^{ij} satisfying

$$C^{ij}{}_{kl} F^{kl} = \lambda F^{ij} \qquad F^{ij} = F^{[ij]}. \tag{5.10.11}$$

Now the map $F^{ij} \mapsto C^{ij}{}_{kl}F^{kl}$, involved in (5.10.11), takes self-dual (resp. anti-self-dual) bivectors into self-dual (resp. anti-self-dual) bivectors. For, if F is self-dual, then

$$^{*}(C^{ij}{}_{kl}F^{kl}) = -\overset{*}{C}{}^{ij}{}_{kl}\overset{**}{F}{}^{kl} = -\overset{*}{C}{}^{ij}{}_{kl}{}^{*}\overset{*}{F}{}^{kl} = -\mathrm{i}(C^{ij}{}_{kl}F^{kl}).$$

So we shall consider only eigenvectors in the three-dimensional complex space of self-dual bivectors. For these

$$F^{ij} = S^{ij}{}_{\mathrm{AB}}F^{\mathrm{AB}}$$

for a two-index spinor F^{AB}; substituting this into (5.10.11) and using (5.7.18) gives:

$$\Psi^{\mathrm{AB}}{}_{\mathrm{CD}}F^{\mathrm{CD}} = \frac{1}{2}\lambda F^{\mathrm{AB}}.$$

Thus classifying the self-dual eigen-bivectors of C corresponds to finding symmetric spinor eigenvectors for the map:

$$\alpha : [F^{\mathrm{AB}}] \longmapsto [\Psi^{\mathrm{AB}}{}_{\mathrm{CD}}F^{\mathrm{CD}}].$$

In the space of symmetric two-index spinors, we take as a basis the spinors:

$$\underset{1}{E}{}^{\mathrm{AB}} = \delta_1^{\mathrm{A}}\delta_1^{\mathrm{B}},\ \underset{2}{E}{}^{\mathrm{AB}} = \sqrt{2}\delta_1^{(A}\delta_2^{B)},\ \underset{3}{E}{}^{\mathrm{AB}} = \delta_2^{\mathrm{A}}\delta_2^{\mathrm{B}}.$$

Then $F^{\mathrm{AB}} = F^a \underset{a}{E}{}^{\mathrm{AB}}$ where the components F^a of F in this basis are:

$$F^1 = F^{11},\ F^2 = \sqrt{2}F^{12},\ F^3 = F^{33}$$

i.e. single indices are associated with pairs according to the scheme

a	AB
1	11
2	12 or 21
3	22

and there is a multiplication by $\sqrt{2}$ for an occurrence of 12 or 21. The matrix of the map α in this basis turns out to be derivable by the same rule, so that we have

$$\begin{aligned}
\alpha^1{}_1 &= \Psi^{11}{}_{11} = \Psi_{2211} \equiv \Psi_2 \\
\alpha^1{}_2 &= \sqrt{2}\Psi^{11}{}_{12} = \sqrt{2}\Psi_{1212} \equiv \Psi_3 \\
\alpha^1{}_3 &= \Psi^{11}{}_{22} = \Psi_{2222} \equiv \Psi_4 \\
\alpha^2{}_1 &= \sqrt{2}\Psi^{12}{}_{11} = -\sqrt{2}\Psi_{2111} \equiv -\sqrt{2}\Psi_1 \\
\alpha^2{}_2 &= 2\Psi^{12}{}_{12} = -2\Psi_{2112} \equiv -2\Psi_2 \\
\alpha^2{}_3 &= \sqrt{2}\Psi^{12}{}_{22} = -\sqrt{2}\Psi_{2122} \equiv -\sqrt{2}\Psi_3 \\
\alpha^3{}_1 &= \Psi^{22}{}_{11} = \Psi_{1111} \equiv \Psi_0 \\
\alpha^3{}_2 &= \sqrt{2}\Psi^{22}{}_{12} = \sqrt{2}\Psi_{1112} \equiv \sqrt{2}\Psi_1 \\
\alpha^3{}_3 &= \Psi^{22}{}_{22} = \Psi_{1122} \equiv \Psi_2 \,.
\end{aligned} \tag{5.10.12}$$

We now consider the eigenvectors of this matrix in the various cases. If there exists a pair of independent spinors (i.e. in all cases except N and O), then we can choose them as $\eta^{(1)}$ and $\eta^{(2)}$ and perform a Lorentz transformation to give:

$$\eta_1^{(1)} = 1; \quad \eta_2^{(2)} = 0; \quad \eta_1^{(2)} = 0; \quad \eta_2^{(2)} = 1.$$

(If we have case III, we choose $\eta^{(2)}$ to be the spinor with multiplicity 3.) Then, except for case D, we can scale the remaining spinors and multiply Ψ by an overall factor β , so that:

$$\Psi'_{\mathrm{ABCD}} = \beta\Psi_{\mathrm{ABCD}} = \eta^{(1)}_{(\mathrm{A}}\eta^{(2)}_{\mathrm{B}}\eta^{(3)}_{\mathrm{C}}\eta^{(4)}_{\mathrm{D})}$$

with

$$\eta_1^{(3)} = x, \quad \eta_2^{(3)} = 1, \quad \eta_1^{(4)} = y, \quad \eta_2^{(4)} = 1.$$

The factor β is irrelevant to the eigenvector structure (though it alters the numerical values of the eigenvalues) and so we shall ignore it from now on, identifying Ψ' and Ψ.

Case II corresponds to one of $x = y$, $x = 0$ or $y = 0$; Case III corresponds to $x = y = 0$ and cases N, D and O have been excluded and will be considered later. We now have

$$\Psi_0 = 0, \quad \Psi_1 = \frac{1}{4}xy, \quad \Psi_2 = \frac{1}{6}(x+y), \quad \Psi_3 = \frac{1}{4}, \quad \Psi_4 = 0.$$

Inserting these values in (5.10.12) gives for α

$$[\alpha^a{}_b] = \frac{1}{6}\begin{bmatrix} x+y & \frac{3}{\sqrt{2}} & 0 \\ -\frac{3}{\sqrt{2}}xy & -2(x+y) & -\frac{3}{\sqrt{2}} \\ 0 & \frac{3}{\sqrt{2}}xy & x+y \end{bmatrix}.$$

The eigenvalues and eigenvectors are then found to be

$$\begin{array}{ll} \frac{1}{6}(x+y) & (1,0,-xy)^{\dagger} \\ \frac{1}{6}(x-2y) & (1,\sqrt{2}y,xy)^{\dagger} \\ \frac{1}{6}(y-2x) & (1,\sqrt{2},xy)^{\dagger}\ . \end{array}$$

Thus in case II two of the eigenvalues with their corresponding eigenvectors coincide; while in case III all the eigenvectors and eigenvalues coincide.

It is customary to denote the eigenvector structure of an automorphism α by the Segré symbol: a list of positive integers enclosed in (square) brackets, disjoint sublists of which may be enclosed in (round) parentheses, with the sum of the integers equalling the dimensions of the space. For example, in dimension 17 a Segré symbol might be

$$[2(14)3(511)].$$

Each unparenthesized integer or parenthesized sublist corresponds to a distinct eigenvalue (so in the above example there are four distinct eigenvalues). Each integer n for a given eigenvalue λ corresponds to a single eigenvector x such that $(\alpha-\lambda I)^{n-1}y=x$, for some vector y, but this is untrue for any larger n.

Thus for case II, the Segré symbol is [12] (there are two distinct eigenvectors and hence two integers adding up to 3, and two distinct eigenvalues so there are no parentheses) while for case III the symbol is [3]. Case O is trivial (every vector is an eigenvector with eigenvalue 0); the Segré symbol is [(111)]. So it remains to consider the cases D and N. In case D, take $\eta^{(1)}_{A}=\eta^{(2)}_{A}=\delta^{1}_{A}, \eta^{(3)}_{A}=\eta^{(4)}_{A}=\delta^{2}_{A}$ (by a Lorentz transformation). Then

$$\Psi_0=\Psi_1=\Psi_3=\Psi_4=0 \qquad \Psi_2=\frac{1}{6}$$

$$\alpha=\frac{1}{6}\begin{bmatrix}1 & 0 & 0\\ 0 & -2 & 0\\ 0 & 0 & 1\end{bmatrix}$$

which immediately shows that there are three distinct eigenvectors with values $\frac{1}{6}$, $\frac{1}{6}$ and $-\frac{1}{3}$, giving a Segré symbol of [(11)1].

In case N we take

$$\eta^{(1)}_{A} = \eta^{(2)}_{A} = \eta^{(3)}_{A} = \eta^{(4)}_{A} = \delta^{1}_{A}$$

$$\Psi_0 = \frac{1}{24}, \ \Psi_1 = \Psi_2 = \Psi_3 = \Psi_4 = 0$$

$$[\alpha] = \frac{1}{24}\begin{bmatrix} 0 & 0 & 0 \\ 0 & 0 & 0 \\ 1 & 0 & 0 \end{bmatrix}$$

giving all the eigenvalues 0 and two eigenvectors, with a Segré symbol of [(21)].

6

COUPLING BETWEEN FIELDS AND GEOMETRY

6.1 Newtonian fluids

The novel and to some extent unexplained idea underlying the general relativistic theory of gravity is that energy generates curvature in space-time; conversely curvature interacts with whatever energy distribution exists in space-time establishing with it an equilibrium which is mathematically described by Einstein's field equations.

The previous chapters have been devoted to the analysis of the geometrical properties of a manifold; hereafter we shall turn our attention to the properties of the physical fields in a given manifold. The most important of these fields represent perfect fluids, and so we shall enter the subject referring first to the properties of a fluid in the context of Newtonian mechanics. The evolution of an adiabatic Newtonian fluid is governed by the conservation equations of energy and momentum. The expression of the total energy per unit volume of the fluid is given by:

$$\mathcal{E} = \frac{1}{2}\rho_0 v^2 + \rho_0 \epsilon \tag{6.1.1}$$

where ρ_0 is the (baryonic) mass density, $\vec{v}$ the velocity of the unit volume element and ϵ is the specific (per unit mass) internal energy. We introduce the convective derivative

$$\frac{D}{Dt} \equiv \frac{\partial}{\partial t} + \vec{v} \cdot \vec{\mathrm{grad}},$$

(which we recognise as the four-vector expression ∇_v, or $v^a \partial_a$, in the Newtonian limit where v has components $(1, \vec{v})$). Then the continuity equation is:

$$\frac{D\rho_0}{Dt} = -\rho_0 \vec{\text{div}}\vec{v} \qquad (6.1.2)$$

and the Euler equation is:

$$\frac{D\vec{v}}{Dt} = -\frac{1}{\rho_0}\vec{\text{grad}}p \qquad (6.1.3)$$

where p is the (isotropic) pressure of the fluid.

From the first law of thermodynamics and the assumed adiabaticity of the fluid we also have:

$$\frac{D\epsilon}{Dt} = \frac{p}{\rho_0^2}\frac{D\rho_0}{Dt} = -\frac{p}{\rho_0}\vec{\text{div}}\vec{v} \qquad (6.1.4)$$

from (6.1.2).

Thus taking the convective derivative of (6.1.1) and using (6.1.2-4) we find

$$\frac{D\mathcal{E}}{Dt} = \frac{\partial}{\partial t}\mathcal{E} + \vec{v}\cdot\vec{\text{grad}}\mathcal{E} = -\left(\frac{1}{2}v^2 + \epsilon + \frac{p}{\rho_0}\right)\left(\rho_0\vec{\text{div}}\vec{v}\right) - \vec{v}\cdot\vec{\text{grad}}\,p$$

$$= -\mathcal{E}\vec{\text{div}}\vec{v} - \vec{\text{div}}(p\vec{v}) \qquad (6.1.5)$$

or, collecting terms and inserting the definition (6.1.1) of $\mathcal{E}$ on the right

$$\frac{\partial\mathcal{E}}{\partial t} = -\vec{\text{div}}\left[\left(\epsilon + \frac{p}{\rho_0} + \frac{1}{2}v^2\right)\rho_0\vec{v}\right]. \qquad (6.1.6)$$

This is a continuity equation which assures energy conservation; here the quantity $(\epsilon + p/\rho_0 + \frac{1}{2}v^2)\rho_0\vec{v}$ is a flux of energy density. A similar relation holds for the linear momentum. The momentum per unit volume of the fluid is $\rho_0\vec{v}$, hence its variation with time reads, for each component:

$$\frac{\partial}{\partial t}(\rho_0 v^\alpha) = -\frac{\partial}{\partial x^\beta}\left(p\delta^{\alpha\beta} + \rho_0 v^\alpha v^\beta\right) \qquad \alpha,\ \beta\ = 1,2,3 \qquad (6.1.7)$$

using again Eqs. (6.1.2) and (6.1.3). Here the conservation of momentum is measured by the terms:

$$\pi^{\alpha\beta} = p\delta^{\alpha\beta} + \rho_0 v^\alpha v^\beta \qquad (6.1.8)$$

which describe the momentum flux density and are the components of a three-dimensional tensor. The fluid then is fully described by four sets of quantitites which we summarize as follows:

$\overset{\text{N}}{T}{}^{00} \equiv \frac{1}{2}\rho_0 v^2 + \rho_0 \epsilon$	total energy density (scalar)
$\overset{\text{N}}{T}{}^{\alpha 0} \equiv \rho_0 v^\alpha$	α-component of the momentum density (vector)
$\overset{\text{N}}{T}{}^{0\alpha} \equiv \left(\rho_0 \epsilon + p + \frac{1}{2}\rho_0 v^2\right) v^\alpha$	α-component of the energy flux density (vector)
$\overset{\text{N}}{T}{}^{\alpha\beta} \equiv p\delta^{\alpha\beta} + \rho_0 v^\alpha v^\beta$	α-component of the flux of the β-component of the momentum density (tensor)

(6.1.9)

These quantities are related by the continuity equation which we rewrite as

$$\frac{\partial}{\partial t}\overset{\text{N}}{T}{}^{00} + \frac{\partial}{\partial x^\alpha}\overset{\text{N}}{T}{}^{0\alpha} = 0$$
$$\frac{\partial}{\partial t}\overset{\text{N}}{T}{}^{\alpha 0} + \frac{\partial}{\partial x^\beta}\overset{\text{N}}{T}{}^{\alpha\beta} = 0. \tag{6.1.10}$$

Here $\overset{\text{N}}{T}{}^{\alpha\beta} = \overset{\text{N}}{T}{}^{\beta\alpha}$ and $\overset{\text{N}}{T}{}^{\alpha 0} \neq \overset{\text{N}}{T}{}^{0\alpha}$; the non-symmetry of the 4×4-matrix

$$\overset{\text{N}}{T} \equiv \begin{pmatrix} \overset{\text{N}}{T}{}^{00} & \overset{\text{N}}{T}{}^{0\alpha} \\ \overset{\text{N}}{T}{}^{\alpha 0} & \overset{\text{N}}{T}{}^{\alpha\beta} \end{pmatrix}$$

is due (as one realizes a posteriori) to the non-relativistic assumption we have made.

6.2 Generalization to special relativity

To extend the previous arguments to special relativity one needs to take into account the basic result of the theory, namely that mass and energy are equivalent entities. Hence momentum density and energy flux density must be equal, up to a factor c^2 ($c =$ velocity of light), so the special relativistic form of the matrix T would need $\overset{\text{S}}{T}{}^{0\alpha} = \overset{\text{S}}{T}{}^{\alpha 0}$, while the space-space components $\overset{\text{S}}{T}{}^{\alpha\beta}$ remain symmetric in order to ensure the torque balance of any isolated fluid configuration. A relativistic fluid, then, is entirely described by a symmetric 4×4-matrix whose elements are the components of a tensor, a fact which ensures the Lorentz invariance of the fluid description. This tensor is termed the energy-momentum tensor; for a perfect fluid* for instance it reads:

$$T^{ab} = (p + \rho_0 c^2 + \rho_0 \epsilon)\, u^a u^b + p\eta^{ab} \qquad a,\ b = 0, 1, 2, 3 \qquad (6.2.1)$$

where $u^a \equiv \left[-v^\alpha/(c\sqrt{1 - v^2/c^2})\ ,\ 1/\sqrt{1 - v^2/c^2}\right]$ is the fluid four-velocity (relative to a given inertial frame) and $\rho = \rho_0 c^2$ is the density of the rest-mass energy. The components of the energy momentum tensor of the relativistic fluid (6.2.1) reduce to the Newtonian form (6.1.9) when the non-relativistic limits: $v/c \ll 1$, $p/\rho \ll 1$ hold.

Conservation of energy and momentum, a prerequisite that any isolated system has to satisfy, implies as a generalization of (6.1.10), that

$$\frac{\partial T^{ab}}{\partial x^a} = 0 \qquad a,\ b = 0, 1, 2, 3 \qquad (6.2.2)$$

i.e. a single tensorial equation which, in its space and time decomposition, reduces to (6.1.10) in the non-relativistic limit.

The arguments which have been here presented for a fluid (in a simple case) apply to any other energy distribution in space-time such as an electromagnetic field or a non-perfect fluid. The simplifying assumption, however, which limits the validity of the above arguments resides in the fact that the energy distribution is here assumed not to affect the background geometry which therefore remains globally pseudo-Euclidean. In a physically realistic situation the fluid is a source of curvature, and so the question arises how to describe the matter-energy distribution when the background geometry is non-flat, and hence how the dynamic equations (6.2.2) generalize to that case.

6.3 Coupling between fields and geometry: the field action

The requirement that a matter system be described by a symmetric tensor which satisfies equations of energy and momentum

*A discussion on perfect fluids is given in Sects. (6.5), (9.10) and (9.11).

conservation suggests that it could be derived from a principle of least action. This method is very convenient (when applicable) for it allows a natural relation between conservation laws and symmetry properties. In our case the symmetry requirement is general covariance. Let an energy field be specified by a Lagrangian L_f which depends not only on the field variables but also on the metric coefficients g_{ij}. The Lagrangian must be a scalar to ensure a description of the field independently of whatever coordinate system we use; we call *action* the integral of the Lagrangian over an open set Ω of the space-time manifold, with compact closure:

$$S_f(\Omega) = \int_\Omega L_f d\Omega \tag{6.3.1}$$

where $d\Omega = \sqrt{-g}\boldsymbol{d}^4x$ is the invariant volume element. The field, or a collection of fields, is assumed to be in equilibrium with the background geometry; in searching for this condition, neither of the two systems, the field and the geometry, should be considered separately but only as one system in which the dynamical properties of the geometry are also described by a Lagrangian L_g. We define a total action:

$$S(\Omega) = S_g(\Omega) + S_f(\Omega) = \int_\Omega (L_f + L_g) d\Omega. \tag{6.3.2}$$

The metric is here considered as a dynamical field, thus the equation which establishes the equilibrium between the geometry and the external field is obtained by imposing the condition of stationarity of S under an arbitrary variation of the metric with support in Ω:

$$\underset{g}{\delta} S = 0. \tag{6.3.3}$$

The variation of the individual actions, S_f and S_g, under $\underset{g}{\delta}$ provide tensorial expressions which covariantly characterize the respective fields and also allow for conservation equations. Let us consider the field action first. The assumption here is that the Lagrangian L_f depends only on $g_{ij}(x)$ and not on its derivatives, because this condition is satisfied by all known reasonable matter distributions. Variation with respect to g_{ij} then yields:

$$\underset{g}{\delta} S_f = \int \left[\frac{\partial L_f}{\partial g_{ij}} \underset{g}{\delta} g_{ij} d\Omega + L_f \, \underset{g}{\delta}(\sqrt{-g})\boldsymbol{d}^4x \right]. \tag{6.3.4}$$

Recalling from (2.10.2) that $\delta g = g g^{ij} \delta g_{ij}$, this equation becomes

$$\underset{g}{\delta} S_f = \frac{1}{2} \int_\Omega T^{ik} \underset{g}{\delta} g_{ik} d\Omega \tag{6.3.5}$$

where

$$T^{ik} \equiv 2 \frac{\partial L_f}{\partial g_{ik}} + L_f g^{ik} \tag{6.3.6a}$$

is a symmetric tensor. Since:

$$\delta g = g g^{ij} \delta g_{ij} = -g g_{ij} \delta g^{ij}$$

then

$$-T_{ij} = 2 \frac{\partial L_f}{\partial g^{ij}} - L_f g_{ij}. \tag{6.3.6b}$$

We identify the tensor T^{ij} with the energy-momentum tensor of the field; it provides in fact the right expression for the known fields and satisfies conservation equations. Suppose we have a one-parameter family of diffeomorphisms $\varphi_t \colon M \to M$, with φ_0 =Id, such that $\varphi_t(x) = x$ unless $x \in U$ with $\overline{U} \subset \Omega$. Any of these diffeomorphisms can be thought of as simply a relabeling of the points in U, and so does not alter the values of S_f, in the sense that, if Ψ denotes the matter fields, then we have:

$$S_f(U;\ g,\ \Psi) = S_f(U;\ \varphi_t^* g;\ \varphi_t^* \Psi). \tag{6.3.7}$$

(If we have coordinates in U, then the effect of φ_t on the expressions for g and Ψ in terms of coordinates is precisely a change in coordinates, under which S_f is unchanged.) Imposing this requirement on S_f is the most general form of the principle of covariance. In practice it is usually met by ensuring that the Lagrangian is a scalar, which is in turn guaranteed by constructing it out of appropriate tensorial quantities (covariant derivatives and the like). Differentiating (6.3.7) with respect to t and setting $t = 0$ gives:

$$\left[\left(\frac{\delta}{\delta g} S_f \right) \left(\frac{d\varphi_t^* g}{dt} \right) + \left(\frac{\delta}{\delta \Psi} S_f \right) \left(\frac{d\varphi_t^* \Psi}{dt} \right) \right]_{t=0} = 0.$$

If Ψ satisfies the field equations, then by definition of the variational principle,

$$\Big((\delta / \delta \Psi) S_f \Big) = 0.$$

Thus we have

$$\left[\left(\frac{\delta}{\delta g}S_f\right)\left(\frac{d\varphi_t^* g}{dt}\right)\right]_{t=0} = \left(\frac{\delta S_f}{\delta g}\right) L_\xi g = 0$$

where ξ is the vector field generated by the family φ_t. From (6.3.5) and Sect. 2.2, this reads:

$$\frac{1}{2}\int_\Omega T^{ij}\nabla_{(i}\xi_{j)}d\Omega = 0. \qquad (6.3.8)$$

Using the symmetry of T^{ij} and Gauss' theorem, we obtain

$$\int_\Omega \left(\nabla_i T^{ij}\right)\xi_j d\Omega - \int_{\partial\Omega} T^{ij}\xi_j\sqrt{-g}d\Sigma_i = 0. \qquad (6.3.9)$$

Since ξ vanishes on $\partial\Omega$, (6.3.9) yields, by the arbitrariness of ξ:

$$\nabla_i T^{ij} = 0\ . \qquad (6.3.10)$$

This generalizes equation (6.2.2) to a curved space-time. The vanishing here of a covariant divergence does not imply a set of proper conservation equations (for which one would require the vanishing of a partial divergence); nevertheless these could be recovered by taking into account the contributions to energy and momentum by the background geometry to which the field is necessarily coupled. There is however some ambiguity in the definition of the energy content of a given space-time domain, but this problem will be considered after we have discussed the gravitational action.

6.4 The gravitational action and the Einstein equations

Since the space-time geometry responds to an external field by developing curvature, according to the amount of energy density stored in the field, it will have dynamic equations of its own. A suitable action would be a scalar function, depending only on g_{ij} and their derivatives; in this case g_{ij} are the only dynamical variables of the system.

The aim is to obtain equations which, in the non-relativistic limit, provide the Newtonian equation of gravity; this is a nonlinear second-order partial differential equation in the Newtonian potential, hence a minimal requirement is that the general relativistic gravitational equations should also be of the same degree and linear in the second partial derivatives of the metric coefficients which play the role of potentials. This is ensured by the action being the integral of a Lagrangian L_g which is a function of g_{ij}, $\partial_k g_{ij}$ and $\partial_k \partial_l g_{ij}$ only; is linear in the latter and is such that terms like $\partial L_g / \partial g_{ij,k}$, do not contain $\partial_k \partial_l g_{ij}$. The simplest action which satisfies all these requirements was found by Hilbert (Hilbert, 1917) to be:

$$S_g = -\frac{c^4}{16\pi G} \int R\sqrt{-g}\, \boldsymbol{d}^4 x \tag{6.4.1}$$

where R is the scalar curvature , and the factor in front is needed to match correctly with the Newtonian limit. The variation of S_g now reads

$$\begin{aligned} \delta_g S_g &= -\frac{c^4}{16\pi G} \int \delta \left(g^{ij} R_{ij} \sqrt{-g} \right) \boldsymbol{d}^4 x \\ &= -\frac{c^4}{16\pi G} \int \boldsymbol{d}^4 x \Big[R_{ij} \sqrt{-g} \delta g^{ij} + g^{ij} \delta R_{ij} \sqrt{-g} \\ &\quad - \frac{1}{2} g^{ij} R_{ij} g_{km} \delta g^{km} \sqrt{-g} \Big] . \end{aligned} \tag{6.4.2}$$

From (3.4.2), the Ricci tensor, namely:

$$R_{ij} = \partial_a \Gamma^a_{ij} - \partial_j \Gamma^a_{ia} + \Gamma^b_{ij} \Gamma^a_{ba} - \Gamma^b_{ia} \Gamma^a_{bj} \tag{6.4.3}$$

undergoes a variation

$$\delta R_{ij} = \partial_a \delta \Gamma^a_{ij} - \partial_j \delta \Gamma^a_{ia} + \delta \Gamma^b_{ij} \Gamma^a_{ba} + \Gamma^b_{ij} \delta \Gamma^a_{ba} - \delta \Gamma^b_{ia} \Gamma^a_{bj} - \Gamma^b_{ia} \delta \Gamma^a_{bj} . \tag{6.4.4}$$

The variation of the affine connection is a tensor; in fact:

$$\begin{aligned} \delta \Gamma^i_{jk} &= -\delta(g_{ma}) g^{im} \Gamma^a_{jk} + \frac{1}{2} g^{ia} \\ &\quad \times \left(\partial_j \delta g_{ak} + \partial_k \delta g_{aj} - \partial_a \delta g_{jk} \right) \\ &= \frac{1}{2} g^{ia} \left(\nabla_j \delta g_{ak} + \nabla_k \delta g_{aj} - \nabla_a \delta g_{jk} \right) , \end{aligned} \tag{6.4.5}$$

hence the variation of the Ricci tensor can be written*

$$\delta R_{ij} = \nabla_a (\delta \Gamma^a_{ij}) - \nabla_j (\delta \Gamma^a_{ia}) . \tag{6.4.6}$$

*Eq. (6.4.6) is known as the *Palatini identity* and is obtained in a natural way from (6.4.4) if we take the Γ's, as Palatini suggested, as independent dynamical variables together with the g^{ij} (Palatini, 1919).

Equation (6.4.2) becomes:

$$\underset{g}{\delta} S_g = -\frac{c^4}{16\pi G}\int \boldsymbol{d}^4x \Big[\Big(R_{ij} - \frac{1}{2}g_{ij}R\Big)\delta g^{ij}\sqrt{-g} + \sqrt{-g} \times \Big(\nabla_a(g^{ij}\delta\Gamma^a_{ij}) - \nabla^i(\delta\Gamma^a_{ia})\Big)\Big]\,; \tag{6.4.7}$$

the second term under the integration can be rewritten in terms of divergences

$$\int \boldsymbol{d}^4x \left[\partial_a\left(g^{ij}\delta\Gamma^a_{ij}\sqrt{-g}\right) - \partial^i\left(\sqrt{-g}\delta\Gamma^a_{ia}\right)\right]$$

and therefore gives no contribution to $\underset{g}{\delta} S$ which finally remains as

$$\underset{g}{\delta} S_g = -\frac{c^4}{16\pi G}\int \boldsymbol{d}^4x\sqrt{-g}\left(R_{ij} - \frac{1}{2}g_{ij}R\right)\delta g^{ij}. \tag{6.4.8}$$

Recalling (3.5.6), i.e. $G_{ij} \equiv R_{ij} - \frac{1}{2}g_{ij}R$ (the Einstein tensor), the requirement that S_g is a scalar leads with arguments similar to those for T_{ij}, to the equations $\nabla_j G^{ij} = 0$ which simply reproduce the Bianchi identities (3.5.3) and (3.5.5). Going back to (6.3.3), we finally deduce the equations which establish the equilibrium between fields and geometry:

$$\underset{g}{\delta} S = \int \left[-\frac{c^4}{16\pi G}\left(R_{ij} - \frac{1}{2}g_{ij}R\right) + \frac{1}{2}T_{ij}\right]\underset{g}{\delta}\sqrt{-g}\;\boldsymbol{d}^4x = 0; \tag{6.4.9}$$

with the arbitrariness of $\underset{g}{\delta} g^{ij}$ yielding the Einstein equations:

$$R_{ij} - \frac{1}{2}g_{ij}R = \frac{8\pi G}{c^4}T_{ij}. \tag{6.4.10}$$

It is customary to write $8\pi G/c^4 = \kappa$ and call it the Einstein (coupling) constant (Einstein, 1915a, 1915b).

Einstein's field equations are ten, coupled, second order partial non-linear differential equations in the metric coefficients. Since both sides of (6.4.10) have vanishing covariant divergence ($\nabla_j G^{ij} = \nabla_j T^{ij} = 0$), then the number of independent equations reduces to six. These entirely suffice to determine the space-time metric because four of the metric coefficients can be arbitrarily assigned by the freedom one has in making coordinate transformations, which amount to four conditions. The interaction between the matter-energy fields and the space-time geometry occurs with no explicit specification of the field properties but only through the energy-momentum tensor; therefore the background geometry would not distinguish between physically different fields provided they have the same energy-momentum distribution. This can be regarded as the general relativistic version of the equivalence between the inertial and the gravitational masses.

A convenient way to write Einstein's equations is to take the trace of (6.4.10):

$$R - 2R = \kappa T = -R \tag{6.4.11}$$

and inserting this back into (6.4.10):

$$R_{ij} = \kappa \left(T_{ij} - \frac{1}{2} g_{ij} T \right) ; \tag{6.4.12}$$

in this form, one recognizes that the Ricci tensor is contributed to by the local distribution of matter-energy. When this is absent, namely in the vacuum, the Einstein equations read:

$$R_{ij} = 0. \tag{6.4.13}$$

This equation not only provides solutions which describe the geometry of space-time outside a given matter-energy distribution, but also gives solutions which differ from the trivial Minkowski solution even when there is no matter-energy distribution at all. These solutions are somehow implied by the non-linearity of the gravitational interaction, according to which curvature (i.e. gravity), being itself endowed with energy and momentum, generates more curvature itself; these solutions therefore describe *self-sustaining gravitational fields.*

Let us now consider the next implication of the Einstein equations, namely how the non-local distribution of matter-energy contributes to the local definition of curvature. As we knew from (3.5.7), the curvature tensor is locally defined as the composition of its trace-free part (the Weyl tensor) and its trace which is made of terms containing the Ricci tensor and the curvature scalar. In vacuum, where the Ricci tensor vanishes, the curvature tensor is only contributed to by the Weyl tensor (see Eq. (3.5.7)); this however will be different from zero only if the space-time is endowed with non-homogeneous distribution of energy. From

(3.5.10), (3.5.8) and (6.4.12) we have in fact:

$$\nabla_i C_{rs}{}^{ji} = \frac{1}{2}\left[\nabla_s \left(R^j_r - \frac{1}{6}R\delta^j_r\right) - \nabla_r \left(R^j_s - \frac{1}{6}R\delta^j_s\right)\right]$$
$$= \kappa \left(\nabla_{[s}T^j_{r]} + \frac{1}{3}\delta^j_{[s}\partial_{r]}T\right)\,; \qquad (6.4.14)$$

hence sources for the Weyl tensor are the gradients of the energy-momentum tensor. We will show in Sect. 8.1 (cf. Eqs. (8.1.25) to (8.1.29)) that these equations can be interpreted as being analogous to Maxwell's with the gradient of T acting as a source.

6.5 The energy-momentum tensor of a perfect fluid

A fluid configuration is said to be a *perfect fluid* when it has no viscosity and no heat conduction (see Sects. 9.10 and 9.11). It is characterized by a normalized four-velocity u (i.e. $(u|u) = -1$) and by a set of scalar quantities: a density of total mass-energy ρ, an isotropic pressure p, a specific entropy s, a temperature T and a specific relativistic enthalpy $w = (\rho + p)/n$, n being the baryon number density. These scalars are defined in an instantaneous Lorentz frame carried by the fluid; only two of them are independent, the others can be determined from them by an equation of state and the usual thermodynamic relations. The aim here is to show that the energy-momentum tensor for a perfect fluid can be derived from a Lagrangian according to the general rule (6.3.6) (Schutz, 1970). The description of a fluid by variational methods has been the object of an extensive investigation by several authors; the main difficulty resides in the impossibility of deriving the equation of motion of the fluid in terms of its five Eulerian* variables, namely three components of its four-velocity and the parameters of the equation of state, from a variational principle based on a free variation of these variables. In the case of a perfect fluid, the variation *must* be constrained by the requirement that the rates of entropy and particle production be conserved. Let us

*A variable is termed Eulerian when it is referred to an arbitrary frame; it is termed Lagrangian when it is referred to a comoving (with the fluid) frame.

then introduce a baryon number flux vector density

$$n^i = nu^i\sqrt{-g} \tag{6.5.1}$$

so that

$$n = \left(\frac{n^i n^j g_{ij}}{g}\right)^{\frac{1}{2}} . \tag{6.5.2}$$

We require a variation of the variables, satisfying the following constraints

$$a) \;\; \delta s = 0 \qquad b) \;\; \delta n^i = 0. \tag{6.5.3}$$

Here we include as variables also the metric coefficients because we shall later consider variations with respect to the metric only. To understand the constraints (6.5.3) better from a physical point of view, and to demonstrate that they do indeed preserve entropy and particle production rates, let us recall from (2.10.7)a:

$$\nabla_i(nu^i) = (-g)^{-\frac{1}{2}}\partial_i n^i. \tag{6.5.4}$$

Thus, taking the gradient of (6.5.3)b and using the relation $\partial_i\delta = \delta\partial_i$ we obtain:

$$\begin{aligned}\delta(\partial_i n^i) &= \delta\left((-g)^{-\frac{1}{2}}\partial_i n^i\right)(-g)^{\frac{1}{2}} + (-g)^{-\frac{1}{2}}\partial_i n^i \delta(-g)^{\frac{1}{2}} \\ &= (-g)^{\frac{1}{2}}\left[\delta\left(\nabla_i(nu^i)\right) + \frac{1}{2}\nabla_i(nu^i)g^{ab}\delta g_{ab}\right] = 0 \end{aligned} \tag{6.5.5}$$

This equation ensures that the rate of particle production in the fluid (if any) is preserved under the variation δ, this being a weaker demand than assuming a priori that the number density is conserved, i.e. $\nabla_i(nu^i) = 0$. Similarly taking the gradient of (6.5.3)a and contracting with n^i we obtain, from (6.5.3)b :

$$\delta(n^i\partial_i s) = 0$$

showing again that it is the rate of entropy production in the fluid which is here assumed to be preserved*. Under these conditions, the appropriate Lagrangian for a one-component perfect fluid is:

$$L_f = -\rho \, . \tag{6.5.6}$$

*This implies that under variation with compact support, the entropy and the particle number density is not changed.

Then for a variation δ subject to these constraints the field action

$$S_f = -\int \rho\sqrt{-g}\, \mathbf{d}^4x \tag{6.5.7}$$

undergoes a variation:

$$\delta S_f = -\int [\delta\rho \sqrt{-g} + \rho\, \delta(\sqrt{-g})]\, \mathbf{d}^4x \ . \tag{6.5.8}$$

Let an equation of state be given as $\rho = \rho(n,\ s)$. Then from the thermodynamic relation

$$\left(\frac{\partial \rho}{\partial n}\right)_s = w$$

and from (6.5.3)a, we have:

$$\delta\rho = w\delta n. \tag{6.5.9}$$

From (6.5.3)b and (6.5.2) we obtain:

$$\delta n = \frac{n}{2}(-g)u^iu^j\left(\frac{\delta g_{ij}}{g} - \frac{g_{ij}}{g^2}\delta g\right) = -\frac{n}{2}\left(u^iu^j + g^{ij}\right)\delta g_{ij} \tag{6.5.10}$$

so (6.5.8) becomes:

$$\begin{aligned}\delta S_f &= \frac{1}{2}\int \left[wn(u^iu^j + g^{ij}) - \rho g^{ij}\right]\sqrt{-g}\ \delta g_{ij}\ \mathbf{d}^4x \\ &= \frac{1}{2}\int \left[(\rho + p)u^iu^j + pg^{ij}\right]\sqrt{-g}\ \delta g_{ij}\ \mathbf{d}^4x.\end{aligned} \tag{6.5.11}$$

This yields an energy-momentum tensor for a perfect fluid:

$$T^{ij} = (\rho + p)u^iu^j + pg^{ij} \tag{6.5.12}$$

which formally generalizes the special relativistic expression given in (6.2.1).

A different expression for the Lagrangian of a fluid, albeit equivalent to (6.5.6), which allows for a more powerful description of the system in terms of velocity potentials, is given by

$$L'_f = p\ . \tag{6.5.13}$$

Expressing the equation of state in the form $p = p(w',\ s)$ where $w' = (\rho + p)/\rho_0$, ρ_0 being the rest-mass density, the following relation holds:

$$dp = \rho_0 dw' - \rho_0 T ds.$$

The basic assumption which leads to (6.5.13) is that the four-velocity of the fluid can be expressed as:

$$u^i = g^{ij}\frac{1}{w'}\left(\varphi_{,j} + \alpha\beta_{,j} + \theta s_{,j}\right)$$

where φ, α, β, θ, s are velocity-potentials (scalar fields) which together with w' and g_{ij} are the dynamical variables of the fluid to be varied freely. Varying with respect to the metric, and from the normalization condition of u^i, we obtain

$$\delta w' = -\left(\frac{1}{2w'}\right)\delta g^{ij}\left(\varphi_{,i} + \alpha\beta_{,i} + \theta s_{,i}\right)\left(\varphi_{,j} + \alpha\beta_{,j} + \theta s_{,j}\right) \tag{6.5.14}$$

but

$$\delta g^{ij} = -g^{ir}g^{js}\delta g_{rs} \tag{6.5.15}$$

so

$$\delta w' = \frac{w'}{2}u^r u^s \delta g_{rs}, \tag{6.5.16}$$

hence the action

$$S_f = \int p\sqrt{-g}\,\boldsymbol{d}^4x \tag{6.5.17}$$

undergoes the variation:

$$\delta S_f = \frac{1}{2}\int \left(\rho_0 w' u^r u^s + p g^{rs}\right)\sqrt{-g}\delta g_{rs}\,\boldsymbol{d}^4x \tag{6.5.18}$$

which leads again to the energy-momentum tensor (6.5.12).

6.6 The energy-momentum tensor of a single particle

The description of a physical system which is made of a single fluid element (a particle) closely resembles that of a perfect fluid but with some additional subtleties (see Tulczyjew, 1957). Let γ be the particle world-line, with parameter s and $\dot{z}^i$ the components of its tangent field. Consider a neighbourhood W of γ, constituted by the union of normal neighbourhoods of the points of γ, then call $\Sigma(s)$ the space-like section of W which is spanned by the space-like geodesics which stem from $\gamma(s)$ orthogonally to γ. We can now expand the vector field $\dot{z}^i$ on γ into a vector field in W, as follows:

$$\breve{u}^i(x)|_\gamma = \Gamma(\gamma(s),\ x;\ \gamma|_x)_j{}^i \dot{z}^j(s) \tag{6.6.1}$$

where $\gamma|_x$ is the unique geodesic joining a general point x^i on $\Sigma(s)$ and the point $\gamma(s)$ on γ with coordinates $z^i(s)$. Within W, the vector field $\breve{u}^i$ identifies a congruence of curves (the integral curves of $\breve{u}^i$) so that one can define an energy flux vector density as:

$$n^i = \int \mu \breve{u}^i \delta^4(x - \gamma(s))ds \tag{6.6.2}$$

where μ is the particle rest-mass energy and δ^4 is the four-dimensional Dirac delta. The latter has the character of a density because

$$\int \delta^4(x - \underset{0}{x}) f(x) \mathbf{d}^4 x = f(\underset{0}{x}) \tag{6.6.3}$$

is a well-defined number, independent of coordinates, for any $f \in \mathcal{F}(M)$. It is also clearly independent of the metric and the field variables, and so

$$\delta \left(\delta^4(x - \underset{0}{x}) \right) = 0\ . \tag{6.6.4}$$

From (2.8.1) it follows that on $\Sigma(s)$:

$$(\breve{u}^i \breve{u}_i)\,|_{\Sigma(s)} = (\dot{z}^i \dot{z}_i)\,|_{\gamma(s)}$$

hence a well-defined energy density can be defined in W as:

$$\rho(x)\sqrt{-g} = (-n^i n_i)^{\frac{1}{2}} = \mu \int \delta^4(x - \gamma(s))\,(-\breve{u}^i \breve{u}_i)^{\frac{1}{2}}\, ds \tag{6.6.5}$$

and the action of a single particle becomes:

$$S_f = -\int_W \rho(x)\sqrt{-g}\, \mathbf{d}^4 x = -\mu \int \int_W \delta^4(x - \gamma(s))\,(-\breve{u}^i \breve{u}_i)^{\frac{1}{2}}\ \mathbf{d}^4 x ds\ . \tag{6.6.6}$$

To deduce the energy-momentum tensor, as in the case of a perfect fluid, we impose the constraint:

$$\delta n^i = 0. \tag{6.6.7}$$

From (6.6.4) and (6.6.2), this implies:

$$\delta \breve{u}^i = 0, \tag{6.6.8}$$

hence, from (6.6.6), (6.6.4) and (6.6.8) we have

$$\delta S_f = \frac{1}{2}\int\int_W \delta^4(x-\gamma(s))\frac{\mu}{(-\breve{u}^r\breve{u}_r)^{\frac{1}{2}}}\delta g_{ij}\breve{u}^i\breve{u}^j\, \boldsymbol{d}^4x ds. \qquad (6.6.9)$$

From (6.3.5) and within W we finally obtain:

$$T^{ij}(x) = \mu\int\frac{\delta^4(x-\gamma(s))}{(-\dot{z}^i\dot{z}_i)^{\frac{1}{2}}} \times \Gamma(\gamma(s),\, x;\ \gamma)_k{}^i\Gamma(\gamma(s),\, x;\ \gamma)_l{}^j\dot{z}^k(s)\dot{z}^l(s)ds\ . \qquad (6.6.10)$$

6.7 The energy-momentum tensor of the electromagnetic field

An important example of a physical field is the electromagnetic field. It is formally characterized by the skew-symmetric tensor $F_{ij} = 2\partial_{[i}A_{j]}$ where $\boldsymbol{A}$ is a one-form, termed the potential, which is defined up to a gauge transformation $\boldsymbol{A}\to\boldsymbol{A}' = \boldsymbol{A}+\boldsymbol{d}\lambda$ with λ an arbitrary scalar function. An invariant description of the field is given in terms of the Lagrangian:

$$L_f = -\frac{1}{16\pi}g^{ab}g^{cd}F_{ac}F_{bd} \qquad (6.7.1)$$

(the simplest scalar function definable from the field variables A_i) which leads to linear second-order partial differential equations in the A_i. In this case if we chose as field variables the F_{ij}, we would not be able to obtain Maxwell's equations from a variational principle in which the F_{ij} were varied without constraints; one needs in fact to impose the condition $\partial_{[j}F_{ij]} = 0$. In order to obtain the field equations from an unconstrained variational principle, one needs to choose as field variables the potentials A_i although, as in the case of the velocity potentials (6.5.14) for fluids, they have no direct physical meaning. It is noteworthy that in the definition of the skew-symmetric covariant tensor F_{ab} no metric appears. So varying (6.7.1) with respect to the metric only as in (6.3.6), we obtain:

$$T^{ij}_{\text{e.m.}} = \frac{1}{4\pi}\left[F^{ia}F^j{}_a - \frac{1}{4}F^{rs}F_{rs}g^{ij}\right]. \qquad (6.7.2)$$

6.8 The energy-momentum pseudotensor

In general relativity the space-time geometry is not conceived as a passive background but as a dynamical field which interacts with the other fields and also with itself. The latter is particularly significant in highly relativistic situations such as in the early phases of the Universe or near the curvature singularities which arise from gravitational collapse. The geometrical (or gravitational) interaction occurs because geometry has itself energy and momentum; but, unlike the other physical fields, these cannot be localized. In fact the local pseudo-Euclidean character of space-time, as dictated by the principle of equivalence, allows one to select a coordinate system in the neighbourhood of each point, such that the connection coefficients vanish at that point. The consequence of this property is that the *interacting* part of the geometry, that is, that component which cannot be made to vanish, only resides in the second (and higher) derivatives of the metric. These terms however become significant over sufficiently extended portions of the space-time (the size depending on the sensitivity of the external fields to curvature effects). Therefore any determination of energy and momentum of the background geometry, having the necessary character of invariance with respect to arbitrary choices of coordinates, must be global in nature. Since the principle of equivalence forbids the finding of a tensor "density" for the gravitational energy and momentum, the best we could hope for, in terms of local quantities, might be a non-tensorial object which, suitably integrated over a large region of space-time, would lead to a quantity that was sufficiently gauge-invariant for practical purposes.

In order to find this object, let us reconsider the geometric action (6.4.1); it is a general rule that the contribution to the variation of any action by a plain divergence is zero once it is assumed that the variations of the dynamical variables vanish near the boundary of the integration domain. This applies directly to S_g in (6.4.1); in fact the integrand of S_g can be written as:

$$R\sqrt{-g} = \partial_k \left[\sqrt{-g}\left(g^{il}\Gamma^k_{il} - g^{kl}\Gamma^i_{il}\right)\right] - \sqrt{-g}g^{ik}\left(\Gamma^r_{ik}\Gamma^l_{rl} - \Gamma^r_{il}\Gamma^l_{kr}\right) \tag{6.8.1}$$

from (3.4.3); therefore the result of varying $R\sqrt{-g}$ is the same as varying the Lagrangian

$$\mathcal{L} \equiv -\sqrt{-g} g^{ik} \left(\Gamma^r_{ik}\Gamma^l_{rl} - \Gamma^r_{il}\Gamma^l_{kr}\right) \tag{6.8.2}$$

which is a function of g^{ij} and $\partial_k g^{ij}$, being quadratic in the latter. Hence we have

$$\delta \int R\sqrt{-g}\, \mathbf{d}^4x = \int \left(R_{ij} - \frac{1}{2} g_{ij} R\right) \sqrt{-g}\delta g^{ij}\, \mathbf{d}^4x = \delta \int \mathcal{L}\, \mathbf{d}^4x. \tag{6.8.3}$$

By standard variation methods, this becomes:

$$\left(R_{ij} - \frac{1}{2} g_{ij} R\right) = (-g)^{-\frac{1}{2}} \left[\frac{\partial \mathcal{L}}{\partial g^{ij}} - \partial_k \left(\frac{\partial \mathcal{L}}{\partial \left(\partial_k g^{ij}\right)}\right)\right]; \tag{6.8.4}$$

thus, from (6.4.10):

$$\kappa T_{ij}\sqrt{-g} = \frac{\partial \mathcal{L}}{\partial g^{ij}} - \partial_k \left(\frac{\partial \mathcal{L}}{\partial \left(\partial_k g^{ij}\right)}\right). \tag{6.8.5}$$

Equation (6.3.10) can now be written in a more convenient form; from the properties of the covariant derivative, equation (6.3.10) becomes:

$$\partial_j \left(\sqrt{-g} T_i^{\ j}\right) + \frac{1}{2} \frac{\partial g^{kl}}{\partial x^i} T_{kl} \sqrt{-g} = 0 \tag{6.8.6}$$

so the coupling between field and geometry can now be taken into account under the form (6.8.5). The second term in (6.8.6) yields:

$$\frac{1}{2\kappa} \frac{\partial g^{kl}}{\partial x^i} \left[\frac{\partial \mathcal{L}}{\partial g^{kl}} - \partial_s \left(\frac{\partial \mathcal{L}}{\partial \left(\partial_s g^{kl}\right)}\right)\right] = \frac{1}{2\kappa} \partial_j \left[\mathcal{L}\delta_i^j - \frac{\partial g^{kl}}{\partial x^i} \frac{\partial \mathcal{L}}{\partial \left(\partial_j g^{kl}\right)}\right] \tag{6.8.7}$$

where use is made of

$$\frac{\partial \mathcal{L}}{\partial x^i} = \frac{\partial \mathcal{L}}{\partial g^{kl}} \partial_i g^{kl} + \frac{\partial \mathcal{L}}{\partial \left(\partial_m g^{kl}\right)} \partial_i (\partial_m g^{kl}).$$

Thus defining:

$$\sqrt{-g} t_i^{\ r} = \frac{1}{2\kappa} \left(\mathcal{L}\delta_i^r - \frac{\partial g^{kl}}{\partial x^i} \frac{\partial \mathcal{L}}{\partial \left(\partial_r g^{kl}\right)}\right) \tag{6.8.8}$$

equation (6.8.6) becomes:

$$\partial_j \left[\sqrt{-g} \left(T_i^{\ j} + t_i^{\ j}\right)\right] = 0. \tag{6.8.9}$$

The quantities $t_i^{\ j}$ in (6.8.8), first introduced by Einstein, do not transform as the components of a tensor, except under linear coordinate transformations; for this reason the set $\left(T^i_{\ j} + t^i_{\ j}\right)$ is termed *the Einstein complex.*

In addition to not being a tensor, $t_i^{\ j}$ has the further inadequacy of providing a non-symmetric quantity when the lower index is raised. However, because the result of the variation of the action is not changed when adding a divergence, one can use this freedom to construct a complex which is symmetric. In order to do so, let us consider in more detail the structure of the Einstein complex. Let us first show that we can write:

$$2\kappa\sqrt{-g}\left(T_i^{\ k} + t_i^{\ k}\right) = \partial_m S_i^{km}. \tag{6.8.10}$$

From (6.8.8) and (6.8.4), we have in fact:

$$\begin{aligned}
2\kappa\sqrt{-g}\left(T_i^{\ k} + t_i^{\ k}\right) &= \sqrt{-g}\left(2G_i^{\ k} + 2\kappa t_i^{\ k}\right) \\
&= 2g^{ik}\left[\frac{\partial \mathcal{L}}{\partial g^{ij}} - \partial_m\left(\frac{\partial \mathcal{L}}{\partial\left(\partial_m g^{ij}\right)}\right)\right] \\
&\quad + \delta_i^k \mathcal{L} - (\partial_i g^{mn})\frac{\partial \mathcal{L}}{\partial\left(\partial_k g^{mn}\right)} \\
&= -\partial_m\left[2g^{jk}\frac{\partial \mathcal{L}}{\partial\left(\partial_m g^{ij}\right)}\right] \\
&\quad + \left[\delta_i^k \mathcal{L} - (\partial_i g^{mn})\frac{\partial \mathcal{L}}{\partial\left(\partial_k g^{mn}\right)}\right. \\
&\quad \left. + 2g^{jk}\frac{\partial \mathcal{L}}{\partial g^{ij}} + 2(\partial_m g^{jk})\frac{\partial \mathcal{L}}{\partial\left(\partial_m g^{ij}\right)}\right].
\end{aligned} \tag{6.8.11}$$

It is possible to prove that (Schrödinger, 1963):

$$\delta_i^k \mathcal{L} - (\partial_i g^{mn})\frac{\partial \mathcal{L}}{\partial\left(\partial_k g^{mn}\right)} + 2g^{jk}\frac{\partial \mathcal{L}}{\partial g^{ij}} + 2(\partial_m g^{jk})\frac{\partial \mathcal{L}}{\partial\left(\partial_m g^{ij}\right)} = 0 \tag{6.8.12}$$

hence:

$$S_i^{km} = -2g^{jk}\frac{\partial \mathcal{L}}{\partial\left(\partial_m g^{ij}\right)}\,. \tag{6.8.13}$$

From (6.8.2), this becomes explicitly:

$$S_i^{km} = 2\sqrt{-g}\left[\Gamma^m_{il} g^{kl} - \frac{1}{2}\left(\delta_i^m \Gamma^n_{ln} + \delta_l^m \Gamma^n_{in}\right) g^{kl} - \frac{1}{2}\delta_i^k \left(g^{rs}\Gamma^m_{rs} - g^{mr}\Gamma^s_{rs}\right)\right]. \quad (6.8.14)$$

A long algebraic manipulation leads to:

$$S_i^{kl} = U_i^{kl} + \partial_m W_i^{klm} \quad (6.8.15)$$

where:

$$U_i^{kl} = U_i^{[kl]} = (-g)^{-\frac{1}{2}} g_{in} \partial_m \left[(-g)(g^{kn}g^{lm} - g^{ln}g^{km})\right] \quad (6.8.16)$$

is a super-potential found by Freud (Freud, 1939), and

$$W_i^{klm} = \left(\delta_i^l g^{km} - \delta_i^m g^{kl}\right)\sqrt{-g} \quad (6.8.17)$$

being antisymmetric in (l, m), gives no contribution to the Einstein complex (6.8.10) which then reads:

$$2\kappa\sqrt{-g}\left(T_i^{\ j} + t_i^{\ j}\right) = \partial_l U_i^{jl}. \quad (6.8.18)$$

Let us now multiply (6.8.18) by $\sqrt{-g}g^{ik}$; we have, after some algebra:

$$2\kappa(-g)\left(T^{kj} + g^{ik}t_i^{\ j}\right) = \partial_l\partial_m\left[(-g)(g^{kj}g^{lm} - g^{jm}g^{lk})\right] - \partial_m\left[(-g)(g^{jn}g^{lm} - g^{jm}g^{ln})\right] \times \left[g_{in}\partial_l g^{ik} - \frac{1}{2}\delta_n^k g_{rs}\partial_l g^{rs}\right]$$

Redefine a new object $\hat{t}^{kj}$ as:

$$2\kappa(-g)\hat{t}^{kj} = 2\kappa(-g)g^{ik}t_i^{\ j} + \partial_m\left[(-g)(g^{jn}g^{lm} - g^{jm}g^{ln})\right] \times \left[g_{in}\partial_l g^{ik} - \frac{1}{2}\delta_n^k g_{rs}\partial_l g^{rs}\right] \quad (6.8.19)$$

so a new complex is obtained:

$$2\kappa(-g)\left(T^{kj} + \hat{t}^{kj}\right) = \partial_l\partial_m\left[(-g)(g^{kj}g^{ml} - g^{mj}g^{kl})\right] \quad (6.8.20)$$

where now a symmetry with respect to the pair of indices (k, j) is apparent. The quantity $\hat{t}^{kj}$ was found by Landau and Lifschitz (Landau and Lifschitz, 1962; see also Cornish, 1964a). From

(6.8.19), (6.8.8) and (6.8.2):

$$\begin{aligned}2\kappa\hat{t}^{ik} = {} & (\partial_l \tilde{g}^{ik})(\partial_m \tilde{g}^{lm}) - (\partial_l \tilde{g}^{il})(\partial_m \tilde{g}^{km}) \\ & + \frac{1}{2} g^{ik} g_{lm} \left(\partial_p \tilde{g}^{ln}\right)(\partial_n \tilde{g}^{pm}) - g^{il} g_{mn} (\partial_p \tilde{g}^{kn})(\partial_l \tilde{g}^{mp}) \\ & - g^{kl} g_{mn} (\partial_p \tilde{g}^{in})(\partial_l \tilde{g}^{mp}) + g_{lm} g^{np} (\partial_n \tilde{g}^{il})(\partial_p \tilde{g}^{km}) \\ & + \frac{1}{8}(2g^{il} g^{km} - g^{ik} g^{lm}) \\ & \times (2g_{kp} g_{qr} - g_{pq} g_{nr})(\partial_l \tilde{g}^{nr})(\partial_m \tilde{g}^{pq}) \qquad (6.8.21)\end{aligned}$$

where $\tilde{g}_{ij} = \sqrt{-g} g_{ij}$. From (6.8.20), the Landau and Lifschitz super-potential

$$\hat{U}^{kmjl} = (-g)(g^{kj} g^{ml} - g^{mj} g^{kl}) \qquad (6.8.22)$$

is antisymmetric in both pairs (k, m) and (j, l); therefore the complex (6.8.20) satisfies the conservation law:

$$\partial_k \left[(-g)\left(T^{kj} + \hat{t}^{kj}\right)\right] = 0. \qquad (6.8.23)$$

The interpretation of $\hat{t}^{ij}$ as the contribution to the energy-momentum complex by the background geometry is only justified a posteriori, in that, when coupled with T^{ij} **and** in some special cases, it provides the right expression for the mass and the angular momentum of a given isolated system when measured at infinity. This will be discussed in the following chapter.

7

DYNAMICS ON CURVED MANIFOLDS

7.1 Conservation laws

A necessary prerequisite to a discussion on the equations of motion in general relativity is a realization of the uncertainties which underlie any global definition of the energy, momentum and angular momentum of a gravitating system. As specified in the previous chapter (see (6.3.2)) a conservation law in a given space-time $(M, \overset{*}{\boldsymbol{g}})$ must take into account, besides the physical fields, which are sources of the background geometry, *also* the geometry itself, since through its dynamical character it contributes to the total energy and momentum.

A quantity which proved to be useful in the definition of global conservation laws is the complex (6.8.20), which satisfies the continuity equation (6.8.23). But this equation leads to conserved quantities physically interpretable as total energy, momentum and angular momentum, *only* if we restrict the space-time to be asymptotically empty and flat; that is, if T_{ij} goes to zero sufficiently fast along any space-like direction to allow the space-time curvature to become arbitrarily small at large enough distances from the source. This will be specified more precisely in the following discussion. Equation (6.8.23) is covariant only with respect to linear coordinate transformations; similarly a one-form $\boldsymbol{\xi}$ on M, which is constant in a given coordinate system, i.e.

$$\partial_a \xi_b = 0 \tag{7.1.1}$$

remains so with respect to that group of transformations. Hence the equation

$$\partial_a \left[(-g)\left(T^{ab}+\hat{t}^{ab}\right)\xi_b\right]=0 \tag{7.1.2}$$

is an invariant only under the action of this group. Limited to this then, the volume integral of (7.1.2) over a domain $\Omega \subset M$ of compact closure, covered by a chart with the orientation of M, yields from Gauss' theorem:

$$\int_\Omega \partial_a \left[(-g)\left(T^{ab}+\hat{t}^{ab}\right)\xi_a\right]\, d\Omega = \oint_{\partial\Omega} (-g)\left(T^{ab}+\hat{t}^{ab}\right)\xi_b\, d^3\Sigma_a = 0 \tag{7.1.3}$$

where $\partial\Omega$, the boundary of Ω, is a three-dimensional hypersurface of M. If we choose Ω so that the induced metric on $\partial\Omega$ is nowhere non-degenerate (that is, $\partial\Omega$ is never null) it is possible to express the surface element $d^3\Sigma_a$ as $d^3\Sigma\; n_a$, where $d^3\Sigma$ is the invariant volume element of $\partial\Omega$ and n_a is the unit normal to $\partial\Omega$, chosen to be *outward-pointing* if space-like and *inward-pointing* if time-like. Let us then choose Ω as being confined within a cylindrical world-tube σ, whose surface is everywhere time-like and bounded by two space-like non-intersecting hypersurfaces Σ_1 and Σ_2 (see Fig. 7-1). Assume also that Σ_2 is to the future of Σ_1. Then the integral (7.1.3) can be written as:

$$\begin{aligned}\int_{\Sigma_2\cap\sigma} &(-g)\left(T^{ab}+\hat{t}^{ab}\right)\xi_b n_a\, d^3\Sigma \\ &- \int_{\Sigma_1\cap\sigma} (-g)\left(T^{ab}+\hat{t}^{ab}\right)\xi_b n_a\, d^3\Sigma \\ &= -\int_{\partial\sigma} (-g)\left(T^{ab}+\hat{t}^{ab}\right)\xi_b n_a\, d^3\Sigma.\end{aligned} \tag{7.1.4}$$

As previously stated, a physical interpretation of (7.1.4) is possible only with the assumption that T^{ab} falls off at infinity. For simplicity, we make the stronger assumption that T^{ab} vanishes outside some compact set σ' with $\sigma' \subset \sigma$.

We also require asymptotic flatness, in the sense that the space-time metric in M goes as:

$$g_{ab} = \eta_{ab} + \mathrm{O}\left(\frac{1}{r}\right) \qquad g_{ab,c} = \mathrm{O}\left(\frac{1}{r^2}\right) \tag{7.1.5}$$

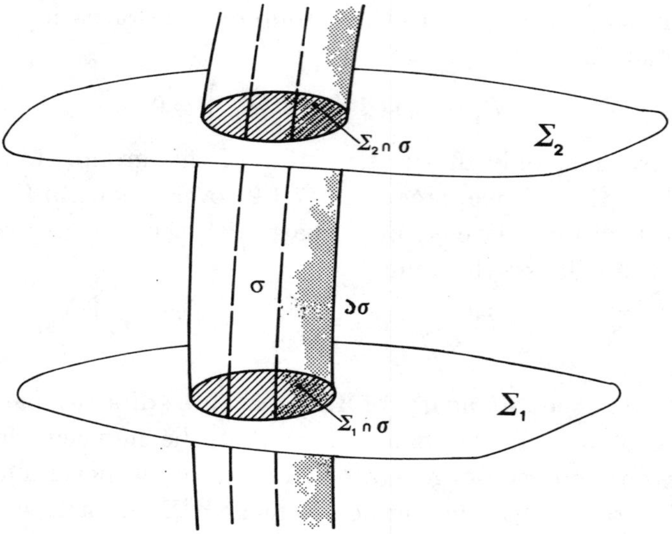

Fig. 7-1 Identifying the four-volume Ω as being confined within σ, Σ_1 and Σ_2.

in the limit* of $r \to \infty$, r being on Σ some coordinate distance from the source. Then we show that the first two integrals in (7.1.4) converge in the limit as $\sigma \to M$; i.e. as $\Sigma_A \cap \sigma \to \Sigma_A$ ($A = 1, 2$). From (6.8.20), (6.8.22) and (7.1.1), the integral on Σ_A can be written as:

$$\frac{1}{2\kappa}\int_{\Sigma_A \cap \sigma} \partial_l \left(\partial_m \hat{U}^{albm} \xi_b\right) n_a \, d^3\Sigma = \frac{1}{2\kappa}\oint_{S_A} \left(\partial_m \hat{U}^{albm}\right) \xi_b \, dS_{al} \tag{7.1.6}$$

where $S_A = \partial(\Sigma_A \cap \sigma)$ is a two-dimensional boundary of $\Sigma \cap \sigma$ and $\boldsymbol{dS}$ is the volume two-form on S. In the limit of $\Sigma \cap \sigma \to \Sigma$, since we can choose the limiting sequence so that S_{ab} goes to infinity as $O(r^2)$, the integrals converge if $\left(\partial_m \hat{U}^{albm}\right) \xi_b$ goes as $O(r^{-2})$ when $r \to \infty$. This is certainly true from (6.8.22) and (7.1.5) (Cornish, 1964). The arbitrariness of $\boldsymbol{\xi}$ and the invariance of (7.1.6) with respect to linear transformations make it convenient to choose as

*In this chapter only, the symbol $O(\frac{1}{r^k})$ is used in a stronger sense than usual, as designating a quantity f such that $r^k f \to$lim. as $r \to \infty$.

a solution of (7.1.1) the set $\xi_i^{\hat{a}} = \delta_i^{\hat{a}}$, with $\hat{a} = 0, 1, 2, 3$, namely the components of an orthonormal coordinate basis at the flat infinity; hence we define

$$P^{\hat{a}}(\Sigma) \equiv \frac{1}{2\kappa} \oint_{S(\infty)} \partial_m \hat{U}^{bl\hat{a}m} \, dS_{bl} \qquad (7.1.7)$$

as the $\hat{a}$-momentum of the system on Σ. Evidently the $P^{\hat{a}}(\Sigma)$'s are invariant with respect to transformations which are asymptotically linear, and they depend only on the asymptotic behaviour of the hypersurface Σ, (see Arnowitt et al., (1962) for an extensive discussion on this topic). Returning to equation (7.1.4): from (7.1.7) and in the limit of $r \to \infty$, it becomes

$$P^{\hat{a}}(\Sigma_2) - P^{\hat{a}}(\Sigma_1) = -\frac{1}{2\kappa} \int_{\partial\sigma} \left(\partial_l \partial_m \hat{U}^{\hat{a}lbm} \right) n_b \, d^3\Sigma. \qquad (7.1.8)$$

A system is said to be *non-radiating* if the integral on the right-hand side vanishes when $\partial\sigma$ is enlarged so that $\Sigma_A \cap \sigma \to \Sigma_A$. In this case $P^{\hat{a}}(\Sigma) = P^{\hat{a}}$ is a conserved quantity in the sense of being independent of Σ.

Condition (7.1.1) is very restrictive, since equation (7.1.2) is also ensured if the one-form $\boldsymbol{\xi}$ satisfies the flat Killing equation:

$$\partial_{(a}\xi_{b)} = 0 \qquad (7.1.9)$$

namely $\boldsymbol{\xi}$ is a Killing one-form at flat infinity. Now since Minkowski space-time admits ten Killing one-forms (it is a maximally symmetric space), (7.1.9) provides ten conservation equations of the type (7.1.8). Four of them are given by $\xi_a^{\hat{a}} = \delta_a^{\hat{a}}$, and six more are given by solutions of the type: $\xi_a^{\hat{a}\hat{b}} = \delta_{ab}^{\hat{a}\hat{b}} x^b$; $\hat{a}, \hat{b} = 0, 1, 2, 3$ where $\{x^b\}$ are orthogonal coordinates. Using the latter in (7.1.6), we define:

$$S^{\hat{a}\hat{b}} = -\frac{1}{\kappa} \oint_{S(\infty)} x^{[\hat{a}} \partial_m \hat{U}^{\hat{b}]mnl} dS_{nl} \qquad (7.1.10)$$

as the angular momentum of the system, measured at a flat (actually Minkowskian) space-like infinity.

The combination, expressed in (7.1.2), of the continuity equation (6.8.23) with an infinitesimal translation, generated by a one-form $\boldsymbol{\xi}$ constrained by (7.1.9), yields conservation laws which are not generally covariant and which can be given a physical interpretation only within a restricted class of coordinates (asymptotic Lorentzian) (Møller, 1958, 1961). However, the symmetry properties of the Landau-Lifschitz superpotential (6.8.22) together with (7.1.9) (despite their non-tensorial character with respect to general coordinate transformations) allow the definition of global conservation laws which are generally covariant. Define the new quantities:

$$\mathcal{P}^{al} \equiv \frac{1}{\kappa}(-g)^{-\frac{1}{2}}\left[\partial_m \hat{U}^{albm}\xi_b - \partial_m\left(\hat{U}^{albm}\xi_b\right)\right]. \tag{7.1.11}$$

From the properties of $\hat{U}^{albm}$ and (7.1.9), these become:

$$\begin{aligned}\mathcal{P}^{al} &= -\frac{1}{\kappa}(-g)^{-\frac{1}{2}}\left(\hat{U}^{albm}\partial_m\xi_b\right)\\ &= -\frac{1}{\kappa}(-g)^{\frac{1}{2}}(g^{ab}g^{lm} - g^{lb}g^{am})\partial_m\xi_b = \frac{2}{\kappa}(-g)^{\frac{1}{2}}\nabla^{[a}\xi^{l]}\end{aligned} \tag{7.1.12}$$

where the symmetry of the connection coefficients is ensured by the use of holonomic coordinates and the assumption of zero torsion. Thus:

$$\mathcal{P}^a := \partial_l\mathcal{P}^{al} = \frac{2}{\kappa}\partial_l\left[(-g)^{\frac{1}{2}}\nabla^{[a}\xi^{l]}\right] = \frac{2}{\kappa}(-g)^{\frac{1}{2}}\nabla_l\nabla^{[a}\xi^{l]} \tag{7.1.13}$$

is a vector density which satisfies the identity:

$$\partial_a\mathcal{P}^a = \frac{2}{\kappa}(-g)^{\frac{1}{2}}\nabla_a\nabla_l\nabla^{[a}\xi^{l]} = 0. \tag{7.1.14}$$

Integrating this over a four-volume Ω:

$$\int_\Omega \partial_a\mathcal{P}^a \boldsymbol{d}^4x = \frac{2}{\kappa}\int_\Omega \nabla_a\nabla_l\nabla^{[a}\xi^{l]}\, d\Omega = 0 \tag{7.1.15}$$

and applying Gauss' theorem, we are led, with similar arguments as before, to the identification of a conserved quantity which is a *true* invariant:

$$P_\kappa = \frac{2}{\kappa}\oint_S(-g)^{\frac{1}{2}}\nabla^{[a}\xi^{b]}dS_{ab} = \frac{2}{\kappa}\oint_S *\boldsymbol{d\xi} \tag{7.1.16}$$

where

$$*\boldsymbol{d\xi} = \frac{1}{2}(-g)^{\frac{1}{2}}\delta_{abrs}\nabla^{[a}\xi^{b]}\boldsymbol{d}x^r\wedge\boldsymbol{d}x^s.$$

This expression is due to Komar (Komar, 1959); it provides a satisfactory definition of total energy and angular momentum, respectively, measured at space-like infinity in the case of an asymptotically flat and empty space-time (namely space-time describing an isolated system) which admits a Killing one-form $\boldsymbol{\xi}$, which is respectively time-like (the space-time has time symmetry) or axial (the space-time is axisymmetric). When $\boldsymbol{\xi}$ is a Killing one-form, then (7.1.16) can be written, going one step back in the derivation:

$$P_{_K} = -\frac{2}{\kappa}\int_{\Sigma\cap\sigma'} \nabla_a\nabla^{[a}\xi^{b]}\; d^3\Sigma_b = \frac{2}{\kappa}\int_{\Sigma\cap\sigma'} R^{bm}\xi_m\; d^3\Sigma_b$$
$$= 2\int_{\Sigma\cap\sigma'} \left(T^b{}_a - \frac{1}{2}\delta^b_a T\right)\xi^a\; d^3\Sigma_b, \qquad (7.1.17)$$

where we have used Einstein equations and the integration is extended to the domain $(\Sigma\cap\sigma)'$ where $T_{ab}\neq 0$. Let us now recall that:

$$d^3\Sigma_b = \frac{1}{3!}\sqrt{-g}\,\delta_{brst}\,\boldsymbol{dx}^r\wedge\boldsymbol{dx}^s\wedge\boldsymbol{dx}^t$$
$$= n_b|h|^{\frac{1}{2}}\frac{1}{3!}\delta_{\alpha\beta\gamma}\,\boldsymbol{dx}^\alpha\wedge\boldsymbol{dx}^\beta\wedge\boldsymbol{dx}^\gamma \qquad \alpha,\ \beta,\ \gamma\neq b$$

where $|h|^{\frac{1}{2}}$ is the determinant of the induced metric on Σ and n^b is a unit normal to Σ. Hence (7.1.17) becomes:

$$P_{_K} = 2\int_{\Sigma\cap\sigma'} \left(T^b{}_a - \frac{1}{2}\delta^b_a T\right)\xi^a n_b|h|^{\frac{1}{2}}\,\boldsymbol{d}^3x. \qquad (7.1.18)$$

For an application of this formula, see Sect. 10.3.

7.2 The equations of motion of an extended body

Equations (7.1.7) and (7.1.10) enable one to determine the total gravitational mass and angular momentum of an isolated system at spatial infinity, provided space-time is asymptotically flat and a Killing one-form exists at infinity (cf. eqn. 7.1.9). The question now arises, how can one deduce the equation of motion of an isolated system from the knowledge of its local properties. The energy-momentum tensor gives a complete description of an uncharged material system and all the information about its motion is contained in the continuity equation $\nabla_i T^{ij} = 0$. The latter however are implied by the field equations $R_{ij} - \frac{1}{2}g_{ij}R = \kappa T_{ij}$ where the properties of the system and those of the field are deeply intertwined. What is needed for a complete description of the system is the separation, in the above equations, of the parameters which relate to the material system from those which relate to the gravitational field. This is a difficult task which can be tackled either by resorting to some overall approximation procedure, or by attempting to find an exact solution to the problem (Dixon, 1970a, b 1979). While the latter program is still far from completion, the elegance and rigour of the approach is strongly suggestive of a successful end and therefore we shall outline some of its main aspects.

Let the energy-momentum tensor of the system be at least C^1 and non-zero in a world-tube W of M such that the intersection of W with any space-like hypersurface Σ is bounded and lies in some open set U of M which is a normal neighbourhood of each of its points. This ensures that the space-like geodesics in $W \cap \Sigma$ are well behaved (they do not intersect) and the body is spatially finite. Assume now that space-time admits an isometry described by a Killing vector field ξ. Thus from (6.3.10) and the Killing equation it follows that:

$$\nabla_i (T^i{}_j \xi^j) = 0. \tag{7.2.1}$$

An integration of (7.2.1) over a four-volume, as shown in Fig. 7-1, and application of Gauss' theorem, leads to

$$\int_\Omega \nabla_i (T^i{}_j \xi^j) d\Omega = \int_{\partial\Omega} T^{ij} \xi_j \, d^3\Sigma_i = 0, \tag{7.2.2}$$

where $d^3\Sigma_i$ is a surface element.

If we extend the volume of integration to asymptotic infinity where we assume $T^{ij} = 0$, then (7.2.2) leads to a conservation equation of the type

$$\int_{\Sigma\cap W} T^{ik} \xi_k \, d^3\Sigma_i = \text{ const.} \tag{7.2.3}$$

for any space-like hypersurface Σ whose normal n is taken to be future-oriented. The integration of (7.2.3) requires the knowledge of the Killing one-form everywhere over the body's section, besides that of the internal structure. It is known from Sect. 3.8 that a Killing field is completely determined once ξ_i and $\nabla_i \xi_j$ are known at some point. Let p_0 and p_1 be two such points inside $\Sigma \cap W$ and let $\Omega^{i_0}(p_0, p_1)$ be as defined in Sect. 3.7, namely (minus) the components of the vector at p_0 tangent to the unique space-like geodesic from p_0 to p_1, with modulus equal to the geodesic distance between them. The existence of a space-time symmetry implies naturally that:

$$\pounds_{\xi}\Omega^{i_0}(p_0, p_1) = 0$$

and, recalling that $\Omega^{i_0}(p_0, p_1)$ behaves like a vector at p_0 and like a scalar at p_1, this becomes explicitly:

$$\xi^{j_0}\nabla_{j_0}\Omega^{i_0} - \Omega^{j_0}\nabla_{j_0}\xi^{i_0} + \xi^{j_1}\partial_{j_1}\Omega^{i_0} = 0.$$

From Sect. 3.7, lowering the index i_0, we have:

$$\xi_{j_0}\Omega_{i_0}{}^{j_0} - \Omega^{j_0}\nabla_{j_0}\xi_{i_0} + \xi_{j_1}\Omega_{i_0}{}^{j_1} = 0. \tag{7.2.4}$$

If p_0 is sufficiently close to p_1, the matrix $\Omega_{i_0}{}^{j_1}$ is non-singular since it approaches the identity. Hence defining its inverse as $\overset{-1}{\Omega}{}_{k_1}{}^{i_0}$ and multiplying (7.2.4) with this, we have:

$$\xi_{k_1} = -\overset{-1}{\Omega}{}_{k_1}{}^{i_0}\Omega_{i_0}{}^{j_0}\xi_{j_0} + \overset{-1}{\Omega}{}_{k_1}{}^{i_0}\Omega^{j_0}\nabla_{j_0}\xi_{i_0}\,. \tag{7.2.5}$$

Using this in (7.2.3), we have:

$$\xi_{j_0}P^{j_0} + \frac{1}{2}\nabla_{[j_0}\xi_{i_0]}S^{j_0 i_0} = \text{const.} \tag{7.2.6}$$

where:

$$P^{j_0} = -\int_{\Sigma} T^{i_1 k_1}\overset{-1}{\Omega}{}_{k_1}{}^{s_0}\Omega_{s_0}{}^{j_0}n_{i_1}\, d\Sigma \tag{7.2.7}$$

$$S^{i_0 j_0} = -2\int_{\Sigma} T^{s_1 k_1}\overset{-1}{\Omega}{}_{k_1}{}^{[i_0}\Omega^{j_0]}n_{s_1}\, d\Sigma. \tag{7.2.8}$$

The character of the world-function as a two-point function ensures that P^{i_0} and $S^{i_0 j_0}$ are respectively a vector and an antisymmetric tensor at p_0. This being understood, we shall drop the subscript $_0$ from the coordinate indices.

In the case where the space-time is maximally symmetric, the independence of the relations (7.2.3) from Σ *for all choices of* ξ also ensures that the quantities defined in (7.2.7) and (7.2.8) do not depend on the way the section Σ is chosen.

They are termed (*proper*) *momentum* and *angular momentum* of the body relative to p_0.

At this stage there is some arbitrariness in the choice of Σ and of the point p_0 in Σ. We need a one-parameter family of space-like cross sections $\Sigma(\lambda)$ of W and a point p in each which will be the base-point for the above construction. Both these are achieved by first fixing the smooth curve γ_0, say, and then taking $\Sigma(\lambda)$ to be the surface generated by the space-like geodesics orthogonal to γ_0 at $\gamma_0(\lambda)$. In the analysis above, $p_0 = \gamma_0(\lambda_0)$, say, and for general λ the evaluation of $P^k(\gamma_0, \Sigma(\lambda))$ and $S^{kl}(\gamma_0, \Sigma(\lambda))$ now gives well-defined tensor fields on γ_0. No special requirements are needed for the curve γ_0 (apart from being smooth) or for the parameter λ; however it will be physically important to impose some restrictions so to identify it uniquely as the centre-of-mass world-line. We shall delay to the following section a discussion on this point.

We can make the definitions (7.2.7) and (7.2.8) quite general. But when there exists a Killing field, relation (7.2.6) holds identically along any arbitrary world-line γ (determining the choice of $\Sigma(\lambda)$) and constrains the way in which P^i and S^{ij} vary on γ. Differentiating (7.2.6) along γ_0, we obtain:

$$\xi_k \left[\frac{DP^k}{D\lambda} + \frac{1}{2} R^k{}_{lij} v^l S^{ij} \right] + \frac{1}{2} \nabla_{[k} \xi_{l]} \left[\frac{DS^{kl}}{D\lambda} - 2P^{[k} v^{l]} \right] = 0 \quad (7.2.9)$$

where use was made of the condition (3.8.5), namely:

$$\nabla_i \nabla_j \xi_k = R^m{}_{ijk} \xi_m$$

for a Killing field and we have set $v^k \equiv d\gamma_0^k / d\lambda$. Equation (7.2.9) holds with no approximations; it expresses the connection between the integrals of motion and the isometries.

Since only momentum and angular momentum enter explicitly in (7.2.9) one is led to ask whether that equation extends to give a definition of general laws of motion in any multipole approximation for the body under consideration. The answer is positive and in fact we can give an intuitive argument which suggests the form of the equations of motion. If space-time $(M, \overset{*}{\boldsymbol{g}})$ is of constant curvature, then the contributions to the force and the torque on the body arising from the interaction between the curvature and the moments of the body higher than the dipole (quadrupole, etc...) vanish identically. However the curvature isotropy about every point ensures that the space-time is maximally symmetric. Therefore (see Sect. 3.8) ξ_k and $\nabla_{[k}\xi_{l]}$ may be given arbitrary values at p_0, and so from (7.2.9)*:

$$\frac{DP^k}{D\lambda} + \frac{1}{2}R^k{}_{lij}v^l S^{ij} = 0 \qquad (7.2.10)$$

$$\frac{DS^{kl}}{D\lambda} - 2P^{[k}v^{l]} = 0. \qquad (7.2.11)$$

In the special case here considered these equations give complete information, but in a general case the higher multipoles of the body, unless neglected on a physical ground, give rise to a (self)-force and torque which will contribute to the right-hand side of equations (7.2.10) and (7.2.11). Thus we have:

$$\frac{DP^k}{D\lambda} + \frac{1}{2}R^k{}_{lij}v^l S^{ij} = f^k \qquad (7.2.12)$$

$$\frac{DS^{kl}}{D\lambda} - 2P^{[k}v^{l]} = \tilde{G}^{kl} \qquad (7.2.13)$$

where f^k, $\tilde{G}^{kl}$ are functions of the position $(\gamma_0(\lambda))$ on the curve, of the metric, and of the multipole $(n \geq 4)$ moments $J^{a_1,a_2\ldots a_n}$ of T^{ij}. The latter depend functionally on the metric field. It is reasonable to generalize the above arguments by requiring that equations (7.2.12) and (7.2.13) hold also in the case when there are no specific symmetries. The latter would have the effect of constraining the parameters of the system as in (7.2.6) and (7.2.9). In this case it can be proven that the self-force and torque are also constrained by the equation:

$$\xi_k f^k + \frac{1}{2}\nabla_{[k}\xi_{l]}\tilde{G}^{kl} = 0. \qquad (7.2.14)$$

This ensures the consistency of the identification made in (7.2.12) and (7.2.13).

*Equations (7.2.10) and (7.2.11) were derived first by Mathisson (1937) and Papapetrou (1951); see also Tulczyjew (1959).

7.3 The centre-of-mass description

The countably many variables which carry all the information about the body, i.e. P^k, S^{kl} and the higher multipole moments that fix f^k and $\tilde{G}^{kl}$, depend on the curve γ with respect to which the moment expansion of T^{ab} is performed. In order to specify the curve γ as the world-line of the centre-of-mass of the system, let us first recall its Newtonian definition. Using Cartesian coordinates x^α ($\alpha = 1, 2, 3$) and denoting by z^α those of the centre-of-mass, we have:

$$z^\alpha \int_V \rho \, \boldsymbol{d}^3 x = \int_V \rho x^\alpha \, \boldsymbol{d}^3 x \tag{7.3.1}$$

where ρ is the density of the body and V its three-volume.

The relativistic generalization of (7.3.1) requires some care. Suppose first that we are in flat space-time (special relativity). Then $\rho = T^{00}$ and the natural generalization of (7.3.1) reads:

$$z^\alpha n_a \int_{x^0=\text{const.}} T^{ab} n_b \, d\Sigma = \int_{x^0=\text{const}} T^{ab} n_a n_b x^\alpha \, d\Sigma$$

with $n_a = \delta^0_a$; or

$$n_a \int_{x^0=\text{const.}} (z^\alpha - x^\alpha) T^{ab} n_b \, d\Sigma = 0. \tag{7.3.2}$$

We can rewrite this in terms of S^{ab} as follows. The forms of (7.2.7) and (7.2.8) are easily deduced by noticing that in flat space with p_0 the point (z^α, z^0) we have $\Omega^i(z, x') = z^i - x'^i$ and $\Omega^i_j = \delta^i_j$, $\Omega^{i'}_j = \delta^{i'}_j$. Hence:

$$\begin{aligned} a) \quad & P^k(\Sigma) = \int_\Sigma T^{kr} n_r \, d\Sigma \\ b) \quad & S^{kl}(\Sigma) = 2 \int_\Sigma \left(z^{[k} - x'^{[k} \right) T^{l]r} n_r \, d\Sigma \end{aligned} \tag{7.3.3}$$

where, as a consequence of the continuity equations $\partial_a T^{ab} = 0$ and of the symmetry of flat space, the momentum P^k and the angular momentum S^{ab} about the point z on Σ are independent of the particular choice of Σ. Using the fact that $z^0 = x^0$, (7.3.2) can be written, from (7.3.3)b, as

$$n_a S(z^l)^{ab} = 0 \tag{7.3.4}$$

where we have explicitly written the dependence of S on z. The vector n^a can be identified physically as defining the Lorentz frame in which the body is at rest, i.e. in which the momentum is zero, in the sense that $P^k = Mn^k$, where n^k is a unit time-like vector and M is a positive constant regarded as the total mass of the body.

These conditions uniquely identify the event z on Σ as the body's centre of mass. Its covariant form allows generalization to arbitrary coordinates. In a curved space-time, it can be proved that, provided the curvature is not too high, at each point $z \in \Sigma \cap W$ there always exists a future-directed time-like unit vector u, such that

$$P^{[i}u^{k]} = 0, \qquad u^i u_i = -1. \tag{7.3.5}$$

In the above conditions, the arbitrariness in the choice of z can be removed by selecting that point z_0 in $\Sigma \cap W$ where:

$$u_a(z_0)S^{ab}(z_0) = 0. \tag{7.3.6}$$

(It can be shown that this is unique; Beiglböck, 1967.)

As Σ varies in the one-parameter family of cross sections $\Sigma(\lambda)$, $z_0(\lambda)$ describes a world-line γ_0 in W which, by analogy with (7.3.4), will be defined as the centre-of-mass line of the body. The vector field u on γ_0 is in general distinct from the vector field v^a which, being tangent to γ_0, describes properly the four-velocity of the body. Along γ_0 then we have from (7.3.5):

$$P^k = Mu^k \tag{7.3.7}$$

where $M(\lambda)$ is a positive quantity which is given the interpretation of the total mass of the body. In general it is not constant; in fact from (7.2.12), (7.3.7) and the second equation of (7.3.5) we have:

$$\frac{dM}{d\lambda} = \frac{1}{2}R^{abcd}u_a v_b S_{cd} - f^a u_a. \tag{7.3.8}$$

Evidently $dM/d\lambda$ vanishes in a flat space-time but we shall show that it also vanishes in a curved space-time whenever one neglects the force f^a and the torque $\tilde{G}^{ab}$ which arise from the multipole moments of the body. Contracting (7.2.12) with v_k, we have:

$$\frac{dM}{d\lambda} v_a u^a + M v_a \frac{Du^a}{D\lambda} = f^a v_a \ . \tag{7.3.9}$$

Differentiating (7.3.6) along γ_0 we also have:

$$u_a \frac{DS^{ab}}{D\lambda} = -\frac{Du_a}{D\lambda} S^{ab}. \tag{7.3.10}$$

Contraction with $Du_b/D\lambda$ yields:

$$u_a \frac{Du_b}{D\lambda} \frac{DS^{ab}}{D\lambda} = 0 \tag{7.3.11}$$

but from (7.2.13), this becomes :

$$M v^b \frac{Du_b}{D\lambda} = \frac{Du_b}{D\lambda} u_a \tilde{G}^{ab}. \tag{7.3.12}$$

Using this equation in (7.3.9), we finally obtain:

$$\frac{dM}{d\lambda} = (v_a u^a)^{-1} \left[\frac{Du_c}{D\lambda} u_b \tilde{G}^{cb} + f^c v_c \right]. \tag{7.3.13}$$

The important result here is that the first multipole moment of the body which causes a loss or absorption of gravitational energy by the body is the quadrupole. To this order the explicit expressions of f^a and $\tilde{G}^{ab}$ are (Dixon, 1970a):

$$f^a = \frac{1}{6}(u^r v_r) J^{lmnp} \nabla^a R_{lmnp} \tag{7.3.14}$$

$$\tilde{G}^{ab} = \frac{4}{3}(u^r v_r) J_{lmn}{}^{[a} R^{lmnb]} \tag{7.3.15}$$

where J^{abcd} is the quadrupole tensor.

An important task in studying the equations of motion is to express the centre-of-mass four-velocity v^a in terms of the other parameters of the body as the momentum, spin and higher multipole terms, (Ehlers and Rudolf, 1977). That requires a solution of (7.2.12) and (7.2.13) for v^i, with the conditions (7.3.5) and (7.3.6). To this purpose it will be useful to recall some properties of v^i in the simpler situations. In a flat space-time, the three-velocity of the centre-of-mass with respect to the zero-momentum observer vanishes. To show this, let us consider the flat space-time form of equations (7.2.10) and (7.2.11):

$$\frac{DP^a}{D\lambda} = 0 \qquad \frac{DS^{ab}}{D\lambda} = P^a v^b - P^b v^a. \tag{7.3.16}$$

Contracting the second of (7.3.16) with u^a, we obtain:

$$u_a \frac{DS^{ab}}{D\lambda} = -M\left[v^b + (u_a v^a)u^b\right] = -Mv^a\left(\delta^b_a + u^b u_a\right). \quad (7.3.17)$$

Here the tensor $\underset{u}{h}{}^b{}_a \equiv \delta^b_a + u^b u_a$ projects orthogonally to u, so the quantity $\underset{u}{h}{}^b{}_a v^a$ is the velocity of the mass centre as seen in the rest frame of the observer with four-velocity u (the zero-momentum frame; an extensive discussion on frames and measurements will be given in chapter 9). Differentiating (7.3.6) along γ_0 and taking into account the first of (7.3.16), we obtain:

$$u_a \frac{DS^{ab}}{D\lambda} = 0 \quad (7.3.18)$$

and so:

$$\underset{u}{h}{}^a{}_b v^b = 0 \quad (7.3.19)$$

as wanted. By definition, moreover, the latter implies:

$$v^a = -(v^r u_r)u^a, \quad (7.3.20)$$

showing that v and u are parallel. Since M is constant, this implies:

$$\frac{Dv^a}{D\lambda} = 0 \qquad \frac{Du^a}{D\lambda} = 0 \qquad \frac{DS^a}{D\lambda} = 0 \quad (7.3.21)$$

where $S^a = -\frac{1}{2}\eta^{abcd}u_b S_{cd}$ is the spin vector. The above quantities mean that the centre-of-mass world line γ_0 is a geodesic and S is parallely propagated along it.

In a curved space-time these results are in general not true. In order to deduce the velocity of the centre-of-mass relative to the zero-momentum frame, we need some more algebra. Expressing equation (7.2.13) in terms of the spin vector S, we obtain:

$$S_{[a}\left(\frac{D}{D\lambda}u_{b]}\right) + \left(\frac{D}{D\lambda}S_{[a}\right)u_{b]} = \left(P_{[a}v_{b]}\right)^* + \frac{1}{2}\overset{*}{\tilde{G}}_{ab} \quad (7.3.22)$$

where $*$ means dualization, (see Sect. 1.17). Contraction of (7.3.22) with u^a yields:

$$\underset{u}{h}{}^a{}_b \frac{DS_a}{D\lambda} = u^a \overset{*}{\tilde{G}}_{ab} \quad (7.3.23)$$

then combining this with (7.3.22) leads to:

$$S_{[a}\left(\frac{D}{D\lambda}u_{b]}\right) = \left(P_{[a}v_{b]}\right)^* + \frac{1}{2}\left[\overset{*}{\tilde{G}}_{ab} - 2u_{[a}\overset{*}{\tilde{G}}_{b]c}u^c\right]. \quad (7.3.24)$$

Now take the dual of (7.3.24) to obtain:

$$\left[\left(\frac{D}{D\lambda}u_{[a}\right)S_{b]}\right]^* = P_{[a}v_{b]} + u_{[a}\tilde{G}_{b]c}u^c. \tag{7.3.25}$$

The term on the left-hand side can also be written as:

$$\left[\left(\frac{D}{D\lambda}u^{[a}\right)S^{b]}\right]^* = -u^{[a}\eta^{b]}{}_{cdr}u^c S^d \frac{Du^r}{D\lambda} \tag{7.3.26}$$

which combined with (7.3.25) gives:

$$u^{[a}\left(Mv^{b]} + \tilde{G}^{b]c}u_c + \eta^{b]}{}_{cdr}u^c S^d \frac{Du^r}{D\lambda}\right) = 0 \tag{7.3.27}$$

or equivalently

$$Mv^b + \tilde{G}^{bc}u_c + \eta^b{}_{cdr}u^c S^d \frac{Du^r}{D\lambda} = -(u^s v_s)Mu^b \tag{7.3.28}$$

and hence

$$M\underset{u}{h}{}^a{}_b v^b = -\eta^a{}_{bcd}u^b S^c \frac{Du^d}{D\lambda} - \tilde{G}^{ab}u_b. \tag{7.3.29}$$

The vector v enters implicitly in $Du^c/D\lambda$, so let us calculate the first term on the right-hand side explicitly. From (7.2.12) and after lengthy calculations, it becomes:

$$\eta^a{}_{bcd}u^b S^c \frac{Du^d}{D\lambda} = \frac{1}{M}\left[v^a \overset{*\,*}{R}{}^{bcrs}u_b S_c u_r S_s - (v^b u_b)\overset{*\,*}{R}{}^{acrs}S_c u_r S_s \right.$$
$$\left. + (v^b S_b)\overset{*\,*}{R}{}^{acrs}u_c u_r S_s + \eta^a{}_{bcd}u^b S^c f^d\right]. \tag{7.3.30}$$

Contraction of (7.3.29) with S_a gives:

$$S^a v_a = -\frac{1}{M}\tilde{G}^{ab}S_a u_b \tag{7.3.31}$$

and hence, using this in (7.3.30) and substituting in (7.3.29) we finally obtain*

$$\left(1 + \frac{1}{M^2}\overset{*\,*}{R}{}^{abcd}u_a S_b u_c S_d\right)\underset{u}{h}{}^r{}_s v^s = \frac{1}{M^2}(u^d v_d)\underset{u}{h}{}^r{}_s \overset{*\,*}{R}{}^{sabc}S_a u_b S_c$$
$$+ \frac{1}{M}\left(\delta^r_d - \frac{1}{M^2}\overset{*\,*}{R}{}^{rabc}u_a u_b S_c S_d\right)\tilde{G}^{ed}u_e$$
$$- \frac{1}{M^2}\eta^r{}_{abc}u^a S^b f^c. \tag{7.3.32}$$

---

*Equation (7.3.32) in the pole-dipole approximation only was first derived by Tod et al., (1976).

The curve γ_0 can be parametrized so that

$$v^a u_a = -1. \tag{7.3.33}$$

thus equation (7.3.32) provides the velocity of the centre-of-mass of the system, in terms of the body's multipole moments, P^a, S^a, $J^{a_1, a_2 \ldots a_n}$ which are, together with γ_0, intrinsically determined by the energy-momentum tensor T^{ij}. If the self-force and torque vanish ($f^a = \tilde{G}^{ab} = 0$) the velocity of the mass centre relative to the zero-momentum frame is determined by the space-time curvature and the angular momentum of the body S. The latter is in general not parallely transported along the world-line γ_0, as can be deduced from (7.3.23) which can be written more conventionally as:

$$\frac{DS^a}{D\lambda} = u^a \frac{Du^b}{D\lambda} S_b. \tag{7.3.34}$$

The four-velocities v and u can be made to coincide under specific symmetry conditions, when (7.3.23) would imply that the spin vector is Fermi-Walker transported (see Sect. 4.4) along γ_0. This deviation from parallel transport gives rise to the Thomas precession.

7.4 Motion of a point particle

A body will be called *point-like* if the typical scale of the cross section of its world-tube W with any space-like hypersurface Σ is much less than the scale over which the space-time curvature changes appreciably. These conditions are achieved by taking the limit in which $\Sigma(\lambda) \cap W \to \gamma_0(\lambda)$ and in which the gravitational strength tends to zero. Then we can neglect all terms containing curvature gradients and multipole terms in (7.3.32) and (7.2.12), yielding immediately

$$v = u \quad , \quad \frac{Du}{D\lambda} = 0 \ ;$$

namely the centre-of-mass world line appears to be an affine geodesic of the background metric. The result is nonetheless independent of the space-time symmetries and can be derived in general from equation (6.3.10) using expression (6.6.10) for the energy-momentum tensor of a single particle. We have in this case (putting $\varphi = (-\breve{u}^i\breve{u}_i)^{\frac{1}{2}} \neq 0$):

$$0 = \nabla_i T^{ij} = \int \frac{\mu}{\varphi} \Big[\nabla_i \left(\delta_g^4(x - \gamma(\lambda))\breve{u}^i\right) \breve{u}^j$$
$$+ \breve{u}^i \nabla_i \breve{u}^j \delta_g^4(x - \gamma(\lambda)) - \frac{1}{\varphi}\breve{u}^i\breve{u}^j\partial_i\varphi\delta_g^4(x - \gamma(\lambda))\Big] \, d\lambda \quad (7.4.1)$$

where $\delta_g^4(x) = \delta^4(x)/\sqrt{-g}$ and all the other quantities here have the meaning specified in Sect. 6.6. The first term on the right hand side can be calculated as follows. If ψ is a test function (a smooth function equal to zero outside some compact set) then

$$\int \sqrt{-g}\mathbf{d}^4x\psi(x) \int \frac{\mu}{\varphi}\nabla_i \left(\delta_g^4(x - \gamma(\lambda))\breve{u}^i\right) \breve{u}^j d\lambda$$
$$= \int \mathbf{d}^4x\psi(x) \int \frac{\mu}{\varphi}\partial_i \left(\delta^4(x - \gamma(\lambda))\breve{u}^i\right) \breve{u}^j d\lambda$$
$$= - \int \int \mathbf{d}^4x\partial_i \left(\psi(x)\frac{\mu}{\varphi}\breve{u}^j\right) \delta^4(x - \gamma(\lambda))\breve{u}^i d\lambda$$
$$= - \int d\lambda \dot{z}^i \partial_i \left(\frac{\psi(x)\mu\breve{u}^j}{\varphi(x)}\right)_{x=\gamma(\lambda)} = 0$$

using the fact that $\breve{u}^i = \dot{z}^i$ on $\gamma(\lambda)$ and noting that the integrand in the last integral is a total derivative with respect to λ.

The remaining terms in (7.4.1) then give

$$0 = \int \frac{\mu}{\varphi} \left[\breve{u}^i\nabla_i\breve{u}^j - \frac{1}{\varphi}\breve{u}^i\breve{u}^j\partial_i\varphi\right] \delta_g^4(x - \gamma(\lambda))d\lambda \, .$$

As before, multiplying by a test function $\psi(x)$ and integrating, we obtain

$$0 = \int \frac{\mu}{\varphi} \left[\breve{u}^i\nabla_i\breve{u}^j - \frac{1}{\varphi}\breve{u}^i\breve{u}^j\partial_i\varphi\right] \psi(\gamma(\lambda))d\lambda$$

which, if it is to hold for any ψ, implies

$$\breve{u}^i\nabla_i\breve{u}^j - \frac{1}{\varphi}\breve{u}^i\breve{u}^j\partial_i\varphi = 0$$

or

$$\dot{z}^i\nabla_i\dot{z}^k = f(\lambda)\dot{z}^k \quad (7.4.2)$$

which is the expected equation of a geodesic (in general not affinely parametrized). A particle which moves according to the law

(7.4.2) is termed *free falling* since it is acted upon by no other fields but gravity itself.

A direct approach, starting from the action itself rather than the energy-momentum tensor, yields a simple geometrical interpretation for the geodesic equation in terms of variation of length. The action S_f in (6.6.6) can be written more suitably as

$$S_f = -\int_\gamma d\lambda \int_W \mu\varphi\delta_g^3(x - \gamma(\lambda))\, d^3\Sigma = -\int_\gamma \mu(-\dot{z}^i\dot{z}_i)^{\frac{1}{2}}\, d\lambda.$$

We recognize this expression as the length of the curve traced by the particle according to the definition (1.19.4), and so the requirement that the action be extremised becomes the requirement that the length is extremized, which is a well-known property of a geodesic (Levi-Civita, 1918).

To verify the equivalence of the variational definition of a geodesic with the definition in terms of the geodesic equation, consider a variation of the curve subject to the conditions that the function $\varphi^2 = |(\dot{\gamma}^k\dot{\gamma}_k)|$ is never zero and the curve is fixed at the parameter values λ_1 and λ_2 between which the action is evaluated. For convenience we suppose that the curve and its variation are contained in a given chart of M; then (we drop the subscript $_f$ for convenience) varying

$$S(\lambda_1, \lambda_2; \gamma) = -\mu \int_{\lambda_1}^{\lambda_2} |(\dot{\gamma}^j\dot{\gamma}_j)|^{\frac{1}{2}}\, d\lambda \tag{7.4.3}$$

with respect to γ^i we require:

$$\delta_\gamma S = -\mu \int_p^q \frac{1}{2}|(\dot{\gamma}|\dot{\gamma})|^{-\frac{1}{2}} \left[(\delta g_{ij})\dot{\gamma}^i\dot{\gamma}^j + 2g_{ij}(\delta\dot{\gamma})\dot{\gamma}^j\right]\, d\lambda = 0. \tag{7.4.4}$$

Let $\dot{\gamma}^i = d\gamma^i/d\lambda$; then from the condition $\delta(d/d\lambda) = (d/d\lambda)\delta$ and integrating by parts we obtain:

$$\begin{aligned}\delta_\gamma S = -\frac{\mu}{2}\int_p^q &\Big\{|(\dot{\gamma}|\dot{\gamma})|^{-\frac{1}{2}}(\partial_k g_{ij})\dot{\gamma}^i\dot{\gamma}^j \\ &- 2\frac{d}{d\lambda}\left[|(\dot{\gamma}|\dot{\gamma})|^{-\frac{1}{2}} g_{kj}\dot{\gamma}^j\right]\Big\}\,\delta\gamma^k\, d\lambda \\ = -\frac{\mu}{2}\int_p^q &|(\dot{\gamma}|\dot{\gamma})|^{-\frac{1}{2}}\end{aligned}$$

$$\times \left\{ (\partial_k g_{ij})\dot{\gamma}^i\dot{\gamma}^j - 2(\partial_r g_{kj})\dot{\gamma}^r\dot{\gamma}^j - 2g_{kj}\frac{d\dot{\gamma}^j}{d\lambda} + 2f(\lambda)g_{kj}\dot{\gamma}^j \right\} \delta\gamma^k \; d\lambda = 0 \qquad (7.4.5)$$

where $f(\lambda) = d \; \ln|(\dot{\gamma}|\dot{\gamma})|^{\frac{1}{2}}/d\lambda$. The quantity in braces in (7.4.5) can be rewritten as:

$$(\partial_k g_{rj} - \partial_r g_{kj} - \partial_j g_{kr})\dot{\gamma}^r\dot{\gamma}^j - 2g_{kj}\frac{d\dot{\gamma}^j}{d\lambda} + 2f(\lambda)g_{kj}\dot{\gamma}^j$$
$$= -2g_{kj}\left(\frac{d\dot{\gamma}^j}{d\lambda} + \Gamma^j_{rs}\dot{\gamma}^r\dot{\gamma}^s - f(\lambda)\dot{\gamma}^j\right) \qquad (7.4.6)$$

from (2.8.7); hence, from the arbitrariness of $\delta\gamma^i$, we have the geodesic equation

$$\frac{d\dot{\gamma}^i}{d\lambda} + \Gamma^i_{rs}\dot{\gamma}^r\dot{\gamma}^s = f(\lambda)\dot{\gamma}^i \qquad (7.4.7)$$

where the notation for geodesic lines has been adopted. As already discussed in Sect. 2.4, the geodesic curve can be given an affine parametrization $\sigma = \sigma(\lambda)$ such that $f'(\sigma) = 0$; in this case the modulus $|(\dot{\gamma}^i\dot{\gamma}_i)_\sigma|$ is constant along the curve, so a reparametrization

$$d\tau = |(\dot{\gamma}^i\dot{\gamma}_i)_\sigma|^{\frac{1}{2}} d\sigma \qquad (7.4.8)$$

will preserve the affine character of τ and make the transformed tangent field of unitary length: $(\dot{\gamma}^i\dot{\gamma}_i)_\tau = -1$. The parameter τ is termed *proper time* and its physical significance will be elucidated in chapter 9.

Let us now assume that a point particle moves from p to a point q' slightly away from q along a different geodesic (Fig. 7-2). If the motion is confined within a normal neighbourhood of p, the end points q, q', etc. are uniquely determined by the tangent to the geodesic at p.

To each geodesic then corresponds a different action $S(\gamma)$, hence the difference between $S(\gamma)$ and $S(\gamma + \delta\gamma)$ will be from (7.4.4):

$$\delta_\gamma S = \mu \frac{g_{ij}\dot{\gamma}^j}{(-\dot{\gamma}^r\dot{\gamma}_r)^{\frac{1}{2}}}\delta\gamma^i. \qquad (7.4.9)$$

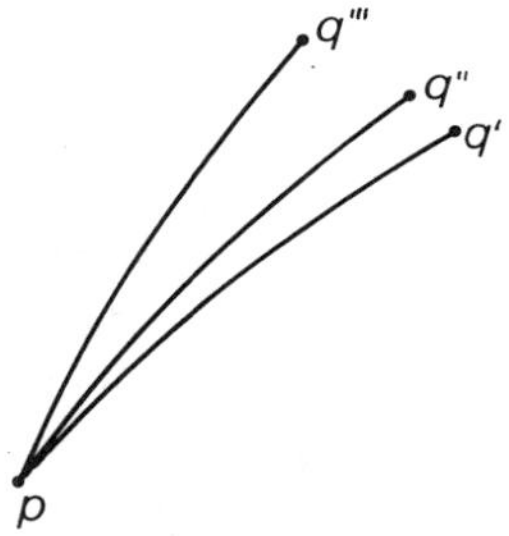

Fig. 7-2 The geodesic followed by a free particle from p depends on its four-momentum.

We define the four-momentum of the particle to be the quantity

$$P_i = \mu \frac{g_{ij}\dot{\gamma}^j}{(-\dot{\gamma}^r\dot{\gamma}_r)^{\frac{1}{2}}} = \partial_i S \tag{7.4.10}$$

which satisfies the equation:

$$g^{ij}(\partial_i S)(\partial_j S) + \mu^2 = 0. \tag{7.4.11}$$

This is known as the Hamilton-Jacobi equation for a point particle since it can be derived in the context of a general Hamiltonian formalism (Tulczyjew, 1977).

7.5 Constants of motion

If space-time admits a Killing vector field, we know from (2.13.5) that

$$(\xi|\dot{\gamma}) = \text{ const.} \tag{7.5.1}$$

along the geodesics. This is a constant of the motion, and there are as many of them as there are one-parameter families of isometries of the space-time. From the variational approach followed in the previous section, we know that a constant of motion is associated to an ignorable coordinate of the metric. In fact from (7.4.5) and (7.4.10) we have:

$$(-\dot{\gamma}^r\dot{\gamma}_r)^{-\frac{1}{2}} \frac{\mu}{2} \frac{\partial g_{ij}}{\partial x^k} \dot{\gamma}^i\dot{\gamma}^j - \frac{d}{d\lambda} P_k = 0 \tag{7.5.2}$$

and so to any ignorable x^α there corresponds an equation

$$P_\alpha = \text{const.} \tag{7.5.3}$$

on $\gamma(\tau)$. We shall now clarify the connection between (7.5.1) and (7.5.3). Evidently they describe the same physical quantity; we need to show that to any ignorable coordinate for the metric, there corresponds locally a Killing field and vice versa. Suppose now that x^α is such an ignorable coordinate. Then consider, in a coordinate neighbourhood of M, the field of vectors ∂_α which identify the local coordinate direction; they form a Killing vector field. In fact by definition we have:

$$\left(\pounds_{\partial_\alpha} \overset{*}{g}\right)_{ij} = (\partial_k g_{ij})\delta^k_\alpha = 0. \tag{7.5.4}$$

Thus from (7.5.1) and (7.4.10) with $\dot{\gamma}^r \dot{\gamma}_r = -1$ we have

$$(\partial_\alpha | \dot{\gamma}) = g_{ij}\delta^i_\alpha \frac{dx^j}{d\lambda} = g_{\alpha j}\frac{dx^j}{d\lambda} = \frac{1}{\mu}P_\alpha = \text{const.} \tag{7.5.5}$$

The ignorable character of a coordinate is clearly a non-covariant property of the space-time metric; nevertheless the existence of a constant of motion is necessarily coordinate-independent, for it bears a precise physical meaning. This is made manifest by the correspondence between constants of motion and isometries, the latter being clearly an invariant property of a manifold. The converse of the above statement, namely given a set of Killing vector fields ξ_A there always exists a coordinate system (adapted to the symmetries) such that $\xi_A(x^i) = \delta^i_A$, is ensured by the Frobenius theorems in Sect. 2.12. In that case the coordinates x^A are ignorable coordinates for the metric.

7.6 Maxwell's equations for a free electromagnetic field

In the absence of charges and currents, the behaviour of an electromagnetic field in a curved space-time $(M, \overset{*}{g})$ is deducible from an action

$$S_{\rm em} = \int_\Omega L_{\rm em}\sqrt{-g}\ \boldsymbol{d}^4x \tag{7.6.1}$$

where

$$L_{\rm em} = -\frac{1}{16\pi}g^{ir}g^{js}F_{ij}F_{rs} \tag{7.6.2}$$

and Ω is an open set of M. The antisymmetric tensor F_{ij} (known as the Faraday tensor) satisfies the constraint equations:

$$\partial_{[k}F_{ij]} = 0 \tag{7.6.3}$$

which also provide the first group of Maxwell's equations. They imply, as already mentioned in Sect. 6.7, that there exists in M a one-form $\boldsymbol{A}$, such that

$$F_{ij} = 2\partial_{[i}A_{j]}. \tag{7.6.4}$$

The components of the four-potential $\boldsymbol{A}$ will be considered as the dynamical variables of the field; but since it is defined only up to a gauge transformation ($A \to A' = A + d\lambda\ ,\ \lambda \in \mathcal{F}(M)$), it will not enter the Lagrangian except via the gauge-invariant form (7.6.2).

The Lagrangian (7.6.2) is in fact the only scalar which is quadratic in the field components F_{ij}. It leads, after varying the action with respect to A, to first-order linear differential equations in F. Assuming that $\delta\boldsymbol{A} = 0$ on $\partial\Omega$ and requiring that $S_{\rm em}$ is stationary under variation of $\boldsymbol{A}$, we obtain:

$$\underset{A}{\delta} S_{\rm em} = \int_\Omega \frac{\partial L_{\rm em}}{\partial(\partial_k A_i)}\delta(\partial_k A_i)\sqrt{-g}\ \boldsymbol{d}^4x = 0. \tag{7.6.5}$$

Form the standard procedure of the variational method, equation (7.6.5) becomes:

$$-\int_\Omega \partial_k\left[\frac{\partial L_{\rm em}}{\partial(\partial_k A_i)}\sqrt{-g}\right]\delta A_i\ \boldsymbol{d}^4x = 0 \tag{7.6.6}$$

which implies:

$$\frac{1}{\sqrt{-g}}\partial_k(\sqrt{-g}F^{ik}) = \nabla_k F^{ik} = 0. \tag{7.6.7}$$

These are the second group of Maxwell's equations. In terms of the vector potential they become:

$$g^{ir}g^{ks}\nabla_k F_{rs} = g^{ir}g^{ks}\nabla_k(\partial_r A_s - \partial_s A_r)$$
$$= -\Box A^i + g^{ir}\nabla_r(\nabla_s A^s) + R^i_j A^j = 0 \quad (7.6.8)$$

where $\Box = g^{rs}\nabla_r\nabla_s$ and R^i_j is the Ricci tensor. Imposing the Lorentz condition:

$$\nabla_s A^s = 0 \quad (7.6.9)$$

which is allowed by the gauge freedom on $\boldsymbol{A}$, we obtain the wave equation:

$$\left(\Box\delta^i_j - R^i_j\right) A^i = 0 \quad (7.6.10)$$

where $\Box\delta^i_j - R^i_j$, the de Rham wave operator, generalizes to a curved space-time the d'Alembertian operator of special relativity (Friedlander, 1975).

A noteworthy property of Maxwell's equations is that of being conformally invariant. Under a transformation of the type $\tilde{g}_{ij} = \Omega^2 g_{ij}$, $\tilde{F}^{ij} = \Omega^{-4}F^{ij}$ we have $\sqrt{-\tilde{g}} = \Omega^4\sqrt{-g}$; hence equation (7.6.7) is trivially invariant, while equation (7.6.3) is identically invariant being independent of the background metric.

7.7 Maxwell's equations in the presence of charges and currents

The equations which govern an electromagnetic field in the presence of its sources, namely charges and currents, can be derived from an action of the type:

$$S = S_{\rm sour} + S_{\rm int} + S_{\rm em}. \quad (7.7.1)$$

Here $S_{\rm sour}$ describes the distribution of charges which we assume to be point particles smoothly distributed through space-time; $S_{\rm int}$ describes the interaction between the charges and the electromagnetic field which they generate and $S_{\rm em}$ describes the free electromagnetic field. The first and the last term of the action are known, being respectively:

$$S_{\rm sour} = -\int \rho\sqrt{-g}\, \boldsymbol{d}^4x \quad (7.7.2)$$

and (7.6.1). The interaction term should depend on the dynamical variables of both the charges and the field, hence we shall assume that the action will depend on the potential $\boldsymbol{A}$ as in special relativity, that is:

$$S_{\text{int}} = \int_\Omega L_{\text{int}}\,(x^i, u^i; \boldsymbol{A})\,\sqrt{-g}\;\boldsymbol{d}^4 x \tag{7.7.3}$$

where u^i is the four-velocity of the charge. Let us suppose that the equations of motion for the charges are satisfied, so we vary S only with respect to $\boldsymbol{A}$:

$$\underset{A}{\delta} S = \underset{A}{\delta} S_{\text{em}} + \int_\Omega \frac{\partial L_{\text{int}}}{\partial A_i}\,\delta A_i\; d\Omega = 0\ . \tag{7.7.4}$$

From (7.6.1) and (7.6.2) we have:

$$-\frac{1}{4\pi}\partial_k\,(\sqrt{-g}F^{ik}) + \sqrt{-g}J^i = 0 \tag{7.7.5a}$$

or equivalently

$$\nabla_k F^{ik} = 4\pi J^i \tag{7.7.5b}$$

where

$$J^i = \frac{\partial L_{\text{int}}}{\partial A_i} \tag{7.7.6}$$

is a four-vector, termed the *current density.* It satisfies the conservation equation:

$$\partial_i(\sqrt{-g}J^i) = 0 \tag{7.7.7}$$

from (7.7.5) and the antisymmetry of F^{ik}. Integrating (7.7.7) over a four-volume and applying Gauss' theorem we have:

$$\int_\Omega \partial_i(\sqrt{-g}J^i)\;\boldsymbol{d}^4 x = \oint_{\partial\Omega} J^i\; d^3\Sigma_i = 0. \tag{7.7.8}$$

If we define Ω as being bounded by two space-like hypersurfaces Σ_1 and Σ_2 which cross a world-tube of time-like lines (see Fig. 7-1) and by a lateral cylindrical hypersurface $\partial\sigma$, then if we can extend Ω so that $J^i = 0$ on $\partial\sigma$, equation (7.7.8) leads to:

$$\int_\Sigma J^i\; d^3\Sigma_i \equiv Q = \text{ const.} \tag{7.7.9}$$

Q is the total charge contained in Σ and as expected is a conserved quantity. If n^i is a unit normal to Σ, then $d^3\Sigma_i = n_i d^3\Sigma$ hence:

$$J^i n_i = q_\Sigma \tag{7.7.10}$$

where q_Σ is the charge density in the rest frame of the three-volume $d^3\Sigma$. If the source is a single point particle with rest mass μ and charge e, moving with four-velocity u, an analysis similar to that for a chargeless particle made in Sect. 6.6 leads us to define a charge density current:

$$J^i = e \int \breve{u}^i \delta_g^4(x - \gamma(\lambda))\sqrt{-g}\ d\lambda. \tag{7.7.11}$$

From the discussion and (7.7.6) it follows that:

$$L_{\text{int}} = J^i A_i \tag{7.7.12}$$

hence S_{int} for a single charge reads:

$$S_{\text{int}} = e \int_\gamma d\lambda \int_\Omega \breve{u}^i A_i \delta^4(x - \gamma(\lambda))\ \boldsymbol{d}^4 x = e \int_\gamma u^i A_i\ d\lambda. \tag{7.7.13}$$

The purpose is now to derive the equations of motion of a (in general) radiating charge interacting with an electromagnetic field F_{ij}. Let us then vary the action S in (7.7.1) with respect to the particle variables, to yield from (7.7.13) and (7.4.3):

$$\begin{aligned} \delta_\gamma S &= -\mu \delta_\gamma \int_{\lambda_1}^{\lambda_2} (-u^i u_i)^{\frac{1}{2}}\ d\lambda + e\delta_\gamma \int_{\lambda_1}^{\lambda_2} u^i A_i\ d\lambda \\ &= -\mu \int_{\lambda_1}^{\lambda_2} u^i \nabla_i u_k \delta\gamma^k\ d\lambda + e \int_{\lambda_1}^{\lambda_2} F_{ki} u^i \delta\gamma^k d\lambda = 0. \end{aligned} \tag{7.7.14}$$

Hence we have:

$$u^i \nabla_i u^k = \frac{e}{\mu} F^k{}_i u^i. \tag{7.7.15}$$

This equation should be coupled to (7.7.5), with (7.6.3), to give a consistent description of the classical electrodynamics. Expressed in terms of the vector potential and assuming the Lorentz gauge, equation (7.7.5) becomes a non-homogeneous de Rham wave equation (cf. 7.6.10), whose solution in a curved space-time is known in terms of advanced or retarded covariant Green's functions:

$$A_i^{\text{adv/ret}} = 4\pi e \int_{-\infty}^{+\infty} G_{ij}^{\text{adv/ret}}(x; z(\tau)) u^j(\gamma(\tau))\ d\tau. \tag{7.7.16}$$

The integration here is extended over the entire particle's history. The field strength is obtained from (7.7.16) after differentiation, but in order to determine the effect of the radiation reaction on the particle itself, one has to calculate the balance of energy and momentum between the particle and the field. This turns out to be a highly non-trivial calculation which goes beyond the scope of this book; the result gives the following equation of motion for a charge moving in an arbitrary Riemannian space-time subject to an externally imposed electromagnetic field $(F^{\text{ext}})_{ij}$ (De Witt and De Witt, 1964; De Witt and Brehme, 1960):

$$\begin{aligned}\mu_0 c\frac{Du^i}{D\tau} &= \frac{e}{c}(F^{\text{ext}})^i{}_k u^k + \frac{2}{3}\frac{e^2}{c^3}\left[\frac{D^2u^i}{D\tau^2} - \frac{u^i}{c^2}\left(\frac{Du^k}{D\tau}\frac{Du_k}{D\tau}\right)\right] \\ &\quad - \frac{e^2}{3c}\left(R^i{}_j u^j + \frac{1}{c^2}u^i R_{jk}u^j u^k\right) \\ &\quad + \frac{e^2}{c}u^k\int_{-\infty}^{\tau} f^i{}_{kj}u^j(\tau')\,d\tau'. \end{aligned} \tag{7.7.17}$$

The second term on the right is the classical radiation damping term; the third term, due to Hobbs (Hobbs, 1968), arises from the local distribution of matter-energy and finally the last term gives a non-local contribution which contains the derivatives of the curvature tensor via the term $f^i{}_{kj}$. A charged particle therefore will deviate from a geodesic up to e^2 terms, even in the absence of an external field $((F^{\text{ext}})_{ij})$.

7.8 The radiation field

The propagation of a source-free electromagnetic field in a given space-time $(M, \overset{*}{g})$ is described by the wave equation (7.6.10). It indicates evidently that the velocity of propagation, with respect to any Lorentz frame, is finite. So if we assume that an electromagnetic field is generated by a source activated suddenly at a finite time and confined in a small region of a space-like hypersurface Σ, then there will exist a hypersurface S with equation

$$S(x) = 0 \tag{7.8.1}$$

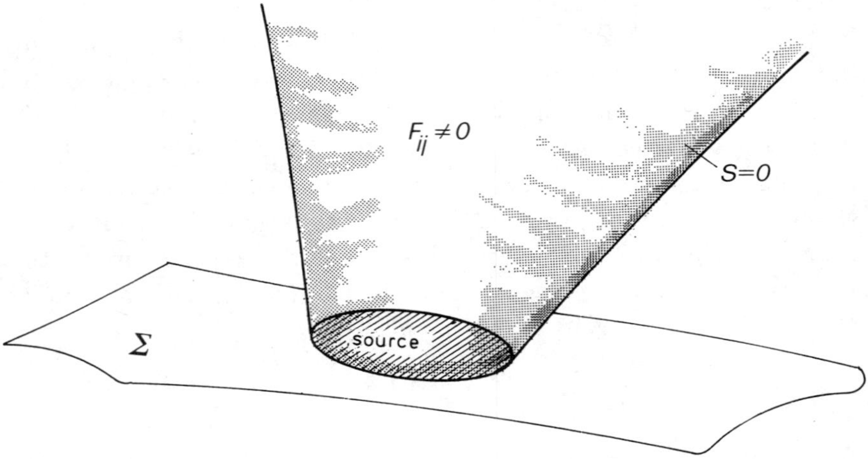

Fig. 7-3 The surface $S = 0$ separates the space-time points which were affected by the electromagnetic perturbation from those which were not.

which separates the events in M which are reached by the electromagnetic disturbances ($S > 0$) from those where it is still zero ($S < 0$) (see Fig. 7-3).

Introducing a step function $g(S)$:

$$g(S) = \begin{cases} 0 & \text{if } S < 0 \\ 1 & \text{if } S \geq 0 \end{cases}, \tag{7.8.2}$$

then the electromagnetic tensor can be written as:

$$F_{ij} = g(S) f_{ij}, \tag{7.8.3}$$

where we assume that the functions f_{ij} are continued smoothly into the region $S < 0$. The hypersurface S is a discontinuity surface for the electromagnetic field and the values the field takes on it are termed the *discontinuities* of the field; these are denoted by $\overline{f}_{ij}$. Our aim is to study the space-time behaviour of the discontinuities. Using (7.8.3) in Maxwell's equations (7.6.3) and (7.6.7), we obtain respectively:

$$\partial_{[i} F_{jk]} = \delta(S) \partial_{[i} S f_{jk]} + g(S) \partial_{[i} f_{jk]} = 0 \tag{7.8.4}$$

$$\nabla_k F^{ik} = \delta(S) \partial_k S f^{ik} + g(S) \nabla_k f^{ik} = 0 \tag{7.8.5}$$

where the differentiation law $\partial_i g(S) = \delta(S)\partial_i S$ is used with $\delta(S)$ being the Dirac delta function. From the properties of this and (7.8.2) we deduce on S:

$$\overline{f}_{[ij}\partial_{k]}S = 0 \tag{7.8.6}$$

$$\overline{f}^{ik}\partial_k S = 0\ . \tag{7.8.7}$$

These are the Maxwell equations for the discontinuities; $\partial_k S$ is the gradient one-form of S. Contracting (7.8.6) with $g^{ij}\partial_j S$ and using equation (7.8.7), we obtain:

$$g^{ij}(\partial_j S)(\partial_j S) = 0 \tag{7.8.8}$$

which shows that S is a null hypersurface. Let us now define the null directions orthogonal to S (the rays) as :

$$k^i \equiv \frac{dx^i}{d\lambda} = g^{ij}\partial_j S \qquad \lambda \in \mathbb{R}. \tag{7.8.9}$$

Since (7.8.8) holds all along the null generators of S, with $k_i = \partial_i S$, we have identically:

$$\frac{d}{d\lambda}(k^r k_r) = 0 \tag{7.8.10}$$

which implies:

$$k_r(k^i \nabla_i k^r) = 0. \tag{7.8.11}$$

The 4-acceleration of the null rays $k^i\nabla_i k^r$, being orthogonal to k, is either a null or a space-like vector. We show that it is a null vector. The one-form $\boldsymbol{k}$ satisfies the condition:

$$\nabla_{[i}k_{j]} = 0$$

and therefore also

$$k_{[l}\nabla_m k_{n]} = 0. \tag{7.8.12}$$

Contracting the latter with k^l we have identically:

$$\left(k_{[m}\nabla_{n]}k_l\right)k^l + k^l\nabla_l k_{[m}k_{n]} = 0. \tag{7.8.13}$$

Alternatively, since $k^a k_a = 0$ holds on S, the tangential derivative $k_{[m}\nabla_{n]}$ of $k^a k_a = 0$ should still be zero (Wald, 1984):

$$\left(k_{[m}\nabla_{n]}k_l\right)k^l = 0 \tag{7.8.14}$$

hence comparing with (7.8.13) we have:

$$k^l \nabla_l k_{[m} k_{n]} = 0 \tag{7.8.15}$$

which shows, from (2.9.5), that the generators of S are null geodesics.

Let us now study more closely the properties of the electromagnetic tensor $\overline{f}_{ij}$. From (7.8.7) the matrix $\{\overline{f}_{ij}\}$ associated to $\overline{f}_{ij}$ is singular ($\mathrm{Det}|\overline{f}_{ij}| = 0$), hence its rank is less than 4. Because of the antisymmetry of $\overline{f}_{ij}$, the rank of $\{\overline{f}_{ij}\}$ can only be 2. So, from the properties of matrices, there exist two one-forms $\boldsymbol{\xi}$ and $\boldsymbol{\eta}$ such that:

$$\overline{f}_{ij} = 2\xi_{[i}\eta_{j]} \tag{7.8.16}$$

on S. In this case $\{\overline{f}_{ij}\}$ is said to be *degenerate*. From (7.8.6) it follows that

$$k_{[i}\xi_j \eta_{k]} = 0 \tag{7.8.17}$$

which implies that any of these one-forms can be expressed as linear combination of the remaining; in particular

$$\boldsymbol{k} = \alpha\boldsymbol{\xi} + \beta\boldsymbol{\eta} \qquad \alpha, \beta \in \mathbb{R}. \tag{7.8.18}$$

Without loss of generality we can choose:

$$\boldsymbol{k} = \boldsymbol{\eta} \tag{7.8.19}$$

so the field of discontinuities will have the form:

$$\overline{f}_{ij} = 2\xi_{[i}k_{j]} \tag{7.8.20}$$

and from (7.8.7):

$$\xi^i k_i = 0. \tag{7.8.21}$$

Since $\boldsymbol{\xi}$ cannot be null (it would be parallel to $\boldsymbol{k}$), it is space-like; therefore:

$$\xi^i \xi_i \equiv \mu > 0. \tag{7.8.22}$$

The positivity of μ will have physical significance in later applications.

Expression (7.8.20) has important consequences; in fact we have identically:

$$a) \qquad \overline{f}^{ij}\overline{f}_{ij} = 0$$

$$b) \qquad \overset{*}{\overline{f}}{}^{ij}\overline{f}_{ij} = 0. \tag{7.8.23}$$

Comparison with special relativity shows that a field of discontinuities propagates locally as a plane-fronted wave. The surface of discontinuity S is a wave front and the function $S(x)$, which characterizes it is the phase of the wave, termed the *eikonal.* Equation (7.8.8) is also known as the *eikonal equation.*

The discontinuity equations coincide with the laws of geometrical optics. This approximation holds in space-time whenever the wave length of the electromagnetic disturbance is very short compared to the background mean curvature radius, as measured in a local Lorentz frame, and also to the mean length scale over which the wavelength, the amplitude and the polarization of the electromagnetic perturbation itself vary. An electromagnetic field which satisfies conditions (7.8.23), (7.8.6) and (7.8.7) is termed a *null field.*

The vector ξ is the *amplitude vector* of the field and to deduce its propagation equation in the direction of k we need to consider the discontinuity in the energy-momentum tensor. From (6.7.2) and (7.8.20) this becomes:

$$\overline{T}^{i}{}_{j} = \frac{1}{4\pi}\mu k^i k_j. \tag{7.8.24}$$

From the continuity equation

$$\nabla_j T^{ij} = 0 \tag{7.8.25}$$

and the geometrical optics approximation (which leads to $\nabla_j \overline{T}^{ij} = 0$) the propagation law of ξ follows:

$$\frac{D\xi}{D\lambda} = -\frac{1}{2}\xi(\nabla_i k^i). \tag{7.8.26}$$

Let us introduce a scalar amplitude $\mathcal{A}$ defined as

$$\xi = \mathcal{A}\mathrm{e} \tag{7.8.27}$$

where e is a unit vector termed the *polarization vector.* From (7.8.27) and (7.8.26) we obtain the propagation equation for the scalar amplitude:

$$\frac{d\mathcal{A}}{d\lambda} = -\frac{1}{2}\mathcal{A}\nabla^i k_i \tag{7.8.28}$$

and the corresponding behaviour of the polarization vector:

$$k^i \nabla_i \mathrm{e} = 0. \tag{7.8.29}$$

The polarization vector is parallely propagated along the rays.

7.9 The light cone

At any point p the set of all null vectors in $T_p(M)$ constitutes the *null cone* at p. The zero vector separates the null cone into two halves, and the physical conditions near p allow one to distinguish these as the future- and past-pointing halves. In the discussion in the previous section we have implicitly assumed that light rays propagate outwards from the source along null geodesics whose tangent vectors are in the future half. Similarly time-like vectors are divided by the null cone into future- and past-directed ones.

If this choice of past and future can be made continuously throughout M, then the space-time is said to be time-orientable. (It can be proved that time-orientability implies the existence of a time-like vector field on M.) In this case, a *causal structure* can be defined on M, using the past/future division.

A smooth curve is said to be *causal* if its tangent vector is everywhere non-zero and is time-like or null; it is future-directed or past-directed according to whether its tangent vector is past- or future-pointing. For any point p in M the set of points q such that there is a future-directed causal curve from p to q is called the *causal future* of p, denoted by $J^+(p)$. The causal past $J^-(p)$ is defined similarly. If we restrict ourselves to time-like curves, then we obtain the chronological future and past $I^+(p)$ and $I^-(p)$. Clearly $I^\pm \subset J^\pm$. All the points of $J^+(p)$ can be causally affected from p. Under the exponential map, those null vectors v for which $\mathrm{Exp}_p v$ is defined, map to points to which a light signal can be sent from p. The set of all such points which are not in $I^+(p)$ or $I^-(p)$ (i.e. the points that can be reached by light but not by massive particles) form a subset of M called *horismos* or *light cone* of p, $E(p)$ (Hawking and Ellis, 1973); it divides into the future horismos $E^+(p)$ and the past horismos $E^-(p)$, according to whether the light signal required is past- or future-directed. These sets need not be disjoint.

Given a point $r \in I^{\pm}(p)$, there always exists an open neighbourhood of r whose points can be connected to p by a time-like curve, so $I^{\pm}(p)$ are open sets. If we denote by $\bar{A}$ the closure of a set A, and define the boundary of A as

$$\dot{A} = \bar{A} \cap \overline{(M - A)} \tag{7.9.1}$$

then, from the definitions, it follows that:

$$\bar{I}^{\pm}(p) = \bar{J}^{\pm}(p), \tag{7.9.2}$$

and therefore:

$$\overline{(M - I^{\pm})} = \overline{(M - J^{\pm})} \tag{7.9.3}$$

hence

$$\dot{I}^{\pm}(p) = \dot{J}^{\pm}(p). \tag{7.9.4}$$

It also follows that

$$J^{\pm}(p) - I^{\pm}(p) = E^{\pm}(p)\ . \tag{7.9.5}$$

If we restrict all the above to a normal neighbourhood $U_{\mathcal{N}}$ of p, denoting the resulting sets by $I^{\pm}(p, U_{\mathcal{N}})$, etc., then one can show that $J^{\pm}$ are closed sets and $\dot{J}^{\pm}(p, U_{\mathcal{N}}) = E^{\pm}(p, U_{\mathcal{N}})$. But this does not hold in the whole of M. Consider, for example, the situation depicted in Fig. 7-4; if a point is removed from the future light cone of p, a point q exists which belongs to $\dot{J}^+(p)$ but not to $E^+(p)$, for it cannot be joined to p by a null geodesic (see Hawking and Ellis, 1973).

Given a point $q \in E^+(p, U_{\mathcal{N}})$ where $U_{\mathcal{N}}$ is a normal neighbourhood of p, no other point $r \in E^+(p, U_{\mathcal{N}})$ has a time-like separation from q. In fact if there exists a future time-like curve γ joining q to r, then $r \in I^+(q, U_{\mathcal{N}})$, but $I^+(q, U_{\mathcal{N}}) \subset I^+(p, U_{\mathcal{N}})$ and so $r \in I^+(p, U_{\mathcal{N}})$, contradicting the hypothesis. A similar argument

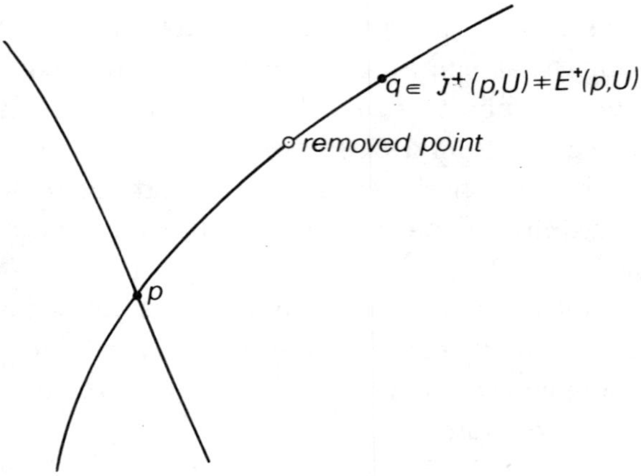

Fig. 7-4 If we remove a point from $\dot{J}^{\pm}(p, U)$, then $\dot{J}^{\pm}(p, U) \neq E^{\pm}(p, U)$.

holds for the case of a curve from r to q. This property can be reformulated by saying that $E = E^{\pm}(p, U_{\mathcal{N}})$ satisfies:

$$I^{\pm}(E) \cap E = \emptyset. \tag{7.9.6}$$

A set E of M which satisfies (7.9.6) with $U_{\mathcal{N}}$ extended to the whole of M is said to be *achronal.* The future (past) horismos of p are achronal sets in $U_{\mathcal{N}}$ but not necessarily in the whole of M.

7.10 Stationary space-times

A space-time manifold $(M, \overset{*}{g})$ is termed (*pseudo*)-*stationary* if there exists a Killing vector field which is time-like in some open set $U \subset M$. Let the vector field be k: $\pounds_k \overset{*}{g} = 0$, then the surfaces $\mathcal{S}|_{_U}: (k|k) \equiv V = \text{const.}$ are time-like; in fact the normal to $\mathcal{S}$, $\partial_i V$, is everywhere space-like in U, being orthogonal to k:

$$k^i \partial_i V = 2k^i k^j \nabla_j k_i = 0. \tag{7.10.1}$$

If U does not coincide with M^*, then in $M - U$, k is space-like or null. Let $\mathcal{S}_0 \equiv \partial U$ be the surface on which k is null, so $\mathcal{S}_0$: $V = 0$. This surface is invariant under the diffeomorphisms corresponding

---

*If U coincides with M, then the space-time is termed *stationary*.

to k. Hence k is tangent to $\mathcal{S}_0$ and so, from Sect. 1.15, it may be time-like or null. We shall prove the following

THEOREM 1

(Vishveshwara, 1968) The surface $\mathcal{S}_0$ is null if and only if

$$\text{a)} \quad k_{[i}\nabla_j k_{k]} = 0 \qquad \text{and} \qquad \text{b)} \quad k^i\nabla_i k^j \neq 0 \quad \text{on} \quad \mathcal{S}_0 \,. \tag{7.10.2}$$

Proof: If the first relation holds, since $\nabla_j k_k = \nabla_{[j}k_{k]}$, we can write it as

$$k_i\nabla_j k_k + k_k\nabla_i k_j + k_j\nabla_k k_i = 0 \,. \tag{7.10.3}$$

Hence the one-form normal to $\mathcal{S}$ has square modulus:

$$\begin{aligned}
\partial^i V \partial_i V &= 4\,(k^i\nabla^j k^k)\left(k_k\nabla_j k_i\right) \\
&= -\,4\left(k_k\nabla_j k_i\right)(k^k\nabla^i k^j + k^j\nabla^k k^i) \\
&= -\,4V(\nabla_j k_i)(\nabla^i k^j) - 4(k_k\nabla_j k_i)(k^j\nabla^k k^i).
\end{aligned}$$

On $\mathcal{S}_0$ this becomes:

$$\partial^i V \partial_i V = 2V(\nabla_j k_i)(\nabla^i k^j) = 0 \tag{7.10.4}$$

but (7.10.2b) is equivalent to $\partial_i V \neq 0$ so $\mathcal{S}_0$ is a null surface. The converse is trivial; if $\mathcal{S}_0$ is a null surface, then

$$k_i = f\partial_i V \qquad f \in \mathcal{F}(M)$$

hence (7.10.2) follows. ∎

A (pseudo)-stationary space-time is said to be *(pseudo)-static* if condition (7.10.2) holds throughout M. Let us now further specialize the space-time properties.

A space-time is said to be *axisymmetric* when the metric is invariant under an action $\pi\colon \mathrm{SO}(2)\times M \to M$ of the one-parameter rotation group SO(2) such that the fixed points of π (points p such that $\pi(g,p) = p$ for all $g \in \mathrm{SO}(2)$), if any, form a two-dimensional surface, imbedded in M. The two-surface (if any) is called the axis of symmetry. The generator of the action π is a Killing vector field m which is always zero on the axis of symmetry. The trajectories of the action are closed (topologically circular) lines. A space-time is *(pseudo)-stationary and axisymmetric* when it is invariant under the action $\pi' \colon R(1) \times \mathrm{SO}(2) \times M \to M$ of the two-parameter Abelian group $R(1) \times \mathrm{SO}(2)$; in this case the generators of the group, namely m and k act simultaneously on M and commute:

$$[k, m] = o. \tag{7.10.5}$$

If a space-time is both (pseudo)-stationary and axisymmetric, and is also asymptotically flat, then it can be shown that condition (7.10.5) is necessarily satisfied. In what follows we shall restrict ourselves to these cases only, since they are of the most importance physically.

Of particular interest is the case when the one-forms associated to m and k satisfy the Frobenius condition (cf. 2.12.14), namely (Carter, 1969):

$$m_{[d}k_c\nabla_b k_{a]} = 0 \qquad m_{[d}k_c\nabla_b m_{a]} = 0. \tag{7.10.6}$$

Under these conditions, we know (cf. Sect. 2.12) that there exists a coordinate system adapted to the symmetries, say $(t, \varphi, x^{\mathrm{A}})_{\mathrm{A}=1,2}$ in terms of which the metric reads:

$$ds^2 = V\ dt^2 + 2W\ dt\ d\varphi + X\ d\varphi^2 + g_{\mathrm{AB}}\ dx^{\mathrm{A}}\ dx^{\mathrm{B}} \tag{7.10.7}$$

where

$$V = (k|k),\ W = (k|m),\ X = (m|m)\ ;$$

t is a time coordinate (time-like sufficiently far from the metric source) in the range $[-\infty, +\infty]$ and φ is an azimuthal angular coordinate in the range $[0, 2\pi]$; furthermore:

$$m = \partial_\varphi\ ;\ k = \partial_t\ ;\ \partial_t g_{ij} = \partial_\varphi g_{ij} = 0\ ;\ g_{t\mathrm{A}} = g_{\varphi\mathrm{A}} = 0\ ;\quad \mathrm{A} = 1, 2. \tag{7.10.8}$$

We prove the following:

PROPOSITION

If conditions (7.10.5) and (7.10.6) hold, then the surface $H \colon V - W^2/X \equiv \sigma = 0$ is a stationary null surface.

Let us first prove the following:

LEMMA

If (7.10.6) holds, then the one-form $\boldsymbol{\ell} = \boldsymbol{k} - (W/X)\boldsymbol{m}$ *satisfies the Frobenius condition:*

$$\ell_{[a}\nabla_b\ell_{c]} = 0 \ . \tag{7.10.9}$$

Proof: From (1.14.12), equation (7.10.6) can be written as:

$$\begin{aligned} a)\quad & 2m_{[d}k_c\nabla_{b]}k_a + m_{[d}\nabla_c k_{b]}k_a - k_{[d}\nabla_c k_{b]}m_a = 0 \\ b)\quad & 2m_{[d}k_c\nabla_{b]}m_a + m_{[d}\nabla_c m_{b]}k_a - k_{[d}\nabla_c m_{b]}m_a = 0 \ . \end{aligned} \tag{7.10.10}$$

Contracting (7.10.10a) with m^a and (7.10.10b) with k^a we obtain

$$\begin{aligned} m_{[d}k_c\nabla_{b]}W &= -Wm_{[d}\nabla_c k_{b]} + Xk_{[d}\nabla_c k_{b]} \\ m_{[d}k_c\nabla_{b]}W &= -Vm_{[d}\nabla_c m_{b]} + Wk_{[d}\nabla_c m_{b]} \end{aligned} \tag{7.10.11}$$

using (7.10.5) and the Killing equations for k and m. Relations (7.10.11) imply:

$$-Wm_{[d}\nabla_c k_{b]} + Xk_{[d}\nabla_c k_{b]} = -Vm_{[d}\nabla_c m_{b]} + Wk_{[d}\nabla_c m_{b]} \tag{7.10.12}$$

so, contracting this with m^b and using (7.10.5) again, we obtain:

$$\begin{aligned} & -Wm_{[d}\partial_{c]}W + Xk_{[d}\partial_{c]}W \\ & \quad = -Vm_{[d}\partial_{c]}X + Wk_{[d}\partial_{c]}X + (W^2 - VX)\nabla_d m_c . \end{aligned} \tag{7.10.13}$$

Let us now consider the one-form $\boldsymbol{\ell}' = X\boldsymbol{k} - W\boldsymbol{m}$; the following relation holds, from (7.10.8) and (7.10.12):

$$\ell'_{[b}\nabla_d\ell'_{c]} = (W^2 - VX)m_{[b}\nabla_d m_{c]} - Xk_{[b}m_c\partial_{b]}W - Wm_{[b}k_c\partial_{d]}X \ . \tag{7.10.14}$$

However, multiplying (7.10.13) with m_b and antisymmetrizing over all the indices, we have:

$$(W^2 - VX)\, m_{[b}\nabla_d m_{c]} = Xk_{[b}m_c\partial_{d]}W + Wm_{[b}k_c\partial_{d]}X \ . \tag{7.10.15}$$

Hence, from (7.10.14), it follows that $\ell'_{[b}\nabla_d\ell'_{c]} = 0$. Thus, since $\ell_a = (1/X)\ell'_a$, we have finally:

$$\ell_{[b}\nabla_d\ell_{c]} = \frac{1}{X^2}\ell'_{[b}\nabla_d\ell'_{c]} + \frac{1}{X^3}\ell'_{[b}(\partial_d X)\ell'_{c]} = 0 \tag{7.10.16}$$

proving the lemma. ∎

We now proceed to prove the proposition. The one-form $\boldsymbol{\ell}$ has square modulus:

$$\ell^a \ell_a = V - \frac{W^2}{X} = \sigma \tag{7.10.17}$$

and it is clearly time-like where k is time-like ($V < 0$) and remains so below the surface $V = 0$ until it becomes null on the surface $H\colon \sigma = 0$. The proof will follow straightforwardly if we could use Theorem 1; to this aim we need to show that $\boldsymbol{\ell}$ is a Killing one-form at least on H. For that to be true we have to show that the quantity

$$\omega \stackrel{\text{def}}{=} -\frac{W}{X} \tag{7.10.18}$$

has zero derivatives on H. By definition we have:

$$X^2 \partial_a \omega = -2\left(X k^b - W m^b\right) \nabla_a m_b \tag{7.10.19}$$

hence, after some algebra:

$$X^2 m_{[b} k_c \partial_{a]} \omega = (XV - W^2) m_{[b} \nabla_c m_{a]} = 0 \qquad \text{on } H. \tag{7.10.20}$$

This implies that on the surface H the one-form $\partial_a \omega$ can be written as a linear combination of $\boldsymbol{m}$ and $\boldsymbol{k}$, viz.

$$\partial_a \omega = \alpha k_a + \beta m_a \qquad \alpha, \beta \in \mathbb{R} \tag{7.10.21}$$

but from (7.10.8) we have:

$$\partial_t \omega = \partial_\varphi \omega = 0$$

and

$$\partial_{\text{A}} \omega = \alpha g_{t\text{A}} + \beta g_{\varphi\text{A}} = 0 \qquad (\text{A} = 1, 2)$$

as we wanted to prove. The one-form $\boldsymbol{\ell}$ is then null and Killing on H so from Theorem 1 H is a stationary null surface, proving the proposition.

In the next section the properties of the surface H will be analyzed in more detail. It seems however appropriate now to illustrate some general features of motion in the space-time (7.10.7).

Consider in $M|_{\sigma<0}$ the family of time-like curves with tangent unit vectors:

$$u_{\Omega} = e^{\text{U}} (k + \Omega m) \qquad \Omega \in \mathcal{F} \tag{7.10.22}$$

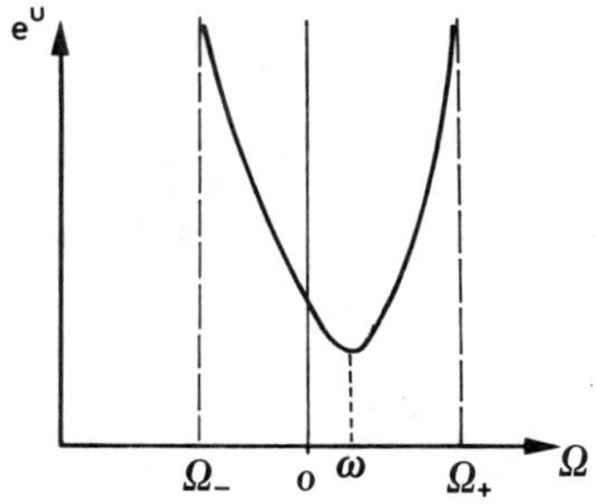

Fig. 7-5 When $V < 0$, Ω_+ and Ω_- have opposite signs so orbits with opposite sense of revolution are possible.

where e^{U} is a normalization factor which, for unit length, is given by

$$e^{\mathrm{U}} = \left[-V - 2\Omega X - \Omega^2 W\right]^{-\frac{1}{2}} \tag{7.10.23}$$

and Ω is a constant parameter. Physically these curves represent the world-lines of point particles moving on stationary (spatially) circular orbits around the metric source, with angular velocity Ω. The identification of Ω as angular velocity (from (7.10.22) and (7.10.8), $\Omega = (d\varphi/dt)$) is particularly meaningful if the space-time is asymptotically flat. The stationary motion described by (7.10.22) is allowed everywhere where e^{U} is real, namely when Ω is in the range

$$\Omega_- < \Omega < \Omega_+$$

with

$$\Omega_\pm = -\frac{W}{X} \pm \frac{1}{X}\left[W^2 - XV\right]^{\frac{1}{2}} . \tag{7.10.24}$$

These extremes would describe light-like trajectories, since $e^{\mathrm{U}} = \infty$ on them. In the region where $V < 0$, Ω_+ and Ω_- have opposite sign and the allowed range of e^{U} is shown in Fig. 7-5.

The values of Ω with opposite sign clearly indicate orbits with opposite sense of revolution around the metric source. When $\Omega = 0$, the vector field $u_{\Omega=0}$ in (7.10.22) becomes proportional to the Killing vector k; everywhere k is time-like (viz. $V < 0$) the integral curves of $u_{\Omega=0}$ describe the world-line of particles which

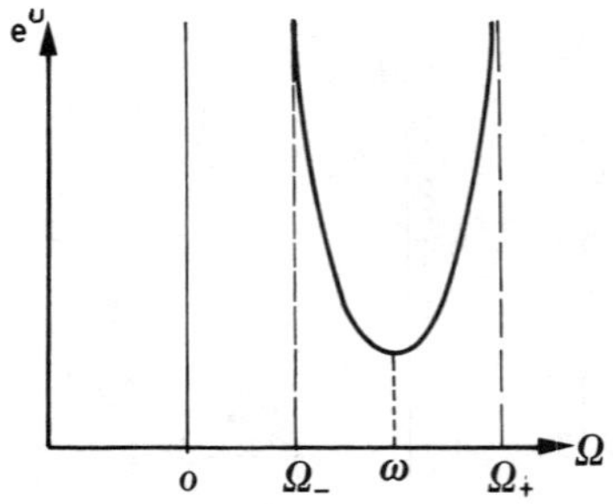

Fig. 7-6 When $V \geq 0$, Ω_+ and Ω_- have the same sign so orbits are allowed only in one sense of revolution.

are termed *static* in M^*. As it will be discussed in Sec.11.6, the curves characterized by the minimum value of e^U describe the trajectories of particles with zero angular momentum (cf. (11.6.2)) despite their angular velocity with respect to infinity being non-zero but equal to $\Omega_{min} = \omega = -(W/X)$. This quantity depends only on the background geometry and arises as result of a relativistic effect termed *gravitational dragging.* On and below the surface $V = 0$, both Ω_- and Ω_+ have the same sign (Fig. 7-6) while their difference shrinks with $|\sigma|$.

In this case, namely when we are between the surfaces $\mathcal{S}_0 \colon V = 0$ and $H \colon \sigma = 0$, the stationary motion is allowed only with one sense of revolution around the metric source. Evidently static particles ($\Omega = 0$) can be found only where $V < 0$. On and below the surface $H \colon \sigma = 0$, no stationary motion is allowed, hence it is named the *limit of stationarity.* However H is a null surface and so, being tangent to the light cone at each of its points, allows crossing in one direction only.

7.11 The geometry of stationary null surfaces

The property of behaving like a one-way membrane, although characteristic of any null surface, is particularly important, for

*In a coordinate system adapted to the symmetries as in (7.10.8), the space-like coordinates are constant on these world-lines.

its astrophysical implications, when the surface is stationary in M. It is then usually the case that this surface is the boundary of events from which it is possible to escape to infinity: such a surface is termed an *event horizon.* We shall refer to stationary null surfaces generally as *horizons.* This means that a particle on a stationary (spatially) circular orbit, with tangent field u_{Ω}, as in (7.10.22) and (7.10.8), stays on a surface $\sigma < 0$ $(u^i \nabla_i \sigma = 0)$ for all values of its proper time, without ever crossing the null surface $H: \sigma = 0$. This surface is therefore bounded within all such circular orbits and divides M into the regions $M_<: \sigma < 0$ containing allowed orbits, and the region $M_>: \sigma > 0$. We shall show below that in general H consists of two separate *branches.* Restrict attention to one of these, H_+ say, choosing the time orientation such that $M_<$ is to the past of the null surface H_+. We note that at least locally the points in $M_<$ cannot be causally connected to the remainder of M by future pointing causal curves.

A stationary and axisymmetric space-time which satisfies (7.10.6), does not necessarily admit a surface with the properties of H. Its existence in fact imposes on the space-time further constraints. The Frobenius condition (7.10.9) reads:

$$\ell_a \nabla_{[b}\ell_{c]} + \ell_c \nabla_{[a}\ell_{b]} + \ell_b \nabla_{[c}\ell_{a]} = 0. \tag{7.11.1}$$

On H we have:

$$\nabla_a \ell_b = \nabla_{[a}\ell_{b]} \tag{7.11.2}$$

so (7.11.1) is equivalent to:

$$\ell_a \nabla_b \ell_c = 2\ell_{[c} \nabla_{b]} \ell_a. \tag{7.11.3}$$

Contracting (7.11.1) with ℓ^a, we have:

$$\sigma \nabla_{[b}\ell_{c]} + \ell^a \nabla_{[a}\ell_{b]}\ell_c - \ell^a \nabla_{[a}\ell_{c]}\ell_b = 0. \tag{7.11.4}$$

On H_+ this becomes:

$$\ell^a \nabla_a \ell_{[b}\ell_{c]} = 0 \tag{7.11.5}$$

showing that the Killing null generators of H are also geodesics, namely:

$$\ell^a \nabla_a \ell_b = \mathrm{k}\ell_b. \tag{7.11.6}$$

Here k is a real function on H_+. From (7.11.6) and (7.11.2) we have:

$$\partial_a \sigma = -2\mathrm{k}\ell_a. \tag{7.11.7}$$

The parameter, λ say, on these geodesics is not in general affine ($\mathrm{k} \neq 0$); an affine reparametrization $\tau = \tau(\lambda)$ is always possible and given by equations (2.4.17), namely:

$$\frac{d\tau}{d\lambda} = \exp\left(\int \mathrm{k}\, d\lambda\right). \tag{7.11.8}$$

The function k can, however, be expressed in terms of ℓ alone. Contracting (7.11.1) with $\nabla^b \ell^c$, we have:

$$\ell_a \nabla_{[b}\ell_{c]} \nabla^b \ell^c = -\ell_c \nabla_{[a}\ell_{b]} \nabla^b \ell^c - \ell_b \nabla_{[c}\ell_{a]} \nabla^b \ell^c; \tag{7.11.9}$$

restricting ourselves to H_+, we are led to:

$$\ell_a \left(\nabla_b \ell_c\right) \left(\nabla^b \ell^c\right) = -2\mathrm{k}^2 \ell_a$$

hence

$$\mathrm{k}^2 = -\frac{1}{2} \left(\nabla_a \ell_b\right) \left(\nabla^a \ell^b\right). \tag{7.11.10}$$

From this it follows identically:

$$\ell^b \partial_b \mathrm{k} = -\frac{1}{2}\ell^b \left(\nabla_d \nabla_b \ell_c\right) \left(\nabla^b \ell^c\right) = -\frac{\nabla^b \ell^c}{2\mathrm{k}} R_{cbde} \ell^d \ell^e = 0 \tag{7.11.11}$$

from (7.11.2) and (3.8.5), so k is constant along the generators of H. Equation (7.11.8) then becomes:

$$\begin{aligned} \frac{d\tau}{d\lambda} \propto \mathrm{e}^{\mathrm{k}\lambda} \quad ; \quad \tau \propto \mathrm{e}^{\mathrm{k}\lambda} \quad ; \quad \mathrm{k} \neq 0 \\ \tau = \lambda \quad ; \quad \mathrm{k} = 0 \qquad . \end{aligned} \tag{7.11.12}$$

When $\mathrm{k} \neq 0$ we see that the complete not-affine null generators of H_+ ($-\infty \leq \lambda \leq +\infty$) correspond to extendible affine generators, since when $\lambda \to -\infty$, $\tau \to 0$. If $\bar{\ell}^a$ is the tangent to the affine generators, such that $\bar{\ell}^a \nabla_a \bar{\ell}^b = 0$ on H_+, then at the point $\tau = 0$ they can be extended to $\tau \to -\infty$ covering a smooth continuation of H in $(M, \overset{*}{g})$.

From (7.11.12) it follows that $\bar{\ell}^i \propto \ell^i \mathrm{e}^{-\mathrm{k}\lambda}$, so when $\lambda \to -\infty$ ($\tau \to 0$) $\bar{\ell}^i$ remains finite hence ℓ^i vanishes at $\tau = 0$. We shall now show that there exists a second branch (H_-, say) of the horizon at $\tau = 0$ intersecting H_+ transversally. Suppose that $\tau = 0$ on the two-dimensional surface Σ bounding H_+, so that $\ell_a = 0$ on Σ. So for any vectors p^a, q^a tangent to Σ we have

$$p^a \nabla_a \ell^b = q^a \nabla_a \ell^b = 0 \ .$$

Let n^a be a null vector perpendicular to p^a and q^a (unique up to a factor) and satisfying $n^a \bar{\ell}_a = 1$ where $\bar{\ell}^a$ is the tangent to the affine generators. Then

$$p_a n^b \nabla_b \ell^a = -p_a n^b \nabla^a \ell_b = 0$$

and $q_a n^b \nabla_b \ell^a = 0$ similarly. Hence $n^b \nabla_b \ell^a = \alpha \bar{\ell}^a + \beta n^a$ for some α and β. Contracting with n_a gives:

$$0 = n_a n^b \nabla_b \ell^a = \alpha$$

and so

$$n^b \nabla_b \ell^a = \beta n^a \ . \tag{7.11.13}$$

Consider now the value of ℓ^a on Σ with initial tangent vector n^a. If we use (3.8.7) with ℓ^a for ξ^a and n^a for X^a and initial conditions $\ell^a|_{s=0} = 0$ and $(D\ell^a/Ds)|_{s=0} = \beta n^a$ (from (7.11.13)) then it is seen that the solution to (3.8.7) has $\ell^a \propto n^a$ for all s: i.e. there exists a null geodesic trajectory of the Killing vector starting at Σ with tangent vector independent of $\bar{\ell}^a$ at Σ. These geodesics, from all of Σ, generate a separate branch of the horizon. The point with $\tau = 0$ is termed the *bifurcation point* and the whole of H is termed a *bifurcated horizon.* Bifurcation disappears when k $= 0$ (Boyer, 1969). In most of the known stationary and axisymmetric solutions of Einstein field equations the surface H coincides with the event horizon, a property of the space-time which will be discussed in Section 10.5.

Let us now assume that the space-time contains a surface with the properties of H. Consider the Ricci identity:

$$\nabla_{[a} \nabla_{b]} \ell_c = \frac{1}{2} R_{abc}{}^d \ell_d;$$

contracting with ℓ^b we obtain:

$$\nabla_a \left(\ell^b \nabla_b \ell_c\right) - \left(\nabla_a \ell^b\right) \left(\nabla_b \ell_c\right) - \ell^b \nabla_b \nabla_a \ell_c = R_{abcd} \ell^b \ell^d. \tag{7.11.14}$$

On H, this becomes:

$$-\frac{1}{2}\nabla_a\nabla_c\sigma + (\nabla_a\ell^b)(\nabla_c\ell_b) = R_{abcd}\ell^b\ell^d \qquad (7.11.15)$$

since, from (3.8.5):

$$\ell^b\nabla_b\nabla_a\ell_c = R_{cabd}\ell^b\ell^d = 0.$$

Hence contracting over a and c and recalling (7.11.10):

$$\nabla_a\nabla^a\sigma + 4\mathrm{k}^2 = -2R_{ab}\ell^a\ell^b. \qquad (7.11.16)$$

The Laplacian of σ on H can be evaluated independently. We have that condition (7.11.7) holds on H. Therefore by differentiating tangentially to H we have:

$$\ell_{[a}\nabla_{b]}\nabla_c\sigma = -2\ell_{[a}\nabla_{b]}(\mathrm{k}\ell_c). \qquad (7.11.17)$$

Contracting over b and c and recalling (7.11.6), this becomes:

$$\ell_a\nabla^b\nabla_b\sigma - \ell_b\nabla_a\nabla^b\sigma = -2\mathrm{k}^2\ell_a. \qquad (7.11.18)$$

But also:

$$\ell^b\nabla_a\nabla_b\sigma = \ell^b\nabla_b\nabla_a\sigma = 2\left(\ell^b\nabla_b\ell^d\right)\nabla_a\ell_d + 2\ell^b\ell^d\nabla_b\nabla_a\ell_d = -2\mathrm{k}^2\ell_a$$

thus on H

$$\nabla^b\nabla_b\sigma = -4\mathrm{k}^2. \qquad (7.11.19)$$

Comparing this with (7.11.16), we have:

$$R_{ab}\ell^a\ell^b = 0. \qquad (7.11.20)$$

If Einstein equations hold, this implies:

$$T_{ab}\ell^a\ell^b = 0. \qquad (7.11.21)$$

On a stationary null surface then, a space-time is either empty ($T_{ab} = 0$) or it is endowed with an electromagnetic field which has ℓ as its principal double null direction ($F_{ab}\ell^b = \overset{*}{F}_{ab}\ell^b = 0$, see Sect. 7.8). The first conclusion is due to the fact that any physically realistic non-electromagnetic contribution to the energy-momentum tensor satisfies the strict inequality

$$T_{ab}X^aX^b < 0$$

for any time-like or null vector field in M, or is zero. The second conclusion is implied by the alternative condition $\ell_{[a}T_{b]c}\ell^c = 0$ on H which leads to:

$$\ell_{[a}R_{b]c}\ell^c = 0. \tag{7.11.22}$$

This, together with (7.11.3), leads to a further restriction on the space-time. Since (7.11.3) holds everywhere on H, then it is preserved under tangential differentiation on H, namely:

$$\ell_{[r}\nabla_{s]}\left(\ell_a\nabla_b\ell_c\right) = 2\ell_{[r}\nabla_{s]}\left(\ell_{[c}\nabla_{b]}\ell_a\right). \tag{7.11.23}$$

From (7.11.2) and (3.8.5), this reduces to:

$$\ell_r\ell_{[a}R_{b]t}\ell^t = \ell^c\ell^t R_{trc[a}\ell_{b]}\,. \tag{7.11.24}$$

But, from (7.11.22) and the decomposition (3.4.7), we deduce:

$$\ell^c\ell^t C_{trc[a}\ell_{b]} = 0 \tag{7.11.25}$$

namely ℓ is a double null principal eigenvector of the Weyl tensor on H (cf. Sect. 5.10). A consequence of (7.11.24) is that k is constant on H. In fact the tangential differentiation of (7.11.6) yields:

$$\ell_b\ell_{[c}\nabla_{d]}\mathrm{k} = R_{abt[d}\ell_{c]}\ell^a\ell^t = 0$$

that is

$$\ell_{[c}\nabla_{d]}\mathrm{k} = 0. \tag{7.11.26}$$

The parameter k, termed the *surface gravity*, plays a key role in the black-hole theory since this is entirely based on the properties of stationary null surfaces.

8

GEOMETRY OF CONGRUENCES

8.1 Tetrad decomposition of the Riemann tensor

In Chapter 4 we introduced the tetrad formalism as a powerful tool for investigating the structure properties of a space-time and introducing the concept of physical measurement on a curved manifold. We saw in Chapter 5 that the possibility of covering such a manifold with a field of tetrads allows one to define on it a spinor structure, an essential requirement for extending the quantum theory of fields to curved manifolds. We shall now further specialize the tetrad formalism to the study of congruences, in view of the fact that these describe in most cases the space-time evolution of physical systems. Let C_X be a time-like congruence of curves and $\{\lambda_{\hat{a}}\}$ a smooth tetrad field on it with the obvious identification of $\lambda_{\hat{0}} = X$, and parametrized so that $(X|X) = -1$ throughout.

The tetrad components of the Riemann tensor can be derived starting from the commutation relations applied to the covariant derivatives of any of the tetrad vectors:

$$\nabla_k \nabla_j \lambda_{\hat{a}i} - \nabla_j \nabla_k \lambda_{\hat{a}i} = R^n{}_{ijk} \lambda_{\hat{a}n}. \tag{8.1.1}$$

Contracting with $\lambda^i_{\hat{b}} \lambda^j_{\hat{c}} \lambda^k_{\hat{d}}$ and recalling (4.2.9a), we have:

$$R_{\hat{a}\hat{b}\hat{c}\hat{d}} = -\lambda_{\hat{d}}(\Gamma_{\hat{a}\hat{b}\hat{c}}) + \lambda_{\hat{c}}(\Gamma_{\hat{a}\hat{b}\hat{d}}) + 2\Gamma_{\hat{a}\hat{n}[\hat{c}}\Gamma^{\hat{n}}{}_{|\hat{b}|\hat{d}]} + 2\Gamma_{\hat{a}\hat{b}\hat{r}}\Gamma^{\hat{r}}{}_{[\hat{c}\hat{d}]}. \tag{8.1.2}$$

These are 36 components which can be divided into three groups and analysed separately, namely $R_{\hat{\alpha}\hat{\beta}\hat{\gamma}\hat{\delta}}$; $R_{\hat{0}\hat{\alpha}\hat{\beta}\hat{\gamma}}$; $R_{\hat{0}\hat{\beta}\hat{0}\hat{\delta}}$. In what follows, we shall for convenience write $(f)_{,\hat{\alpha}}$ instead of $\lambda_{\hat{\alpha}}(f)$.

The symmetry properties of the Riemann tensor which are explicit from (8.1.2) are:

$$R_{\hat{a}\hat{b}\hat{c}\hat{d}} = R_{[\hat{a}\hat{b}][\hat{c}\hat{d}]}. \tag{8.1.3}$$

Further symmetries are assured by the Jacobi identity which, imposed on the $\lambda_{\hat{a}}$'s, gives, from (4.2.12):

$$C_{\hat{a}[\hat{c}\hat{b},\hat{d}]} + C^{\hat{n}}{}_{[\hat{c}\hat{b}}C_{|\hat{a}|\hat{d}]\hat{n}} = 0. \qquad (8.1.4)$$

These imply, from (8.1.2):

$$\begin{aligned} &a) \quad R_{\hat{a}[\hat{b}\hat{c}\hat{d}]} = 0 \\ &b) \quad R_{\hat{a}\hat{b}\hat{c}\hat{d}} = R_{\hat{c}\hat{d}\hat{a}\hat{b}}; \end{aligned} \qquad (8.1.5)$$

we have here 16 constraint equations which algebraically connect the Riemann tensor components (8.1.2). Written in terms of the Ricci rotation coefficients, the identities (8.1.4) loose their compactness somewhat but allow a more immediate geometrical interpretation of their single terms:

$$\begin{aligned} &\frac{3}{2}\Gamma_{\hat{a}[\hat{b}\hat{c},\hat{d}]} + \Gamma^{\hat{\nu}}{}_{[\hat{d}\hat{b}]}\Gamma_{\hat{a}[\hat{\nu}\hat{c}]} + \Gamma^{\hat{\nu}}{}_{[\hat{c}\hat{d}]}\Gamma_{\hat{a}[\hat{\nu}\hat{b}]} \\ &+ \Gamma^{\hat{\nu}}{}_{[\hat{b}\hat{c}]}\Gamma_{\hat{a}[\hat{\nu}\hat{d}]} - \Gamma_{\hat{0}[\hat{d}\hat{b}]}\Gamma_{\hat{a}[\hat{0}\hat{c}]} - \Gamma_{\hat{0}[\hat{c}\hat{d}]}\Gamma_{\hat{a}[\hat{0}\hat{b}]} - \Gamma_{\hat{0}[\hat{b}\hat{c}]}\Gamma_{\hat{a}[\hat{0}\hat{d}]} = 0. \end{aligned} \qquad (8.1.6)$$

Let us now consider the first group of components of the Riemann tensor, namely the 9 completely spatial $R_{\hat{\alpha}\hat{\beta}\hat{\gamma}\hat{\delta}}$. We realize that at least two indices must be equal so these components are of the type $R_{\hat{\alpha}\hat{\beta}\hat{\alpha}\hat{\gamma}}$. Here we assume that equal indices on the same level *must not* be summed. From (8.1.2) we then have:

$$\begin{aligned} R_{\hat{\alpha}\hat{\beta}\hat{\alpha}\hat{\delta}} = & -2\Gamma_{\hat{\alpha}\hat{\beta}[\hat{\alpha},\hat{\delta}]} + 2\Gamma_{\hat{\alpha}\hat{\nu}[\hat{\alpha}}\Gamma^{\hat{\nu}}{}_{|\hat{\beta}|\hat{\delta}]} + 2\Gamma_{\hat{\alpha}\hat{\beta}\hat{\nu}}\Gamma^{\hat{\nu}}{}_{[\hat{\alpha}\hat{\delta}]} \\ & + \Gamma_{\hat{0}\hat{\alpha}\hat{\alpha}}\Gamma_{\hat{0}\hat{\beta}\hat{\delta}} - \Gamma_{\hat{0}\hat{\alpha}\hat{\delta}}\Gamma_{\hat{0}\hat{\beta}\hat{\alpha}} - 2\Gamma_{\hat{\alpha}\hat{\beta}\hat{0}}\Gamma_{\hat{0}[\hat{\alpha}\hat{\delta}]}. \end{aligned} \qquad (8.1.7)$$

Comparing with (4.6.8) and (4.6.11), we recognize here the structure of the Riemann tensor as a composition of terms with a definite geometrical meaning reflecting the choices of the congruence and of the tetrad field. For these indices equations (8.1.6) reduce to only three non-trivial equations which read:

$$\begin{aligned} &\Gamma_{\hat{\alpha}\hat{\beta}[\hat{\alpha},\hat{\delta}]} + \Gamma_{\hat{\alpha}\hat{\delta}[\hat{\beta},\hat{\alpha}]} + 2\Gamma^{\hat{\nu}}{}_{[\hat{\delta}\hat{\beta}]}\Gamma_{\hat{\alpha}[\hat{\nu}\hat{\alpha}]} + 2\Gamma^{\hat{\nu}}{}_{[\hat{\alpha}\hat{\delta}]}\Gamma_{\hat{\alpha}[\hat{\nu}\hat{\beta}]} + 2\Gamma^{\hat{\nu}}{}_{[\hat{\beta}\hat{\alpha}]}\Gamma_{\hat{\alpha}[\hat{\nu}\hat{\delta}]} \\ &\quad - 2\Gamma_{\hat{0}[\hat{\delta}\hat{\beta}]}\Gamma_{\hat{\alpha}[\hat{0}\hat{\alpha}]} - 2\Gamma_{\hat{0}[\hat{\alpha}\hat{\delta}]}\Gamma_{\hat{\alpha}[\hat{0}\hat{\beta}]} - 2\Gamma_{\hat{0}[\hat{\beta}\hat{\alpha}]}\Gamma_{\hat{\alpha}[\hat{0}\hat{\delta}]} = 0. \end{aligned} \qquad (8.1.8)$$

Using (8.1.8) in (8.1.7) and recalling definitions (4.5.7), (4.5.8) and (4.6.11), we obtain:

$$R_{\hat{\alpha}\hat{\beta}\hat{\alpha}\hat{\delta}} = P_{\hat{\alpha}\hat{\beta}\hat{\alpha}\hat{\delta}} + 2\Theta_{\hat{\alpha}[\hat{\alpha}}\Theta_{\hat{\beta}]\hat{\delta}} + 2\Lambda_{\hat{\alpha}(\hat{\beta}}\omega_{\hat{\delta})\hat{\alpha}} + \omega_{\hat{\delta}\hat{\alpha}}\omega_{\hat{\beta}\hat{\alpha}} \qquad (8.1.9)$$

where:

$$P_{\hat{\alpha}\hat{\beta}\hat{\alpha}\hat{\delta}} = -\Gamma_{\hat{\alpha}\hat{\beta}[\hat{\alpha},\hat{\delta}]} - \Gamma_{\hat{\alpha}\hat{\delta}[\hat{\alpha},\hat{\beta}]} + \Gamma_{\hat{\alpha}\hat{\nu}\hat{\alpha}}\Gamma^{\hat{\nu}}{}_{(\hat{\beta}\hat{\delta})} + \Gamma^{\hat{\nu}}{}_{[\hat{\alpha}\hat{\beta}]}\Gamma_{\hat{\alpha}\hat{\delta}\hat{\nu}}$$
$$+ \Gamma_{\hat{\nu}(\hat{\alpha}\hat{\delta})}\Gamma^{\hat{\nu}}{}_{(\hat{\beta}\hat{\alpha})} - \Gamma^{\hat{\nu}}{}_{[\hat{\alpha}\hat{\delta}]}\Gamma_{\hat{\nu}[\hat{\alpha}\hat{\beta}]} + \Gamma_{\hat{\alpha}\hat{\beta}\hat{\nu}}\Gamma^{\hat{\nu}}{}_{[\hat{\alpha}\hat{\delta}]} \qquad (8.1.10)$$

behaves as a Lorentz tensor (i.e. transforms as a tensor when one changes the tetrad) and, as expected, manifestly shows the symmetries of the Riemann tensor and depends entirely on the structure of the spatial triads at each point of the manifold.

Relation (8.1.9) is particularly useful when one is dealing with a hypersurface orthogonal congruence ($\omega_{\hat{\alpha}\hat{\beta}} = 0$). In this case the vector X is the unit normal to a hypersurface Σ, say. The change of X for tangential displacements over Σ measures the curvature of the hypersurface with respect to the imbedding space-time. From (4.5.3) we have:

$$\lambda_{i\hat{\beta}}\lambda^{j}_{\hat{\alpha}}\nabla_j X^i = -\Gamma_{\hat{0}(\hat{\alpha}\hat{\beta})} \equiv K_{\hat{\alpha}\hat{\beta}} \qquad (8.1.11)$$

and this quantity is termed the *extrinsic curvature* of the hypersurface Σ. In terms of this, relation (8.1.9) becomes

$$R_{\hat{\alpha}\hat{\beta}\hat{\alpha}\hat{\delta}} = P_{\hat{\alpha}\hat{\beta}\hat{\alpha}\hat{\delta}} + 2K_{\hat{\alpha}[\hat{\alpha}}K_{\hat{\beta}]\hat{\delta}} \qquad (8.1.12)$$

which is the Gauss equation; (cf. eq. (4.6.8)). In this case the tensor $P_{\hat{\alpha}\hat{\beta}\hat{\alpha}\hat{\delta}}$ is termed the *intrinsic curvature* of Σ (Eisenhart, 1926; Schouten, 1954).

The second group of components of (8.1.2) leads to a larger number of equations, namely 18, which are from (4.5.7) and (4.5.8′):

$$R_{\hat{0}\hat{\beta}\hat{\gamma}\hat{\delta}} = -2\Gamma_{\hat{0}\hat{\beta}[\hat{\gamma},\hat{\delta}]} + 2\Gamma_{\hat{0}\hat{\sigma}[\hat{\gamma}}\Gamma^{\hat{\sigma}}{}_{|\hat{\beta}|\hat{\delta}]} + 2\Gamma_{\hat{0}\hat{\beta}\hat{0}}\Gamma^{\hat{0}}{}_{[\hat{\gamma}\hat{\delta}]} + 2\Gamma_{\hat{0}\hat{\beta}\hat{\sigma}}\Gamma^{\hat{\sigma}}{}_{[\hat{\gamma}\hat{\delta}]}$$
$$= 2\nabla_{[\hat{\delta}}\Theta_{\hat{\gamma}]\hat{\beta}} - 2\nabla_{[\hat{\delta}}\omega_{\hat{\gamma}]\hat{\beta}} - 2\dot{X}_{\hat{\beta}}\omega_{\hat{\gamma}\hat{\delta}}. \qquad (8.1.13)$$

Hereafter, for ease of notation, we write $\nabla_{\hat{a}}\Phi^{\hat{b}\ldots}{}_{\hat{c}\ldots}$ to mean $(\nabla_{\hat{a}}\mathbf{\Phi})^{\hat{b}\ldots}{}_{\hat{c}\ldots}$, for any tensor $\mathbf{\Phi}$. Of the 18 equations (8.1.13), only 8 are independent, this being assured by the constraint equations (8.1.6) which yield 10 equations. These can be divided into two groups; the first is the single equation: $R_{\hat{0}[\hat{\beta}\hat{\gamma}\hat{\delta}]} = 0$ which can also be written as:

$$\omega_{[\hat{\gamma}\hat{\beta},\hat{\delta}]} + 2\omega_{\hat{\nu}[\hat{\gamma}}\Gamma^{\hat{\nu}}{}_{\hat{\beta}\hat{\delta}]} + \omega_{[\hat{\delta}\hat{\beta}}\dot{X}_{\hat{\gamma}]} = 0\ , \qquad (8.1.14)$$

or in a more compact form, from (4.5.8′):

$$\nabla_{[\hat{\delta}}\omega_{\hat{\gamma}\hat{\beta}]} + \omega_{[\hat{\delta}\hat{\beta}}\dot{X}_{\hat{\gamma}]} = 0. \tag{8.1.15}$$

The second group is made by the remaining nine, of the form: $R_{\hat{\beta}[\hat{0}\hat{\gamma}\hat{\delta}]} = 0$ which lead to:

$$\dot{\Gamma}_{\hat{\beta}[\hat{\gamma}\hat{\delta}]} + \nabla_{[\hat{\delta}}\Psi_{|\hat{\beta}|\hat{\gamma}]} - \Psi^{\hat{\nu}}{}_{[\hat{\delta}}\Gamma_{\hat{\gamma}]\hat{\beta}\hat{\nu}} + 2\Psi_{\hat{\beta}[\hat{\gamma}}\dot{X}_{\hat{\delta}]} = 0 \tag{8.1.16}$$

where

$$\dot{\Gamma} = X^i\partial_i\Gamma \tag{8.1.17}$$

and

$$\Psi_{\hat{\alpha}\hat{\beta}} = \Theta_{\hat{\alpha}\hat{\beta}} + \Lambda_{\hat{\alpha}\hat{\beta}} + \omega_{\hat{\alpha}\hat{\beta}}. \tag{8.1.18}$$

In the case of a hypersurface orthogonal congruence, equations (8.1.13) become, from (8.1.11):

$$R_{\hat{0}\hat{\beta}\hat{\gamma}\hat{\delta}} = 2\nabla_{[\hat{\delta}}K_{\hat{\gamma}]\hat{\beta}} \tag{8.1.19}$$

which is known as the Codazzi equation. The restrictions (8.1.16) are now to be considered as evolution equations for the transverse Ricci coefficients; they clearly limit one's freedom of choosing the spatial triads on any hypersurface Σ.

We are now left with the third group of the 9 Riemann tensor components $R_{\hat{0}\hat{\alpha}\hat{0}\hat{\beta}}$. From (8.1.2) these read:

$$R_{\hat{0}\hat{\alpha}\hat{0}\hat{\beta}} = -2\Gamma_{\hat{0}\hat{\alpha}[\hat{0},\hat{\beta}]} + 2\Gamma_{\hat{0}\hat{c}[\hat{0}}\Gamma^{\hat{c}}{}_{|\hat{\alpha}|\hat{\beta}]} + 2\Gamma_{\hat{0}\hat{\alpha}\hat{c}}\Gamma^{\hat{c}}{}_{[\hat{0}\hat{\beta}]}. \tag{8.1.20}$$

The restrictions now yield only three equations which assure the symmetry of (8.1.20) in the pair $(\hat{\alpha},\hat{\beta})$. We then have from (8.1.6):

$$\dot{\omega}_{\hat{\beta}\hat{\alpha}} + \nabla_{[\hat{\beta}}\dot{X}_{\hat{\alpha}]} - 2\omega_{\hat{\sigma}[\hat{\beta}}\Theta_{\hat{\alpha}]}{}^{\hat{\sigma}} + 2\omega_{\hat{\sigma}[\hat{\beta}}\Lambda_{\hat{\alpha}]}{}^{\hat{\sigma}} = 0. \tag{8.1.21}$$

Using this in (8.1.20) we finally obtain:

$$R_{\hat{0}\hat{\alpha}\hat{0}\hat{\beta}} = -\dot{\Theta}_{\hat{\alpha}\hat{\beta}} - \Theta_{\hat{\alpha}\hat{\sigma}}\Theta^{\hat{\sigma}}{}_{\hat{\beta}} - \omega_{\hat{\sigma}\hat{\alpha}}\omega_{\hat{\beta}}{}^{\hat{\sigma}} + \nabla_{(\hat{\beta}}\dot{X}_{\hat{\alpha})} + 2\Theta_{\hat{\rho}(\hat{\alpha}}\Lambda_{\hat{\beta})}{}^{\hat{\rho}} + \dot{X}_{\hat{\alpha}}\dot{X}_{\hat{\beta}}. \tag{8.1.22}$$

The last two equations are of basic importance in relativistic fluidodynamics (Ellis, 1967; Stewart and Ellis, 1968).

Let us now consider the tetrad components of the Weyl tensor which was introduced in (3.4.7). It is of some interest to define the quantities:

$$E^{\hat{\alpha}}{}_{\hat{\beta}} = -C_{\hat{0}}{}^{\hat{\alpha}}{}_{\hat{0}\hat{\beta}} \qquad E^{\hat{\alpha}}{}_{\hat{\alpha}} = 0 \tag{8.1.23}$$

$$H^{\hat{\alpha}}{}_{\hat{\beta}} = \overset{*}{C}{}^{\hat{0}\hat{\alpha}}{}_{\hat{0}\hat{\beta}} \qquad H^{\hat{\alpha}}{}_{\hat{\alpha}} = 0 \tag{8.1.24}$$

which are termed, respectively, the *electric* and the *magnetic* parts of the Weyl tensor. These are traceless three-dimensional Lorentz tensors with respect to the internal rotation group. In terms of these quantities the Bianchi identities take on a form which closely resembles that of Maxwell's equations, justifying their names. In tetrad components, the Bianchi identities (3.5.10) can be written as:

$$\nabla_{\hat{d}} C^{\hat{a}\hat{b}\hat{c}\hat{d}} = \nabla^{[\hat{b}} R^{\hat{a}]\hat{c}} + \frac{1}{6} \eta^{\hat{c}[\hat{b}} \nabla^{\hat{a}]} R. \tag{8.1.25}$$

Equations (8.1.25) lead to four sets of equations:

$$\nabla_{\hat{\mu}} E^{\hat{\beta}\hat{\mu}} = -\frac{1}{2}\left[\nabla_{\hat{\beta}} R_{\hat{0}\hat{0}} + \frac{1}{6}\nabla_{\hat{\beta}} R - \nabla_{\hat{0}} R_{\hat{0}\hat{\beta}}\right] \tag{8.1.26}$$

$$\epsilon^{\hat{\gamma}\hat{\sigma}}{}_{\hat{\rho}} \nabla_{\hat{\sigma}} H^{\hat{\alpha}\hat{\rho}} - \nabla_{\hat{0}} E^{\hat{\alpha}\hat{\gamma}} = \frac{1}{2}\left[\nabla_{\hat{\alpha}} R_{\hat{0}\hat{\gamma}} - \nabla_{\hat{0}}\left(R_{\hat{\alpha}\hat{\gamma}} - \frac{1}{2}\delta_{\hat{\alpha}\hat{\gamma}} R\right)\right] \tag{8.1.27}$$

$$\nabla_{\hat{\mu}} H^{\hat{\alpha}\hat{\mu}} = \epsilon_{\hat{\rho}\hat{\sigma}}{}^{\hat{\alpha}} \nabla^{[\hat{\rho}} R^{\hat{\sigma}]\hat{0}} \tag{8.1.28}$$

$$\epsilon^{\hat{\gamma}\hat{\lambda}}{}_{\hat{\sigma}} \nabla_{\hat{\lambda}} E^{\hat{\alpha}\hat{\sigma}} + \nabla_{\hat{0}} H^{\hat{\alpha}\hat{\gamma}} = \frac{1}{2}\epsilon_{\hat{\rho}\hat{\beta}}{}^{\hat{\alpha}}\left[\nabla^{[\hat{\beta}} R^{\hat{\rho}]\hat{\gamma}} + \frac{1}{6}\eta^{\hat{\gamma}[\hat{\beta}}\nabla^{\hat{\rho}]} R\right] \tag{8.1.29}$$

where

$$\epsilon_{\hat{\alpha}\hat{\beta}\hat{\gamma}} = \eta_{ijkl} X^l \lambda^i_{\hat{\alpha}} \lambda^j_{\hat{\beta}} \lambda^k_{\hat{\gamma}}. \tag{8.1.30}$$

It is more interesting to find the explicit expressions of $E_{\hat{\alpha}\hat{\beta}}$ and $H_{\hat{\alpha}\hat{\beta}}$ in terms of the Riemann tensor and its contractions relative to the assigned tetrad field. From (3.4.7), written in tetrad components, and from (8.1.23), one finds:

$$E_{\hat{\alpha}\hat{\beta}} = -R_{\hat{0}\hat{\alpha}\hat{0}\hat{\beta}} - \frac{1}{2} R_{\hat{\alpha}\hat{\beta}} + \frac{1}{6}\delta_{\hat{\alpha}\hat{\beta}}\left(R + 3R_{\hat{0}\hat{0}}\right) . \tag{8.1.31}$$

The electric part of the Weyl tensor describes the gravitational tidal forces. It has a Newtonian analog which arises in the non-relativistic limit, from terms involving $R_{\hat{0}\hat{\alpha}\hat{0}\hat{\beta}}$ and its trace, and reads:

$$\underset{\text{N}}{E}_{\hat{\alpha}\hat{\beta}} = \frac{\partial^2 \varphi}{\partial x^{\hat{\alpha}} \partial x^{\hat{\beta}}} - \frac{1}{3}\delta_{\hat{\alpha}\hat{\beta}} \nabla^2 \varphi \tag{8.1.32}$$

where φ is the Newtonian potential.

The magnetic part $H_{\hat{\alpha}\hat{\beta}}$ has no Newtonian analog and reads explicitly

$$H_{\hat{\alpha}\hat{\beta}} = -\frac{1}{2}\epsilon_{\hat{\alpha}}{}^{\hat{\rho}\hat{\sigma}}\Big[2\nabla_{[\hat{\sigma}}\Theta_{\hat{\rho}]\hat{\beta}} + \delta_{\hat{\beta}[\hat{\rho}}\nabla^{\hat{\mu}}\Theta_{\hat{\sigma}]\hat{\mu}} - \delta_{\hat{\beta}[\hat{\sigma}}\nabla_{\hat{\rho}]}\Theta + 2\nabla_{[\hat{\sigma}}\omega_{\hat{\rho}]\hat{\beta}}$$
$$+ \delta_{\hat{\beta}[\hat{\rho}}\nabla^{\hat{\mu}}\omega_{\hat{\sigma}]\hat{\mu}} + 2\dot{X}_{\hat{\beta}}\omega_{\hat{\rho}\hat{\sigma}} + 2\dot{X}_{\hat{\mu}}\omega_{[\hat{\sigma}}{}^{\hat{\mu}}\delta_{\hat{\rho}]\hat{\beta}}\Big]. \tag{8.1.33}$$

It has been shown (Ashtekhar and Hansen, 1978) that $H_{\hat{\alpha}\hat{\beta}}$ plays an important role in the problem of defining at spatial infinity (see Sect. 10.4) the angular momentum of a space-time which is asymptotically flat and empty.

8.2 The expansion equation

Equation (8.1.22) governs the evolution of the expansion, the curvature entering directly as a source. Let us here discuss in more detail the significance of the expansion in terms of the properties of the congruence. If ξ is a connecting vector field of the congruence, then by definition it satisfies the equation:

$$\pounds_X \xi = o. \tag{8.2.1}$$

From the properties of the Lie derivative, we have:

$$\pounds_X \xi = \dot{\xi}^{\hat{a}}\lambda_{\hat{a}} + \xi^{\hat{\alpha}} C_{\hat{c}\hat{0}\hat{\alpha}}\lambda^{\hat{c}} = o. \tag{8.2.2}$$

Scalar multiplication with $\lambda_{\hat{\beta}}$ yields:

$$\dot{\xi}_{\hat{\beta}} = 2\Gamma_{\hat{\beta}[\hat{0}\hat{\alpha}]}\xi^{\hat{\alpha}} = \Psi_{\hat{\beta}\hat{\alpha}}\xi^{\hat{\alpha}} \tag{8.2.3}$$

from (8.1.18). Further multiplication with $\xi^{\hat{\beta}}$ leads to:

$$\frac{1}{2}\dot{\ell^2} = \Theta_{\hat{\alpha}\hat{\beta}}\xi^{\hat{\alpha}}\xi^{\hat{\beta}} \tag{8.2.4}$$

where $\ell^2 = \xi_{\hat{\alpha}}\xi^{\hat{\alpha}} = h_{ij}\xi^i\xi^j$ is the *length* of the connecting vector measured in the congruence *rest frame* . Let us now introduce the following decomposition of the expansion $\Theta_{\hat{\alpha}\hat{\beta}}$:

$$\Theta_{\hat{\alpha}\hat{\beta}} = \sigma_{\hat{\alpha}\hat{\beta}} + \frac{1}{3}\delta_{\hat{\alpha}\hat{\beta}}\Theta$$

into a trace-free part $\sigma_{\hat{\alpha}\hat{\beta}}$ and its trace Θ; equation (8.2.4) becomes:

$$\dot{\ell} = \frac{1}{3}\Theta\ell + \sigma_{\hat{\alpha}\hat{\beta}}e^{\hat{\alpha}}e^{\hat{\beta}}\ell \tag{8.2.5}$$

where $e^{\hat{\alpha}}$ are the direction cosines of ℓ in the spatial triad. These directions also change along the congruence, in fact from (8.2.3) and (8.2.5) we find:

$$\dot{e}_{\hat{\beta}} = \left(\Lambda_{\hat{\alpha}\hat{\beta}} + \omega_{\hat{\alpha}\hat{\beta}}\right)e^{\hat{\alpha}} + \left[\sigma_{\hat{\alpha}\hat{\beta}} - \left(\sigma_{\hat{\rho}\hat{\pi}}e^{\hat{\rho}}e^{\hat{\pi}}\right)\delta_{\hat{\alpha}\hat{\beta}}\right]e^{\hat{\alpha}}. \tag{8.2.6}$$

Equation (8.2.5) says that as we move along the congruence, the distance between neighbouring world-lines experiences two changes at the same time: an isotropic variation measured by Θ which is then termed the *isotropic* (or *volumetric*) *expansion*; and a non-isotropic one, measured by $\sigma_{\hat{\alpha}\hat{\beta}}$, which is termed the *shear*. Equation (8.2.6) moreover says that the direction of a connecting vector changes along the congruence, the change again consisting of two parts: a uniform rotation due to the combined effects of the Fermi rotation of the tetrad and the vorticity of the congruence; and a non-uniform rotation due to the shear itself.

The expansion coefficients determine the geometry of a space-like section of the congruence at any time of its evolution. Call $\Sigma(\tau)$ the cross section of a narrow tube of world-lines of the congruence, centred on a given curve $\gamma(\tau)$, and generated by geodesics orthogonal to $\gamma(\tau)$. Assume also that initially, at $\tau = \tau_0$ say, $\Sigma(\tau_0)$ is a sphere. If $\sigma_{\hat{\alpha}\hat{\beta}} = 0$, the volume of this sphere, measured in the space-triad defined in $\Sigma(\tau_0)$ is just: $\underset{0}{V} = (4/3)\pi\underset{0}{\ell}^3$ (subscript $_0$ means at τ_0). After a proper time lapse of $\delta\tau$, the volume is

$$\underset{0}{V}(\tau) = \underset{0}{V}\left(1 + \underset{0}{\Theta}\delta\tau\right)$$

from (8.2.5); we would then have an expansion ($\underset{0}{\Theta} > 0$) or a contraction ($\underset{0}{\Theta} < 0$) with no departure from spherical symmetry. Assume now that $\Theta = 0$ and $\sigma_{\hat{\alpha}\hat{\beta}} \neq 0$; at $\tau = \tau_0$ let $\underset{1}{e}^{\hat{\alpha}}$, $\underset{2}{e}^{\hat{\alpha}}$, $\underset{3}{e}^{\hat{\alpha}}$ be three mutually orthogonal directions along which we consider three connecting vectors stemming from a given world-line of the congruence. If $\Sigma(\tau_0)$ was originally a sphere ($\underset{1}{\ell} = \underset{2}{\ell} = \underset{3}{\ell} = \underset{0}{\ell}$), after a time $\delta\tau$ it will no longer be so, in fact the connecting vectors become:

$$\underset{\alpha}{\ell}(\tau) = \underset{0}{\ell}\left[1 + \left(\sigma_{\hat\rho\hat\pi}\, \underset{\alpha}{e}^{\hat\rho}\, \underset{\alpha}{e}^{\hat\pi}\right)_0 \delta\tau + \cdots\right] \qquad \alpha = 1, 2, 3$$

and their directions in $\Sigma(\tau)$ will no longer be orthogonal; the original sphere now appears as an ellipsoid. To first order in the shear however we can still consider the connecting vectors of length $\underset{\alpha}{\ell}\,(\tau)$ orthogonal to each other, so the volume of $\Sigma(\tau)$ reads:

$$V(\tau) = \frac{4}{3}\pi \underset{0}{\ell}^3 \left[1 + \sigma_{\hat\rho\hat\pi}\left(\underset{1}{e}^{\hat\rho}\, \underset{1}{e}^{\hat\pi} + \underset{2}{e}^{\hat\rho}\, \underset{2}{e}^{\hat\pi} + \underset{3}{e}^{\hat\rho}\, \underset{3}{e}^{\hat\pi} \right)_0 \delta\tau + \mathrm{O}(\sigma^2 d\tau^2)\right] . \tag{8.2.7}$$

From the properties of direction cosines, the quantity which multiplies $\delta\tau$ vanishes, so the shear would not contribute to the volume, to first order.

The coefficients Θ and $\sigma_{\hat\alpha\hat\beta}$ describe the kinematical properties of a congruence of world-lines, while its dynamical behaviour is completely described by their propagation laws. From (8.1.22) we obtain the propagation equation for the isotropic expansion:

$$\dot\Theta = -R_{\hat0\hat0} + 2(\omega^2 - \sigma^2) - \frac{1}{3}\Theta^2 + \nabla_{\hat\alpha}\dot X^{\hat\alpha} + \dot X^2 \tag{8.2.8}$$

where

$$\omega^2 = \frac{1}{2}\omega_{\hat\alpha\hat\beta}\omega^{\hat\alpha\hat\beta} \quad \sigma^2 = \frac{1}{2}\sigma_{\hat\alpha\hat\beta}\sigma^{\hat\alpha\hat\beta} \quad \dot X^2 = \dot X_{\hat\alpha}\dot X^{\hat\alpha}$$

which is known as the *Raychauduri equation* (Raychauduri, 1955).

The trace-free part of (8.1.22) yields the propagation equation of the shear which reads:

$$\begin{aligned}\dot\sigma_{\hat\alpha\hat\beta} = {} & E_{\hat\alpha\hat\beta} + \frac{1}{2}\left(R_{\hat\alpha\hat\beta} - \frac{1}{3}\delta_{\hat\alpha\hat\beta}R_{\hat\sigma}{}^{\hat\sigma}\right) + \tilde A_{\hat\alpha\hat\beta} - 2\Theta_{\hat\sigma(\hat\alpha}\Lambda_{\hat\beta)}{}^{\hat\sigma} \\ & - \frac{2}{3}\sigma_{\hat\alpha\hat\beta}\Theta - \omega_{\hat\mu\hat\alpha}\omega_{\hat\beta}{}^{\hat\mu} - \sigma_{\hat\mu\hat\alpha}\sigma_{\hat\beta}{}^{\hat\mu} - \frac{2}{3}\delta_{\hat\alpha\hat\beta}\left(\omega^2 - \sigma^2\right) .\end{aligned} \tag{8.2.9}$$

where:

$$\tilde A_{\hat\alpha\hat\beta} = \nabla_{(\hat\alpha}\dot X_{\hat\beta)} + \dot X_{\hat\alpha}\dot X_{\hat\beta} - \frac{1}{3}\delta_{\hat\alpha\hat\beta}\left(\nabla_{\hat\sigma}\dot X^{\hat\sigma} + \dot X^2\right) . \tag{8.2.10}$$

An observer who moves along the congruence detects the relative displacement between neighbouring world-lines, but what he actually measures is their relative acceleration. In tetrad form this quantity is easily evaluated; differentiating (8.2.3) with respect to the parameter τ, we have:

$$\ddot{\xi}_{\hat{\alpha}} = \left[-R_{\hat{0}\hat{\alpha}\hat{0}\hat{\beta}} + \nabla_{\hat{\beta}} \dot{X}_{\hat{\alpha}} + \dot{X}_{\hat{\alpha}} \dot{X}_{\hat{\beta}} + \dot{\Lambda}_{\hat{\alpha}\hat{\beta}} \right. \\ \left. + 2 \left(\Theta_{\hat{\sigma}\hat{\beta}} + \omega_{\hat{\sigma}\hat{\beta}} \right) \Lambda_{\hat{\alpha}}{}^{\hat{\sigma}} + \Lambda_{\hat{\sigma}\hat{\alpha}} \Lambda_{\hat{\beta}}{}^{\hat{\sigma}} \right] \xi^{\hat{\beta}}. \quad (8.2.11)$$

In case of a Fermi tetrad ($\Lambda_{\hat{\alpha}\hat{\beta}} = \dot{\Lambda}_{\hat{\alpha}\hat{\beta}} = 0$) and of a geodesic flow ($\dot{X}_{\hat{\alpha}} = 0$), the above equation reduces to the geodesic deviation equation:

$$\ddot{\xi}_{\hat{\alpha}} = -R_{\hat{0}\hat{\alpha}\hat{0}\hat{\beta}} \xi^{\hat{\beta}} \quad (8.2.12)$$

which allows a direct measurement of the background curvature.

8.3 The vorticity equation

The vorticity measures how a world-line of a congruence rotates around a neighbouring one. As can be deduced from (8.2.6), an instantaneous axis of rotation is identified, in the space triad, as:

$$\omega_{\hat{\alpha}} = \frac{1}{2} \epsilon_{\hat{\alpha}\hat{\beta}\hat{\gamma}} \omega^{\hat{\beta}\hat{\gamma}}. \quad (8.3.1)$$

We shall call this quantity the *vorticity* vector. A property of the vorticity vector can be obtained immediately from equation (8.1.15); contracting there with $\epsilon^{\hat{\gamma}\hat{\beta}\hat{\delta}}$ we have:

$$\omega^{\hat{\alpha}}{}_{,\hat{\alpha}} - \Gamma_{\hat{\sigma}}{}^{\hat{\alpha}}{}_{\hat{\alpha}} \omega^{\hat{\sigma}} - \omega^{\hat{\sigma}} \dot{X}_{\hat{\sigma}} = \nabla_{\hat{\alpha}} \omega^{\hat{\alpha}} - \omega^{\hat{\sigma}} \dot{X}_{\hat{\sigma}} = 0 \quad (8.3.2)$$

showing that the vorticity vector field is divergence free when the motion is geodesic.

The propagation law for the vorticity is given by equation (8.1.21); it will be more convenient to write it in terms of the vorticity vector itself. Contracting (8.1.21) with $\epsilon^{\hat{\rho}\hat{\alpha}\hat{\beta}}$ and using (1.17.10) we obtain after some algebra:

$$\dot{\omega}^{\hat{\rho}} = -\Theta \omega^{\hat{\rho}} + \omega^{\hat{\sigma}} \left(\Theta^{\hat{\rho}}{}_{\hat{\sigma}} + \Lambda^{\hat{\rho}}{}_{\hat{\sigma}} \right) + \frac{1}{2} \epsilon^{\hat{\rho}\hat{\alpha}\hat{\beta}} \nabla_{[\hat{\beta}} \dot{X}_{\hat{\alpha}]}. \quad (8.3.3)$$

If $e^{\hat{\rho}}$ are the direction cosines of the vorticity vector, then equation (8.3.3) is equivalent to the two equations:

$$\dot{e}^{\hat{\rho}} = \tilde{h}^{\hat{\rho}}{}_{\hat{\alpha}} \left[(e^{\hat{\sigma}}\Theta^{\hat{\alpha}}{}_{\hat{\sigma}} + e^{\hat{\sigma}}\Lambda^{\hat{\alpha}}{}_{\hat{\sigma}}) + \frac{1}{2}\frac{\epsilon^{\hat{\alpha}\hat{\sigma}\hat{\beta}}}{\omega}\nabla_{[\hat{\beta}}\dot{X}_{\hat{\sigma}]} \right] \tag{8.3.4}$$

$$\dot{\omega} = -\Theta\omega + \omega\Theta_{\hat{\rho}\hat{\sigma}}e^{\hat{\rho}}e^{\hat{\sigma}} + \frac{1}{2}\epsilon^{\hat{\rho}\hat{\beta}\hat{\alpha}}\nabla_{[\hat{\beta}}\dot{X}_{\hat{\rho}]}e_{\hat{\alpha}} \tag{8.3.5}$$

where $\tilde{h}^{\hat{\rho}}{}_{\hat{\alpha}} = \delta^{\hat{\rho}}_{\hat{\alpha}} - e^{\hat{\rho}}e_{\hat{\alpha}}$ is an operator which projects orthogonally to the vorticity vector and $\omega = \left(\omega_{\hat{\alpha}}\omega^{\hat{\alpha}}\right)^{1/2}$.

In these equations, the acceleration provides a source for the vorticity causing a change in both the modulus and the direction. The latter is determined not only by the shear, but also by the Fermi rotation. This causes a precession of the vorticity axis around the Fermi axis ($\zeta^{\hat{\alpha}} = (1/2)\epsilon^{\hat{\alpha}\hat{\beta}\hat{\gamma}}\Lambda_{\hat{\beta}\hat{\gamma}}$), as can be seen by writing equation (8.3.4) as follows:

$$\dot{e}^{\hat{\rho}} = \tilde{h}^{\hat{\rho}}{}_{\hat{\alpha}}\sigma^{\hat{\alpha}}{}_{\hat{\beta}}e^{\hat{\beta}} + \epsilon^{\hat{\rho}\hat{\alpha}\hat{\beta}}e_{\hat{\alpha}}\zeta_{\hat{\beta}} + \frac{1}{2}\tilde{h}^{\hat{\rho}}{}_{\hat{\alpha}}\frac{\epsilon^{\hat{\alpha}\hat{\sigma}\hat{\beta}}}{\omega}\nabla_{[\hat{\beta}}\dot{X}_{\hat{\sigma}]}. \tag{8.3.6}$$

The precession is not *constant* along the congruence in the sense that the precession angle is not preserved even in the simpler case of a geodesic motion ($\dot{X} = 0$) and of a constant Fermi rotation (i.e. when $\dot{\Lambda}_{\hat{\alpha}\hat{\beta}} = 0$); in fact from (8.3.6) we have:

$$(e^{\hat{\alpha}}\zeta_{\hat{\alpha}})^{\cdot} = \tilde{h}_{\hat{\rho}}{}^{\hat{\beta}}\sigma_{\hat{\alpha}\hat{\beta}}e^{\hat{\alpha}}\zeta^{\hat{\rho}}. \tag{8.3.7}$$

8.4 The Einstein equations in tetrad form

The knowledge of the tetrad components of the curvature tensor makes it easy to represent the Einstein equations themselves. The natural splitting into space and time, yielded by a tetrad frame, leads to a physically more intuitive structure of the gravitational field equations.

Let us write down the tetrad components of the Ricci tensor and of the scalar curvature; from (8.1.9), (8.1.13) and (8.1.22) we have:

$$R_{\hat{0}\hat{0}} = -\dot{\Theta} - \Theta_{\hat{\alpha}\hat{\beta}}\Theta^{\hat{\alpha}\hat{\beta}} + \omega_{\hat{\alpha}\hat{\beta}}\omega^{\hat{\alpha}\hat{\beta}} + \nabla_{\hat{\alpha}}\dot{X}^{\hat{\alpha}} + \dot{X}^2 \tag{8.4.1}$$

$$R_{\hat{\alpha}\hat{\beta}} = P_{\hat{\alpha}\hat{\beta}} + \dot{\Theta}_{\hat{\alpha}\hat{\beta}} + \Theta\Theta_{\hat{\alpha}\hat{\beta}} - 2\Theta_{\hat{\sigma}(\hat{\alpha}}\Lambda_{\hat{\beta})}{}^{\hat{\sigma}} + 2\omega_{\hat{\sigma}(\hat{\alpha}}\Lambda_{\hat{\beta})}{}^{\hat{\sigma}} - \nabla_{(\hat{\alpha}}\dot{X}_{\hat{\beta})} - \dot{X}_{\hat{\alpha}}\dot{X}_{\hat{\beta}} \tag{8.4.2}$$

$$R_{\hat{0}\hat{\beta}} = \nabla_{\hat{\sigma}}\left(\Theta_{\hat{\beta}}{}^{\hat{\sigma}} - \delta^{\hat{\sigma}}_{\hat{\beta}}\Theta - \omega_{\hat{\beta}}{}^{\hat{\sigma}}\right) - 2\dot{X}_{\hat{\sigma}}\omega_{\hat{\beta}}{}^{\hat{\sigma}} \tag{8.4.3}$$

$$R = P - \frac{2}{3}\Theta^2 + 2\left(\sigma^2 - \omega^2\right) - 2\omega_{\hat{\alpha}\hat{\beta}}\Lambda^{\hat{\alpha}\hat{\beta}} + 2\nabla_i\left(\Theta X^i - \dot{X}^i\right). \tag{8.4.4}$$

The Einstein's equation in tetrad form is simply:

$$G_{\hat{a}\hat{b}} = R_{\hat{a}\hat{b}} - \frac{1}{2}\eta_{\hat{a}\hat{b}}R = \kappa T_{\hat{a}\hat{b}} \tag{8.4.5}$$

which now decomposes from (8.4.1)–(8.4.4) as

$$G_{\hat{0}\hat{0}} = \frac{1}{2}\left[P + 2\omega_{\hat{\alpha}\hat{\beta}}\Lambda^{\hat{\alpha}\hat{\beta}} + \frac{2}{3}\Theta^2 + 2\left(\omega^2 - \sigma^2\right)\right] = \kappa T_{\hat{0}\hat{0}} \tag{8.4.6}$$

$$G_{\hat{0}\hat{\beta}} = \nabla_{\hat{\sigma}}\left(\sigma_{\hat{\beta}}{}^{\hat{\sigma}} - \frac{2}{3}\Theta\delta^{\hat{\sigma}}_{\hat{\beta}} - \omega_{\hat{\beta}}{}^{\hat{\sigma}}\right) - 2\dot{X}_{\hat{\sigma}}\omega_{\hat{\beta}}{}^{\hat{\sigma}} = \kappa T_{\hat{0}\hat{\beta}} \tag{8.4.7}$$

$$\begin{aligned} G_{\hat{\alpha}\hat{\beta}} &= P_{\hat{\alpha}\hat{\beta}} - 2\omega_{\hat{\sigma}(\hat{\alpha}}\Lambda^{\hat{\sigma}}{}_{\hat{\beta})} - 2\Theta_{\hat{\sigma}(\hat{\alpha}}\Lambda_{\hat{\beta})}{}^{\hat{\sigma}} \\ &\quad - \frac{1}{2}\delta_{\hat{\alpha}\hat{\beta}}\left[P - 2\omega_{\hat{\sigma}\hat{\rho}}\Lambda^{\hat{\sigma}\hat{\rho}} - \frac{2}{3}\Theta^2 + 2\left(\sigma^2 - \omega^2\right)\right] - \dot{X}_{\hat{\alpha}}\dot{X}_{\hat{\beta}} \\ &\quad + \lambda^j_{\hat{\alpha}}\lambda^k_{\hat{\beta}}\nabla_i\left[X^i\Theta_{jk} - \delta^i_{(j}\dot{X}_{k)} - g_{jk}\left(X^i\Theta - \dot{X}^i\right)\right] \\ &= \kappa T_{\hat{\alpha}\hat{\beta}}. \end{aligned} \tag{8.4.8}$$

Here we see that only the equation involving $G_{\hat{\alpha}\hat{\beta}}$ contains second *time* derivatives of the metric tensor, which is the dynamical variable, through terms like $X^i\nabla_i\Theta_{\hat{\alpha}\hat{\beta}}$ and $X^i\nabla_i\Theta$.

We have shown that $\Theta_{\hat{\alpha}\hat{\beta}}$ can be written from (4.5.7) as:

$$\Theta_{\hat{\alpha}\hat{\beta}} = \lambda_{i(\hat{\beta}}\partial_{\hat{\alpha})}X^i + \frac{1}{2}\lambda^i_{(\hat{\alpha}}\lambda^j_{\hat{\beta})}\partial_{\hat{0}}g_{ij}. \tag{8.4.9}$$

So equations (8.4.8) are properly the dynamical field equations while the remaining ones are constraint equations (for an extensive treatment of the Einstein equations of evolution, see Fisher and Marsden, 1972).

To understand better their role and importance let us assume that we are given a hypersurface-orthogonal congruence of time-like curves in a vacuum; then $T_{\hat{a}\hat{b}} = 0$, $\omega_{\hat{\alpha}\hat{\beta}} = 0$. The constraint equations (8.4.6) and (8.4.7), expressed in terms of the extrinsic curvature $K_{\hat{\alpha}\hat{\beta}} \equiv \Theta_{\hat{\alpha}\hat{\beta}}$ on any such hypersurface, Σ say, read (Arnowitt et al., 1962):

$$G_{\hat{0}\hat{0}}|_{\Sigma} = \frac{1}{2}\left[P + K^2 - K_{\hat{\rho}\hat{\sigma}}K^{\hat{\rho}\hat{\sigma}}\right]_{\Sigma} = 0 \qquad (8.4.10)$$

$$G_{\hat{0}\hat{\beta}}|_{\Sigma} = \nabla_{\hat{\sigma}}\left(K_{\hat{\beta}}{}^{\hat{\sigma}} - \delta^{\hat{\sigma}}_{\hat{\beta}}K\right)|_{\Sigma} = 0. \qquad (8.4.11)$$

These equations assure that the initial values of the metric and its derivatives on a given hypersurface $\Sigma|_{\text{in}}$ cannot be assigned arbitrarily. They constitute a set of four independent constraints on the ten metric coefficients while the six remaining equations (8.4.8) guarantee the *time* evolution of the metric tensor along the congruence. The constraint equations however need to be preserved on any subsequent hypersurface Σ and this is assured by the Bianchi identities. In the contracted form, these read, from (3.5.5):

$$\dot{G}^{\hat{0}\hat{0}} + 2\dot{X}_{\hat{\sigma}}G^{\hat{\sigma}\hat{0}} + \nabla_{\hat{\beta}}G^{\hat{0}\hat{\beta}} = 0 \qquad (8.4.12)$$

$$\dot{G}^{\hat{0}\hat{\alpha}} + \dot{X}^{\hat{\alpha}}G^{\hat{0}\hat{0}} + \dot{X}_{\hat{\sigma}}G^{\hat{\sigma}\hat{\alpha}} + \Lambda_{\hat{\sigma}}{}^{\hat{\alpha}}G^{\hat{\sigma}\hat{0}} + \nabla_{\hat{\beta}}G^{\hat{\alpha}\hat{\beta}} = 0. \qquad (8.4.13)$$

It follows from these that if the dynamical equations $G^{\hat{\alpha}\hat{\beta}} = 0$ hold at all times, then the constraint equations hold at all times if they hold initially. In fact since, by assumption, on Σ_{in}: $G^{\hat{\alpha}\hat{\beta}} = \partial_{\hat{\beta}}G^{\hat{\alpha}\hat{\beta}} = 0$, and also $G^{\hat{0}\hat{0}} = G^{\hat{0}\hat{\alpha}} = \partial_{\hat{\beta}}G^{\hat{0}\hat{\beta}} = 0$, then $\dot{G}^{\hat{0}\hat{0}} = \dot{G}^{\hat{0}\hat{\alpha}} = 0$ from (8.4.12) and (8.4.13).

8.5 The geometry of null rays

In Section 8.2 we showed how to analyse the rate of change of a small three-dimensional surface element generated by geodesics orthogonal to a curve of the congruence and then Lie-propagated by the congruence. If the congruence is null, however, then the space

$${}^{\perp}T_p(M) = \{Y \mid (Y|\underset{p}{X}) = 0\}$$

(where X is the tangent field to the congruence) contains $\underset{p}{X}$ itself. If we take $\{\underset{p}{X}, a, b\}$ as a basis for ${}^{\perp}T_p$, then we can find a two-parameter family of curves of the congruence $\gamma_{t,u}(s)$ with $\partial\gamma^i_{t,u}/\partial t = a^i$, $\partial\gamma^i_{t,u}/\partial u = b^i$, which expresses the full range of possible variations of γ orthogonal to X. Thus the geometry of a null congruence is concerned with the way a two-dimensional element, generated by a and b, changes as it is Lie-propagated.

In fact the connecting vector a, connecting $\gamma_{0,0}$ with an "infinitesimally neighbouring" curve $\gamma_{\delta t,0}$ is not fixed (independently of the parametrisation of $\gamma_{\delta t,0}$) by the requirement that $(a|\underset{p}{X}) = 0$, as it is in the time-like case, since $a + \alpha\underset{p}{X}$ is an equally possible choice for any α. Thus the space of neighbouring curves is parametrized not by ${}^{\perp}T_p(M)$ but by ${}^{\perp}T_p(M)/[\underset{p}{X}]$ in which two vectors are regarded as equivalent if they differ by a multiple of $\underset{p}{X}$.

In practice, however, we choose an arbitrary a and b and parametrise neighbouring curves by linear combinations of these. But we have to remember the possibility of adding an arbitrary multiple of X. We note that, since all the geometry reduces to a two-parameter family of curves whose connecting vectors are orthogonal to X, the full three-parameter congruence is not needed. In particular our analysis holds equally well for the two-parameter family of curves defined by a null surface.

$$S \equiv \{p|f(p) = 0\}\ , \qquad \boldsymbol{d}f \text{ null}$$

where the integral curves of $\overset{-1}{g}(\boldsymbol{d}f) = X$ are null geodesics in S, with connecting vectors ξ lying in S and so satisfying $\boldsymbol{g}(X,\xi) = \boldsymbol{d}f(\xi) = \xi(f) = 0$.

It will be convenient to standardize the choice of the basis $\{X, a, b\}$ by making a generalization of the concept of a tetrad appropriate to congruences of null rays (such as arise with electromagnetic and gravitational radiation). This involves generalising $T_p(M)$ to a complex space. We can turn any real vector space V into a complex vector space $V^{\mathbb{C}}$ by forming the set $V \times V$ of all pairs (u, v) with $u, v \in V$ and then writing (u, v) as $u + \mathrm{i}v$. The scalar multiplication by real numbers can then be extended to complex numbers if we define $\mathrm{i}(u, v)$ by

$$\mathrm{i}(u, v) = (-v, u) = -v + \mathrm{i}u.$$

With this definition $V^{\mathbb{C}}$ becomes a complex vector-space.

Let $\{\lambda_{\hat{\alpha}}\}$ be a tetrad; then construct a set of vectors $\{\ell, k, m, \overline{m}\}$ in $T_p^{\mathbb{C}}(M)$ defined as

$$\ell = \frac{1}{\sqrt{2}}(\lambda_{\hat{0}} + \lambda_{\hat{1}}) \qquad k = \frac{1}{\sqrt{2}}(\lambda_{\hat{0}} - \lambda_{\hat{1}})$$
$$m = \frac{1}{\sqrt{2}}(\lambda_{\hat{2}} + \mathrm{i}\lambda_{\hat{3}}) \qquad \overline{m} = \frac{1}{\sqrt{2}}(\lambda_{\hat{2}} - \mathrm{i}\lambda_{\hat{3}}) \ . \quad (8.5.1)$$

These satisfy the following conditions:

$$(\ell|\ell) = (m|m) = (k|k) = (\bar{m}|\bar{m}) = 0$$
$$(\ell|k) = -1 \qquad (m|\bar{m}) = 1$$
$$(\ell|m) = (\ell|\bar{m}) = (k|m) = (k|\bar{m}) = 0 \ . \quad (8.5.2)$$

The set $\{\ell, k, m, \bar{m}\}$ forms a basis of $T_p^{\mathbb{C}}(M)$; in fact any relation of the type:

$$u = A\ell + Bk + Cm + D\bar{m} = o \qquad u \in T_p(M) \qquad A, B, C, D \in \mathbb{C}$$

implies, from (8.5.2), that $A = B = C = D = 0$. Any set $\{\lambda_{\hat{\alpha}}\}$ satisfying (8.5.2) is called a *null tetrad* (Newman and Penrose, 1962). From the general definition of a basis, we have:

$$\overset{*}{\eta}_{\hat{a}\hat{b}} = (\lambda_{\hat{a}}|\lambda_{\hat{b}}) = -2\delta_{\hat{0}(\hat{a}}\delta_{\hat{b})\hat{1}} + 2\delta_{\hat{2}(\hat{a}}\delta_{\hat{b})\hat{3}} \quad (8.5.3)$$

hence, recalling (4.2.6):

$$g_{ij} = -2\ell_{(i}k_{j)} + 2m_{(i}\bar{m}_{j)}. \quad (8.5.4)$$

As already mentioned, null tetrads are most useful in the study of null congruences. Consider a narrow beam of affinely parametrised null rays and let ℓ be their tangent vector, then augment ℓ to a null tetrad $(\ell, k, m, \bar{m})$; we shall investigate how an original circular section of this beam propagates along the congruence. At some point p on the congruence define a new pair of space-like, real and orthonormal vectors:

$$a = \frac{1}{\sqrt{2}}(m + \bar{m}) \ , \quad b = \frac{1}{\sqrt{2}}\mathrm{i}(m - \bar{m}) \quad (8.5.5)$$

with $(a|\ell) = (b|\ell) = 0$. They span a two-plane orthogonal to ℓ, Σ_0 say, through p. Take them as a basis in $T_p(\Sigma_0)$ and write the unit connecting vector ζ, joining p to an infinitesimally close null ray of the beam, as

$$\zeta = a\ \cos\varphi + b\ \sin\varphi = \frac{1}{\sqrt{2}}\left(e^{i\varphi}m + e^{-i\varphi}\bar{m}\right) \tag{8.5.6}$$

for some real azimuth φ on Σ_0. Together with ζ, define another connecting vector ι which at p is orthogonal to ζ and has the form:

$$\iota = -a\ \sin\varphi + b\ \cos\varphi = \mathrm{i}\frac{1}{\sqrt{2}}\left(e^{i\varphi}m - e^{-i\varphi}\bar{m}\right). \tag{8.5.7}$$

ζ and ι are interpreted as generating a unit circular (infinitesimal) section of the beam, centred at p.

Let us now propagate the circle from p to p', a parameter distance $\delta\lambda$ from p along the null rays of the beam. The properties of the beam cross section at the point p' are fully characterized by the rate of change of the connecting vector, which satisfies the equation:

$$\frac{D\zeta^i}{D\lambda} = \ell^j\nabla_j\zeta^i = \zeta^j\nabla_j\ell^i. \tag{8.5.8}$$

The latter is due to the property of ζ of being Lie-transported along the congruence of the beam. Because the vectors ζ and ι form a (real) basis in $T_p(\Sigma_0)$, the components of (8.5.8) with respect to these vectors tell us the behaviour of the connecting vector in its parallel and transverse directions. Hence, from (8.5.6), (8.5.7) and (8.5.8) we have:

$$\zeta^i\zeta^j\nabla_i\ell_j = \frac{1}{2}\left(e^{2i\varphi}\sigma + e^{-2i\varphi}\bar{\sigma} + \rho + \bar{\rho}\right) \tag{8.5.9}$$

$$\iota^j\zeta^i\nabla_i\ell_j = \frac{\mathrm{i}}{2}\left(e^{2i\varphi}\sigma - e^{-2i\varphi}\bar{\sigma} + \rho - \bar{\rho}\right) \tag{8.5.10}$$

where we put:

$$\sigma = m^i m^j\nabla_i\ell_j\ , \quad \rho = m^i\overline{m}^j\nabla_j\ell_i. \tag{8.5.11}$$

The terms in (8.5.9) and (8.5.10) which do not depend on φ show an isotropic deformation of the beam cross section: in particular $\Theta = (\rho + \bar{\rho})$ measures the isotropic expansion and $\omega = (\mathrm{i}/2)(\rho - \bar{\rho})$ measures the relative rotation (vorticity) of the rays in the beam. The terms which depend on φ describe a non-isotropic

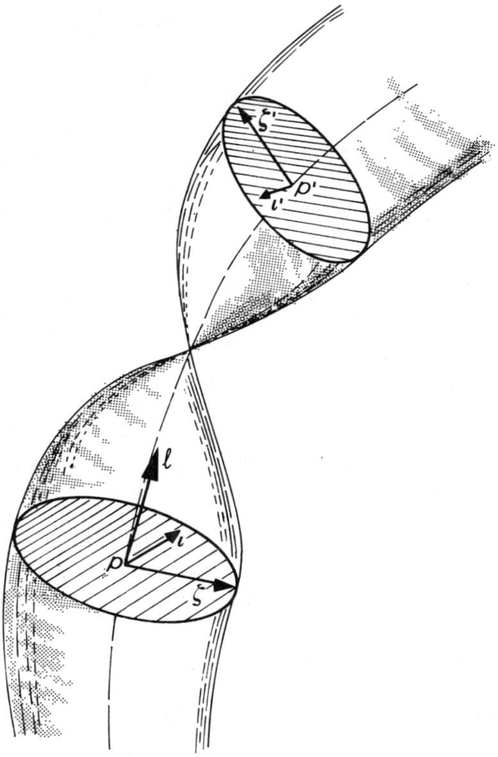

Fig. 8-1 A circular cross section to a beam of null rays is propagated into a rotated ellipse.

deformation of the beam cross section, so the original circular section evolves into an ellipse (Fig. 8-1).

To first order in $\delta\lambda$, we have $\zeta' = \zeta + D\zeta$ so:

$$|\zeta'|^2 = (\zeta'|\zeta') = 1 + \left(e^{2i\varphi}\sigma + e^{-2i\varphi}\bar{\sigma} + \Theta\right)\delta\lambda + \mathrm{O}(\delta\lambda^2). \quad (8.5.12)$$

The measures of the semiaxes of the ellipse are given by the φ-extremes of $|\zeta'|$ and these occur at:

$$e^{2i\varphi_\pm} = \pm\left(\frac{\bar{\sigma}}{\sigma}\right)^{\frac{1}{2}}$$

and so the lengths of the semiaxes are:

$$|\zeta'|_{\pm} = 1 + \left(\frac{1}{2}\Theta \pm |\sigma|\right)\delta\lambda + \mathrm{O}(\delta\lambda^2). \tag{8.5.13}$$

Here the quantity $|\sigma| = \sqrt{\bar{\sigma}\sigma}$ is the *shear* and directly determines the eccentricity of the ellipse:

$$e^2 = 4|\sigma|\delta\lambda + \mathrm{O}(\delta\lambda^2). \tag{8.5.14}$$

To the same order, the area of the ellipse reads:

$$\mathcal{A} = \pi\zeta'_+\zeta'_- = \pi\left[1 + \Theta\delta\lambda + \left(\frac{\Theta^2}{4} - |\sigma|^2\right)\delta\lambda^2 + \cdots\right]. \tag{8.5.15}$$

We notice that the shear contributes to the area only to higher-order terms in λ and with opposite sign with respect to Θ; moreover, as expected, no contribution comes to the area from the vorticity ω. The rotation of the beam cross section is described by equation (8.5.10) where the deviation from the uniform rotation of the beam, due to the shear, is clearly shown:

$$\iota^i\zeta^j\nabla_j\ell_i = \omega - |\sigma|\ \sin 2(\varphi - \varphi_+), \tag{8.5.16}$$

where the choice of signs is made so that φ_+ designates the position of the major axis. The real quantities Θ, ω, $|\sigma|$ are termed *optical scalars* of the beam and determine the behaviour of the null congruences, (Sachs, 1962; Jordan et al., 1961).

A comparison of (8.5.11) with (8.5.4) allows us to express the optical scalars in coordinate components:

$$\begin{aligned}\Theta &= (\rho + \bar{\rho}) = 2m^{(i}\bar{m}^{j)}\nabla_j\ell_i = \nabla_i\ell^i \\ \omega^2 &= -\frac{1}{4}(\rho - \bar{\rho})^2 = -g^{ir}g^{js}(\nabla_r\ell_s)(\nabla_{[i}\ell_{j]}) \\ |\sigma|^2 &= g^{ir}g^{js}(\nabla_r\ell_s)\nabla_{(i}\ell_{j)} - \Theta^2.\end{aligned} \tag{8.5.17}$$

The properties of a congruence of null rays in a curved manifold are only known once we know the transport equations for the optical scalars. For this purpose let us consider the Ricci identity (3.1.10) applied to the null vector ℓ:

$$\nabla_{[i}\nabla_{j]}\ell^r = \frac{1}{2}R^r{}_{kij}\ell^k \tag{8.5.18}$$

where we assume the torsion to be zero. Contracting this with ℓ^j we obtain

$$\frac{D}{D\lambda}\left(\nabla_i \ell^r\right) = -\left(\nabla_i \ell^j\right)\left(\nabla_j \ell^r\right) - R^r{}_{kij}\ell^k \ell^j \tag{8.5.19}$$

where we have used the fact that ℓ is tangent to an affine geodesic.

Contracting now (8.5.19) with $\bar{m}^i m_r$ and making the further assumption that the vectors of the null tetrad are parallely propagated along the rays, we obtain from (8.5.11):

$$\frac{d\rho}{d\lambda} = -\rho^2 - \sigma\bar{\sigma} - \frac{1}{2}R_{kj}\ell^k \ell^j. \tag{8.5.20}$$

Remembering from (8.5.17) that $\rho = (1/2)(\Theta - 2\mathrm{i}\omega)$, this leads to two equations:

$$\frac{d\Theta}{d\lambda} = -\frac{1}{2}\Theta^2 + 2\left(\omega^2 - |\sigma|^2\right) - R_{ij}\ell^i \ell^j \tag{8.5.21}$$

which is the Raychauduri (1955) equation for a congruence of null rays, and

$$\frac{d\omega}{d\lambda} = -\omega\Theta \tag{8.5.22}$$

which is the vorticity equation for such a congruence. Contracting (8.5.19) with $m^i m_r$, we obtain the differential equation for the shear:

$$\frac{d|\sigma|}{d\lambda} = -|\sigma|\Theta - \frac{1}{2}R_{rkij}\ell^k \ell^j \left(e^{2\mathrm{i}\varphi_+} m^i m^r + e^{-2\mathrm{i}\varphi_+} \bar{m}^i \bar{m}^r\right). \tag{8.5.23}$$

Equations (8.5.21), (8.5.22) and (8.5.23) are the basic equations of a null congruence and play a key role in gravitational collapse as well as in ray optics in observational cosmology.

8.6 Singularities

Most solutions in general relativity contain regions where physical quantities (pressure, tidal forces, etc.) become unbounded and where there are geodesics that end at a finite value of their affine parameter. A space-time where the latter property (or, sometimes, a generalization of it) holds is called singular, or is said to "have a singularity". It is still unclear whether (apart from special cases) singularities are always associated with unbounded physical quantities. In this section we shall give an indication of the way congruences are used in the study of singularities; a large subject on which we can merely touch (Clarke, 1990; Hawking and Ellis, 1973).

Consider first time-like geodesic congruences that are initially non-rotating; for example, a set of time-like geodesics with tangent field u emanating from a fixed point. In this case $\dot{u} = 0$ and the vorticity equation (8.3.3) shows that $\omega = 0$ everywhere. The Raychauduri equation (8.2.8) then becomes

$$\dot{\Theta} = -R_{\hat{0}\hat{0}} - 2\sigma^2 - \frac{1}{3}\Theta^2. \tag{8.6.1}$$

We can now estimate $R_{\hat{0}\hat{0}}$ by noting that a generic energy-momentum $T^i{}_j$ can be decomposed into orthogonal eigenvectors according to

$$T^i{}_j = \rho\lambda'^i_{\hat{0}}\lambda'_{\hat{0}j} + \sum_{\alpha=1}^{3} p_\alpha \lambda'^i_{\hat{\alpha}}\lambda'_{\hat{\alpha}j} \tag{8.6.2}$$

where we chose a tetrad $\{\lambda'_{\hat{\alpha}}\}$ along the eigenvectors and write ρ, p_α for the energy density and principal pressures. Einstein's equations give:

$$R^i{}_j = \kappa\left[\frac{1}{2}(\rho + 3\bar{p})\lambda'^i_{\hat{0}}\lambda'_{\hat{0}j} + \sum_\alpha (p_\alpha + \frac{1}{2}\rho - \frac{3}{2}\bar{p})\lambda'^i_{\hat{\alpha}}\lambda'_{\hat{\alpha}j}\right] \tag{8.6.3}$$

where $\bar{p} = \frac{1}{3}\sum_\alpha p_\alpha$. So, if u is a time-like vector, we can write its tetrad components as $u_{\hat{\alpha}} = (\cosh\chi,\ \vec{n}\sinh\chi)$ ($\sum(n^{\hat{\alpha}})^2 = 1$) giving:

$$\begin{aligned} R_{ij}u^i u^j &\geq \kappa\left[\frac{1}{2}(\rho + 3\bar{p})\cosh^2\chi + \frac{1}{2}(\rho + p_1 - p_2 - p_3)\sinh^2\chi\right] \\ &= \frac{\kappa}{2}\left[(\rho + p_1)\cosh 2\chi + (p_2 + p_3)\right] \end{aligned} \tag{8.6.4}$$

where p_1 is the smallest of p_α ($p_1 < p_2 < p_3$, say). In the case where $u = \lambda'_{\hat{0}}$ we have $\chi = 0$ and

$$R_{ij}\lambda'^i_{\hat{0}}\lambda'^j_{\hat{0}} \geq 0 \qquad \text{if} \qquad \rho + 3\bar{p} \geq 0\,, \tag{8.6.5}$$

while in general

$$R_{ij}u^i u^j \geq 0 \qquad \text{if} \quad \rho + p_1 \geq 0 \qquad \text{and} \qquad \rho + 3\bar{p} \geq 0 \tag{8.6.6}$$

i.e. if $\rho + p_\alpha \geq 0$ $\forall\alpha$ and $\rho + 3\bar{p} \geq 0$. The condition in (8.6.6) is called the *strong energy condition*, and it clearly implies the condition in (8.6.5). If we let $\chi \to \infty$ so that $(\cosh\chi)^{-1}u$ tends to a null vector ℓ then we have from (8.6.4) that:

$$R_{ij}\ell^i\ell^j \geq 0 \qquad \text{if} \;\; \rho + p_1 \geq 0, \tag{8.6.7}$$

i.e. if $\rho + p_\alpha \geq 0$ $\forall\alpha$. This is implied by the conditions $\rho + p_\alpha \geq 0$, $\rho \geq 0$ which constitute the *weak energy condition.*

All these conditions are satisfied by all known classically-defined types of matter having a generic $T^i{}_j$; the non-generic cases are treated by Hawking and Ellis (1973).

As a first application of these results, we give a derivation of the Robertson-Walker "big bang" singularity. We take a congruence generated by the flow lines of the matter in a homogeneous isotropic universe, so that $X = \lambda_{\hat{0}} = \lambda'_{\hat{0}}$. By isotropy the flow is non-rotating (and $\sigma = 0$, though this is not required) and geodesic, so (8.6.5) and (8.6.1) give:

$$\dot{\Theta} \leq -\frac{1}{3}\Theta^2. \tag{8.6.8}$$

Writing in the form $\Theta^{-2}\dot{\Theta} \leq -(1/3)$, integrating between parameter values τ_1 and τ_2 $(\tau_1 < \tau_2)$ and rearranging gives:

$$\Theta(\tau_1) \geq \left[\Theta(\tau_2)^{-1} + \frac{(\tau_1 - \tau_2)}{3}\right]^{-1}; \tag{8.6.9}$$

so if at the present time, τ_2, we have $\Theta > 0$, then at the earlier time, τ_1, Θ was larger, and has to become unboundedly large in the interval

$$\tau_1 \in \left[\tau_2 - 3\Theta(\tau_2)^{-1} \;,\; \tau_2\right].$$

Unboundedness of Θ implies that the volume of a tube of matter shrinks to zero, producing a *matter singularity* in which, for normal matter, $\rho \to \infty$ and $p \to \infty$.

Closely related to this situation (where the matter-volume shrinks to zero) is the concept of a *crushing singularity*, defined* as one where *any* irrotational geodesic congruence has $\liminf_{\tau \to \tau_1} \Theta(\tau) = -\infty$ (where the parametrization is increasing from the singularity, which occurs at $\tau = \tau_1$). To study this, or similar concepts of singularity strength, we define a variable x by solving for x the equation:

$$3\dot{x} = \Theta x \tag{8.6.10}$$

with initial condition $x(\tau_0) = 1$, say, so that (8.6.1) becomes

$$3\frac{\ddot{x}}{x} = -R_{\hat{0}\hat{0}} - 2\sigma^2 \tag{8.6.11}$$

or $\ddot{x} = -f(\tau)x$ where $f(\tau) \geq R_{\hat{0}\hat{0}}/3$.

We shall now show that a crushing singularity occurs provided

$$R_{\hat{0}\hat{0}} > \frac{3}{4(\tau_1 - \tau)^2}$$

i.e. provided, from (8.6.4), that $\rho + p_1 > 0$ and

$$\rho + \sum_\alpha p_\alpha > \frac{3}{2\kappa(\tau_1 - \tau)^2}.$$

For simplicity parametrise so that $\tau_1 = 0$; so we are working in an interval $\tau_0 < \tau < 0$. x satisfies (8.6.11) with

$$f(\tau) < \frac{1}{(4\tau^2)}. \tag{8.6.12}$$

If at τ_0 we have $\dot{x} > 0$, then x is increasing until the first zero of $\dot{x}$, so until then we have, from (8.6.11), $\ddot{x} \leq -(1/4\tau^2)x(\tau_0)$; whence, integrating, we find that $\dot{x}$ becomes zero before $\tau = 0$. So we can without loss of generality take τ_0 such that $\dot{x} \leq 0$. Let $y(\tau)$ be the solution of

$$\ddot{y} = \frac{1}{4\tau^2}y \tag{8.6.13}$$

with $y(\tau_0) = x(\tau_0) = 1$, $\dot{y}(\tau_0) = \dot{x}(\tau_0) \leq 0$. Since $y = \sqrt{|\tau|}\,(A + B\ln|\tau|)$, y becomes zero for some $\tau < 0$. The following standard comparison argument shows that x also becomes 0. Set

*Note, however, that some authors use the term to denote a space-time with a foliation, the trace of whose second fundamental form diverges uniformly.

$h = x/y$ in the interval before y becomes 0, where $y > 0$. Then from (8.6.11), (8.6.12) and (8.6.13)

$$\ddot{h} \leq -2\dot{h}\frac{\dot{y}}{y}. \tag{8.6.14}$$

But from (8.6.13)

$$\frac{d}{d\tau}\left(\frac{\dot{y}}{y}\right) = -\frac{y}{4\tau^2} - \frac{\dot{y}^2}{y^2} \leq 0 \qquad \dot{y}(\tau_0) \leq 0,$$

so $\dot{y}/y \leq 0$. Also $\dot{h} = (\dot{x}y - x\dot{y})/y^2$ so $\dot{h}(\tau_0) = 0$ and so $\dot{h}$ is initially negative (from (8.6.14)). Thus (8.6.14) implies that

$$\frac{\ddot{h}}{\dot{h}} \geq 0$$

until the first zero of $\dot{h}$ or y. Integrating this shows that $\dot{h}$ decreases, so it remains negative and cannot be zero before y is zero. Since h is initially 1, h decreases. Thus $x < y$ before the first zero of y, and hence $x = 0$ at or before the first zero of y. From (8.6.10), together with $\dot{x} < 0$, this shows that at some point $\Theta \to -\infty$, giving a crushing singularity.

We shall next discuss some of the ideas that enter into the singularity theorems of Hawking, Penrose and Geroch showing that singularities (in the sense of incomplete geodesics) can actually occur in general situations. These use null congruences arising from the geometry of the space-time. A central role is played by the concept of a *trapped surface.*

Suppose S is a space-like two-dimensional surface. Thus at each point $p \in S$, $T_p(S)$ is spanned by two vectors a, b which we may take to be orthonormal; we can then take $2^{-\frac{1}{2}}(a + ib) = m$ and $\bar{m}$ to be two members of a null tetrad, which we can complete by future pointing null vectors k and ℓ. The vectors k and ℓ are fixed by $T_p(S)$ up to transformations of the form

$$k \to \alpha k \ , \ \ell \to \alpha^{-1}\ell \tag{8.6.15}$$

for $\alpha > 0$, or interchanges of k and ℓ; hence the pair of null directions determined by $T_p(S)$ is fixed. For p ranging over a neighbourhood U on S, choose $\underset{p}{k}$ and $\underset{p}{\ell}$ arbitrarily but smoothly and construct two congruences by taking all the geodesics having

$\underset{p}{k}$ as initial vector for all $p \in U$ to form one congruence, and all those with $\underset{p}{\ell}$ as the other. We then note from (8.5.17) that the sign of the expansion Θ is unchanged by the transformation (8.6.15); thus if the Θ for both congruences is negative, this is independent of the choice of k and ℓ. In this case we call S a trapped surface.

There is a deep theorem of Schoen and Yau (Schoen and Yau, 1983; Clarke, 1988) which shows that, roughly speaking, if enough matter is concentrated in a region of small enough extent, then a trapped surface forms (the theorem makes precise the meanings of these words!). Thus we know that trapped surfaces can in principle arise in the universe, and we have reason to believe from stellar evolution that they could well form in stellar collapse. Let us then consider a curve γ in a geodesic null congruence with negative Θ. We can repeat the analysis given earlier, which used the Raychaudhuri equation for time-like geodesics, in the null case. From (8.5.11) we have:

$$\omega = (b^j a^i - a^j b^i)\nabla_j \ell_i = \ell_i [a, b]^i = 0$$

on S, by Frobenius' theorem (since a, b are tangent to S and so $[a, b] \in T(S)$). Hence, from (8.5.22), $\omega = 0$ everywhere. Assuming the weak energy condition and using (8.6.7), the Raychaudhuri equation (8.5.2) becomes

$$\frac{d\Theta}{d\lambda} \leq -\frac{1}{2}\Theta^2.$$

Integrating between λ_1 and λ_2 gives (adopting a different rearrangement from (8.6.9))

$$\Theta(\lambda_2) \leq \left[\Theta(\lambda_1)^{-1} + \frac{1}{2}(\lambda_2 - \lambda_1)\right]^{-1}.$$

So if $\Theta(\lambda_1)$ is negative, $\Theta(\lambda_2)$ will become unboundedly negative in the interval $\lambda_2 \in [\lambda_1, \lambda_1 - 2\Theta(\lambda_1)^{-1}]$. This means that the area element defined by neighbouring curves goes to zero in this interval, and hence that one of the connecting vectors of the congruence equal to a or b on S vanishes at some point q. In this case we say that q is *conjugate* to S. (This generalizes the concept of conjugate points introduced in Sec. 3.6.)

The singularity theorems all use some variant of the idea that, in the situation just described, when q is conjugate to S then any point r on the curve beyond q can be connected to S by a time-like curve. (For the proof of this, see Hawking and Ellis, 1973, Sec. 4.5.) One then considers the set $I^+(S)$ of all points that can be connected to S by past-directed time-like curves; $I^+(S)$ is called the chronological future of S. The boundary of $I^+(S)$ is initially formed from the congruences (or congruence) just constructed. But when the congruence reaches a conjugate point it enters into the interior of $I^+(S)$. Speaking very roughly, this means that if S is compact (e.g. a sphere) then either the whole of $I^+(S)$ "wraps in on itself", which produces closed time-like lines (a violation of causality) or else there are other parts of the boundary of $I^+(S)$, which has to be null, generated by null geodesics that cannot be continued back to join up with the congruence. The latter possibility then can be shown to be impossible unless, somewhere in the space-time, there is a singularity making some null geodesics incomplete (Hawking and Ellis, 1973).

9

PHYSICAL MEASUREMENTS IN SPACE-TIME

9.1 The concept of measurement

In the previous chapter we showed how to work in terms of frames, but, although it was clear that the tetrad components of a tensor can be physically significant quantities, we still need a criterion to assign to each of them a physical interpretation. To this purpose we use a tetrad frame to establish an instantaneous inertial frame, in which the laws of special relativity hold. In this case a physical interpretation of a tetrad projected quantity may be found once one is able to cast that quantity in a form immediately recognizable as a known special relativistic expression. We shall apply this criterion to mechanics, electrodynamics and fluid dynamics and shall interpret as genuine relativistic effects (curvature effects) those quantities which have no part in a special relativistic treatment and contain the curvature either explicitly or implicitly.

One aspect of what can be properly termed the *theory of measurements* in space-time is to establish what can be directly measured. It has been pointed out already in Chap. 4 that an observer can interact with a distant system only by means of light signals. The properties of that system can then be deduced only from the measurements which are allowed in his or her rest frame, in particular proper-time readings, angles, frequencies and energy fluxes. Since the recent development of space and laboratory technologies have considerably enhanced one's ability to perform high-precision measurements it has become more and more compelling to have a consistent formalism for a theory of measurements (Martin et al., 1985). There is therefore a need to establish a relation between these measurements and the geometry of space-time. Here we shall not enter into the details of the experimentation but we limit ourselves to relations between the observable quantities and geometrical terms like curvature components, spatial distances and proper recession velocities. In Sect. 9.2 we shall find how the geometrical concept of spatial distance is related to the observer's proper time when the curvature does not vanish; similarly in Sect. 9.4 we connect the spatial distance and the curvature to the measurements of angles and finally in Sect. 9.6 we discuss the relationship between the Doppler velocity of the system under investigation and its proper recession velocity (a geometrical non-observable quantity), the difference being explicitly written in terms of the curvature. In addition to the above types of measurements an observer interacting locally with a gravitational wave can also measure shear stresses induced on a mechanical system.

9.2 The measurement of time intervals and space distances

Let γ be the world-line of a physical observer; the parameter s on it is taken to be the proper time, so the tangent vector field $\dot{\gamma}$ is normalized as:

$$(\dot{\gamma}|\dot{\gamma}) = -1. \tag{9.2.1}$$

Here we shall analyse the concepts of *space* and *time* distances between two events, relative to the given observer, referring closely to the geometrical analogue in Euclidean geometry.

Consider an event p not belonging to γ but sufficiently close to it that a normal neighbourhood $U_{\mathcal{N}}(p)$ exists which contains the intersections A_1 and A_2 of γ with the generators of the light-cone in p. Referring to the setting of Fig. 9-1, we see that all points on γ between A_1 and A_2 can be connected to p by a unique space-like geodesic, which we shall denote by $\boldsymbol{\zeta}$ with parameter σ. Let $\mathrm{A}_1 = \gamma(s_{\mathrm{A}_1})$ and $\mathrm{A}_2 = \gamma(s_{\mathrm{A}_2})$. Then, from (1.19), the world-function $\Omega(\mathrm{A}, p)$ along the geodesic connecting a general point $\mathrm{A} = \gamma(s)$

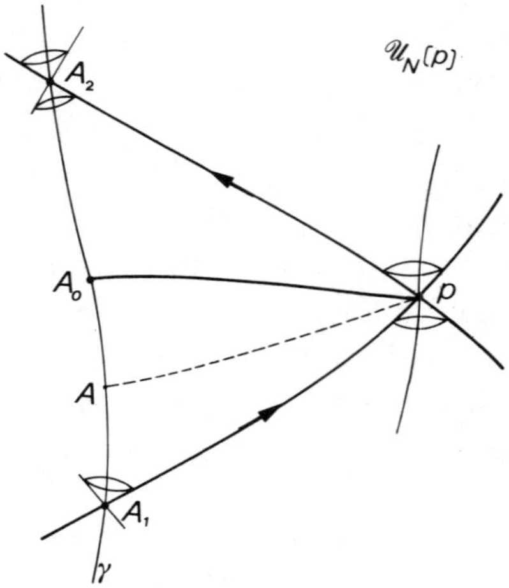

Fig. 9-1 The localization of the point p with respect to γ requires a set of measurements to be confined to a normal neighbourhood of p.

with p, reads:

$$\Omega(\mathrm{A}, p) = \frac{1}{2}\left(\sigma_p - \sigma_{\mathrm{A}}\right)^2 \left(\dot{\zeta}|\dot{\zeta}\right)_{\mathrm{A}}; \tag{9.2.2}$$

the geodesics ζ are assumed affinely parametrised, so $\left(\dot{\zeta}|\dot{\zeta}\right)$ is constant on them.

Because p is kept fixed and A is any point on γ: $s_{\mathrm{A}_1} \leq s \leq s_{\mathrm{A}_2}$, the world-function in (9.2.2) is itself a function of s, hence we write it as $\Omega(s)$ with the constraints:

$$\Omega(s_{\mathrm{A}_1}) = \Omega(s_{\mathrm{A}_2}) = 0. \tag{9.2.3}$$

Since γ is a smooth curve, there exists a value $s_{\mathrm{A}_0}: s_1 < s_{\mathrm{A}_0} < s_2$ where $\Omega(s)$ is extreme, namely at $\mathrm{A}_0 = \gamma(s_{\mathrm{A}_0})$:

$$\left.\frac{d\Omega}{ds}\right|_{\mathrm{A}_0} = 0. \tag{9.2.4}$$

It can be shown that A_0 is unique providing p is sufficiently close to γ. At the point A_0, relation (9.2.4) is equivalent to

$$\dot{\gamma}(\Omega)\Big|_{A_0} = \Omega_{i_0}\dot{\gamma}^{i_0} = 0, \tag{9.2.5}$$

where $\dot{\gamma}^{i_0} = \dot{\gamma}^i(s_{A_0})$ and Ω_{i_0} is the derivative at A_0 of the world-function (involving the geodesic segment ζ_0 joining A_0 to p) and given from (3.7.8), as:

$$\Omega_{i_0} = -(\sigma_p - \sigma_{A_0})\xi_{i_0} \qquad \xi = \dot{\zeta}_0 \tag{9.2.6}$$

where $\xi_{i_0} = \xi_i(\sigma_0)$. From (9.2.3) and (9.2.5) we find:

$$\dot{\gamma}_{i_0}\xi^{i_0} = 0. \tag{9.2.7}$$

We define a measurement of *space distance* between the observer on γ and the event p to be the length of this geodesic segment ζ which strikes γ orthogonally at A_0; namely (Synge, 1960):

$$L(p,\gamma) \equiv \left(2\Omega(s_{A_0})\right)^{\frac{1}{2}} = |(\sigma_p - \sigma_{A_0})|(\xi|\xi)^{\frac{1}{2}}_{A_0}. \tag{9.2.8}$$

The event A_0 on γ is termed *simultaneous* with p, with respect to the observer on γ.

The above definition arises from a mathematical construction which does not allow a direct determination of $L(p,\gamma)$, thus we need to derive this value from the physical quantities which can be directly measured.

Call

$$\delta T_\gamma(A_1, A_2) \equiv \frac{s_{A_2} - s_{A_1}}{c}$$

the proper-time interval between the emission of a signal at A_1 to p and the reading of the reflected echo at A_2; the world-function $\Omega(s)$ relative to a general ζ between p and some $\gamma(s)$ with $s_{A_1} \leq s \leq s_{A_2}$ can be expanded in a power series about s_{A_0}:

$$\Omega(s) = \Omega(s_{A_0}) + \sum_{n=1}^{\infty} \frac{1}{n!}\frac{d^n\Omega}{ds^n}\Big|_{s_{A_0}} (s - s_{A_0})^n \,. \tag{9.2.9}$$

From (3.7.8), (3.7.19), (3.7.20) and (3.7.21) we recall the following relations:

$$\begin{aligned}
\frac{d\Omega}{ds} &= \Omega_{;i}\dot{\gamma}^i \\
\frac{d^2\Omega}{ds^2} &= \Omega_{;ij}\dot{\gamma}^i\dot{\gamma}^j + \Omega_{;i}a^i \\
\frac{d^3\Omega}{ds^3} &= \Omega_{;ijk}\dot{\gamma}^i\dot{\gamma}^j\dot{\gamma}^k + 3\Omega_{;ij}a^i\dot{\gamma}^j + \Omega_{;i}\dot{\gamma}^k\nabla_k a^i \\
\frac{d^4\Omega}{ds^4} &= \Omega_{;ijkl}\dot{\gamma}^i\dot{\gamma}^j\dot{\gamma}^k\dot{\gamma}^l + 5\Omega_{;ijk}a^i\dot{\gamma}^j\dot{\gamma}^k + \Omega_{;ijk}\dot{\gamma}^i\dot{\gamma}^j a^k + 4\Omega_{;ij}\left(\dot{\gamma}^k\nabla_k a^i\right)\dot{\gamma}^j \\
&\quad + 3\Omega_{;ij}a^i a^j + \Omega_{;i}\dot{\gamma}^r\nabla_r\left(\dot{\gamma}^k\nabla_k a^i\right) \\
&\quad \dots
\end{aligned} \tag{9.2.10}$$

where $a^i = \gamma^k\nabla_k\gamma^i$ is the four-acceleration of the given observer.

The coefficients in (9.2.10) are functions of the parameter distance on the curve ζ_0 connecting A_0 to p; assuming that these points are close enough that one can neglect the curvature gradients, we obtain from (3.7.19), (3.7.20) and (3.7.21), to first order in the curvature, the following expressions:

$$\begin{aligned}
\Omega_{;i_0 j_0} &\approx g_{i_0 j_0} + \frac{1}{2}\left[S_{ijkm}\xi^k\xi^m\right]_{\mathrm{A}_0}(\sigma_p - \sigma_{\mathrm{A}_0})^2 \\
\Omega_{;i_0 j_0 k_0} &\approx -\left[S_{ijkm}\xi^m\right]_{\mathrm{A}_0}(\sigma_p - \sigma_{\mathrm{A}_0}) \\
\Omega_{;i_0 j_0 k_0 l_0} &\approx \left[S_{ijkl}\right]_{\mathrm{A}_0} .
\end{aligned} \tag{9.2.11}$$

We recall from (3.7.22):

$$S_{ijkm}\xi^k\xi^m = -\frac{2}{3}R_{ikjm}\xi^k\xi^m. \tag{9.2.12}$$

Within the assumptions which lead to (9.2.11) and from (9.2.10) we recognize that the dominant curvature contribution in the expansion (9.2.9) is contained in the second derivative terms. In fact, noticing that along γ we have

$$a_i\dot{\gamma}^i = 0 \qquad \text{and} \qquad \left.\frac{d\Omega}{ds}\right|_{\mathrm{A}_0} = 0 \tag{9.2.13}$$

equation (9.2.9) can be written as:

$$\begin{aligned}\Omega(s) &= \Omega(s_{A_0}) - \frac{1}{2}(s - s_{A_0})^2 \\ &\times \left[1 + (\sigma_p - \sigma_{A_0})\xi_i a^i - \frac{1}{2} S_{ijkm}\dot{\gamma}^i\dot{\gamma}^j\xi^k\xi^m(\sigma_p - \sigma_{A_0})^2\right]_{A_0} \\ &- \frac{1}{6}(s - s_{A_0})^3 \left[(\sigma_p - \sigma_{A_0})\xi_i(\gamma^k\nabla_k a^i)\right]_{A_0} \\ &- \frac{1}{24}(s - s_{A_0})^4 (a_i a^i)_{A_0} + O(|\Delta\sigma + \Delta s|^5) + O(|\text{Riem}|^2) \qquad (9.2.14)\end{aligned}$$

where $O(|\Delta\sigma + \Delta s|^5)$ means terms containing powers of the parameter distances higher than the fourth. All the other first-order curvature contributions are coupled to the observer's acceleration which we shall later assume to be zero. The values s_{A_1} and s_{A_2} corresponding to the events A_1 and A_2 are obtained from (9.2.14) by imposing the conditions (9.2.3); however due to the complexity of the resulting equation, it will be more instructive to limit ourselves to the case when γ is itself a geodesic ($a = o$). Let $\dot{\gamma} \equiv u$, so (9.2.14) with (9.2.3) reduces to:

$$(s - s_{A_0})^2 \left[1 - \frac{1}{2} S_{ijkm} u^i u^j \xi^k \xi^m (\sigma_p - \sigma_{A_0})^2\right] \approx 2\Omega(s_{A_0}) \qquad (9.2.15)$$

which then admits for s_{A_1} and s_{A_2} the two solutions:

$$\begin{aligned} s_{A_1} &\approx s_{A_0} - \left(2\Omega(s_{A_0})\right)^{\frac{1}{2}} \left[1 - \frac{1}{6} R_{ikjm} u^i \xi^k u^j \xi^m (\sigma_p - \sigma_{A_0})^2\right]_{A_0} \\ s_{A_2} &\approx s_{A_0} + \left(2\Omega(s_{A_0})\right)^{\frac{1}{2}} \left[1 - \frac{1}{6} R_{ikjm} u^i \xi^k u^j \xi^m (\sigma_p - \sigma_{A_0})^2\right]_{A_0} . \end{aligned} \qquad (9.2.16)$$

From the above formulas and definitions we finally have:

$$c\,\delta T_\gamma(A_1, A_2) \approx 2L(p,\gamma) - \frac{1}{3}\left(R_{ikjm} u^i \xi^k u^j \xi^m\right)_{A_0} L^3(p,\gamma) \qquad (9.2.17)$$

where we have reparametrized ζ_0 so that $(\xi|\xi) = 1$. Equation (9.2.17) gives the first-order curvature contribution to the relationship between the round trip time of a bouncing signal and the geometrical distance between γ and p.

Let us now justify the interpretation given to (4.1.11) and (4.1.12) of a time interval and a space distance respectively, between two infinitesimally close events and relative to a given observer. If we work in the infinitesimal domain, we neglect in (9.2.14) the curvature and limit ourselves to terms of second order in the parameter distance. Choose a point A on γ very close to A_0, (here γ is again a general time-like curve); from (9.2.14) we have:

$$\begin{aligned}2\Omega(s_{A_0}) &= 2\Omega(s_A) + (s_{A_0} - s_A)^2 + O(|\Delta\sigma + \Delta s|^3)\\ &= (\sigma_p - \sigma_A)^2 g_{ij}(s_A)\dot{\zeta}^i(\sigma_A)\dot{\zeta}^j(\sigma_A)\\ &\quad + (\sigma_p - \sigma_A)^2(\dot{\gamma}^i\dot{\zeta}_i)^2_A + O(|\Delta\sigma + \Delta s|^3)\\ &= (\sigma_p - \sigma_A)^2 \left[(g_{ij} + \dot{\gamma}_i\dot{\gamma}_j)\dot{\zeta}^i\dot{\zeta}^j\right]_A\\ &\quad + O(|\Delta\sigma + \Delta s|^3); \end{aligned} \tag{9.2.18}$$

where the following relation was used from (9.2.7) and (9.2.10):

$$\begin{aligned}-(\sigma_p - \sigma_A)(\dot{\gamma}^i\dot{\zeta}_i)_A &= \left.\frac{d\Omega}{ds}\right|_A = \left.\frac{d^2\Omega}{ds^2}\right|_A (s_A - s_{A_0}) + O(\Delta s^2)\\ &= -(s_A - s_{A_0}) + O(|\Delta s + \delta\sigma|^2). \end{aligned} \tag{9.2.19}$$

Hence for a point p sufficiently close to γ, we have from (9.2.8):

$$L_{\dot{\gamma}}(A, p) = (h_{ij}\delta x^i \delta x^j)^{\frac{1}{2}} + O(\delta x^2) \tag{9.2.20}$$

where δx^i denotes the coordinate difference between A and p and h_{ij} are the components of the transverse projective tensor:

$$h_{ij} = g_{ij} + \dot{\gamma}^i\dot{\gamma}^j \tag{9.2.21}$$

relative to the observer on γ.

Relations (9.2.19) and (9.2.20) are very useful; they allow one to express the invariant measurements of distance and time interval between any two events (sufficiently close) in terms of coordinates and vector components. From (9.2.19), we interpret the time interval between the event A and the event A_0 on γ simultaneous to p as the time separation between the events A and p relative to the observer on γ; it reads:

$$\delta T_{\dot{\gamma}}(A_0, A) = -\frac{1}{c}(\dot{\gamma}_i\delta x^i)_A + O(\delta x^2). \tag{9.2.22}$$

Comparing (9.2.22) and (9.2.20) with (4.1.11) and (4.1.12) we find that they coincide and this justifies their interpretation.

9.3 Measurements of angles

Angles can be measured with great accuracy, therefore it is very important to relate angles to the measurement of other physical quantities. In the experimental example shown in Fig. 9-1, a light signal emitted at A_1 in a given direction is in general recorded at A_2 (after having been reflected at p) in a different direction, this effect being due to the acceleration of the observer and to the curvature of the background geometry. In order to measure this effect, let us first define the angle between two null directions from a single point, relative to an observer on γ. Consider Fig. 9-2; when a light signal is emitted at A towards a distant target, p say, the direction of emission at A is identified by selecting a point p' on the photon trajectory and infinitesimally close to A (the tip of the photon gun). The *distance* between γ and p' is given by the segment of the space-like geodesic through p' which strikes γ orthogonally at A_0 (Fig. 9-2); from (9.2.20), the length $\overline{A_0 p'}$ reads:

$$L(p', \gamma) \approx (h_{ij} k^i k^j)^{\frac{1}{2}}_{A} \Delta\sigma \tag{9.3.1}$$

where k is the vector tangent to the null ray from A to p' and $\Delta\sigma$ the parameter interval on it between A and p.

From the properties (4.1.4) of the projecting operator h, the length (9.3.1) is the magnitude of a space-like vector $k_\perp = \Delta\sigma h(k)$ which we refer to as the direction of the photon emission towards p. If a signal is sent from A to another point q say, and with a photon gun similar to the previous one, we identify this direction with a vector at A: $k'_\perp = \Delta\sigma h(k')$ where k' is the vector tangent to the null ray from A to q. We then define the angle $\Theta_{k,k'}$ between these two directions at A, as:

$$\cos\Theta_{k,k'} = \frac{(k'_\perp | k_\perp)}{(k_\perp | k_\perp)^{\frac{1}{2}} (k'_\perp | k'_\perp)^{\frac{1}{2}}} \,. \tag{9.3.2}$$

Evidently, the observer can also send signals in the form of particles, so the above arguments apply to time-like directions as well and the formula can be applied geometrically to any direction not tangent to the world line of the observer.

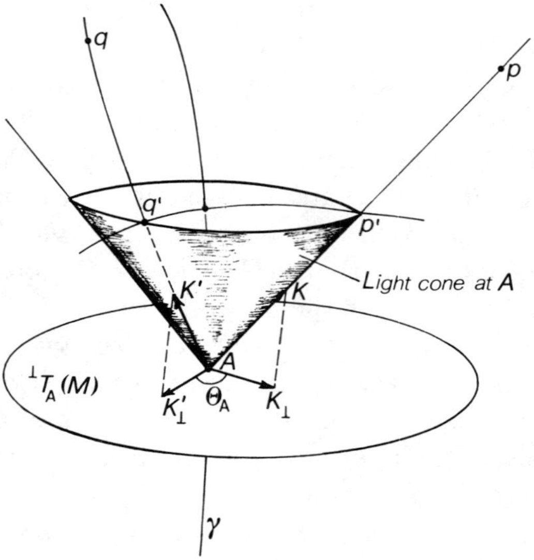

Fig. 9-2 Defining spatial angles in a four-manifold.

9.4 Curvature effects in the measurement of angles

Consider an observer carrying a tetrad frame $\{\lambda_{\hat{a}}\}$ along a world line γ and let $\dot{\gamma} \equiv u$. At the event A, a light signal will be emitted towards a target at p along a direction k which forms an angle $\underset{\hat{\alpha}\,1}{\Theta}$ with respect to a given tetrad direction $\underset{\text{A}\,\hat{\alpha}}{\lambda}$:

$$\cos \underset{\hat{\alpha}\,1}{\Theta}\bigg|_{\text{A}} = \frac{(\lambda_{\hat{\alpha}}|k)}{(k_\perp|k_\perp)^{\frac{1}{2}}}\bigg|_{\text{A}} \tag{9.4.1}$$

from (9.3.2). The signal, after having been reflected at p, is recorded by the same observer at the event B, some time later on γ, forming in general a different angle with respect to the same tetrad leg $\underset{\text{B}\,\hat{\alpha}}{\lambda}$:

$$\cos \underset{\hat{\alpha}\,2}{\Theta}\bigg|_{\text{B}} = \frac{(\lambda_{\hat{\alpha}}|k)}{(k_\perp|k_\perp)^{\frac{1}{2}}}\bigg|_{\text{B}}. \tag{9.4.2}$$

We want to calculate the difference between these two angles. From (3.7.8) we have at a general point $\gamma(s)$ on γ (see Fig. 9-3):

$$\Omega_i = -(\sigma_p - \sigma_{\gamma(s)})k_i \tag{9.4.3}$$

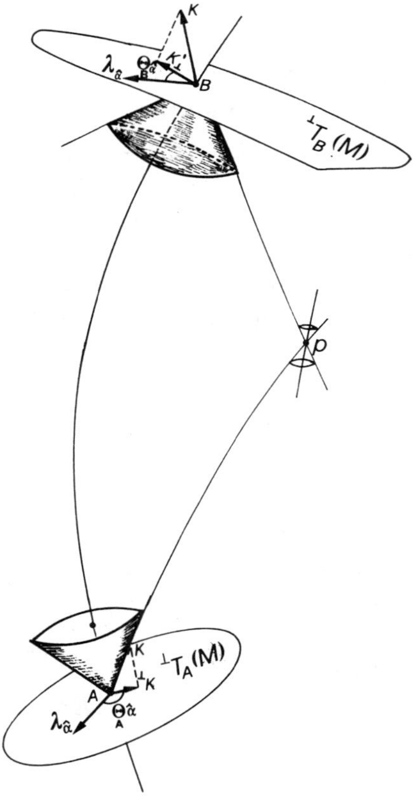

Fig. 9-3 The angle between a fixed direction and a light signal sent to p differs from that between the same direction and the light transmitted from p.

where σ is an affine parameter on the geodesics from $\gamma(s)$ to p.

Applying this to the case $\gamma(s) = \mathrm{A}_1$, equation (9.4.1) reads:

$$\cos \Theta_{\hat\alpha\, 1}\Big|_{\mathrm{A}} = -\frac{\Omega_i \lambda^i{}_{\hat\alpha}}{(\Omega_{\hat\beta}\Omega^{\hat\beta})^{\frac{1}{2}}}\Big|_{\mathrm{A}}. \tag{9.4.4}$$

Similarly we have at B:

$$\cos \Theta_{\hat\alpha\, 2}\Big|_{\mathrm{B}} = \frac{\Omega_i \lambda^i{}_{\hat\alpha}}{(\Omega_{\hat\beta}\Omega^{\hat\beta})^{\frac{1}{2}}}\Big|_{\mathrm{B}}. \tag{9.4.5}$$

Since k is a null vector, we also have:

$$(\Omega_{\hat{\alpha}}\Omega^{\hat{\alpha}})_{A/B} = |\Omega_{\hat{0}}|_{A/B} = |\Omega_i u^i|_{A/B} \tag{9.4.6}$$

so the direction cosines can be written as

$$e_{\hat{\alpha} A/B} \equiv \cos\Theta_{\hat{\alpha}\ A/B} = \mp \left.\frac{\Omega_i \lambda^i{}_{\hat{\alpha}}}{\Omega_i u^i}\right|_{A/B} \tag{9.4.7}$$

where the signs $(-/+)$ apply to A and B, respectively. As the observer moves from A to B, the tetrad frame changes smoothly according to the laws described in Sect. 4.4, so we shall consider the quantities $\Omega_i \lambda^i{}_{\hat{\alpha}} \equiv \Omega_{\hat{\alpha}}$ and $\Omega_i u^i \equiv \Omega_{\hat{0}}$ as smooth functions on γ. Consider the first term and expand it about the point $A_0 = \gamma(s_{A_0})$ which is the event simultaneous to p on γ:

$$\Omega_{\hat{\alpha}}(s) \sim \Omega_{\hat{\alpha}}(s_{A_0}) + \sum_{n=1}^{\infty} \frac{1}{n!}\left(\frac{d^n}{ds^n}\Omega_{\hat{\alpha}}\right)_{A_0} (s - s_{A_0})^n . \tag{9.4.8}$$

From (9.2.10) and assuming for simplicity $a = 0$ and $\Lambda_{\hat{\alpha}\hat{\beta}} = 0$ we obtain to first order in the curvature:

$$\begin{aligned}\Omega_{\hat{\alpha}}(s) = \Omega_{\hat{\alpha}}(s_{A_0}) &+ (s - s_{A_0})\Omega_{ij} u^j \lambda^i{}_{\hat{\alpha}} \\ &+ \frac{1}{2}(s - s_{A_0})^2 \left(\Omega_{ijk} u^j u^k \lambda^i{}_{\hat{\alpha}}\right) \\ &+ \frac{1}{6}(s - s_{A_0})^3 \left(\Omega_{ijkl} u^j u^k u^l \lambda^i{}_{\hat{\alpha}}\right) \\ &+ O(\Delta s^4) + O(|\mathrm{Riem}|^2) .\end{aligned} \tag{9.4.9}$$

Thus from (9.2.11) and the symmetries of the Riemann tensor, we have:

$$\begin{aligned}\Omega_{\hat{\alpha}}(s) = \Omega_{\hat{\alpha}}(s_{A_0}) &+ \frac{1}{2}(s - s_{A_0})(\sigma_p - \sigma_{A_0})^2 \\ &\times \left[S_{ijkm}\xi^k \xi^m u^j \lambda^i{}_{\hat{\alpha}}\right]_{A_0} \\ &- \frac{1}{2}(s - s_{A_0})^2 (\sigma_p - \sigma_{A_0}) \left[S_{ijkm}\xi^m u^j u^k \lambda^i{}_{\hat{\alpha}}\right]_{A_0} \\ &+ O(|\Delta s + \Delta\sigma|^4) + O(|\mathrm{Riem}|^2)\end{aligned} \tag{9.4.10}$$

where now we recall that ξ^i is the tangent to the geodesic $\boldsymbol{\zeta}$ from A_0 to p. The values of the parameter s at the points A and B on γ, are given in (9.2.16) by setting s_{A_1} as s_A and s_{A_2} as s_B.

Substituting these in (9.4.10) we have to the same order:

$$\Omega_{\hat{\alpha}}(s_{A}/s_{B}) \approx \Omega_{\hat{\alpha}}(s_{A_0}) \mp \frac{1}{2}\left[2\Omega(s_{A_0})\right]^{\frac{1}{2}}(\sigma_p - \sigma_{A_0})^2 \times \left[S_{ijkm}\xi^k\xi^m u^j \lambda^i{}_{\hat{\alpha}}\right]_{A_0} - \Omega(s_{A_0})(\sigma_p - \sigma_{A_0})\left[S_{ijkm}\xi^m u^j u^k \lambda^i{}_{\hat{\alpha}}\right]_{A_0}. \qquad (9.4.11)$$

Similarly, for (9.4.6) we have from (9.2.11) and (9.2.7):

$$\Omega_{\hat{0}}(s) \approx (s - s_{A_0})\left[-1 + \frac{1}{2}(\sigma_p - \sigma_{A_0})^2\left(S_{ijkm}u^i u^j \xi^k \xi^m\right)_{A_0}\right], \qquad (9.4.12)$$

so recalling again (9.2.16):

$$\Omega_{\hat{0}}(s_{A}/s_{B}) \approx \mp\left[2\Omega(s_{A_0})\right]^{\frac{1}{2}}\left[-1 + \frac{1}{6}\boldsymbol{K}_{A_0}(\sigma_p - \sigma_{A_0})^2\right] \qquad (9.4.13)$$

where we put

$$\boldsymbol{K} = -R_{ijkm}u^i\xi^j u^k \xi^m.$$

To the desired order, then, equation (9.4.7) reads:

$$\underset{\hat{\alpha}A}{e} \approx -\frac{1}{\left[2\Omega(s_{A_0})\right]^{\frac{1}{2}}}\Big\{\Omega_{\hat{\alpha}}(s_{A_0})\left[1 + \frac{1}{6}\boldsymbol{K}_{A_0}(\sigma_p - \sigma_{A_0})^2\right] - \frac{1}{2}\left[2\Omega(s_{A_0})\right]^{\frac{1}{2}}(\sigma_p - \sigma_{A_0})^2\left(S_{ijkm}\xi^k\xi^m u^j\lambda^i{}_{\hat{\alpha}}\right)_{A_0} - \Omega(s_{A_0})(\sigma_p - \sigma_{A_0})\left(S_{ijkm}\xi^m u^j u^k \lambda^i{}_{\hat{\alpha}}\right)_{A_0}\Big\} \qquad (9.4.14)$$

and

$$\underset{\hat{\alpha}B}{e} \approx -\frac{1}{\left[2\Omega(s_{A_0})\right]^{\frac{1}{2}}}\Big\{\Omega_{\hat{\alpha}}(s_{A_0})\left[1 + \frac{1}{6}\boldsymbol{K}_{A_0}(\sigma_p - \sigma_{A_0})^2\right] + \frac{1}{2}\left[2\Omega(s_{A_0})\right]^{\frac{1}{2}}(\sigma_p - \sigma_{A_0})^2\left(S_{ijkm}\xi^k\xi^m u^j\lambda^i{}_{\hat{\alpha}}\right)_{A_0} - \Omega(s_{A_0})(\sigma_p - \sigma_{A_0})\left(S_{ijkm}\xi^m u^j u^k \lambda^i{}_{\hat{\alpha}}\right)_{A_0}\Big\}. \qquad (9.4.15)$$

The change in the two direction cosines in passing from A to B is given by:

$$\Delta \underset{\hat{\alpha}}{e} = -\left(S_{ijkm}\xi^k\xi^m u^j\lambda^i{}_{\hat{\alpha}}\right)_{A_0}(\sigma_p - \sigma_{A_0})^2. \qquad (9.4.16)$$

This equation gives a measure of how the light gun must be turned relative to a Fermi transported tetrad, in order to detect the reflected light signal at B. From (9.2.12) and (9.2.8), equation (9.4.6) can also be written as:

$$\Delta \underset{\hat{\alpha}}{e} = \frac{2}{3}\left(R_{ikjm}\lambda^{i}{}_{\hat{\alpha}}\xi^{k}u^{j}\xi^{m}\right)_{A_0} L^2(p,\gamma). \tag{9.4.17}$$

A comparison of this with (9.2.17) leads to an *angle–proper-time* relation which, to first order in the curvature, may have a direct observational verification.

9.5 Measurement of frequency

Light signals are the most important carriers of information. Whenever an observer emits or absorbs a light signal, he or she measures its frequency and energy flux. Since this operation takes place within a small neighbourhood of the observer's world-line (assuming a small size for the measuring apparatus) we can limit our considerations to the local inertial frame where special relativity holds.

We have seen in Sect. 7.8 how the discontinuities of an electromagnetic field propagate in space-time. The surface of discontinuity

$$\Phi(x) = 0 \tag{9.5.1}$$

is the *eikonal* of the wave and satisfies the equation (the eikonal equation):

$$g^{ij}k_i k_j = 0 \tag{9.5.2}$$

where $k_i = \partial_i \Phi$.

In the observer's rest frame, reinterpreting Φ as the phase function of the geometrical optics approximation (Sect. 7.8), the instantaneous frequency of the wave is:

$$\nu = -\frac{d\Phi}{dT_\gamma}. \tag{9.5.3}$$

From (9.5.1), the latter is more conveniently written as:

$$\nu = -(\partial_i \Phi)\frac{dx^i}{dT_\gamma} = -ck_i u^i \tag{9.5.4}$$

where c is the velocity of light; hence we have:

$$-k_i u^i = \frac{\nu}{c} \tag{9.5.5}$$

which is the invariant characterization of a frequency.

Let us now consider two observers u on γ and u' on γ', far apart, who exchange a light signal. At the event of emission, e on γ say, the observer u measures a frequency $\nu = -c(u_i k^i)_{\mathrm{e}}$; the *same* signal at the event of reception o on γ', will be measured with a frequency $\nu' = -c(u'_i k^i)_{\mathrm{o}}$. The ratio between these two values is called the *frequency shift* and traditionally denoted by (Trautmann, 1965):

$$(1+z)_{\mathrm{o}} = \left(\frac{\nu}{\nu'}\right)_{\mathrm{o}} = \frac{(u_i k^i)_{\mathrm{e}}}{(u'_i k^i)_{\mathrm{o}}} = \frac{(\check{u}_i k^i)_{\mathrm{o}}}{(u'_i k^i)_{\mathrm{o}}}, \tag{9.5.6}$$

where we parallely propagate along the null geodesic γ joining e to o and recall that along γ we have $(u_i k^i) = (\check{u}_i \check{k}^i)_{\mathrm{o}}$ and $\underset{\mathrm{o}}{\check{k}} = \underset{\mathrm{o}}{k}$, from (2.4.11) and (2.4.19).

When $1+z<1$, we have a blue-shift relative to u', while when $1+z>1$ we have a red-shift.

9.6 Measurement of relative velocities

Let us consider the situation shown in Fig. 9-4; an observer u on a world-line γ with parameter s monitors the position of a particle on a world-line γ' with parameter σ and tangent vector field ℓ (this labeling of the tangent vector to γ' is made for convenience of notation), sending signals (emission events A, B ...) and recording the echoes (at the events A_1, $\mathrm{B}_1 \ldots$).

If the world-lines γ and γ' are sufficiently close, the arguments of Sect. 9.2 apply to each round trip of the light signals, hence, naming A_0 and B_0 the events on γ simultaneous to p and p' on γ', we can state the following: with respect to the observer u, the particle ℓ has covered a distance $\delta L = L(p,\gamma) - L(p',\gamma)$ in

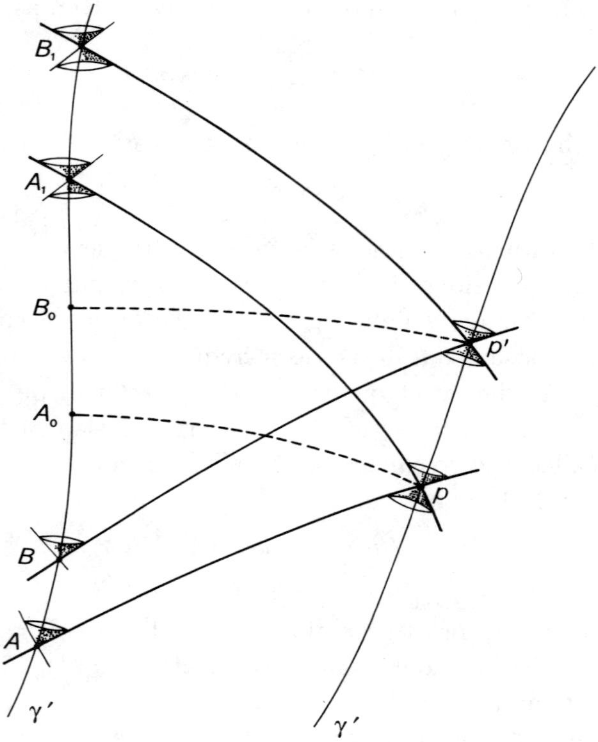

Fig. 9-4 To monitor the motion of a point p on γ', we need a sequence of measurements on the observer's world-line γ.

an interval of proper time $\delta T_\gamma = \delta T_\gamma(\mathrm{B}_0, \mathrm{A}_0)$ measured by u on γ. Thus we define the *radial* velocity (or velocity of recession) of the particle ℓ with respect to the observer u, to be the quantity

$$\tilde{v} = \lim_{p \to p'} \frac{\delta L}{\delta T_\gamma}. \tag{9.6.1}$$

If the particle and the observer (namely the curves γ and γ') are close enough for curvature effects to be neglected and the approximation as in (9.2.17) holds, then it is possible to express (9.6.1) in terms only of the recorded times of emission and reception on γ of the light signals. We have in fact:

$$L(p, \gamma) \approx \frac{1}{2} c\, \delta T_\gamma(\mathrm{A}, \mathrm{A}_1)$$

$$L(p', \gamma) \approx \frac{1}{2} c\, \delta T_\gamma(\text{B}, \text{B}_1)$$

$$\delta T \approx \delta T_\gamma(\text{A}, \text{B}_1) - \frac{1}{2}\left[\delta T_\gamma(\text{A}, \text{A}_1) + \delta T_\gamma(\text{B}, \text{B}_1)\right]. \quad (9.6.2)$$

Then:

$$\tilde{v} \approx \lim_{\text{A/B} \to \text{A}_1/\text{B}_1} \frac{c[\delta T_\gamma(\text{B}, \text{B}_1) - \delta T_\gamma(\text{A}, \text{A}_1)]}{2\delta T_\gamma(\text{A}, \text{B}_1) - [\delta T_\gamma(\text{A}, \text{A}_1) + \delta T_\gamma(\text{B}, \text{B}_1)]} . \quad (9.6.3)$$

If the particle and the observer are so far apart that curvature effects cannot be neglected, then $\tilde{v}$ should include correction terms due to the curvature, in the way that they enter the definition of distance L, given as the solution of (9.2.17).

An indirect measurement of velocity, which naturally stems from the process of exchanging light signals with the particle, is a *Doppler velocity*; that is, the velocity the particle would have if the frequency shift of the exchanged signal were due only to a relative motion. The frequency shift is directly observable and the special relativistic expression of the Doppler shift reads, in the observer's rest frame at any event on γ:

$$\frac{\omega_{\text{ems}}}{\omega_{\text{obs}}} = \left(1 - \frac{v^2}{c^2}\right)^{-\frac{1}{2}} \left[1 - \frac{v}{c} \cos \Theta_{k,\check{\ell}}\right] . \quad (9.6.4)$$

Here subscripts ems and obs mean at the emission (from the particle on γ') and observation (at the observer on γ), respectively; $\Theta_{k,\check{\ell}}$ is the angle between the direction of the light signal and that of the emitting particle, referred to the observer's rest frame by parallel propagation along the light trajectory from the emission point on γ' to the observation point on γ. Let us consider equation (9.5.6); recall that, since the light trajectory is a null geodesic, the following relation holds:

$$(k^i \ell_i)_{p=\gamma'(\sigma)} = (\check{k}_i \check{\ell}^i)_{\gamma(s)} = (k_i \check{\ell}^i)_{\gamma(s)}$$

hence (9.5.6) becomes at $\gamma(s)$:

$$\frac{\omega_{\text{ems}}}{\omega_{\text{obs}}} = \frac{(k_i \check{\ell}^i)}{(k_i u^i)} = \frac{[h_{ij} k^i \check{\ell}^j - (u_i k^i)(u_j \check{\ell}^j)]}{(u_i k^i)}$$

$$= -\frac{(h_{ij} k^i \check{\ell}^j)(h_{rs} \check{\ell}^r \check{\ell}^s)^{\frac{1}{2}}}{(h_{mn} k^m k^n)^{\frac{1}{2}} (h_{pq} \check{\ell}^p \check{\ell}^q)^{\frac{1}{2}}} - (u_j \check{\ell}^j)$$

remembering that $(u_i k^i)_{\gamma(s)} < 0$. From (9.4.2) however, this becomes:

$$\frac{\omega_{\rm ems}}{\omega_{\rm obs}} = -\left[\cos\Theta_{k,\breve{\ell}}(h_{rs}\breve{\ell}^r\breve{\ell}^s)^{\frac{1}{2}} + u_j\breve{\ell}^j\right]$$

$$= -(u_j\breve{\ell}^j)\left[1 - \cos\Theta_{k,\breve{\ell}}\frac{(h_{rs}\breve{\ell}^r\breve{\ell}^s)^{\frac{1}{2}}}{(-u_j\breve{\ell}^j)}\right]. \quad (9.6.5)$$

If we define the velocity of ℓ with respect to u at $\gamma(s)$ as the quantity:

$$\frac{v}{c} = -\left.\frac{(h_{rs}\breve{\ell}^r\breve{\ell}^s)^{\frac{1}{2}}}{(u_j\breve{\ell}^j)}\right|_{\gamma(s)}, \quad (9.6.6)$$

we obtain after simple algebra:

$$-u_j\breve{\ell}^j = \left(1 - \frac{v^2}{c^2}\right)^{-\frac{1}{2}}, \quad (9.6.7)$$

so (9.6.4) is recovered.

The instantaneous Doppler velocity of the particle on γ' relative to the observer on γ can be written in terms of the components v^i of a four-vector orthogonal to u as

$$\frac{v}{c} = (v^i v_i)^{\frac{1}{2}} \quad (9.6.8)$$

where

$$v^i = -(u_r\breve{\ell}^r)^{-1}(h^i{}_j\breve{\ell}^j) = -(u_r\breve{\ell}^r)^{-1}\left[\breve{\ell}^i + u^i(u_s\breve{\ell}^s)\right]. \quad (9.6.9)$$

Of course $v^i u_i = 0$.

The relative Doppler velocity (9.6.6), being based on a measurement of a Doppler shift, is really an *equivalent* velocity. A frequency shift in fact can be thought of as induced both by gravitational effects and by the proper motion of the source relative to the observer. In general then, the observable Doppler velocity (9.6.6) expresses on the same footing a proper relative motion of the source (the scattering particle in our case) and the effects due to the background geometry. The question arises whether one could in fact single out the contribution to the Doppler velocity by the background geometry establishing a relationship between the Doppler shift and the proper recession velocity (9.6.1). In principle the latter could be determined if we were able to know L as a

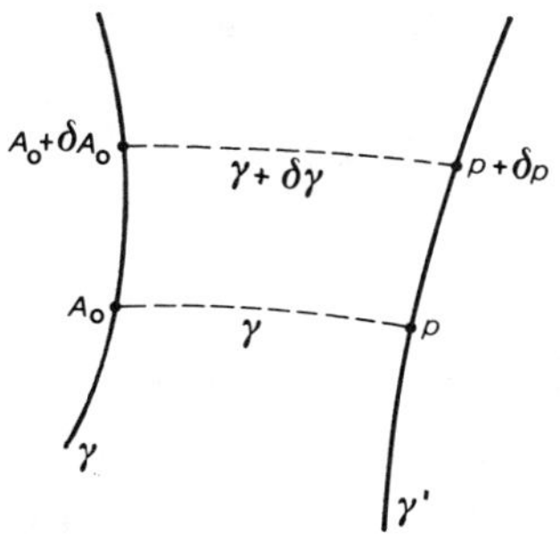

Fig. 9-5 The varied point $A_0 + \delta A_0$ is the simultaneous to $p + \delta p$ relative to the observer on γ.

function of the proper time of the observer on γ. In this case we have, from (9.6.1) and (9.2.8):

$$\tilde{v} = \frac{dL}{ds} = \frac{d}{ds}(2\Omega)^{\frac{1}{2}} = \frac{1}{L}\left(\frac{d\Omega}{ds}\right) \tag{9.6.10}$$

where the differentiation with respect to s is here meant as the result of varying A_0 on γ (varying s) and also varying the point p on γ' (varying σ) in such a way that the corresponding varied geodesic segment joining them is extremal with respect to γ, in the sense of (9.2.4) (see Fig. 9-5); hence:

$$\tilde{v} = \frac{1}{L}\left(\Omega_{i_0} u^{i_0} + \frac{d\sigma}{ds}\Omega_{j_p}\ell^{j_p}\right) = \frac{1}{L}\frac{d\sigma}{ds}\Omega_{j_p}\ell^{j_p}, \tag{9.6.11}$$

where $\Omega_{i_0} u^{i_0} = 0$. From (3.7.8) we have on $\zeta_{\text{A}_0 \to p}$:

$$\Omega_{j_p}\ell^{j_p} = -\Omega_{j_0}\breve{\ell}^{j_0} \ ,$$

thus (9.6.11) becomes:

$$\tilde{v} = -\frac{1}{L}\left(\frac{d\sigma}{ds}\right)\Omega_{i_0}\breve{\ell}^{i_0} \tag{9.6.12}$$

where $\breve{\ell}^{i_0} = \Gamma(p, \text{A}_0; \zeta_{p \to \text{A}_0})_{j_p}{}^{i_0}\, \ell^{j_p}$.

In order to express $\tilde{v}$ as the modulus of a four-vector defined at A_0, let us exploit the property (3.7.9) of the world-function, i.e.:

$$2\Omega = \Omega_{i_0}\Omega^{i_0} = \Omega_{i_p}\Omega^{i_p} \ . \tag{9.6.13}$$

Defining:

$$\tilde{v}_{i_0} = \frac{1}{2\Omega}\left(\frac{d\sigma}{ds}\right)\Omega_{i_0}\Omega_{j_0}\check{\ell}^{j_0} , \tag{9.6.14}$$

equation (9.6.12) can be written as:

$$\tilde{v} = -\left(\tilde{v}_{i_0}\tilde{v}^{i_0}\right)^{\frac{1}{2}} . \tag{9.6.15}$$

Let us now calculate the ratio $d\sigma/ds$. From the properties of the world-function and recalling (3.7.10) and (3.7.19), we obtain:

$$\begin{aligned}\frac{D\Omega_{i_0}}{Ds} &= \Omega_{i_0 j_0}u^{j_0} + \frac{d\sigma}{ds}\Omega_{i_0 j_p}\ell^{j_p} \\ &= u_{i_0} - \frac{d\sigma}{ds}\check{\ell}_{i_0} + \chi_{i_0 j_0}u^{j_0} + \frac{d\sigma}{ds}\zeta_{i_0 j_p}\ell^{j_p}\end{aligned} \tag{9.6.16}$$

where $\chi_{i_0 j_0}$ and $\zeta_{i_0 j_p}$ are curvature-dependent terms which to first order in the curvature are given, as in (3.7.19), by:

$$\begin{aligned}\chi_{i_0 j_0} &= (\lambda_p - \lambda_0)^{-1} g_{i_0 k_0}\int_{\lambda_0}^{\lambda_p}(\lambda_p - \lambda)^2 R^r{}_{mns}\xi^m\xi^n\Gamma_r{}^{k_0}\Gamma_{j_0}{}^s\, d\lambda \\ &\quad + \mathrm{O}(|\mathrm{Riem}|^2) \\ \zeta_{i_0 j_p} &= (\lambda_p - \lambda_0)^{-1} g_{i_0 s_0}\int_{\lambda_0}^{\lambda_p}(\lambda_p - \lambda)(\lambda - \lambda_0) R^r{}_{mnk}\xi^m\xi^n\Gamma_{j_p}{}^k\Gamma_r{}^{s_0} d\lambda \\ &\quad + \mathrm{O}(|\mathrm{Riem}|^2).\end{aligned} \tag{9.6.17}$$

Here λ is the parameter on the space-like extremal geodesics ζ_0's which join γ' to γ, and whose tangent vector field is ξ. Contracting (9.6.16) with u^{i_0} and recalling that

$$\frac{D\Omega_{i_0}}{Ds}u^{i_0} = -\Omega_{i_0}\dot{u}^{i_0}$$

we have:

$$-\Omega_{i_0}\dot{u}^{i_0} = -1 - \frac{d\sigma}{ds}\left(u_{i_0}\check{\ell}^{i_0}\right) + \chi_{i_0 j_0}u^{i_0}u^{j_0} + \frac{d\sigma}{ds}\zeta_{i_0 j_p}u^{i_0}\ell^{j_p} \tag{9.6.18}$$

and so to first order in the curvature we have, relative to the observer on γ:

$$\frac{d\sigma}{ds} = -\left(u_{i_0}\check{\ell}^{i_0}\right)^{-1}\left(1 - \Omega_{i_0}\dot{u}^{i_0} - \chi_{i_0 j_0}u^{i_0}u^{j_0} + \frac{\zeta_{i_0 j_p}u^{i_0}\ell^{j_p}}{u_{k_0}\check{\ell}^{k_0}}\right) + \mathrm{O}(|\mathrm{Riem}|^2). \tag{9.6.19}$$

In the limit of zero curvature and negligible distance between the curves ($\Omega_{i_0} \approx 0$), the above relation reduces to the familiar special relativistic one:

$$\frac{d\sigma}{ds} = -\left(u_{i_0}\check{\ell}^{i_0}\right)^{-1} \equiv \bar{\gamma}^{-1} \tag{9.6.20}$$

from (9.6.7). To that limit in fact the recession velocity and the Doppler velocity coincide. Let us use (9.6.19) in (9.6.14) to get:

$$\begin{aligned}\tilde{v}_{i_0} &= -\left(u_{k_0}\check{\ell}^{k_0}\right)^{-1}\left(1 - \Omega_{i_0}\dot{u}^{i_0} + \mathcal{R}'_0 + \mathrm{O}(|\mathrm{Riem}|^2)\right)\frac{\Omega_{i_0}\Omega_{j_0}\check{\ell}^{j_0}}{2\Omega} \\ &= -\xi_{i_0}\xi_{j_0}\check{\ell}^{j_0}\left(u_{k_0}\check{\ell}^{k_0}\right)^{-1} \\ &\quad\times\left(1 - \Omega_{i_0}\dot{u}^{i_0} + \mathcal{R}'_0 + \mathrm{O}(|\mathrm{Riem}|^2)\right)\end{aligned} \tag{9.6.21}$$

where we used definition (9.2.6) of Ω_{i_0} with $(\xi|\xi) = 1$, and

$$\mathcal{R}'_0 = \left(u_{k_0}\check{\ell}^{k_0}\right)^{-1}\zeta_{i_0 j_p}u^{i_0}\ell^{j_p} - \chi_{i_0 j_0}u^{i_0}u^{j_0}.$$

Since $\xi_{i_0}u^{i_0} = 0$, (9.6.21) can also be written to first order in $|\mathrm{Riem}|$:

$$\tilde{v}_{i_0} \approx -\xi_{i_0}\xi_{j_0}\left[u^{j_0} + \check{\ell}^{j_0}\left(u^{k_0}\ell_{k_0}\right)^{-1}\right]\left(1 - \Omega_{j_0}\dot{u}^{j_0} + \mathcal{R}'_0\right). \tag{9.6.22}$$

The quantity

$$v'_{i_0} = -\left[u_{i_0} + \check{\ell}_{i_0}\left(u_{k_0}\check{\ell}^{k_0}\right)^{-1}\right]$$

although similar to (9.6.9), is not a Doppler velocity four-vector since it is written in terms of a vector $\check{\ell}$ at A_0 parallely propagated along a space-like geodesic (from p to A_0) and not along a null ray (as in the definition (9.6.9)). We can write however

$$\begin{aligned}\check{\ell}_{i_0} &= \Gamma(p, \mathrm{A}_0; \zeta)^{j_p}{}_{i_0}\ell_{j_p} \\ &= \Gamma(p, \mathrm{A}_0; \zeta)^{j_p}{}_{i_0}\Gamma(\mathrm{A}_1, p; \gamma)^{k_1}{}_{j_p}\Gamma(p, \mathrm{A}_1; \gamma)^{s_p}{}_{k_1}\ell_{s_p} \\ &= \mathcal{S}^{k_1}{}_{i_0}\check{\ell}_{k_1}\end{aligned} \tag{9.6.23}$$

where $\check{\ell}_{k_1}$ is the parallel at the point of observation A_1 on γ to ℓ in p along the null ray γ from p, and $\mathcal{S}^{k_1}{}_{i_0}$ is a new quantity which depends on three points and can be expressed as:

$$\mathcal{S}^{k_1}{}_{i_0} = \Gamma(\mathrm{A}_1, \mathrm{A}_0; \gamma)^{k_1}{}_{i_0} + \tilde{S}(p, \mathrm{A}_1, \mathrm{A}_0)^{k_1}{}_{i_0} \tag{9.6.24}$$

the last term being a three-point tensor, whose derivation we omit here but which has the property of being zero if any two points coincide or when the curvature is zero; (Synge, 1960). Thus from (9.6.24), (9.6.23) can be written as:

$$\check{\ell}_{i_0} = \Gamma(A_1, A_0; \gamma)^{k_1}{}_{i_0}\check{\ell}_{k_1} + \tilde{\mathcal{L}}_{i_0} \qquad (9.6.25)$$

where $\tilde{\mathcal{L}}_{i_0}$ is the result (at A_0) of acting on $\check{\ell}_{k_1}$ (at A_1) with $\tilde{S}^{k_1}{}_{i_0}$. The significance of $\tilde{\mathcal{L}}_{i_0}$ is deducible directly from (9.6.25); it is in fact the result of a comparison at A_0 of the images there of the vector ℓ at p under parallel transport along two different paths, one along the space-like geodesic from p to A_0 and the other along the null geodesic from p to A_1 and then from A_1 back to A_0 along γ. From the non-integrability of the connection, $\tilde{\mathcal{L}}_{i_0}$ is a measure of the integrated curvature over the enclosed area. If we assume, for simplicity, that γ is a geodesic ($\dot{u} = 0$), then

$$u_{i_0} = \check{u}_{i_0} = \Gamma(A_1, A_0; \gamma)^{k_1}{}_{i_0}u_{k_1}\ . \qquad (9.6.26)$$

Hence, inserting this and (9.6.25) into v'_{i_0}, we obtain:

$$v'_{i_0} = -\Big\{\check{u}_{i_0} + \Big(\Gamma(A_1, A_0; \gamma)^{k_1}{}_{i_0}\check{\ell}_{k_1} + \tilde{\mathcal{L}}_{i_0}\Big) \times \Big[\check{u}^{j_0}\Big(\Gamma(A_1, A_0; \gamma)^{k_1}{}_{j_0}\check{\ell}_{k_1} + \tilde{\mathcal{L}}_{j_0}\Big)\Big]^{-1}\Big\}. \qquad (9.6.27)$$

Now since $\check{u}^{j_0}\Gamma(A_1, A_0; \gamma)^{k_1}{}_{j_0}\check{\ell}_{k_1} = u^{k_1}\check{\ell}_{k_1} = -\bar{\gamma}_1$ is the *Doppler* Lorentz factor of the particle ℓ relative to the observer u at A_1, we can write (9.6.27) as follows:

$$v'_{i_0} = -\Big[\check{u}_{i_0} + \Big(\Gamma(A_1, A_0; \gamma)^{k_1}{}_{i_0}\check{\ell}_{k_1} + \tilde{\mathcal{L}}_{i_0}\Big)\Big(-\bar{\gamma}_1 + \tilde{\mathcal{L}}_{j_0}\check{u}^{j_0}\Big)^{-1}\Big]. \qquad (9.6.28)$$

Thus to first order in |Riem| we finally have:

$$v'_{i_0} = \check{v}_{i_0} + \bar{\gamma}_1^{-1}\tilde{\mathcal{L}}_{i_0} + \bar{\gamma}_1^{-2}\tilde{\mathcal{L}}_{j_0}\check{u}^{j_0}\Gamma(A_1, A_0; \gamma)^{k_1}{}_{i_0}\check{\ell}_{k_1} + O(|\mathrm{Riem}|^2) \qquad (9.6.29)$$

where:

$$\check{v}_{i_0} = -\Big[\check{u}_{i_0} - \bar{\gamma}_1^{-1}\Gamma(A_1, A_0; \gamma)^{k_1}{}_{i_0}\check{\ell}_{k_1}\Big] \qquad (9.6.30)$$

is the parallel at A_0 of the Doppler velocity four-vector defined at A_1 according to (9.6.9). Then inserting (9.6.29) in (9.6.22), with the assumption that $\dot{u} = 0$, we have

$$\tilde{v}_{i_0} = \xi_{i_0}\xi^{j_0}\left[\check{v}_{j_0} + \tilde{\mathcal{L}}_{k_0}\left(\delta^{k_0}_{j_0}\bar{\gamma}_1^{-1} + \bar{\gamma}_1^{-2}u^{k_0}\Gamma(\text{A}_1, \text{A}_0; \gamma)^{r_1}{}_{j_0}\check{\ell}_{r_1}\right)\right] + \xi_{i_0}\xi^{j_0}\check{v}_{j_0}\mathcal{R}'_0 + \text{O}(|\text{Riem}|^2). \quad (9.6.31)$$

Since all curvature terms are contained in $\tilde{\mathcal{L}}_{j_0}$ (Lathrop, 1973), in $\mathcal{R}'_0$ and the neglected terms, we establish a relationship between the radial component of the Doppler velocity $\check{v}_\xi = \check{v}_{j_0}\xi^{j_0}$ and that of the recession velocity $\tilde{v}_\xi$ as follows:

$$\tilde{v}_\xi = \check{v}_\xi\,(1 + \mathcal{R}'_0) + \bar{\gamma}_1^{-1}\tilde{\mathcal{L}}_\xi + \bar{\gamma}_1^{-2}\tilde{\mathcal{L}}_{\hat{0}}\Gamma(\text{A}_1, \text{A}; \gamma)^{k_1}{}_\xi\check{\ell}_{k_1} + \text{O}(|\text{Riem}|^2). \quad (9.6.32)$$

Here we have selected a tetrad frame adapted to u on γ so that $\xi = \dot{\zeta}$ coincides at A_0 with one space-like tetrad direction, since we have there: $\xi^{i_0}u_{i_0} = 0$. Though it is not clear whether Eq. (9.6.32), as it stands, is suitable for a direct observational test, it has the virtue of showing explicitly how the space-time curvature contributes to the frequency shift of the exchanged signal.

9.7 The velocity composition law

A distinctive feature of relativity is that velocities do not sum up linearly but according to a law which prevents them from appearing larger than the velocity of light, whatever observer we refer to.

Let us have a particle on a world-line with tangent field k and two observers on world-lines having respectively tangent fields u and ℓ. We ask how to express the spatial velocity of k relative to ℓ, say, in terms of the velocity of k relative to u and that of u relative to ℓ. For simplicity let us assume that at the event A of the measurement the three world-lines coincide (Fig. 9-6). This assumption amounts to the neglect of curvature terms, and so the relative velocities are well defined as Doppler velocities.

If we term v and v the instantaneous spatial velocities of k relative to u and ℓ respectively and V that of u relative to ℓ, we have from (9.6.9) at A:

$$v^i = -(u^r k_r)^{-1}[k^i + u^i(u_s k^s)] \quad (9.7.1)$$

$$\text{v}^i = -(\ell^r k_r)^{-1}[k^i + \ell^i(\ell_s k^s)] \quad (9.7.2)$$

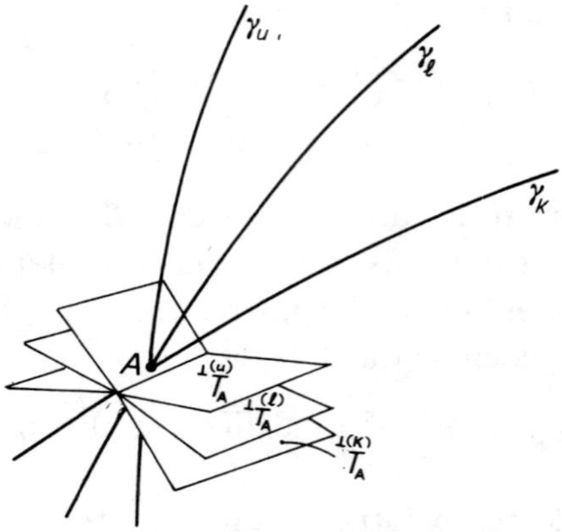

Fig. 9-6 If the three world-lines coincide at A, one can determine the velocity composition law projecting into the respective rest frames at A.

$$V^i = -(u^r\ell_r)^{-1}[u^i + \ell^i(u_s\ell^s)]. \tag{9.7.3}$$

We shall now express v^i in terms of v^i and V^i; from the above equations we have:

$$\begin{aligned}\mathrm{v}^i &= -(k^s\ell_s)^{-1}\left[-(u^rk_r)(v^i+u^i)+\ell^i(\ell^sk_s)\right]\\ &= -(k^s\ell_s)^{-1}\left[-(u^rk_r)(v^i-(u^s\ell_s)(V^i+\ell^i))+\ell^i(\ell^sk_s)\right]\\ &= \frac{(u^rk_r)}{(k^s\ell_s)}v^i - \frac{(u^rk_r)(u^s\ell_s)}{(k^j\ell_j)}V^i - \frac{\ell^i}{(k^s\ell_s)}\left[(u^rk_r)(u^i\ell_i)+k^j\ell_j\right].\end{aligned} \tag{9.7.4}$$

The last term in (9.7.4) can be written as

$$\begin{aligned}-\frac{\ell^i}{(k^s\ell_s)}\left[(u^rk_r)(u^n\ell_n)+k^j\ell_j\right] &= \ell^i\frac{(u^rk_r)}{(k^s\ell_s)}\left[-\frac{k^s+u^s(u^tk_t)}{(u^jk_j)}\right]\ell_s\\ &= \ell^i\frac{(u^rk_r)}{(k^s\ell_s)}v^s\ell_s\end{aligned} \tag{9.7.5}$$

from (9.7.1); hence using this result in (9.7.4) we have:

$$\mathrm{v}^i = \frac{(u^rk_r)}{(k^s\ell_s)}v^j(\delta^i_j+\ell^i\ell_j) - \frac{(u^rk_r)(u^s\ell_s)}{(k^t\ell_t)}V^i\ ; \tag{9.7.6}$$

recalling (4.1.4), we obtain the final result (de Felice, 1979):

$$\mathrm{v}^i = \frac{(u^r k_r)}{(k^s \ell_s)} \left[v^j h^i_j - (u^t \ell_t) V^i \right] \tag{9.7.7}$$

where $h^i_j = \delta^i_j + \ell^i \ell_j$.

We can now calculate the modulus of (9.7.7); from (9.6.8) we have:

$$\frac{\mathrm{v}^2}{c^2} = \left(\frac{1 - \mathrm{v}^2/c^2}{1 - v^2/c^2} \right) \left[h_{ij} v^i v^j + \frac{2 h_{ij} v^j V^i}{\sqrt{1 - V^2/c^2}} + \frac{V^2/c^2}{1 - V^2/c^2} \right]. \tag{9.7.8}$$

Solving for v/c we obtain after some algebra:

$$\frac{\mathrm{v}}{c} = \frac{\beta_u + \frac{V/c}{\sqrt{1 - V^2/c^2}}}{\left[1 - v^2/c^2 + \left(\beta_u + \frac{V/c}{\sqrt{1 - V^2/c^2}} \right)^2 \right]^{\frac{1}{2}}} \tag{9.7.9}$$

where we define:

$$\beta_u = \left(h_{ij} v^i v^j \right)^{\frac{1}{2}} \tag{9.7.10}$$

as the velocity of k with respect to u, as it would be measured by the observer ℓ. We notice that in (9.7.9) the principle of relativity is satisfied; in fact when $v = c$ also $\mathrm{v} = c$, whatever is the velocity of u with respect to ℓ.

9.8 Energy and momentum of a particle

The most important measurements, after the spatial instantaneous velocity of a particle, are its energy and momentum. An observer u who monitors a particle moving along a world-line with tangent k can easily deduce its instantaneous energy and spatial momentum from the knowledge of its Doppler velocity and rest mass. Call these v and μ_0 respectively; then in the rest frame of u we have from special relativity:

$$\mathcal{E} = \frac{\mu_0 c^2}{(1 - \frac{v^2}{c^2})^{\frac{1}{2}}} \tag{9.8.1}$$

$$P = \frac{\mu_0 v}{(1 - \frac{v^2}{c^2})^{\frac{1}{2}}}. \tag{9.8.2}$$

Using (9.6.7), the above relations can be expressed formally in terms of the four-velocities which identify the observer and the particle in space-time. In fact (9.8.1) becomes:

$$\mathcal{E} = -\mu_0 c^2 (u_i \check{k}^i) \tag{9.8.3}$$

which implies:

$$u_i \check{k}^i = -\frac{\mathcal{E}}{\mu_0 c^2} \; . \tag{9.8.4}$$

Similarly from (9.8.2) we have:

$$P = -\mu_0 c^2 (u_i \check{k}^i)(h_{ij} v^i v^j)^{\frac{1}{2}} = \mu_0 c \left(h_{ij} \check{k}^i \check{k}^j \right)^{\frac{1}{2}} \tag{9.8.5}$$

where $h_{ij} = g_{ij} + u_i u_j$. Definitions (9.8.3) and (9.8.5) are meaningful only when one is dealing with a point particle. This however is a rather good approximation in many cases.

9.9 Measurement of electric and magnetic fields

Assume that an observer u moves in a space-time endowed with an electromagnetic field F_{ij}. A natural question is how a measurement of electric and magnetic fields, made in the frame of u, is related to F_{ij}. The observer realizes that an electromagnetic field exists only by studying the behaviour of a charged particle; let the four-velocity of the particle be k. We know from (7.7.15) that the trajectory of the particle deviates from a geodesic because of the Lorentz force which stems from the electromagnetic interaction, namely:

$$k^r \nabla_r k^i = \frac{e}{\mu} F^i{}_j k^j \; . \tag{9.9.1}$$

Here e is the particle's charge and μ its rest energy. Assuming for simplicity that the particle and the observer coincide at the event of measurement, let us see how the force term is measured by u. First we decompose the four-vector $f^i = F^i{}_j k^j$ into a parallel and a transverse component relative to u, namely:

$$f^i = h^i{}_j f^j + \pi^i{}_j f^j . \tag{9.9.2}$$

Each component has a well defined physical meaning which we shall now analyse. The transverse component reads:

$$\tilde{f}^i \equiv h^i{}_j f^j = h^i{}_j F^j{}_r k^r = h^i{}_j F^j{}_r (h^r{}_s k^s - u^r u_s k^s)$$
$$= h^i{}_j h^r{}_s F^j{}_r k^s - h^i{}_j F^j{}_r u^r (u_s k^s) \quad (9.9.3)$$

from (4.1.4) and (4.1.8). Let us now define (Misner et al., 1973):

$$E^j = F^j{}_r u^r \ ; \quad (9.9.4)$$

from the skew-symmetry of $F^i{}_j$ it follows that:

$$E^j u_j = 0 \qquad h^i{}_j E^j = E^i \ . \quad (9.9.5)$$

Hence, using this in (9.9.3), that equation becomes:

$$\tilde{f}^i = h^i{}_j h^r{}_s F^j{}_r k^s - E^i (u_s k^s) \ . \quad (9.9.6)$$

Now we show that, if we define

$$H^i = \overset{*}{F}{}_r{}^i u^r \quad \text{(so that } H^i u_i = 0) \quad (9.9.7)$$

where:

$$\overset{*}{F}{}^{ij} = \frac{1}{2} \eta^{ijmn} F_{mn} \quad (9.9.8)$$

is the dual of F^{ij} (cf. (1.17.18)), then

$$h^i{}_j h^{rs} F^j{}_r = -\eta^{isab} H_a u_b . \quad (9.9.9)$$

From (9.9.7), (9.9.8) and (1.17.9) (we have here $\epsilon = -1$), we have in fact

$$\eta^{iabc} H_i = \frac{1}{2} \delta^{abc}_{jmn} F^{mn} u^j = u^a F^{bc} + u^c F^{ab} + u^b F^{ca} . \quad (9.9.10)$$

Contracting the latter with u_a, we obtain, from (9.9.4):

$$-\eta^{iabc} H_i u_a = F^{bc} + u^c E^b - u^b E^c \ , \quad (9.9.11)$$

but from the skewsymmetry of F_{ij} we also have:

$$F^{bc} = h^b{}_m h^c{}_n F^{mn} + u^b E^c - u^c E^b , \quad (9.9.12)$$

therefore, combining (9.9.12) with (9.9.11) we obtain (9.9.9) as wanted. Using (9.9.9) in (9.9.6) we finally have:

$$\tilde{f}^i = -\eta^{isab} H_a u_b k_s - E^i (u_r k^r) . \quad (9.9.13)$$

Since the first term on the right-hand side of this equation is totally transverse to u, we can also write it as:

$$\eta^{isab} H_a u_b h_{sn} k^n = -\eta^{isab} H_a u_b v_s (u_r k^r)$$

from (9.6.9). Hence we finally have:

$$\tilde{f}^i = -(u_r k^r) \left[-\eta^{isab} H_a u_b v_s + E^i \right] . \tag{9.9.14}$$

This is manifestly the Lorentz force. Its modulus takes on the familiar three-dimensional form; we have in fact:

$$\begin{aligned} \tilde{f} &= \left(\tilde{f}^i \tilde{f}_i \right)^{\frac{1}{2}} \\ &= -(u_r k^r) \left[E^2 - 2\eta^{isba} H_b u_a v_s E_i + \frac{v^2}{c^2} H^2 - \left(v^i H_i \right)^2 \right]^{\frac{1}{2}} \end{aligned} \tag{9.9.15}$$

where $E^2 = E^i E_i$, $H^2 = H^i H_i$.

In the three-dimensional space spanned by the vectors orthogonal to u, the quantity $\eta^{isba} u_a = \epsilon^{isb}$ is the three-dimensional anticommuting tensor; therefore Equation (9.9.15) can be written as:

$$\tilde{f} = -(u_r k^r) \left| \left(\vec{E} - \frac{\vec{v}}{c} \times \vec{H} \right) \right| . \tag{9.9.16}$$

This shows that the moduli of the quantities E^i and H^i defined respectively, in (9.9.4) and (9.9.7) measure the electric and the magnetic fields in the rest frame of the observer u.

In terms of the space-like vectors E^i and H^i, the electromagnetic tensor F_{ij} admits the following decomposition:

$$F_{ij} = -\eta_{ijrs} H^r u^s + 2u_{[i} E_{j]} . \tag{9.9.17}$$

From this it is straightforward to show that:

$$\frac{1}{2} F_{ij} F^{ij} = H^2 - E^2 \tag{9.9.18}$$

$$\frac{1}{2} F_{ij} \overset{*}{F}{}^{ij} = H_a E^a . \tag{9.9.19}$$

9.10 The properties of a fluid

A physical observer who studies the behaviour of a relativistic fluid, which is in general away from thermodynamic equilibrium, has to specify, in his or her own rest frame, the parameters which invariantly characterize the fluid and determine its evolution. For this purpose one would need a full thermodynamical treatment (Israel, 1963; Anile, 1989) which would however bring us well beyond the scope of this book. We shall instead only outline some of the results in the case of a simple fluid, namely when the deviations from local equilibrium are small and the self-gravity of the fluid is neglected.

If T_{ij} is the energy-momentum tensor of the fluid (cf. Sect. 6.5) and u is the four-velocity of the observer, the following decomposition holds (Ellis, 1971):

$$T_{ij} = \tilde{T}_{ij} + 2u_{(i}\tilde{P}_{j)} + u_i u_j \rho \tag{9.10.1}$$

where:

$$\tilde{T}_{ij} \equiv h_i{}^r h_j{}^s T_{rs} \tag{9.10.2}$$

$$\tilde{P}_i \equiv -h_i{}^r T_{rs} u^s \tag{9.10.3}$$

$$\rho \equiv T_{rs} u^r u^s . \tag{9.10.4}$$

An insight into the physical interpretation of these quantities stems from the definition (7.2.7) of the four-momentum of an extended body. If we choose the space-like hypersurface Σ in such a way that its unit normal is parallel to u, then the quantity

$$P^i = -T^i{}_j u^j \tag{9.10.5}$$

turns out to be the *four-momentum density* of the fluid. With respect to the observer u this four-vector can then be decomposed as $P^i = h^i{}_j P^j + \pi^i{}_j P^j$. The first term, namely:

$$\tilde{P}^i = -h^i{}_j T^j{}_r u^r \tag{9.10.6}$$

is the *three-dimensional energy flux density* and describes processes like convection, radiation, heat transfer (thermoconduction) (Landau and Lifschitz, 1959; Novikov and Thorn, 1972). The parallely projected quantity:

$$\pi^i{}_j P^j = u^i T^{rs} u_r u_s \tag{9.10.7}$$

describes an energy density current and its modulus

$$\left(-\pi_{ij} P^i P^j\right)^{\frac{1}{2}} = T^{rs} u_r u_s = \rho \tag{9.10.8}$$

is the *energy density* of the fluid relative to u. The remaining transverse quantity in (9.10.2), being a symmetric tensor in a three-dimensional space, can be written as:

$$\tilde{T}_{ij} = \tilde{\Psi}_{ij} + \frac{1}{3} h_{ij} \tilde{T} \tag{9.10.9}$$

where $\tilde{\Psi}_{ij}$ is a traceless tensor and $\tilde{T} = \tilde{T}^i{}_i$ is the trace of $\tilde{T}_{ij}$. The quantity $\tilde{\Psi}_{ij}$ is termed the *viscous stress tensor* and describes non isotropic dissipative processes; to the assumed accuracy (namely small deviations from local equilibrium) the viscous stresses enter the energy-momentum tensor as linear perturbations to the equilibrium configuration and are given in terms of the fluid shear with a coefficient called the shear viscosity. The trace $\tilde{T}$ is only related to uniform properties of the fluid; in our case it can be written as:

$$\tilde{T} = 3(p + \tilde{\varphi}) \tag{9.10.10}$$

where p is the hydrostatic pressure and $\tilde{\varphi}$ is a relativistic contribution to the pressure arising from the appearance of a volume (or bulk) viscosity. This property vanishes in the non-relativistic limit. Thus the energy-momentum of a fluid can be decomposed with respect to a given observer u as:

$$T_{ij} = \tilde{\Psi}_{ij} + 2u_{(i}\tilde{P}_{j)} + h_{ij}(p + \tilde{\varphi}) + u_i u_j \rho. \tag{9.10.11}$$

We define as *comoving* with the fluid that observer with respect to whom the quantities $\tilde{P}^i$ describe only processes of thermoconduction, the fluid elements having zero momentum, (Landau and Lifschitz, 1959). His or her four-velocity will be hereafter identified as the four-velocity of the fluid and the four-vector $\tilde{P}$ will be denoted by q as is customarily done in the specialized literature. In the case of a perfect fluid (i.e. $\tilde{\Psi}_{ij} = 0$, $\tilde{\varphi} = 0$, $q_i = 0$) the energy-momentum tensor becomes:

$$T_{ij} = (p + \rho) u_i u_j + g_{ij} p. \tag{9.10.12}$$

A perfect fluid is termed a *dust* if, in the comoving frame, $p = 0$.

Let us show, as an application of the projecting operators, how a fluid of dust appears to an arbitrary observer described by a congruence of curves with tangent field ℓ. From (9.10.4) and (9.6.7) we have, in the frame of ℓ:

$$\rho' = T_{ij}\ell^i\ell^j = \rho(u_i\ell^i)^2 = \left(1 - \frac{v^2}{c^2}\right)^{-1}\rho \tag{9.10.13}$$

where v is the instantaneous velocity of the fluid relative to ℓ. From (9.10.3) and (9.6.9) we also have:

$$\tilde{P}'^i = -h'^i{}_j T^j{}_r \ell^r = -h'^i{}_j \rho u^j u_r \ell^r = (u^r\ell_r)^2 v^i \rho = \frac{\rho v^i}{(1 - \frac{v^2}{c^2})} \tag{9.10.14}$$

where $h'_{ij} = g_{ij} + \ell_i\ell_j$, and finally from (9.10.2) and (9.6.9):

$$\tilde{T}_{ij} = h'_i{}^r h'_j{}^r \rho u_r u_s = (u_r\ell^r)^2 v_i v_j \rho \tag{9.10.15}$$

whose trace is from (9.6.8):

$$\tilde{T} = \rho\left(1 - \frac{v^2}{c^2}\right)^{-1}\frac{v^2}{c^2}. \tag{9.10.16}$$

A non-zero pressure is then measured in the frame of ℓ and given by

$$p' = \frac{1}{3}\rho\left(1 - \frac{v^2}{c^2}\right)^{-1}\frac{v^2}{c^2} \tag{9.10.17}$$

which, combined with (9.10.13), gives an equation of state:

$$p' = \frac{1}{3}\rho'\frac{v^2}{c^2}. \tag{9.10.18}$$

The limit $v \to c$ provides the expected result of $p' \to \frac{1}{3}\rho'$; this limit is nonetheless independent of the equation of state of the fluid in its rest frame. In fact, assuming some relationship $p = p(\rho)$, we have in the frame of the observer ℓ:

$$\rho' = \gamma^2(\rho + \beta^2 p) \quad , \quad p = \frac{1}{3}\beta^2\gamma^2\rho + \frac{1}{3}p(\gamma^2\beta^2 + 3) \quad ,$$

$$\beta = \frac{v}{c} \quad , \quad \gamma = \left(1 - \frac{v^2}{c^2}\right)^{-\frac{1}{2}}.$$

From these we have straightforwardly:

$$\frac{p'}{\rho'} = \frac{1}{3}\frac{\beta^2(\rho + p) + 3p\gamma^{-2}}{\rho + \beta p} \xrightarrow{v \to c} \frac{1}{3}.$$

An interesting application is to the energy-momentum tensor of an electromagnetic field. From (6.7.2) we have:

$$T_{ij} = \frac{1}{4\pi}(F_{ir}F_j{}^r - \frac{1}{4}g_{ij}F_{rs}F^{rs}). \tag{9.10.19}$$

An arbitrary observer u will measure:

i) an energy density:

$$\epsilon = T_{ij}u^i u^j = \frac{1}{8\pi}(E^2 + H^2) \tag{9.10.20}$$

from (9.9.17) and (9.9.18);

ii) a momentum density:

$$\tilde{P}^i = -h^i{}_j T^j{}_r u^r = -\eta^{iabc}E_a H_b u_c = -S^i \tag{9.10.21}$$

where S^i is the Poynting four-vector;

iii) a uniform pressure:

$$p = \frac{1}{3}\tilde{T} = \frac{1}{3}\epsilon \tag{9.10.22}$$

from (9.9.18) and (9.10.20) as expected.

9.11 The equations of motion of a fluid

The dynamics of a fluid (see Lichnerowicz, 1967) are relatively simple when one assumes quasi-stationary conditions: in this case the space-time gradients of the bulk viscosity, the heat conduction and the shear viscosity are considered to be quadratic in the deviations from equilibrium and therefore neglected. In the case of transient phenomena however this assumption is not justified as it would lead to unphysical results such as non-hyperbolic equations for the propagation of heat (Israel, 1976). A completely self-consistent description of a non-perfect fluid in general relativity is still a matter of debate so we shall not embark here on a discussion on this topic, but we shall only deduce the equations of motion of a fluid, referring for simplicity to a comoving observer, and leave to the interested reader the implications according to different views (Landau and Lifschitz, 1959; Israel and Stewart, 1979).

From (9.10.11) we have:

$$T_{ij} = \tilde{\Psi}_{ij} + 2u_{(i}q_{j)} + g_{ij}(p + \tilde{\varphi}) + (\rho + p + \tilde{\varphi})u_i u_j . \qquad (9.11.1)$$

Besides the parameters which have already been introduced, other parameters need to be given in order to characterize the behaviour of the fluid in the space-time. These are:

n: the baryon number density*.

m_b: mean rest mass of the baryonic constituents of the fluid assuming that a fluid element is not a single particle (a hydrogen atom, say) but an average collection of the particle components of the fluid.

$\rho_0 \equiv m_b n$: mean rest-mass density of the fluid.

s: local entropy per baryon.

An important assumption is that in the fluid there are no chemical reactions, so m_b is constant. The thermodynamic properties of the fluid are determined by equations of state which are relations of the type:

$$\rho = \rho(n, s) \qquad p = p(n, s) \; . \qquad (9.11.2)$$

The dynamic equations of the fluid are given by:

$$\nabla_j T_i^{\ j} = 0 \qquad (9.11.3)$$

$$\nabla_i (n u^i) = 0 \; . \qquad (9.11.4)$$

Equation (9.11.3) decomposes into a transverse and parallel part, relative to the comoving observer; let us then calculate the transverse part:

$$h^{ij}\nabla_k T^k{}_j = h^{ij}\nabla_k\left(\tilde{\Psi}^k{}_j\right) + q^k(\sigma_k{}^i + \omega_k{}^i)$$

$$+ h^{ij}\dot{q}_j + \frac{4}{3}q^i\Theta + h^{ij}\partial_j(p + \tilde{\varphi}) + (\rho + p + \tilde{\varphi})\dot{u}^i = 0 \; ; \qquad (9.11.5)$$

here the kinematic parameters of the fluid congruence, Θ (volume expansion) and σ_{ij} (shear), are defined from (4.5.7′) as:

$$\Theta = \nabla_k u^k \qquad \sigma_{ij} = h^r{}_i h^s{}_j \nabla_{(s} u_{r)} - \frac{1}{3} h_{ij}\Theta \; . \qquad (9.11.6)$$

*Note that n is a scalar, not a density in the mathematical sense. It is defined as the density in a comoving frame.

In the case of a perfect fluid, namely when we have:

$$\tilde{\Psi}_{ij} = 0, \ \tilde{\varphi} = 0, \ q_i = 0$$

equation (9.11.5) simplifies considerably and becomes (the Euler equation):

$$\dot{u}^i = -\frac{h^{ij}\partial_j p}{\rho + p} \ . \tag{9.11.7}$$

In a general relativistic perfect fluid, the pressure contributes to overcoming the pressure gradient as much as the energy density, enhancing the gravitational pull towards collapse as compared to the Newtonian case. An obvious consequence of (9.11.7) is that a perfect fluid made of incoherent matter (dust: $p = 0$) moves along geodesics ($\dot{u} = 0$).

Let us now consider the parallely projected component of Eqn. (9.11.3):

$$u^i \nabla_j T_i{}^j = u^i \nabla_j \tilde{\Psi}_i{}^j - q_i \dot{u}^i - \nabla_i q^i - \dot{\rho} - (\rho + p + \tilde{\varphi})\Theta = 0. \tag{9.11.8}$$

This equation gives the rate of change of the energy density of the fluid as it evolves. Combining this with the first law of thermodynamics which can be expressed in the form

$$\dot{\rho} = \frac{\rho + p}{n}\dot{n} + nT\dot{s} \tag{9.11.9}$$

and Equation (9.11.4), we obtain the rate of change of the local entropy per baryon as the fluid evolves:

$$nT\dot{s} = -\tilde{\varphi}\Theta - q_i \dot{u}^i - \nabla_i q^i + u^i \nabla_j \tilde{\Psi}_i{}^j. \tag{9.11.10}$$

Hence heat conduction and viscosity are responsible for entropy generation in the fluid.

9.12 The small curvature limit

A situation of great interest is when the curvature of the background geometry is sufficiently small that one can neglect non-linear terms up to a certain degree of approximation. This degree is only fixed by the physics of the problem under consideration according to a general approximation scheme where curvature effects of a given order of magnitude can be singled out and expressed in a way suitable for experimental verification.

Let the space-time be generated by a physical system which is gravitationally bound and weakly relativistic. The first requirement implies, from the virial theorem, that all forms of energy density within the system must not exceed the maximum value of the Newtonian potential in it. If L is a typical size of the system and ρ_0 an average mass density, then the above condition can be written as:

$$\Phi \equiv \rho_0 L^2 G \geq \; v^2 \;, \; \frac{p}{\rho_0} \;, \; \text{etc.}$$

where v is the order of magnitude of the velocities of the matter within the system, p the uniform pressure and so on. When the system is gravitationally bound, then the background curvature tends to zero sufficiently far from the source (i.e. $g_{ij} \rightarrow \eta_{ij}$), the leading term in the difference from η_{ij} corresponding to a radiative component which may arise from gravitational waves being emitted by the system itself.

The second requirement, namely that of the source being weakly relativistic, is ensured when the length scale of the curvature is everywhere small compared to L, i.e.

$$\epsilon^2 \equiv \frac{\rho_0 L^2 G}{c^2} \ll 1. \tag{9.12.1}$$

Hence this implies that $v^2/c^2 \ll 1$; that is only slow motion (as compared to light) is permitted within the system. The approximation scheme which follows under this requirement is consequently termed the *weak field and slow motion.*

Our task is to find a solution for the space-time metric generated by a source in this approximation; this is equivalent to restricting our analysis to space-time points which are at distances from the source small compared to the typical wavelength of the gravitational radiation emitted by the source itself (*the near zone*, that is $L \ll \lambda_{\text{G.W.}} \sim (c/v)L$). Within this region in fact, the time scales of appreciable variations of the system, i.e. $\sim L/v$, are much larger than the light-crossing time, i.e. $\sim L/c$, therefore the time derivatives of any quantity, being related to the motion within the system, can be neglected, to first order in ϵ, with respect to the space ones; i.e.:

$$\partial_0 = \mathrm{O}(\epsilon) \qquad \partial_\alpha = \mathrm{O}(\epsilon^0).$$

Assuming that all physical quantities describing the system are of the order of ϵ^2, it is convenient to expand the terms entering Einstein's equations in a power series of ϵ and analyse the contribution to the curvature by terms of increasing order in that parameter. To zeroth order in ϵ, the background metric is simply Minkowski's η_{ij} since the first contribution to the gravitational potential is the Newtonian $\sim \epsilon^2$; hence in a suitable coordinate system, we can write (Einstein et al., 1938):

$$g_{ij} = \eta_{ij} + h_{ij} \tag{9.12.2}$$

where h_{ij} are functions of the coordinates and are at most of the order of ϵ^2. These perturbation terms will contribute to the metric g_{ij} with even and odd powers of ϵ. Now it turns out, after investigating the behaviour in ϵ of the energy-momentum tensor (to be seen shortly), that to sufficiently low order, h_{ij} contribute with terms even in ϵ to g_{00} and $g_{\alpha\beta}$, and with terms odd in ϵ to $g_{0\alpha}$. This is true only up to fifth order in ϵ when radiation reaction terms come into play. These in fact were found to contribute evenly in ϵ to $g_{0\alpha}$ and oddly to the others (Misner et al., 1973); we shall limit our analysis only to the lowest order in ϵ. In this case the metric tensor can be expanded as follows:

$$\begin{aligned} g_{00} &= -1 + \underset{2}{h}_{00} + \cdots \\ g_{0\alpha} &= \underset{3}{h}_{0\alpha} + \cdots \\ g_{\alpha\beta} &= \delta_{\alpha\beta} + \underset{2}{h}_{\alpha\beta} + \cdots \end{aligned} \tag{9.12.3}$$

where $\underset{\mathrm{N}}{h}_{ij} = \mathrm{O}(\epsilon^{\mathrm{N}})$.

The h_{ij}'s in (9.12.2) do not behave like a tensor except under Lorentz transformations. The inverse metric reads, to first order in h_{ij}:

$$g^{ik} = \eta^{ik} + \mathcal{F}^{ikrs} h_{rs} + \mathrm{O}(h^2_{mn}) \tag{9.12.4}$$

where, from the condition $g^{ik} g_{kj} = \delta^i_j$, we have:

$$\mathcal{F}^{ikrs} h_{rs} = -\eta^{ir}\eta^{ks} h_{rs} \equiv -h^{ik}. \tag{9.12.5}$$

To lowest order in ϵ then, the indices of h_{ij} will be raised and lowered by the Minkowski metric η^{ij}. From (9.12.2) and (9.12.3) it follows that:

$$\sqrt{-g} = 1 + \frac{1}{2} \underset{2}{h} + \mathrm{O}(\epsilon^3) \tag{9.12.6}$$

where $\underset{2}{h} = \underset{2}{h}{}^i{}_i = \mathrm{O}(\epsilon^2)$. The metric forms (9.12.2), (9.12.4) and (9.12.6) are preserved by infinitesimal coordinate transformations of the type:

$$x'^i = x^i + \xi^i \tag{9.12.7}$$

where $|\xi^i|$ and $|\partial_k \xi^i|$ are at most of the order of h_{ij}. We have in fact:

$$g'_{ij} = \eta_{ij} + h_{ij} - 2\partial_{(j}\xi_{i)} + \mathrm{O}(h^2_{rs}) \equiv \eta_{ij} + h'_{ij} + \mathrm{O}(h^2_{rs}); \tag{9.12.8}$$

we shall later use this arbitrariness to impose suitable *gauge* conditions on the field equations.

Assume now that the source is a perfect fluid with an energy-momentum tensor given by:

$$T_i^{\ j} = (\rho + p)u_i u^j + p\delta_i^j \tag{9.12.9}$$

where ρ, p and u^i are respectively the energy density, uniform pressure and four-velocity of the fluid. Let $\{\lambda_{\hat{\alpha}}\}$ be a smooth field of tetrad frames attached to a family of observers $X \equiv \lambda_{\hat{0}}$ gravitationally bound to the system. From the previous discussion and the various definitions of physical measurements so far introduced (see Sects. 9.6 and 9.10) we have, to the lowest order in ϵ, and assuming that v^2/c^2 is of the order of ϵ^2 (it may be in fact much less):

$$\begin{aligned}
T_{\hat{0}\hat{0}} &= (\rho + p)(u^i \lambda_{i\hat{0}})^2 - p = (\rho + p)\left(1 - \frac{v^2}{c^2}\right)^{-1} - p = \rho + \mathrm{O}(\epsilon^2) \\
T_{\hat{0}\hat{\beta}} &= (\rho + p)(u^i \lambda_{i\hat{0}})^2 \frac{v^{\hat{\beta}}}{c}\left(1 + \frac{p}{\rho} + \frac{v^2}{c^2}\right) = \rho \frac{v^{\hat{\beta}}}{c} + \mathrm{O}(\epsilon^3) \\
T_{\hat{\alpha}\hat{\beta}} &= (\rho + p)(u^i \lambda_{i\hat{0}})^2 \frac{v^{\hat{\alpha}}}{c}\frac{v^{\hat{\beta}}}{c} + p\delta_{\hat{\alpha}\hat{\beta}} \\
&= \rho\left(\frac{v^{\hat{\alpha}}}{c}\frac{v^{\hat{\beta}}}{c} + \frac{p}{\rho}\delta^{\hat{\alpha}\hat{\beta}}\right) + \mathrm{O}(\epsilon^4)
\end{aligned} \tag{9.12.10}$$

where we recall that $u^i \lambda_{i\hat{0}} = -(1 - v^2/c^2)^{-\frac{1}{2}}$ from (9.8.4) and $p/\rho = \mathrm{O}(\epsilon^2)$. Furthermore, since $\kappa = 8\pi G/c^4$, we have $\kappa\rho = \mathrm{O}(\epsilon^2)$, hence, the lowest-order contributions from the source term in Einstein's equations are:

$$\kappa T_{\hat{0}\hat{0}} = \mathrm{O}(\epsilon^2) \quad , \quad \kappa T_{\hat{0}\hat{\beta}} = \mathrm{O}(\epsilon^3) \quad , \quad \kappa T_{\hat{\alpha}\hat{\beta}} = \mathrm{O}(\epsilon^4). \tag{9.12.11}$$

Let us now restrict the observers to be quasi-stationary and put:

$$\lambda^i_{\hat{0}} = (-g_{00})^{-\frac{1}{2}} \delta^i_0 = \delta^i_0 \left(1 + \frac{1}{2} \underset{2}{h}_{00} + \cdots\right)$$

$$\lambda_{i\hat{0}} = \begin{cases} \lambda_{0\hat{0}} = -1 + \dfrac{1}{2} \underset{2}{h}_{00} + \cdots \\ \lambda_{\rho\hat{0}} = \underset{3}{h}_{0\rho} + \cdots \end{cases} \tag{9.12.12}$$

$$\lambda^i_{\hat{\alpha}} = \delta^i_\alpha \left(1 - \frac{1}{2} \underset{2}{h}_{\alpha|\alpha|} + \cdots\right)$$

$$\lambda_{i\hat{\alpha}} = \begin{cases} \lambda_{0\hat{\alpha}} = 0 \\ \lambda_{\rho\hat{\alpha}} = \delta_{\rho\alpha} \left(1 - \dfrac{1}{2} \underset{2}{h}_{|\alpha|\alpha}\right) + \underset{2}{h}_{\rho\alpha} + \cdots \end{cases} \tag{9.12.13}$$

where no summation over α is meant in (9.12.13).

From definitions (4.4.2a), (4.5.7), (4.5.8) and from (4.5.3) and (9.12.13), we have:

$$\Theta = \nabla_i \lambda^i_{\hat{0}} = \frac{1}{\sqrt{-g}} \partial_i (\sqrt{-g} \lambda^i_{\hat{0}}) = \frac{1}{2} \partial_0 \underset{2}{h}_\rho{}^\rho + \mathrm{O}(\epsilon^5) \tag{9.12.14}$$

(because $\partial_0 = \mathrm{O}(\epsilon)$), hence $\Theta = \mathrm{O}(\epsilon^3)$. Similarly we have $\dot{\lambda}_{\hat{0}} = \mathrm{O}(\epsilon^3)$ and

$$\begin{aligned} \Theta_{\hat{\beta}\hat{\sigma}} &= \lambda_{\hat{0}i;j} \lambda^i_{(\hat{\beta}} \lambda^j_{\hat{\sigma})} = \lambda_{\hat{0}(\beta,\sigma)} + \Gamma^0_{\beta\sigma} + \mathrm{O}(\epsilon^5) \\ &= \frac{1}{2} \partial_0 \underset{2}{h}_{\beta\sigma} + \mathrm{O}(\epsilon^5) \end{aligned} \tag{9.12.15}$$

$$\omega_{\hat{\beta}\hat{\sigma}} = \lambda_{\hat{0}i;j} \lambda^i_{[\hat{\beta}} \lambda^j_{\hat{\sigma}]} = \underset{3}{h}_{0[\beta,\sigma]} + \mathrm{O}(\epsilon^5). \tag{9.12.16}$$

We now write Einstein's equations (8.4.6) to (8.4.8) to the lowest order in ϵ, as:

$$\underset{2}{G}_{\hat{0}\hat{0}} = \frac{1}{2} \underset{2}{P} = \kappa\rho \quad , \quad \underset{2}{G}_{\hat{\alpha}\hat{\beta}} = \underset{2}{P}_{\hat{\alpha}\hat{\beta}} - \frac{1}{2} \delta_{\hat{\alpha}\hat{\beta}} \underset{2}{P} = \mathrm{O}(\epsilon^4), \tag{9.12.17}$$

$$\underset{3}{G}_{\hat{0}\hat{\beta}} = \partial_{\hat{\sigma}} \left(\underset{3}{\Theta}_{\hat{\beta}}{}^{\hat{\sigma}} - \underset{3}{\Theta} \delta^{\hat{\sigma}}_{\hat{\beta}} - \underset{3}{\omega}_{\hat{\beta}}{}^{\hat{\sigma}}\right) = \kappa\rho \frac{v^{\hat{\beta}}}{c} \tag{9.12.18}$$

$$\underset{4}{G}_{\hat{\alpha}\hat{\beta}} = \underset{4}{P}_{\hat{\alpha}\hat{\beta}} - \frac{1}{2} \delta_{\hat{\alpha}\hat{\beta}} \underset{4}{P} + \lambda^j_{\hat{\alpha}} \lambda^k_{\hat{\beta}} \nabla_i \left[\lambda^i_{\hat{0}} \underset{3}{\Theta}_{jk} - \delta^i_{(j} \underset{3}{\dot{\lambda}}_{k)\hat{0}} - \eta_{jk} \left(\lambda^i_{\hat{0}} \underset{3}{\Theta} - \underset{3}{\dot{\lambda}}{}^i_{\hat{0}}\right)\right]$$

$$= \kappa\rho\left(\frac{v^{\hat{\alpha}}v^{\hat{\beta}}}{c^2} + \frac{p}{\rho}\delta_{\hat{\alpha}\hat{\beta}}\right). \tag{9.12.19}$$

From (9.12.17) and (8.4.8) it follows trivially that:

$$\underset{2}{P}_{\hat{\alpha}\hat{\beta}} = \delta_{\hat{\alpha}\hat{\beta}}\kappa\rho. \tag{9.12.20}$$

The field equations become visibly more and more complicated as the order of the approximation considered becomes higher. We shall write explicitly equations (9.12.17) and (9.12.18) because their solutions are respectively the Newtonian and the first post-Newtonian contributions to the background metric.

From (8.1.10), (4.5.3), (9.12.12) and (9.12.13) we have:

$$\begin{aligned}\underset{2}{P} =& -2\underset{2}{\Gamma}^{\hat{\alpha}\hat{\beta}}{}_{[\hat{\alpha},\hat{\beta}]} = -\delta^{\beta\gamma}\underset{2}{\Gamma}_{\beta}{}^{\alpha}{}_{\alpha,\gamma} + \delta^{\beta\gamma}\underset{2}{\Gamma}_{\beta}{}^{\alpha}{}_{\gamma,\alpha}\\ =& -\nabla^2\left(\underset{2}{h}_{00} + \frac{1}{2}\underset{2}{h}\right) + \partial_\alpha\partial_\beta\left(\underset{2}{h}^{\alpha\beta} - \frac{1}{2}\delta^{\alpha\beta}\underset{2}{h}\right)\end{aligned} \tag{9.12.21}$$

where $\nabla^2 = \delta^{\alpha\beta}\partial_\alpha\partial_\beta$ is the three-dimensional Laplacian operator, and

$$\begin{aligned}\Gamma^{\alpha}_{\beta\alpha} &= \Gamma^{i}_{\beta i} - \Gamma^{0}_{\beta 0} = \frac{1}{2}\partial_\beta\underset{2}{h} + \frac{1}{2}\partial_\beta\underset{2}{h}_{00} + \mathrm{O}(\epsilon^4)\\ \delta^{\beta\gamma}\Gamma^{\alpha}_{\beta\gamma} &= \eta^{ij}\Gamma^{\alpha}_{ij} + \Gamma^{\alpha}_{00} = \partial_\beta\left(\underset{2}{h}^{\alpha\beta} - \frac{1}{2}\delta^{\alpha\beta}\underset{2}{h}\right)\\ &\quad - \frac{1}{2}\partial_\alpha\underset{2}{h}_{00} + \mathrm{O}(\epsilon^4).\end{aligned} \tag{9.12.22}$$

Thus defining the new quantities:

$$\Psi^{ij} = h^{ij} - \frac{1}{2}h\eta^{ij} \tag{9.12.23}$$

equation (9.12.17) becomes:

$$-\frac{1}{2}\nabla^2\underset{2}{\Psi}_{00} + \partial_\alpha\partial_\beta\underset{2}{\Psi}^{\alpha\beta} = \kappa\rho. \tag{9.12.24}$$

Using expressions (9.12.14), (9.12.15) and (9.12.16) in (9.12.18), we obtain:

$$\begin{aligned}&\partial_{\hat{\sigma}}\left(\underset{3}{\Theta}_{\hat{\beta}}{}^{\hat{\sigma}} - \underset{3}{\omega}_{\hat{\beta}}{}^{\hat{\sigma}} - \underset{3}{\Theta}\delta^{\hat{\sigma}}_{\hat{\beta}}\right)\\ &= \partial_\sigma\left(\frac{1}{2}\partial_0\underset{2}{h}_{\beta\sigma} - \frac{1}{2}\partial_\sigma\underset{3}{h}_{0\beta} + \frac{1}{2}\partial_\beta\underset{3}{h}_{0\sigma}\right) - \frac{1}{2}\partial_\beta\partial_0\underset{2}{h}^{\rho}{}_{\rho}\\ &= -\frac{1}{2}\nabla^2\underset{3}{h}_{0\beta} - \frac{1}{4}\partial_0\partial_\beta\underset{2}{h}_{00} + \frac{1}{2}\partial_0\partial_\sigma\underset{2}{\Psi}_{\beta}{}^{\sigma} + \frac{1}{2}\partial_\beta\left(\partial_\sigma\underset{3}{h}_0{}^{\sigma} - \frac{1}{2}\partial_0\underset{2}{h}_{\rho}{}^{\rho}\right)\end{aligned}$$

$$= \kappa\rho\frac{v_\beta}{c} \ . \qquad (9.12.25)$$

Einstein's equations (9.12.24) and (9.12.25) can be simplified if we exploit the freedom in the choice of h_{ij} given by the infinitesimal coordinate transformation (9.12.7), and the corresponding invariance of the curvature tensor. Since

$$h'_{ij} = h_{ij} - 2\partial_{(i}\xi_{j)} \qquad (9.12.26)$$

we have:

$$\xi_\alpha = \underset{2}{\xi}_\alpha + \mathrm{O}(\epsilon^4) \quad , \quad \xi_0 = \underset{3}{\xi}_0 + \mathrm{O}(\epsilon^5) \qquad (9.12.27)$$

and hence to the lowest order in ϵ:

$$\begin{aligned} &\underset{2}{h}'_{00} = \underset{2}{h}_{00} \quad , \quad \underset{3}{h}'_{0\alpha} = \underset{3}{h}_{0\alpha} - \partial_0\underset{2}{\xi}_\alpha - \partial_\alpha\underset{3}{\xi}_0 \\ &\underset{2}{h}'_{\alpha\beta} = \underset{2}{h}_{\alpha\beta} - 2\partial_{(\alpha}\underset{2}{\xi}_{\beta)} \quad , \quad \underset{2}{h}'{}_\rho{}^\rho = \underset{2}{h}{}_\rho{}^\rho - 2\partial_\rho\underset{2}{\xi}^\rho . \end{aligned} \qquad (9.12.28)$$

Taking into account that:

$$\partial_{0'} = \partial_0 + \mathrm{O}(\epsilon^3) \quad , \quad \partial_{\alpha'} = \partial_\alpha + \mathrm{O}(\epsilon^2) \qquad (9.12.29)$$

we have:

$$\begin{aligned} \partial_{0'}\underset{2}{h}'{}_\rho{}^\rho &= \partial_0\underset{2}{h}{}_\rho{}^\rho - 2\partial_0\partial_\rho\underset{2}{\xi}^\rho \\ \partial_{\rho'}\underset{3}{h}'{}_0{}^\rho &= \partial_\rho\underset{3}{h}{}_0{}^\rho - \partial_0\partial_\rho\underset{2}{\xi}^\rho - \nabla^2\underset{3}{\xi}_0 \ . \end{aligned}$$

Hence:

$$\partial_{\rho'}\underset{3}{h}'{}_0{}^\rho - \frac{1}{2}\partial_{0'}\underset{2}{h}'{}_\rho{}^\rho = \partial_\rho\underset{3}{h}{}_0{}^\rho - \frac{1}{2}\partial_0\underset{2}{h}{}_\rho{}^\rho - \nabla^2\underset{3}{\xi}_0 . \qquad (9.12.30)$$

Similarly from:

$$\begin{aligned} \partial_{\rho'}\underset{2}{h}'{}^{\rho\sigma} &= \partial_\rho\underset{2}{h}{}^{\rho\sigma} - \nabla^2\underset{2}{\xi}^\sigma - \partial^\sigma\partial_\rho\underset{2}{\xi}^\rho \\ \partial_{\sigma'}\underset{2}{h} &= \partial_\sigma\underset{2}{h} - 2\partial_\sigma\partial_\rho\underset{2}{\xi}^\rho \end{aligned}$$

it follows that:

$$\partial_{\rho'}\left(\underset{2}{h}'{}^{\rho\sigma} - \frac{1}{2}\delta^{\rho\sigma}\underset{2}{h}\right) = \partial_\rho\left(\underset{2}{h}{}^{\rho\sigma} - \frac{1}{2}\delta^{\rho\sigma}\underset{2}{h}\right) - \nabla^2\underset{2}{\xi}^\sigma \qquad (9.12.31)$$

therefore a gauge $\left(\underset{3}{\xi}^0, \underset{2}{\xi}^\alpha\right)$ can be found which leads to the following conditions (we drop primes):

$$\begin{aligned} &a) \qquad \partial_\rho\underset{3}{h}{}_0{}^\rho - \frac{1}{2}\partial_0\underset{2}{h}{}_\rho{}^\rho = 0 \\ &b) \qquad \partial_\beta\underset{2}{\Psi}{}^{\alpha\beta} = 0 \end{aligned} \qquad (9.12.32)$$

from (9.12.30) and (9.12.31), (Chandrasekhar, 1965). Using these, Einstein's equations, to the orders $O(\epsilon^2)$ and $O(\epsilon^3)$, become:

$$-\frac{1}{2}\nabla^2 \underset{2}{\Psi}_{00} = \kappa\rho \tag{9.12.33}$$

$$-\frac{1}{2}\nabla^2 \underset{3}{h}_{0\beta} - \frac{1}{4}\partial_0\partial_\beta \underset{2}{h}_{00} = \kappa\rho\frac{v^\beta}{c} \,. \tag{9.12.34}$$

The first of these admits as a solution which vanishes at infinity:

$$\underset{2}{\Psi}_{00} = -\frac{4}{c^4}\Phi \tag{9.12.35}$$

where

$$\Phi(\vec{x}, x^0) = -\frac{G}{c^2}\int_{\Sigma^3} \frac{\rho(x')\, d^3x'}{|\vec{x} - \vec{x}'|} \tag{9.12.36}$$

is the Newtonian potential at the point $\vec{x}$ and at the time x^0, the integration being extended over the source at the same time x^0.

The contributions of the order $O(\epsilon^2)$ to the metric, are then:

$$\underset{2}{h}_{00} = -\frac{2\Phi}{c^2} \quad ; \quad \underset{2}{h}_{\alpha\beta} = -\frac{2\Phi}{c^2}\delta_{\alpha\beta}. \tag{9.12.37}$$

Of these, only the term $\underset{2}{h}_{00}$ gives rise to a purely Newtonian mechanics; therefore the metric form:

$$\underset{\mathrm{N}}{g}_{00} = -1 + \underset{2}{h}_{00} \quad , \quad \underset{\mathrm{N}}{g}_{0\alpha} = \underset{\mathrm{N}}{g}_{\alpha\beta} = 0 \tag{9.12.38}$$

is termed the *Newtonian limit* of (9.12.2). The term $\underset{2}{h}_{\alpha\beta}$ is the first post-Newtonian correction to the metric, while the solutions $\underset{3}{h}_{0\alpha}$, which vanish at infinity, are given by:

$$\underset{3}{h}_{0\alpha} = -\frac{1}{2c^3}(\, 7V_\alpha + W_\alpha \,) \tag{9.12.39}$$

where

$$V_\alpha(\vec{x}, x^0) = G\int_{\Sigma^3} \frac{\rho(\vec{x}', x^0)v^\alpha(\vec{x}', x^0)}{|\vec{x} - \vec{x}'|} d^3x'$$
$$W_\alpha(\vec{x}, x^0) = G\int_{\Sigma^3} \frac{\rho(\vec{x}', x^0)}{|\vec{x} - \vec{x}'|^3}\,[(\vec{x} - \vec{x}')\cdot\vec{v}] \times (x^\alpha - x'^\alpha)d^3x' \tag{9.12.40}$$

are obtained with standard methods of solution (Landau and Lifschitz, 1962).

9.13 Gravitational radiation

Sufficiently far from the source (when $T_{ij} = 0$) and at distances greater than $\lambda_{\text{G.W.}}$ (the radiation zone), the time variations cannot be neglected with respect to space variations, so the study of the curvature should be reconsidered.

Let us restrict the analysis to linear terms in h_{ij}, since these give the leading contribution to the curvature. From (9.12.2) and (3.2.3) it follows to first order in h_{ij} (Pirani, 1965):

$$\begin{aligned} a) \quad & R_{ijkl} = \partial_l \partial_{[i} h_{j]k} - \partial_k \partial_{[i} h_{j]l} \\ b) \quad & R_{ik} = \frac{1}{2}\left[-\Box_\eta h_{ik} + \partial_i \partial_l \Psi^l{}_k + \partial_k \partial_l \Psi^l{}_i\right] \\ c) \quad & R = -\Box_\eta h + \partial_k \partial_l h^{kl} \end{aligned} \tag{9.13.1}$$

where $\Box_\eta = \eta^{ij}\partial_i\partial_j$ is the flat D'Alembertian wave operator, $h = \eta^{ij}h_{ij}$ and Ψ_{ij} is given by (9.12.23). Einstein's equations in this case read:

$$R_{ij} = 0. \tag{9.13.2}$$

The freedom in the choice of the coordinates, assured by (9.12.7), allows us to simplify (9.13.2) considerably. The quantities Ψ_{ij} transform as follows:

$$\Psi'^{ij} = \Psi^{ij} - \eta^{ik}\partial_k\xi^j - \eta^{jk}\partial_k\xi^i + \eta^{ij}\partial_k\xi^k + \mathrm{O}(h^2_{rs}) \tag{9.13.3}$$

and hence their divergence transforms, to the same accuracy, as:

$$\partial_{j'}\Psi'^{ij} = \partial_j\Psi^{ij} - \Box_\eta\xi^i. \tag{9.13.4}$$

We can then specialize the four functions ξ^i to be solutions of the non-homogeneous wave equation $\Box_\eta\xi^i = \partial_j\Psi^{ij}$ so that

$$\partial_{j'}\Psi'^{ij} = 0. \tag{9.13.5}$$

Working in this gauge (termed the *Lorentz gauge*) and dropping primes, equation (9.13.2) becomes:

$$\Box_\eta h_{ij} = 0. \tag{9.13.6}$$

This is a flat wave equation showing that small metric perturbances propagate with a wave character at the velocity of light.

Relation (9.13.4) assures that more freedom is left in the choice of the coordinates, since (9.13.5) and (9.13.6) are left unchanged by coordinate transformations of the type of (9.12.7) where ξ^i are homogeneous harmonic functions, that is solutions of the wave equation:

$$\Box_\eta \xi^i = 0. \tag{9.13.7}$$

These considerations imply naturally that the ten functions h_{ij}, solution of (9.13.6), can be constrained by eight supplementary conditions, (9.13.5) and (9.13.7), which leave the metric of a gravitational wave depending only on two parameters.

Since we are in the radiation zone of a material system, we can seek a particular solution of (9.13.6) in terms of retarded plane waves, namely depending on one space coordinate (say, x^1):

$$h_{ij} = f_{ij}(x^0 - x^1). \tag{9.13.8}$$

Similarly $\Psi_i{}^j$ has the same behaviour, hence (9.13.5) implies:

$$\partial_0 \Psi_i{}^0 + \partial_1 \Psi_i{}^1 = 0; \tag{9.13.9}$$

but now we have $\partial_0 = -\partial_1$, hence:

$$\partial_0 \left(\Psi_i{}^0 - \Psi_i{}^1\right) = 0. \tag{9.13.10}$$

Neglecting the non-radiative part of the metric, the latter implies:

$$\Psi_i{}^0 = \Psi_i{}^1 \tag{9.13.11}$$

that is:

$$\Psi_0{}^0 = \Psi_0{}^1 \quad \Psi_1{}^0 = \Psi_1{}^1 \quad h_2^0 = h_2^1 \quad h_3^0 = h_3^1. \tag{9.13.12}$$

From the first two relations in (9.13.12), we deduce:

$$h_{22} + h_{33} = 0 \quad h_{01} = -\frac{1}{2}(h_{00} + h_{11}). \tag{9.13.13}$$

We make a further transformation constrained by (9.13.7), that is, with $\xi^i = f^i(x^0 - x^1)$, so the relevant h_{ij}'s transform as follows:

$$\begin{aligned} h'_{00} &= h_{00} - 2\partial_0\xi_0 \ ; \ h'_{20} = h_{20} - \partial_0\xi_2 \\ h'_{11} &= h_{11} + 2\partial_0\xi_1 \ ; \ h'_{30} = h_{30} - \partial_0\xi_3 \\ h'_{22} &= h_{22} \ ; \ h'_{33} = h_{33} \\ h'_{23} &= h_{23}. \end{aligned} \tag{9.13.14}$$

We then see that all the metric coefficients can be made to vanish except:

$$h_{22} = -h_{33} \ , \quad h_{23} \ . \tag{9.13.15}$$

Thus the metric of a plane gravitational wave propagating in the x^1- direction is given by:

$$ds^2 = -c^2dt^2 + (dx^1)^2 + (1 + h_{22})(dx^2)^2 + (1 - h_{22})(dx^3)^2 + 2h_{23}\, dx^2 dx^3. \tag{9.13.16}$$

The metric perturbance is confined to the (x^2, x^3)-plane, which, due to the almost Euclidean background geometry, can be taken orthogonal to the x^1-direction of the wave propagation. Furthermore the trace h is zero, hence the gauge in which (9.13.16) holds is termed *transverse traceless.* In this case a gravitational wave has two degrees of polarization. Let us write the solution (9.13.8) as:

$$h_{ab} = \mathcal{A}_{ab} \mathrm{e}^{\mathrm{i}k_r x^r} \tag{9.13.17}$$

where k_r is the wave vector one-form and $\mathcal{A}_{ab}$ is a constant amplitude, termed the polarization tensor. The traceless condition on h_{ij} implies $\mathcal{A}^r{}_r = 0$ while condition (9.13.5) leads to:

$$\mathcal{A}^{ab} k_b = 0. \tag{9.13.18}$$

The field equation (9.13.6) moreover is equivalent to:

$$\mathrm{i}\Theta_{\text{G.W.}} \mathcal{A}_{ab} - k_r k^r \mathcal{A}_{ab} = 0 \tag{9.13.19}$$

where $\Theta_{\text{G.W.}} = \partial_r k^r$ is the isotropic expansion of the beam of curves along which the gravitational plane wave propagates. Clearly (9.13.19) implies:

$$\Theta_{\text{G.W.}} = 0 \ , \quad k_r k^r = 0 \tag{9.13.20}$$

hence our approximation leads to a propagation along null rays. This latter condition together with (9.13.18) makes the curvature tensor, which in our case ($R_{ij} = 0$) coincides with the Weyl tensor, equal to:

$$R_{ijkl} = k_{[k} \mathcal{A}_{l][j} k_{i]} \tag{9.13.21}$$

from (9.13.1a). Thus it follows (cf. Sect. 5.10) that:

$$R_{ijkl}k^l = 0 \ , \tag{9.13.22}$$

that is the curvature tensor is algebrically special of Petrov type N characterising the pure radiative character of the solution (9.13.16).

If the gravitational wave is generated by a system of moving sources confined well within the near zone as discussed in the previous section, then we need to consider Einstein's equations with sources. A general solution is difficult to handle, but restricting ourselves to sources which are weakly relativistic in the sense specified in the previous section, we achieve a considerable simplification. If in (9.12.17), (9.12.18) and (9.12.19) we consider to the desired order only linear terms in the metric perturbations h_{ij} and absorb the non-linear terms into the source together with the energy-momentum tensor, we have for the field equations:

$$-\frac{1}{2}\Box_\eta \Psi_{\hat{a}\hat{b}} = \kappa \tilde{T}_{\hat{a}\hat{b}} \tag{9.13.23}$$

from (9.13.1b) and (9.13.1c), and the condition (9.13.5). Here $\tilde{T}_{\hat{a}\hat{b}}$ is a complex, that is the sum of the energy-momentum tensor with the non-linear terms of $G_{\hat{a}\hat{b}}$. From (9.13.5) it follows that:

$$\partial_{\hat{a}} \tilde{T}^{\hat{a}\hat{b}} = 0$$

but now no more freedom is left as in the sourceless case and given by the radiation gauge (9.13.7), because the latter does not preserve (9.13.5). Hence the wave equation (9.13.23) admits six independent solutions which we take as the $\Psi_{\hat{\alpha}\hat{\beta}}$'s.

A standard solution is that of retarded potentials:

$$\Psi_{\hat{\alpha}\hat{\beta}}(\vec{x}, x^0) = \frac{\kappa}{2\pi} \int_{\Sigma^3} \frac{\tilde{T}_{\hat{\alpha}\hat{\beta}}(x^0 - |\vec{x} - \vec{x}'|; \vec{x}')}{|\vec{x} - \vec{x}'|} d^3x' \tag{9.13.24}$$

where $(\vec{x}, x^0)$ are the coordinates of the event where the metric coefficients are calculated and $(\ \vec{x},\ x^0 - |\vec{x} - \vec{x}'|\ )$ are those of the source; the integration is carried over the three-dimensional hypersurface Σ^3 at the retarded time $x^0 - |\vec{x} - \vec{x}'|$. From (9.12.17) and (9.12.18) we see that, to the second and third order in ϵ, $\tilde{T}_{\hat{0}}{}^{\hat{0}}$ and $\tilde{T}_{\hat{0}}{}^{\hat{\alpha}}$ coincide with the energy-momentum tensor, while to fourth order in ϵ, $\tilde{T}_{\hat{\alpha}\hat{\beta}}$ contains non-linear terms from $G_{\hat{\alpha}\hat{\beta}}$. At this order then, equation (9.13.24) is in fact an integral equation since $\Psi_{\hat{\alpha}\hat{\beta}}$ are contained also in $\tilde{T}_{\hat{\alpha}\hat{\beta}}$, however in our case it still provides a solution; in fact exploiting (9.13.24) and applying Gauss' theorem, we obtain to the order $O(\epsilon^4)$:

$$\int_{\Sigma^3} \underset{4}{\tilde{T}}_{\hat{\alpha}\hat{\beta}}\Big|_{\text{ret}} d^3x' = \frac{1}{2}\partial_0^2 \int_{\Sigma^3} \underset{2}{T}_{00}\Big|_{\text{ret}} x^\alpha x^\beta \, d^3x'. \qquad (9.13.25)$$

If we restrict (9.13.24) to the radiation zone, we can approximate $|\vec{x} - \vec{x}'|^{-1} \sim \mathrm{R}^{-1}$, R being an average distance from some origin of the coordinates taken inside the source, giving:

$$\underset{4}{\Psi}_{\alpha\beta}(\mathrm{R}, t) = \frac{2G}{c^4\mathrm{R}} \frac{\partial^2}{\partial t^2} \int_{\Sigma^3(x^0 - \mathrm{R})} \rho_0 x^\alpha x^\beta \, d^3x + O(\epsilon^5/\mathrm{R}^2) \quad (9.13.26)$$

where we put $x^0 = ct$. The integral on the right-hand side is directly related to the mass quadrupole moment of the source, hence the lowest-order contribution to the dynamical components of the metric comes from the rate of change with time of the quadrupole moment of the system and is of the order $O(\epsilon^4)$.

Detectable curvature effects induced by gravitational radiation on material systems (gravitational antennas) are at least $O(\epsilon^6)$. Nevertheless detectable effects on the source due to radiation reaction (for example a period decrease in a binary system) appear as corrections of the order $O(\epsilon^5)$ (Chandrasekhar and Esposito, 1970; Thorne, 1969).

Consider a family of stationary observers in Minkowski space-time, $X^i = \delta^i_0$. Their world-lines are geodesics of the metric η_{ij} but remain so also when a plane gravitational wave impinges on them. In fact, from (9.13.16) we have:

$$\dot{X}^i = \Gamma^i_{00} = 0. \qquad (9.13.27)$$

If $\{\lambda_{\hat{a}}\}$ is a smooth tetrad field attached to this congruence, we find, from the geodesic deviation equations (8.2.12) and to first order in h_{ij}:

$$\ddot{\xi}^{\hat{\alpha}} = E^{\hat{\alpha}}{}_{\hat{\beta}}\xi^{\hat{\beta}} \qquad \hat{\alpha}\ ,\ \hat{\beta} = 2, 3 \qquad (9.13.28)$$

where $E^{\hat{\alpha}}{}_{\hat{\beta}}$ is the electric part of the Weyl tensor (8.1.23) given by:

$$E_{\hat{3}\hat{3}} = -E_{\hat{2}\hat{2}} = -\frac{1}{2}\frac{\partial^2 h_{22}}{\partial x^{0^2}} \quad ; \quad E_{\hat{2}\hat{3}} = -\frac{1}{2}\frac{\partial^2 h_{23}}{\partial x^{0^2}} \tag{9.13.29}$$

from (9.13.1a) and (9.13.16).

If the congruence of curves under consideration is representative of a test material system (a collection of particles, say), then (9.13.28) tells us that the body undergoes unisotropic stresses in the (x^2, x^3)-plane, determined by the rate of change with time of the shear, since from (8.2.9) and to first order in h_{ij}, we have:

$$\partial_0 \sigma_{\hat{\alpha}\hat{\beta}} = E_{\hat{\alpha}\hat{\beta}}. \tag{9.13.30}$$

Thus a gravitational wave excites, to the lowest order, the shear anisotropy of the system. A much more subtle and extensive theory of gravitational radiation measurements allows us to establish a close relationship between the shear $\sigma_{\alpha\beta} \sim O(\epsilon^5)$ induced in the detecting device and the power radiated away by the source. The complexity of this subject, however, suggests that any further discussion here would add nothing to the already vast specialized literature (e.g., Thorne, 1980; Misner et al., 1973).

10

SPHERICALLY SYMMETRIC SOLUTIONS

10.1 The spherically symmetric line element

Spherical symmetry has played an important role in the development of general relativity. The exact solutions of Einstein's field equations which provided the decisive experimental verification of the theory, namely the Schwarzschild external solution and the Robertson-Walker-Friedmann cosmological solutions, were found under the assumption that space-time was spherically symmetric.

A space-time is said to be *spherically symmetric* if, colloquially expressed, it is possible to rotate it leaving its metric (and any other non-metric fields) unchanged. In more precise terms, for every rotation R (a 3×3 rotation matrix) in the rotation group SO(3), there is an isometry of the space-time $\varphi(\mathrm{R})$ and the isometries constitute what is called an *action* of the group, meaning that the composition of the isometries corresponds to the composition of the corresponding rotations:

$$\varphi(\mathrm{R})\varphi(\mathrm{S}) = \varphi(\mathrm{RS}). \tag{10.1.1}$$

Moreover, we require that the action "looks like" the action of the rotation group in three dimensions, in that each of the sets: $\mathrm{O}_x = \{\varphi(\mathrm{R})(x) \mid \mathrm{R} \in \mathrm{SO}(3)\}$, consisting of the *orbit* of all points into which x can be rotated, is a two-dimensional sphere or possibly a point (the latter corresponding to the centre of rotation). If we just consider the *natural action* of SO(3) in three-dimensional space, where R maps $x \mapsto \mathrm{R}x$, then we can distinguish three one-parameter families of rotations, whose members we denote by $\mathrm{R}_{u,x}$, $\mathrm{R}_{u,y}$ and $\mathrm{R}_{u,z}$ where these are rotations through u about the x (resp. y, z) axes and u parametrizes the family. To each family corresponds a Killing vector (cf. Sec. 2.13) which turn out to be:

$$\underset{1}{\xi} = z\partial_y - y\partial_z \qquad \underset{2}{\xi} = x\partial_z - z\partial_x \qquad \underset{3}{\xi} = y\partial_x - x\partial_y \ . \qquad (10.1.2)$$

It is more convenient to use polar coordinates r, θ, φ given by

$$x = r\sin\theta\cos\varphi \quad , \quad y = r\sin\theta\sin\varphi \quad , \quad z = r\cos\theta \ . \qquad (10.1.2')$$

From these, (10.1.2) becomes:

$$\begin{aligned} \underset{1}{\xi} &= \sin\varphi\partial_\theta + \cot\theta\cos\varphi\partial_\varphi \\ \underset{2}{\xi} &= -\cos\varphi\partial_\theta + \cot\theta\sin\varphi\partial_\theta \\ \underset{3}{\xi} &= -\partial_\varphi \ . \end{aligned} \qquad (10.1.3)$$

If we now take the corresponding families $\varphi(R_{u,x})$, etc., of isometries of the space-time, then these too will have their Killing vectors; call them $\underset{i}{\xi}'$. It can be shown that (10.1.1) implies that the commutators (Lie derivatives) of the Killing vectors correspond: if $[\underset{i}{\xi}, \underset{j}{\xi}] = C^k_{ij}\underset{k}{\xi}$, then: $[\underset{i}{\xi}', \underset{j}{\xi}'] = C^k_{ij}\underset{k}{\xi}'$. Now in the case of the rotation group acting with spherical orbits, it turns out that we can actually choose coordinates (t, r, θ, φ) in the space-time (perhaps for a limited range of r and t) such that the functional form of the isometry $\varphi(R)$, when expressed in terms of these coordinates, is identical to the form of the action of R on three-dimensional Euclidean space in spherical polar coordinates, together with t remaining constant. In this case the coordinate expressions of $\underset{i}{\xi}'$ are the same as the expression (10.1.3) for their Euclidean counterparts. Since this is so, we drop the prime from them.

From (10.1.3) and the isometry condition:

$$\left(\pounds_{\underset{\alpha}{\xi}}\overset{*}{\boldsymbol{g}}\right)_{ij} = \underset{\alpha}{\xi}^k\partial_k g_{ij} + g_{ki}\partial_j\underset{\alpha}{\xi}^k + g_{kj}\partial_i\underset{\alpha}{\xi}^k = 0 \ ; \quad \alpha = 1, 2, 3 \quad (10.1.4)$$

we have, after some algebra:

$$\partial_\varphi g_{ij} = 0 \qquad (10.1.5)$$

$$g_{r\theta} = g_{r\varphi} = g_{\theta\varphi} = g_{0\theta} = g_{0\varphi} = 0 \qquad (10.1.6)$$

$$\partial_\theta g_{rr} = \partial_\theta g_{\theta\theta} = \partial_\theta g_{0r} = \partial_\theta g_{00} = 0 \qquad (10.1.7)$$

$$g_{\varphi\varphi} = g_{\theta\theta}\sin^2\theta \qquad (10.1.8)$$

Thus the only non-vanishing components of the metric tensor are: g_{00}, g_{rr}, $g_{\theta\theta}$, $g_{\varphi\varphi}$, g_{0r} and they are functions of r and t only. The line element becomes:

$$ds^2 = g_{00}dt^2 + g_{0r}drdt + g_{rr}dr^2 + g_{\theta\theta}\left(d\theta^2 + \sin^2\theta d\varphi^2\right) \ . \quad (10.1.9)$$

There is some arbitrariness in the choice of the coordinates r and t; a coordinate transformation of the type

$$r' = r'(r,t) \ , \quad t' = t'(r,t) \ , \quad \theta = \theta' \ , \quad \varphi = \varphi' \qquad (10.1.10)$$

enables us to transform away the metric coefficient g_{0r} in (10.1.9); moreover from (10.1.8), the last of (10.1.3) and recalling that $\underset{\alpha}{\xi}$ are space-like, we have always $g'_{\theta\theta}(r',t') = g_{\theta\theta}(r,t) > 0$, hence there is no loss of generality if we name

$$r' = \sqrt{g_{\theta\theta}(r,t)} \qquad (10.1.11)$$

and this leads to the canonical spherically symmetric line element:

$$ds^2 = -e^{\nu(r,t)}dt^2 + e^{\lambda(r,t)}dr^2 + r^2\left(d\theta^2 + \sin^2\theta d\varphi^2\right) \qquad (10.1.12)$$

where we have dropped the prime and defined $g_{00} = -e^{\nu}$, $g_{rr} = e^{\lambda}$. The two-space $t =$const.,$r =$const., with metric

$$d\Sigma_{_2}^2 = g_{\theta\theta}\left(d\theta^2 + \sin^2\theta d\varphi^2\right)$$

in (10.1.12) has Gaussian curvature which agrees with the Gaussian curvature of a two-sphere of radius $g_{\theta\theta}^{1/2}$ in $\mathbb{R}^3$, so the identification (10.1.11) of $g_{\theta\theta}^{1/2}$ as a radial coordinate is justified.

10.2 The external Schwarzschild solution

The choice of the coordinates which lead to (10.1.12) reduces to two the metric coefficients to be determined by the field equations, namely $\nu(r,t)$ and $\lambda(r,t)$. Let us assume that the source of the space-time (10.1.12) is a bounded matter-energy distribution with energy-momentum tensor T_{ij} which, as stated previously, is spherically symmetric, that is:

$$\left(\pounds_{\underset{\alpha}{\xi}}T\right)_{ij} = 0 \qquad \alpha = 1,2,3. \qquad (10.2.1)$$

The same arguments which lead to (10.1.6) can be applied here to deduce that T_{ij} are only functions of r and t and the only non-vanishing components are T_{rr}, T_{00}, $T_{\theta\theta}$, $T_{\varphi\varphi}$, T_{0r}. Moreover we require the space-time to be asymptotically flat, in the sense that, at large r, the metric (10.1.12) should take on the special relativistic form which is deduced from the Minkowski metric after performing the transformation (10.1.2′). Thus we impose:

$$\lim_{r\to\infty} \nu(r,t) = \lim_{r\to\infty} \lambda(r,t) = 0. \qquad (10.2.2)$$

The non-zero connection coefficients for the metric (10.1.12) are:

$$\Gamma^r_{rr} = \frac{\lambda'}{2} , \; \Gamma^0_{00} = \frac{\dot\nu}{2} , \; \Gamma^0_{0r} = \frac{\nu'}{2} , \; \Gamma^\theta_{\varphi\varphi} = -\sin\theta\cos\theta ,$$

$$\Gamma^0_{rr} = \frac{\dot\lambda}{2}\mathrm{e}^{\lambda-\nu} , \; \Gamma^r_{\theta\theta} = -r\mathrm{e}^{-\lambda} , \; \Gamma^r_{00} = \frac{\nu'}{2}\mathrm{e}^{\nu-\lambda} , \; \Gamma^\theta_{r\theta} = \Gamma^\varphi_{r\varphi} = \frac{1}{r} ,$$

$$\Gamma^\varphi_{\theta\varphi} = \cot\theta , \; \Gamma^r_{0r} = \frac{\dot\lambda}{2} , \; \Gamma^r_{\varphi\varphi} = -r\sin^2\theta\mathrm{e}^{-\lambda} ,$$

where $(') \equiv \partial_r$ and $(\dot{}) \equiv \partial_t$. Einstein's equations now read:

$$a) \qquad \mathrm{e}^{-\lambda}\left(\frac{\nu'}{r} + \frac{1}{r^2}\right) - \frac{1}{r^2} = \kappa T^r{}_r(r,t)$$

$$b) \qquad \mathrm{e}^{-\lambda}\left(\frac{1}{r^2} - \frac{\lambda'}{r}\right) - \frac{1}{r^2} = \kappa T^0{}_0(r,t)$$

$$c) \qquad \mathrm{e}^{-\lambda}\frac{\dot\lambda}{r} = \kappa T^r{}_0(r,t)$$

$$d) \qquad \frac{1}{2}\mathrm{e}^{-\lambda}\left(\nu'' + \frac{\nu'^2}{2} + \frac{\nu'-\lambda'}{r} - \frac{\nu'\lambda'}{2}\right) - \frac{1}{2}\mathrm{e}^{-\nu}\left(\ddot\lambda + \frac{\dot\lambda^2}{2} - \frac{\dot\nu\dot\lambda}{2}\right) = \kappa T^\theta{}_\theta = \kappa T^\varphi{}_\varphi , \qquad (10.2.3)$$

where $\kappa = 8\pi G/c^4$. Let us call r_0 the smallest value of the radial coordinate such that:

$$T_{ij} = 0 \qquad \text{at} \;\; r > r_0 \qquad (10.2.4)$$

(r_0 existing since we are assuming the matter to be bounded). Hence, from (10.2.3c) we have, when $r > r_0$:

$$\dot\lambda = 0 \qquad (10.2.5)$$

namely outside the matter source $\lambda(r,t)\Big|_{r>r_0} \equiv \lambda(r)$ is independent of time. Subtracting (10.2.3b) from (10.2.3a) we obtain:

$$\nu' = -\lambda' + \kappa r e^{\lambda}({T_r}^r - {T_0}^0). \tag{10.2.6}$$

Integrating over r from $r \geq r_0$ to infinity and using (10.2.2) and (10.2.4) gives:

$$\nu(r,t) = -\lambda(r) \qquad r \geq r_0 \tag{10.2.7}$$

showing that, outside the matter distribution, ν as well as λ is time-independent, and hence so is the whole metric. This result is known as Birkoff's theorem: *if a space-time contains a region which is spherically symmetric, asymptotically flat and empty ($T_{ij} = 0$) for $r > r_0$, then the metric in this region is time-independent and hence independent of the dynamical properties of its source*, (Birkoff, 1923).

If the distribution of matter is regular down to $r = 0$, the centre of symmetry (which, as we shall see, excludes, among other things, the case of a black-hole), then we can integrate (10.2.6) from 0 to r, obtaining:

$$\nu(r,t) + \lambda(r,t) = \nu(0,t) + \lambda(0,t) + \kappa \int_0^r x e^{\lambda(x,t)}({T_r}^r - {T_0}^0)\, dx. \tag{10.2.8}$$

Setting $r = r_0$ and comparing with (10.2.7) gives:

$$\nu(0,t) + \lambda(0,t) = -\kappa \int_0^{r_0} x e^{\lambda(x,t)}({T_r}^r - {T_0}^0)\, dx. \tag{10.2.9}$$

The boundary conditions at $r = 0$ can be derived from the following *local regularity* requirement. Consider the loop defined by:

$$r = \epsilon \ \text{(small)} \ , \quad t = \text{const.} \ , \quad \theta = \frac{\pi}{2}$$

with φ parametrizing the loop, running from 0 to 2π. The metric length of this loop is $2\pi\epsilon$, which tends to zero as ϵ tends to zero. Hence, if this loop is to be in a regular region of space-time, the Lorentz transformation defined by parallely propagating a tetrad around the loop must tend to the identity. (The alternative, when the transformation tends to a non-identity limit, could in principle be caused by a singular thread of matter at $r = 0$.) Calculating the transformation from (10.2.3) shows that we need:

$$\lambda(r) \to 0 \qquad (r \to 0) ; \tag{10.2.10a}$$

then, from (10.2.9) together with the regularity of T, we deduce that:

$$\nu(r) \to 0 \qquad (r \to 0) . \tag{10.2.10b}$$

We can then solve equation (10.2.3b) with respect to λ; it can be written as:

$$\left(\mathrm{e}^{-\lambda}\right)' + \frac{1}{r}\left(\mathrm{e}^{-\lambda}\right) - \frac{1}{r} - \kappa r T_0{}^0 = 0, \tag{10.2.11}$$

thus, taking into account (10.2.10), a general solution at $r > r_0$ is:

$$\mathrm{e}^{-\lambda(r)} = 1 + \frac{\kappa}{r}\int_0^{r_0} x^2 T_0{}^0\, dx \qquad r > r_0 . \tag{10.2.12}$$

Let us put:

$$M = -\frac{4\pi}{c^2}\int_0^{r_0} x^2 T_0{}^0\, dx \tag{10.2.13}$$

so (10.2.12) becomes:

$$\mathrm{e}^{-\lambda(r)} = 1 - \frac{2MG}{c^2 r}. \tag{10.2.14}$$

The quantity M clearly does not depend on r, nor does it depend on t, from (10.2.5), hence it is a constant. From (10.2.14) and (10.2.7), the metric outside a spherically symmetric source becomes:

$$\begin{aligned} ds^2 = &- \left(1 - \frac{2GM}{c^2 r}\right) dt^2 \\ &+ \left(1 - \frac{2GM}{c^2 r}\right)^{-1} dr^2 + r^2\left(d\theta^2 + \sin^2\theta\, d\varphi^2\right). \end{aligned} \tag{10.2.15}$$

This is the external Schwarzschild solution, (Schwarzschild, 1916a). It depends only on the parameter M, whose physical interpretation is obtained if we consider the small curvature (weak field) limit. In that limit, in fact, we have:

$$ds^2 \approx - \left(1 - \frac{2GM}{c^2 r}\right) dt^2 + \left(1 + \frac{2GM}{c^2 r}\right) dr^2 + r^2\left(d\theta^2 + \sin^2\theta\, d\varphi^2\right). \tag{10.2.16}$$

Comparing this with (9.12.37), we recognize M as being the total mass which enters the Newtonian potential at asymptotically large distances from the source.

10.3 The internal Schwarzschild solution

Inside the metric source, when $T^i{}_j \neq 0$, the considerations which lead to Birkoff's theorem do not hold. The metric coefficients depend in general on t and r and one needs a specific knowledge of $T^i{}_j$ to solve Einstein's equations. A relatively simple solution for the space-time interior of a spherically symmetric configuration is attained when the latter is a perfect fluid and static. In this case:

$$T_i^{\ j} = (\rho + p)u_i u^j + p\delta_i^j \qquad \rho = \rho(r) \qquad p = p(r) \ , \tag{10.3.1}$$

where ρ and p are the energy density and the uniform pressure as measured in the rest frame of the fluid, and u is the fluid four-velocity which we assume to be:

$$u = \mathrm{e}^{-\frac{\nu}{2}}\partial_t . \tag{10.3.2}$$

From (10.3.1) and (10.3.2) we have:

$$T_0^{\ 0} = -\rho \qquad T_r^{\ r} = T_\theta^{\ \theta} = T_\varphi^{\ \varphi} = p \ , \tag{10.3.3}$$

all the remaining components being zero. From (10.2.3c), it then follows that:

$$\lambda(r,t) \equiv \lambda(r)\Big|_{r<r_0} \tag{10.3.4}$$

and so, integrating (10.2.6) from $r = 0$ to some $r < r_0$, we obtain:

$$\nu(r,t) + \lambda(r) = \nu(0,t) + \lambda(0) + \kappa \int_0^r r' \mathrm{e}^{\lambda(r')}[p(r') + \rho(r')]\, dr' \ . \tag{10.3.5}$$

Hence, from the boundary condition (10.2.9) it follows that under conditions (10.3.1) and (10.3.2), ν is also time-independent and is given by:

$$\nu(r) = -\lambda(r) - \kappa \int_r^{r_0} r' \mathrm{e}^{\lambda(r')}[p(r') + \rho(r')]\, dr' \qquad r < r_0 . \tag{10.3.6}$$

Equation (10.2.11) can be solved for $\lambda(r)$ and gives:

$$e^{-\lambda(r)} = 1 - \frac{\kappa}{r}\int_0^r r'^2 \rho(r')\, dr' \ . \qquad (10.3.7)$$

Thus, to find a complete solution for the interior metric, one needs to choose an equation of state for the fluid as $p = p(\rho)$, and to know the behaviour of either ρ or p as a function of r. It is convenient for this purpose to consider equation (9.11.7), which derives from the equation of motion of the fluid: $\nabla_i T^i{}_j = 0$; using equations (10.3.1) to (10.3.3) we have:

$$(\rho + p)\frac{\nu'}{2} + p' = 0. \qquad (10.3.8)$$

Equations (10.3.6) to (10.3.8) together with an equation of state, give a complete set of equations which one needs to solve for the interior metric. The solution originally found by Schwarzschild (1916b) corresponds to the assumptions:

$$\rho = \rho_0 = \text{const.} \ ; \ p = p(r) \ ; \ p(r_0) = 0. \qquad (10.3.9)$$

Using these, equation (10.3.7) becomes:

$$e^{-\lambda} = 1 - \frac{\kappa\rho_0}{3}r^2. \qquad (10.3.10)$$

Differentiating (10.3.6) (i.e. using equation (10.2.6)) and using (10.3.10) we obtain:

$$\nu'\left(1 - \frac{\kappa\rho_0}{3}r^2\right) + \frac{2}{3}\kappa\rho_0 r = \kappa r(p + \rho_0). \qquad (10.3.11)$$

Equation (10.3.8) with (10.3.9) gives:

$$p + \rho_0 = A e^{-\frac{\nu}{2}} \qquad (10.3.12)$$

where A is an integration constant which can be determined imposing the continuity of the interior metric coefficient e^{ν} with the external one, on the fluid boundary at $r = r_0$. From (10.2.15) this leads to:

$$A = \rho_0\left(1 - \frac{\kappa\rho_0 r_0^2}{3}\right)^{\frac{1}{2}} . \qquad (10.3.13)$$

Combining (10.3.12) with (10.3.11) yields:

$$\nu'\left(1 - \frac{\kappa\rho_0}{3}r^2\right) + \frac{2}{3}\kappa\rho_0 r - \kappa r A e^{-\frac{\nu}{2}} = 0 \qquad (10.3.14)$$

which admits as solution:

$$e^{\frac{\nu}{2}} = \left(1 - \frac{\kappa\rho_0}{3}r^2\right)^{\frac{1}{2}} B + \frac{3A}{2\rho_0}. \tag{10.3.15}$$

Here B is an integration constant which, by continuity with the external metric on $r = r_0$, can be fixed as $B = -\frac{1}{2}$. We finally obtain the Schwarzschild solution for the space-time inside a static, constant energy density, perfect fluid sphere:

$$\begin{aligned} ds^2 = & -\left[\frac{3}{2}\left(1 - \frac{\kappa\rho_0 r_0^2}{3}\right)^{\frac{1}{2}} - \frac{1}{2}\left(1 - \frac{\kappa\rho_0 r^2}{3}\right)^{\frac{1}{2}}\right]^2 c^2 dt^2 \\ & + \left(1 - \frac{\kappa\rho_0 r^2}{3}\right)^{-1} dr^2 \\ & + r^2\left(d\theta^2 + \sin^2\theta\, d\varphi^2\right). \end{aligned} \tag{10.3.16}$$

The behaviour of the pressure inside the body is now deduced from equations (10.3.15) and (10.3.12); we have in fact:

$$p(r) = \rho_0 \left[\frac{\left(1 - \frac{\kappa\rho_0}{3}r^2\right)^{\frac{1}{2}} - \left(1 - \frac{\kappa\rho_0}{3}r_0^2\right)^{\frac{1}{2}}}{3\left(1 - \frac{\kappa\rho_0}{3}r_0^2\right)^{\frac{1}{2}} - \left(1 - \frac{\kappa\rho_0}{3}r^2\right)^{\frac{1}{2}}}\right]. \tag{10.3.17}$$

At the center of the body, the pressure diverges when:

$$r_0 = \frac{9}{8}\frac{2GM}{c^2}, \tag{10.3.18}$$

where we put $M = \frac{4}{3}\pi r_0^3 \rho_0/c^2$. This is a singularity for the fluid configuration and also for the space-time since the curvature invariants diverge as well. The constant density assumption, although not very realistic for a normal star, is reasonably good for very dense objects, such as neutron stars where $\rho_0/c^2 \approx 10^{14}\mathrm{g/cm}^3$ and $r_0 \approx 10(2GM/c^2)$. A variety of static interior solutions have been studied but the appearance of a singularity at the centre with values of r_0 larger than $2GM/c^2$, seems to be a common feature indicating the impossibility of producing a sequence of equilibrium static configurations with arbitrarily high densities (Buchdahl, 1959, 1967).

Before concluding this section, let us apply the Komar integral, in the form (7.1.18), to the interior Schwarzschild solution. Since the latter is invariant under time translations, a time Killing one-form $\boldsymbol{\xi}$ exists which, from (10.3.16), has components $\xi^a = \delta^a_0$. Let us then choose the space-like hypersurface Σ, transverse to the fluid flow, as the one with t=const. and let the unit normal to it be $n_a = -\sqrt{-g_{00}}\delta^0_{\ a}$. From (10.3.3), (10.3.16) and (10.3.17), the integral (7.1.18) becomes:

$$\begin{aligned} P_k &= 2\int_{\Sigma\cap\sigma}\left(T^0{}_0 - \frac{1}{2}T\right)\sqrt{-g_{00}}\,|h|^{\frac{1}{2}}\,d^3x \\ &= 4\pi\int_0^{r_0}(1-\beta r^2)^{-\frac{1}{2}}\sqrt{-g_{00}}(\rho_0+3p)r^2\,dr \\ &= 4\pi\int_0^{r_0}(1-\beta r^2)^{-\frac{1}{2}}(1-\beta r^2)^{\frac{1}{2}}\rho_0 r^2\,dr \\ &= 4\pi\frac{\rho_0 r_0^3}{3} = Mc^2 \end{aligned} \tag{10.3.19}$$

from (10.2.13) with $\beta = \kappa\rho_0/3$.

10.4 The global structure of spherically symmetric space-times

In order to pursue our study of the Schwarzschild metric further we need to enlarge our viewpoint somewhat, so we revert to the general situation of Section 10.1. To enable us to give a single treatment for all cases, we shall in the case where some of the orbits of the rotation group are points (as the centre of symmetry) work with the set $M\backslash C$, where C is the set of all such points, in place of M. Thus for this section we assume that all the group orbits are spheres. The missing points can be restored later. Then through any point p there is a two-dimensional group orbit O_p. Locally one can choose coordinates (with, say, p the point $r = r_0$, $t = t_0$, $\theta = 0$) so that near p the metric has the form (10.1.12). Then setting $\theta \equiv 0$ defines a two-surface $\bar{S}_p$ parametrized by r and t orthogonal to all the other group orbits near p. It turns out that $\bar{S}_p$ can be extended to be a maximal surface S_p with the property of cutting all orbits orthogonally (Clarke, 1987).

In the simplest case each orbit intersects S exactly once. In this case, for each $q \in M$ there is a unique $p_1 \in O_p$ with $q \in S_{p_1}$ and a unique $p_2 \in S_p$ such that $S_p \cap O_q = \{p_2\}$. The map $q \mapsto (p_1, p_2)$ is then a diffeomorphism between M and $O_p \times S_p$; since each orbit is an S^2, we express this by saying that $M = S^2 \times S$, the direct (or Cartesian) product of S^2 with $S \equiv S_p$.

In the only other case each orbit intersects S exactly twice and S_q is doubly connected: there is a two–to–one map from a simply connected surface $\tilde{S}_p$ to S_p. This extends to a two–to–one map from a covering space $\tilde{M}$ to M, and $\tilde{M}$ is a spherically symmetric space-time falling within the previous case.

Now any two-dimensional space-time is locally conformal to flat space (see (3.4.24)); i.e. locally there are coordinates u, v such that ${}^2\overset{*}{g}$, the metric induced on S, is:

$$ {}^2\overset{*}{g}(x) = \Omega(x)^{-2} \boldsymbol{dudv} \tag{10.4.1} $$

for some function Ω. Hence the metric on M is locally:

$$ \overset{*}{g} = \Omega(x)^{-2} \boldsymbol{dudv} + r(u,v)^2 \left(\boldsymbol{d}\theta^2 + \sin^2\theta \, \boldsymbol{d}\varphi^2\right). \tag{10.4.2} $$

Since the curves u =const., v =const. are null geodesics, we can set up the following prescription for defining the coordinates u, v.

i) Pick a point p in S and find the null geodesics through p. In terms of the local coordinate representation (10.1.12) this involves solving:

$$ \frac{dt}{dr} = \pm e^{\lambda(r,t)-\nu(r,t)}. \tag{10.4.3} $$

There will be two solutions, corresponding to the two signs: pick one of them, and denote the geodesic by $u \mapsto \gamma(u)$ where u is an affine parameter on γ such that $\gamma(0) = p$.

ii) Similarly for each u, find the null geodesic $u \mapsto \boldsymbol{\lambda}_u(v)$ through $\gamma(u)$ transverse to γ, choosing the parameter v on $\boldsymbol{\lambda}_0$ to be affine and such that $\lambda_0(0) = p$.

iii) For each v, find the null geodesic $u \mapsto \boldsymbol{\mu}_v(u)$ through $\lambda_0(v)$ transverse to $\boldsymbol{\lambda}_0$. (Note that $\boldsymbol{\mu}_0 = \gamma$ providing we parametrize $\boldsymbol{\mu}_0$ appropriately).

Then near to p, provided we restrict the ranges of u, v so that $\lambda_u(v)$ and $\mu_v(u)$ lie in a normal neighbourhood, we find that each point lies on a unique pair of null geodesics $\boldsymbol{\lambda}_u, \boldsymbol{\mu}_v$. If we assign to that point the coordinates u, v, then the metric takes the form (10.4.1).

One can use this procedure systematically to give a canonical global form for S. In many applications it turns out that, locally, one can find coordinates u', v' such that the corresponding functions Ω and r in (10.4.2) are real analytic functions of u', v'. In this case, if u and v are defined by the procedure just outlined, then since analytic functions are involved at each step, the resulting Ω and r will be analytic for u, v as well as for u', v'. Now if we are given a real analytic function on a simply connected subset of $\mathbb{R}^2$ (or more generally, on any simply connected analytic manifold), then it is uniquely determined by its value on any open subset. We can use this to adapt the construction just given to provide a way of extending a given space-time to a larger one.

Suppose we are given a spherically symmetric analytic space-time M, with a corresponding S. Suppose, moreover, that, if we carry out the procedure just outlined, then we obtain coordinates covering the whole of M with Ω and r being analytic. We can then continue with the following steps to extend M.

iv) If there exists a null geodesic in S which cannot be extended to arbitrarily large values of its affine parameter, then choose it as the geodesic γ for the construction of a new set of coordinates u', v' by steps i)-ii).

v) On the $\mathbb{R}^2$ defined by u', v' construct an analytic extention of $\Omega(u', v')$, $r(u', v')$ over a simply connected domain U.

vi) Join U onto S by identifying points with the corresponding u', v'.

vii) Go back to iv) and repeat until no further extension is possible.

For many space-times it turns out that it is possible to do this in such a way as to form a *maximum analytic extension* of M. This is defined as an analytic Hausdorff space-time M' containing M which is simply connected and satisfies the property:

if $\varphi: U \to \mathbb{R}^4$ is a coordinate neighbourhood with $\overline{\varphi(U)}$ compact and $(\overset{-1}{\varphi})^*(\overset{*}{g})$ and its inverse analytically extendible to a neighbourhood of $\overline{\varphi(U)}$, then φ can be extended to an open set $U' \subset U$ with $\overline{\varphi(U)} \subset \varphi(U')$.

There is no guarantee that any given space-time has a maximal analytic extension; but if it does have one, it is unique, in the sense that for any two such extensions M' and M'' there is a unique isometry mapping M' onto M'' and equal to the identity on M.

In the following sections we shall see how this construction, or a modification of it, is used in practice.

10.5 The extended external Schwarzschild solution

When the external Schwarzschild solution is written in the form (10.2.15), either a coefficient of the metric or a coefficient of the inverse metric diverges at $r = r_g \equiv 2GM/c^2$ and at $r = 0$. While the latter is a true singularity, the former is not because it only arises from the choice of the coordinates. A symptom of this is the fact that the curvature invariants diverge on $r = 0$ but remain finite on r_g; the simplest of them reads:

$$I(r) = R^{ijkl}R_{ijkl} = \frac{48G^2M^2}{c^4r^6} . \qquad (10.5.1)$$

On the coordinate singularity r_g, I is equal to:

$$I(r_g) = \frac{3c^8}{4G^4M^4} \qquad (10.5.2)$$

and this suggests that a suitable choice of a coordinate system could remove the apparent singularity.

With this aim we apply the extension procedure of the previous section (simplified by taking advantage of the particular features of the solution), which will yield its maximum analytical extension. The null geodesics in $S = \{\theta = \pi/2\}$ are radial, and so let us consider the family of the radial null geodesics in the metric (10.2.15) with tangent vector $k = k^t\partial_t + k^r\partial_r$, where $k^i = dx^i/d\lambda$ and $\lambda \in \mathbb{R}$. By definition we have:

$$(k|k) = -\left(1 - \frac{r_g}{r}\right)\left(\frac{dt}{d\lambda}\right)^2 + \left(1 - \frac{r_g}{r}\right)^{-1}\left(\frac{dr}{d\lambda}\right)^2 = 0. \quad (10.5.3)$$

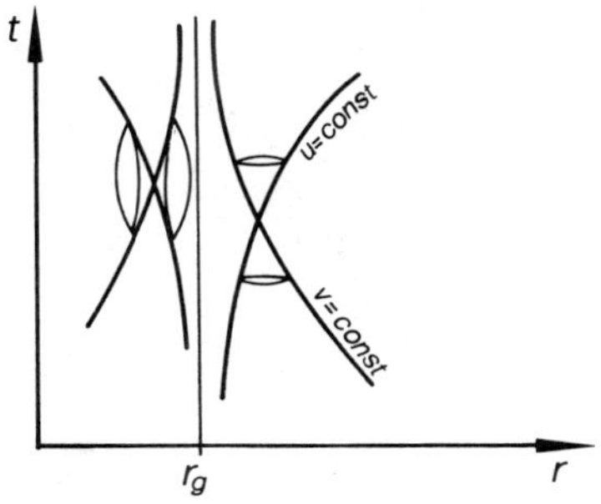

Fig. 10-1 In the Schwarzschild coordinates the description of light rays is singular at $r = r_g$.

Integration of this gives the equations of motion of the null trajectories:

$$u = t - r - r_g \ln\left(\frac{r}{r_g} - 1\right) = \text{const.}$$

$$r > r_g$$

$$v = t + r + r_g \ln\left(\frac{r}{r_g} - 1\right) = \text{const.} \qquad (10.5.4)$$

$$u = t - r - r_g \ln\left(1 - \frac{r}{r_g}\right) = \text{const.}$$

$$r < r_g$$

$$v = t + r + r_g \ln\left(1 - \frac{r}{r_g}\right) = \text{const.} \qquad (10.5.5)$$

Here v and u are constants of integration which label each null trajectory. The behaviour of the light cones in the (t, r) plane is shown in Fig. 10-1 and one easily recognizes the inadequacy of the coordinates in use, for it is impossible to follow a signal through the surface $r = r_g$, since $\lim_{r \to r_g} t = \pm\infty$; although, as we shall see, a continuation is physically possible.

Since u and v form a set of null coordinates, we can perform a coordinate transformation to any of the following sets: (v, r, θ, φ) ; (u, r, θ, φ) ; (u, v, θ, φ). Using (v, r, θ, φ) (ingoing

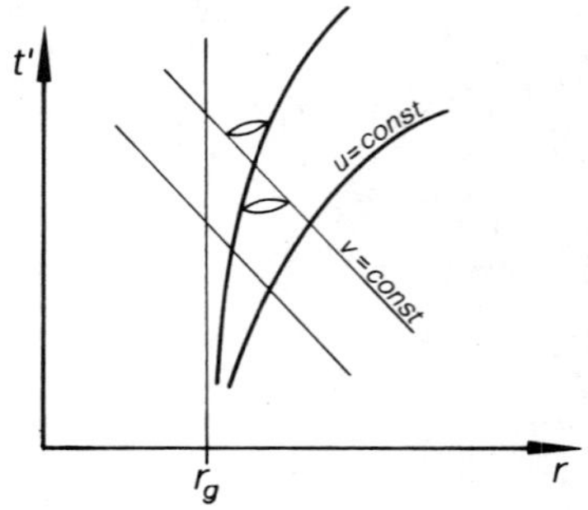

Fig. 10-2 In the ingoing Eddington-Finkelstein coodinates the description of ingoing light rays is regular at $r = r_g$; it remains singular for the outgoing.

Eddington-Finkelstein coordinates (Eddington, 1922; Finkelstein, 1958)) the space-time metric (10.2.15) becomes:

$$ds^2 = -\left(1 - \frac{r_g}{r}\right) dv^2 + 2dvdr + r^2 \left(d\theta^2 + \sin^2\theta \; d\varphi^2\right) . \quad (10.5.6)$$

Here the singularity at $r = r_g$ has been removed, but while the description of the null rays is regular for the ingoing ones, it remains singular for the outgoing:

$$\begin{aligned} v &= \text{const.} && \text{(ingoing)} \\ v &= 2\left(r + r_g \ln\left|\frac{r}{r_g} - 1\right|\right) + \text{const.} && \text{(outgoing)} . \quad (10.5.7) \end{aligned}$$

If we label $t' = v - r$, the behaviour of the light cone in the plane (t', r) is shown in Fig. 10-2.

If we use the set (u, r, θ, φ) (outgoing Eddington-Finkelstein coordinates) the metric (10.2.15) becomes:

$$ds^2 = -\left(1 - \frac{r_g}{r}\right) du^2 - 2dudr + r^2 \left(d\theta^2 + \sin^2\theta \; d\varphi^2\right) ; \quad (10.5.8)$$

in this case the null rays behave oppositely than before, namely the outgoing ones are regular through $r = r_g$ but the ingoing are not:

$$\begin{aligned} u &= -2\left(r + r_g \ln\left|\frac{r}{r_g} - 1\right|\right) + \text{const.} && \text{(ingoing)} \\ u &= \text{const.} && \text{(outgoing)} . (10.5.9) \end{aligned}$$

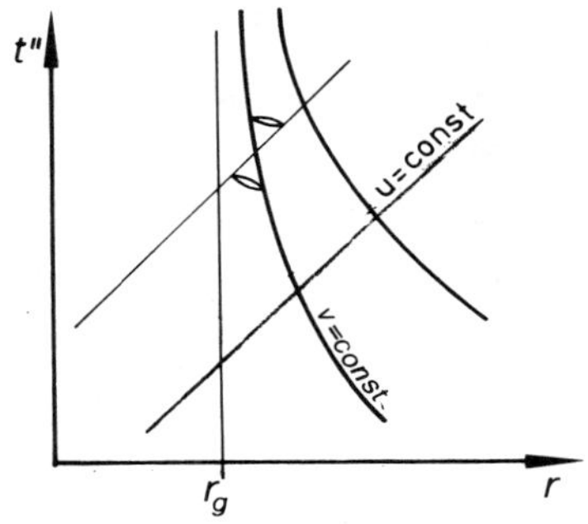

Fig. 10-3 In the outgoing Eddington-Finkelstein coordinates the description of the light rays at $r = r_g$ is opposite to that of Fig. 10-2.

If we label $t'' = u + r$, their behaviour is shown in Fig. 10-3.

Since $dt = dt' = dt''$, the time orientation is the same in all three diagrams, hence we see that, while in both the coordinate patches (t', r) and (t'', r) the surface r_g behaves as a semipermeable membrane, in the two cases it can be crossed in opposite directions. This seems to indicate that light rays propagate differently in the vicinity of the coordinate surface $r = r_g$ according to the coordinates in use. This would be a non-desirable behaviour of the space-time unless we interpret this as a manifestation of the fact that a coordinate system exists in which both situations are possible.

This is assured by the coordinate system (u, v, θ, φ), which (up to rescaling of the form $u \to u'(u)$, $v \to v'(v)$) is of the form discussed in the previous section. In these coordinates, the metric (10.2.15) becomes:

$$ds^2 = -\left(1 - \frac{r_g}{r}\right) du dv + r^2 \left(d\theta^2 + \sin^2\theta \, d\varphi^2\right) \qquad (10.5.10)$$

where $r = r(u, v)$ is the solution of the equation:

$$\frac{1}{2}(v - u) = r + r_g \ln\left|\frac{r}{r_g} - 1\right| \qquad (10.5.11)$$

from (10.5.5). The null rays are along the coordinate lines u =const., v =const. but now the divergence on $r = r_g$ can easily be removed. Any coordinate transformation $v'(v)$, $u'(u)$, where v' and u' are real, differentiable and monotonic functions of their arguments, would not modify the light-cone structure of the space-time (i.e. u =const., v =const.$\Rightarrow u'$ =const., v' =const.). Hence the logarithmic divergence of $v-u$ as a function of r (cf. (10.5.11)) on $r=r_g$ is removed with the following choice:

$$\left.\begin{aligned} u' &= -\exp\left(-\frac{u}{2r_g}\right) \\ v' &= \exp\left(\frac{v}{2r_g}\right) \end{aligned}\right\} \quad r > r_g \tag{10.5.12a}$$

$$\left.\begin{aligned} u' &= \exp\left(-\frac{u}{2r_g}\right) \\ v' &= \exp\left(\frac{v}{2r_g}\right). \end{aligned}\right\} \quad r < r_g \tag{10.5.12b}$$

From these solutions (10.5.11) becomes:

$$-u'v' = \left(\frac{r}{r_g}-1\right)\mathrm{e}^{r/r_g} \tag{10.5.13}$$

which allows one to determine $r=r(u',v')$ uniquely, the metric then taking the form:

$$ds^2 = -\frac{4r_g^3}{r}\mathrm{e}^{-(r/r_g)}du'dv' + r^2\left(d\theta^2+\sin^2\theta\, d\varphi^2\right), \tag{10.5.14}$$

with no divergences at $r=r_g$. It can be cast into a more familiar (t,r)-form if we now define $\bar{t}=1/2(v'+u')$, $\bar{r}=1/2(v'-u')$. Then from (10.5.4) and (10.5.5) we have the coordinate transformation due to Kruskal (1960):

$$\left.\begin{aligned} \bar{r} &= \left(\frac{r}{r_g}-1\right)^{\frac{1}{2}}\mathrm{e}^{r/2r_g}\cosh\left(\frac{t}{2r_g}\right) \\ \bar{t} &= \left(\frac{r}{r_g}-1\right)^{\frac{1}{2}}\mathrm{e}^{r/2r_g}\sinh\left(\frac{t}{2r_g}\right) \end{aligned}\right\} \quad r > r_g \tag{10.5.15a}$$

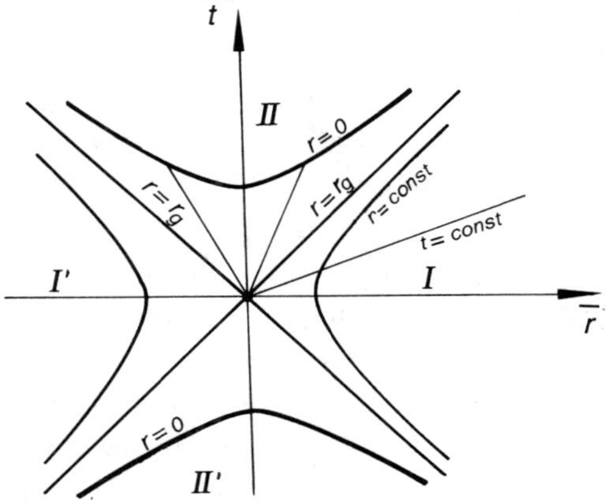

Fig. 10-4 The Kruskal diagram of the Schwarszschild space-time.

$$\bar{r} = \left(1 - \frac{r}{r_g}\right)^{\frac{1}{2}} \mathrm{e}^{r/2r_g} \sinh\left(\frac{t}{2r_g}\right)$$

$$r < r_g$$

$$\bar{t} = \left(1 - \frac{r}{r_g}\right)^{\frac{1}{2}} \mathrm{e}^{r/2r_g} \cosh\left(\frac{t}{2r_g}\right). \qquad (10.5.15\text{b})$$

In these coordinates the Schwarzschild metric becomes:

$$ds^2 = -4\frac{r_g^3}{r}\mathrm{e}^{-(r/r_g)}\left(d\bar{t}^2 - d\bar{r}^2\right) + r^2\left(d\theta^2 + \sin^2\theta\, d\varphi^2\right) \quad (10.5.16)$$

where $r = r(u', v')$. In the plane $(\bar{t}, \bar{r})$ the null rays are straight lines at 45° with respect to the coordinate lines, making explicit the fact (noted in (10.5.1)) that the light-cone structure of the two-dimensional $(\bar{t}, \bar{r})$-plane is that of two-dimensional Minkowski space-time. This is illustrated in Fig. 10-4 where the coordinate lines r =const. and t =const. are also shown.

The condition $r > r_g$ is now equivalent to $|\bar{r}| > |\bar{t}|$, so regions I and I′ are isometric to $r > r_g$ in the original (r, t) coordinate patch. Similarly regions II and II′, where $|\bar{r}| < |\bar{t}|$, but bounded by the hyperbola $\bar{t}^2 - \bar{r}^2 = 1$ which corresponds to $r = 0$, are isometric to $r < r_g$. It is now clear that, if we define the future direction, in the $(\bar{t}, \bar{r})$-patch, to be that in which $\bar{t}$ increases, future-directed light rays can cross the surface $r = r_g$ only if they are outgoing in the region II′ and only if ingoing in the region II, (where by *outgoing* we mean that $|\bar{r}|$ is increasing).

Hence we recognize that region I∪II is isometric to the space-time (10.5.6) where the coordinates (v, r, θ, φ) were used while region I∪II′ is isometric to (10.5.8) where the coordinates (u, r, θ, φ) were used. The surfaces $r = r_g$ are stationary* null surfaces and are *event horizons.* The surface $r = r_g$ which separates regions I and II is termed the *future event horizon* because no event in II can ever be to the past (and so be observed by) an event in I. The region II cannot causally influence any event in I and it is termed a *black-hole*, in view of the fact that no light can come out of it (Misner et al., 1973). The surface $r = r_g$ which separates regions I and II′ is termed the *past event horizon* and region II′ is called a *white-hole.* It is generally believed that black-holes are physically realistic since they arise from the gravitational collapse of massive configurations; on the contrary white-holes could only be of primordial origin since the singularity at $r = 0$ would have been existing prior to every event in the Universe (Lake, 1978). But there are strong arguments suggesting that if they were perturbed by inclusion of radiation, then they would turn into a black-hole due to the gravitational action of the radiation which appears, when emerging from the horizon, infinitely blue-shifted.

10.6 Penrose diagrams

Having obtained the maximal analytic extension of a spherically symmetric space-time, it turns out that it is possible to make a further transformation of the (u, v)-coordinates used to express the metric so as to make their range finite. Set:

$$\tilde{u} = \tan^{-1} u \ ; \quad \tilde{v} = \tan^{-1} v \, . \tag{10.6.1}$$

Applying this to Minkowski space, with metric:

$$ds^2 = -dudv + r^2 \left(d\theta^2 + \sin^2\theta \, d\varphi^2\right)$$

*In the sense of Sect. 7.11.

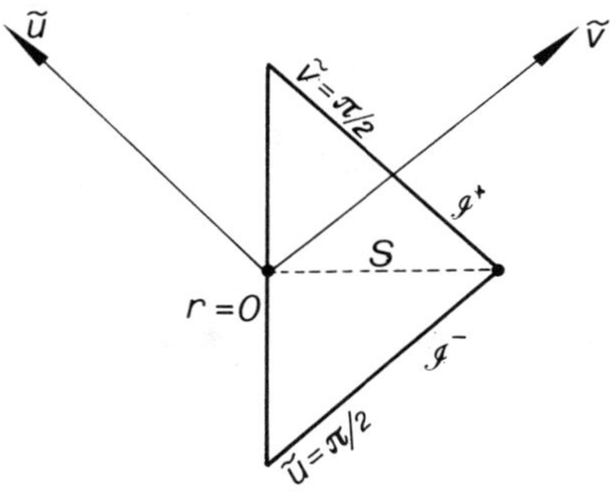

Fig. 10-5 The Penrose diagram of Minkowski space-time.

where $t = (1/2)(v+u)$, $r = (1/2)(v-u)$ gives $\tilde{u}$, $\tilde{v}$ in the range

$$-\frac{\pi}{2} < \tilde{u} < \frac{\pi}{2} \ , \quad \tilde{u} < \tilde{v} < \frac{\pi}{2}$$

with the metric on S $(\theta = 0)$ taking the form

$$(\cos\tilde{u})^{-2}(\cos\tilde{v})^{-2}\, d\tilde{u}d\tilde{v} = \Omega_0^{-2}(\tilde{u},\tilde{v})d\tilde{u}d\tilde{v}, \qquad (10.6.2)$$

say. The domain of $\tilde{u}$,$\tilde{v}$ on S is then as depicted in Fig. 10-5, where we have rotated the $(\tilde{u},\tilde{v})$-axes to maintain the convention that light rays are at 45° to the horizontal.

The coordinate boundaries at $\tilde{u} = \frac{\pi}{2}$, $\tilde{v} = \frac{\pi}{2}$ now provide "end points" for radial null geodesics in S running to infinity, the behaviour $\Omega_0 \to 0$ indicating the way that space-time has been shrunk by a factor Ω_0 near the boundaries. In interpreting the diagram, note that, according to the conventions of Sect. 10.4, the line $r = 0$ has been cut out of the space-time for the purpose of constructing S, so that S is bounded by $r = 0$ and $r = \infty$. Every point of Fig. 10-5 corresponds to a sphere of radius r in the full space-time.

To carry out the analogous transformation for the Kruskal metric we must take note of the fact that the (u,v)-coordinates are analogous to the (u,v)-coordinates in Minkowski space, and so using these will produce a conformal factor Ω similar to Ω_0. On the other hand it is the (u',v')-coordinates that cover the whole extended metric. As a result it is not possible to do a transformation which covers the whole space, produces an appropriate conformal factor and is mathematically simple. Since we will not be using the result for calculations, we sacrifice simplicity and set:

$$\tilde{u} = k(u') \quad \tilde{v} = k(v') \tag{10.6.3}$$

where

$$k(x) = \tan^{-1}\left(\frac{x}{\sqrt{1+x^2}}\ln(1+x^2)\right).$$

Note that for large v' this gives:

$$\tilde{v} \approx \tan^{-1}(2\ln v') = \tan^{-1}\left(\frac{v}{r_g}\right)$$

and similarly for large negative u', so that this corresponds to (10.6.1). Note also that for large x we have

$$xk'(x) \approx \frac{2}{[\tan(k(x))]^2} + \text{ lower order terms }. \tag{10.6.4}$$

Applying (10.6.3) to (10.5.14) gives for the metric of S:

$$\frac{4r_g^3}{r}\left(\frac{r}{r_g}-1\right)\frac{d\tilde{u}d\tilde{v}}{u'v'k'(u')k'(v')} = \Omega_s(\tilde{u},\tilde{v})^{-2}d\tilde{u}d\tilde{v}$$

(where we have used (10.5.13)) which at large values of $|u'|$ or $|v'|$ is determined, from (10.6.4), by

$$\Omega_s(\tilde{u},\tilde{v})^{-2} \approx (\tan(\tilde{u})\tan(\tilde{v}))^2 r_g^2 . \tag{10.6.5}$$

The domain of the $\tilde{u}$,$\tilde{v}$ coordinates is bounded by $-\frac{\pi}{2} < \tilde{u} < \frac{\pi}{2}$, $-\frac{\pi}{2} < \tilde{v} < \frac{\pi}{2}$ (corresponding to $-\infty < u', v' < \infty$ in (10.5.12)) and by $r = 0$, which now becomes, from (10.5.13), $u'v' = +1$ or $\tilde{v} = \frac{\pi}{2} - \tilde{u} + n\pi$. In terms of (t, r)-type coordinates defined by

$$\tilde{t} + \tilde{r} = \tilde{v} = \tan^{-1}(\bar{t} + \bar{r})$$
$$\tilde{t} - \tilde{r} = \tilde{u} = \tan^{-1}(\bar{t} - \bar{r})$$

we have

$$-\frac{\pi}{2} < \tilde{t} < \frac{\pi}{2}$$
$$-\pi + |\tilde{t}| < \tilde{r} < \pi - |\tilde{t}|.$$

The null geodesics with $r \to \infty$ terminate on the boundaries $\tilde{u} = \pm\pi/2$, $\tilde{v} = \pm\pi/2$ where, as for the Minkowski case, $\Omega_s \to 0$ with

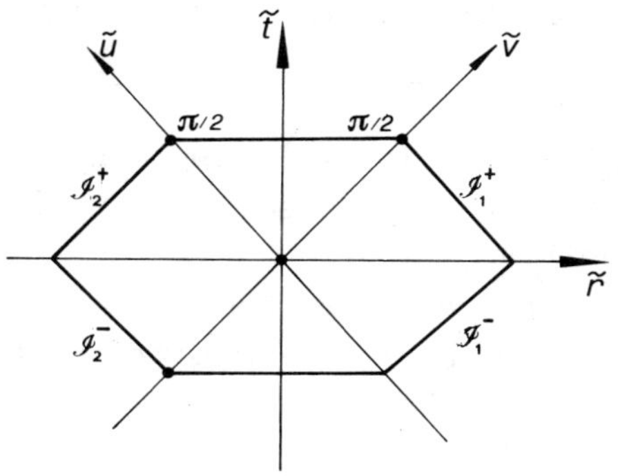

Fig. 10-6 The Penrose diagram of the Schwarzschild space-time.

$d\Omega_s$ regular and non-zero. (These boundaries are labelled $\mathcal{J}$ in Fig. 10-6).

We shall see in Sect. 10.13 that in the Reissner-Nordström solution as well one obtains boundaries at infinity which are null (45°) lines in the diagram, suggesting that this is a general feature. Indeed one can generalize the construction to the case where one no longer has spherical symmetry: just as we have been depicting S as a region in the two-dimensional plane with metric related to the Minkowski metric by a conformal factor Ω^{-2}, so one can consider realising a whole 4-dimensional space-time as a subset of a larger space-time $\tilde{M}$ with metric $\overset{*}{\tilde{g}}$, the two metrics being related by:

$$\overset{*}{\tilde{g}} = \Omega^2 \overset{*}{g}$$

M having a smooth boundary $\mathcal{J}$ in $\tilde{M}$ on which $\Omega \to 0$ (but $d\Omega \neq 0$). Such a construction is called a *conformal compactification* of M.

One can then show, by a direct calculation of the field equation, reexpressing the Riemann tensor in terms of $\tilde{g}$ and Ω, that if the space-time is vacuum near the boundary $\mathcal{J}$ (or if, as in the Reissner-Nordström solution, $T_{ij} \to 0$ sufficiently rapidly), then $\mathcal{J}$ is a null surface. The part of $\mathcal{J}$ lying to the future of M (if any) is called $\mathcal{J}^+$, while the part to the past (if any) is called $\mathcal{J}^-$.

For the cases we have studied in this section, if we extend the construction from S to the full 4-dimensional space-time M, then for Minkowski space the conformally rescaled metric is

$$\tilde{ds}^2 = \Omega_0^2 ds^2 = d\tilde{u} d\tilde{v} + r^2 \Omega_0^2 \left(d\theta^2 + \sin^2\theta \, d\varphi^2\right),$$

where we note that $r^2\Omega_0^2$ is regular and non-zero at $\mathcal{J}$. We have the same expression for the Kruskal metric, with Ω_0 replaced by Ω_s, and again $r^2\Omega_s^2$ is regular.

The asymptotic properties of M are then expressed in terms of the geometry of $(\tilde{M}, \overset{*}{\tilde{g}})$ at $\mathcal{J}$, an extensive topic for which we refer the reader to Penrose (1964).

10.7 Time-like geodesics in the external Schwarzschild solution

The Schwarzschild vacuum solution is a good description of the space-time generated by a non-rotating (spherically symmetric) star. Due to the Birkoff theorem, this extends to the case when the star is either in a very dense state (a neutron star) or is collapsed beyond the event horizon (a black-hole). The knowledge of the behaviour of the free particles in the gravitational field of these objects is of interest in astrophysics but it also illuminates the nature of the gravitational field itself, in particular in the strong field limit.

We consider here time-like geodesic motion in the metric (10.2.15), starting from the Hamilton-Jacobi equation (7.4.11):

$$-\left(1-\frac{r_g}{r}\right)^{-1}\left(\frac{\partial S}{\partial x^0}\right)^2 + \left(1-\frac{r_g}{r}\right)\left(\frac{\partial S}{\partial r}\right)^2 + \frac{1}{r^2}\left(\frac{\partial S}{\partial \theta}\right)^2 + \frac{1}{r^2\sin^2\theta}\left(\frac{\partial S}{\partial \varphi}\right)^2 = -\mu_0^2 c^2 \quad (10.7.1)$$

where μ_0 is the rest mass of the particle and c is the velocity of light. Besides the Killing vector fields (10.1.3), the Schwarzschild (vacuum) space-time admits one more Killing vector, $\underset{0}{\xi}$, which is time-like where the space-time is static, that is for $r > r_g$. With the coordinates in use, $\underset{0}{\xi}$ has components:

$$\underset{0}{\xi}^i = \delta_0^i \quad (10.7.2)$$

hence, considering this and $\underset{3}{\xi}$ in (10.1.3), it follows from the discussion in Sect. 7.5 that time-like geodesics admit the following constants of motion:

$$\underset{0}{\xi}^i P_i = P_0 = \frac{\partial S}{\partial x^0} = -\frac{E}{c} \tag{10.7.3}$$

$$\underset{\varphi}{\xi}^i P_i = P_\varphi = \frac{\partial S}{\partial \varphi} = \ell \,. \tag{10.7.4}$$

Here E and ℓ have the physical significance, as will be clarified later, of total energy and azimuthal angular momentum, respectively, as measured by a static observer at infinity. Taking into account (10.7.3) and (10.7.4), equation (10.7.1) can be solved by separation of variables; putting $S = \sum_i S_i$ where $S_i = S_i(x^{|i|})$ and using (10.7.3), (10.7.4), we have:

$$r^2\left[-\left(1-\frac{r_g}{r}\right)^{-1}\frac{E^2}{c^2}+\left(1-\frac{r_g}{r}\right)\left(\frac{\partial S_r}{\partial r}\right)^2\right] + \left(\frac{\partial S_\theta}{\partial \theta}\right)^2 + \frac{\ell^2}{\sin^2\theta} + \mu_0^2 c^2 r^2 = 0 \,. \tag{10.7.5}$$

A separation constant can then be introduced such that:

$$\left(\frac{\partial S_\theta}{\partial \theta}\right)^2 + \frac{\ell^2}{\sin^2\theta} = L^2 \tag{10.7.6}$$

$$\left(\frac{\partial S_r}{\partial r}\right)^2\left(1-\frac{r_g}{r}\right) - \frac{E^2}{c^2}\left(1-\frac{r_g}{r}\right)^{-1} + \mu_0^2 c^2 = -\frac{L^2}{r^2}. \tag{10.7.7}$$

Here L has the physical meaning of the total angular momentum. At larger distances the expression (10.7.6) (which is independent of r) can be compared with the Newtonian limit and this confirms the interpretation of L and ℓ, stated earlier.

The relation between the four-momentum one-form $P_i = \partial S/\partial x^i$ and the geometric tangent vector field to the geodesic along which the particle moves is:

$$k^i \equiv \dot{x}^i = \frac{1}{\mu_0 c} g^{ij} P_j \quad , \quad k^i k_i = -1 \,, \tag{10.7.8}$$

hence, from (10.7.3) to (10.7.7) we obtain the equations of motion:

$$a) \qquad \dot{x}^0 = \frac{E}{\mu_0 c^2}\left(1 - \frac{r_g}{r}\right)^{-1}$$

$$b) \qquad \dot{\varphi} = \frac{\ell}{\mu_0 c}\,\frac{1}{r^2 \sin^2\theta}$$

$$c) \qquad \dot{\theta} = \pm\left[\left(\frac{L}{\mu_0 c}\right)^2 - \left(\frac{\ell}{\mu_0 c}\right)^2 \frac{1}{\sin^2\theta}\right]^{\frac{1}{2}}$$

$$d) \qquad \dot{r} = \pm\left[\left(\frac{E}{\mu_0 c^2}\right)^2 - 1 + \frac{r_g}{r} - \frac{L^2}{\mu_0^2 c^2 r^2}\left(1 - \frac{r_g}{r}\right)\right]^{\frac{1}{2}}. \qquad (10.7.9)$$

Let us first justify the interpretation of E as the total energy of the particle. Given a field of static observers in the space-time with $r > r_g$, i.e. $u = (-g_{00})^{-\frac{1}{2}}\,\partial_0$, we know, from (9.8.3), that the kinetic energy of the particle relative to the observer u is, in units of the rest-mass energy:

$$\mathcal{E} = \left(1 - \frac{v^2}{c^2}\right)^{-\frac{1}{2}} = -u^i k_i = \frac{E}{\mu_0 c^2}\left(1 - \frac{r_g}{r}\right)^{-\frac{1}{2}}, \qquad (10.7.10)$$

hence in the limit of large r ($r \gg r_g$) and non-relativistic velocities ($v/c \ll 1$), equation (10.7.10) leads to:

$$E = \mu_0 c^2 + \frac{1}{2}\mu_0 v^2 - \frac{GM\mu_0}{r} + \mathrm{O}\left(\frac{v^2}{c^2}, \left(\frac{r_g}{r}\right)^2\right). \qquad (10.7.11)$$

Equation (10.7.9b) states that the sense of the azimuthal rotation is fixed by the sign of ℓ, given as an initial condition on the motion, independently of the values of θ.

Equation (10.7.9c) identifies the inclination of the plane of the orbit with respect to the polar axis $\theta = 0$. This inclination is determined by ℓ, for any given value of L, with the obvious constraint

$$-L \leq \ell \leq L.$$

The equality sign corresponds to the motion with $\theta = \pi/2$. When $\ell = 0$ and $L \neq 0$, the orbital plane contains the axis $\theta = 0$, $\theta = \pi$, while purely radial motion corresponds to $\ell = L = 0$.

Equation (10.7.9d) describes, in each orbital plane, the behaviour of the radial coordinate for any given set of values of the constants of motion L and E^*. Let us search for the condition that $\dot{r} = 0$. Equating the right-hand side of (10.7.9d) to zero and solving for $\gamma = E/\mu_0 c^2$ shows that the condition can be expressed as $\gamma = V(r)$, where the "potential" V is given by:

$$V(r) = \left(1 + \frac{\Lambda^2}{\varrho^2}\right)^{1/2} \left(1 - \frac{1}{\varrho}\right)^{1/2} \equiv W(\Lambda^2, \varrho) \qquad (10.7.12)$$

where $\Lambda = L/(\mu_0 c r_g)$, $\varrho = r/r_g$. The equation $\gamma^2 = [W(\Lambda^2, \varrho)]^2$ describes a one-parameter (Λ^2) family of curves in the plane (γ^2, ϱ) which specify where r has extreme values. The behaviour of a general curve of the family, as a function of the parameter Λ^2, is easily understood if we consider two auxiliary equations:

$$\Lambda^2 \equiv \Lambda^2_{\text{ext}} = \frac{\varrho^2}{2\varrho - 3} \qquad (10.7.13)$$

which is the locus where V^2 has extreme values, and

$$\Lambda^2 \equiv \underset{(1)}{\Lambda}{}^2 = \frac{\varrho^2}{\varrho - 1} \qquad (10.7.14)$$

which is the locus where V^2 is equal to one. These are shown in Fig. 10-7a, and a general curve of the family is shown in Fig. 10-7b.

An inspection of Fig. 10-7b shows that particles with a $\gamma < 1$ are bound in the gravitational field while the circular orbits are found at those values of r where γ has extreme values which are given by the function:

$$\gamma^2 \equiv \gamma^2_{\text{ext}} = \frac{2(\varrho - 1)^2}{\varrho(2\varrho - 3)}. \qquad (10.7.15)$$

This is shown in Fig. 10-8;

The minimum of γ^2_{ext} is found at $\varrho = 3$ with $\underset{0}{\gamma}_{\text{ext}} = 2\sqrt{2}/3$. The branch of the γ^2_{ext} curve with $\varrho \geq 3$ describes stable circular orbits, the remaining part of the curve ($3/2 < \varrho < 3$) describes circular unstable orbits. A stable circular orbit cannot be found in

*The independence of that equation from ℓ and θ manifestly shows that the motion is confined to a plane.

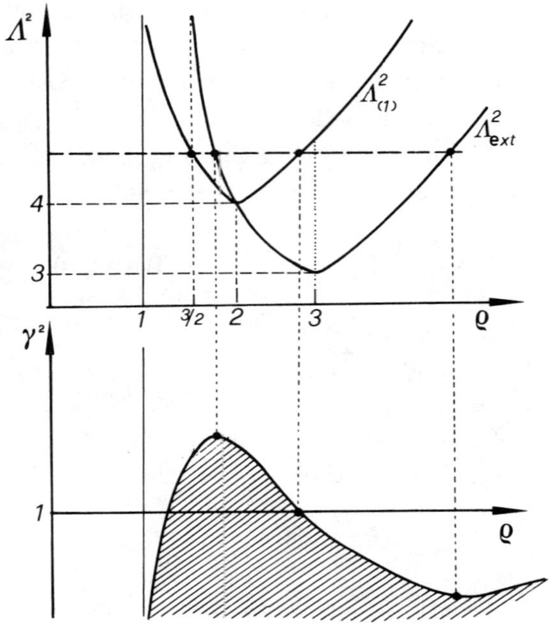

Fig. 10-7 a) The functions $\Lambda^2_{(1)}(\rho)$ and $\Lambda^2_{\text{ext}}(\rho)$ are respectively the loci where $\gamma^2 = 1$ and $\gamma^2_{,\rho} = 0$. b) The general behaviour of a potential curve corresponding to some value of Λ^2 (dashed line in a)). The dashed area is forbidden to the (classical) motion.

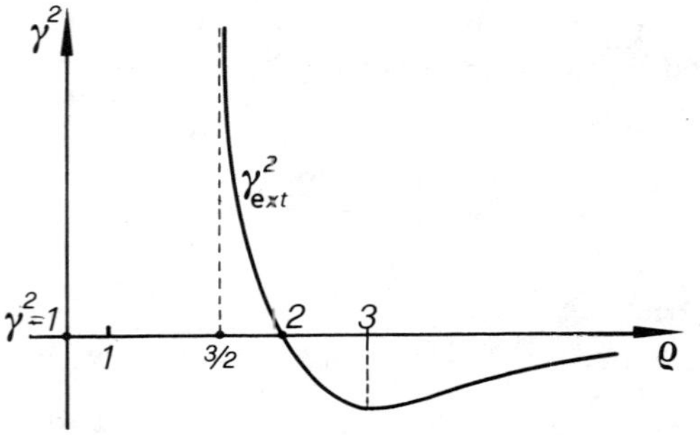

Fig. 10-8 Behaviour of the extreme values of γ^2.

the vacuum Schwarzschild space-time at a coordinate radius less than $\varrho = 3$ ($r = 6GM/c^2$) and with a binding energy higher that $\sim 5.7\%$ of its rest-mass energy (Misner et al., 1973).

10.8 The precession of the apsidal points

The orbit analysis of the previous section is only qualitative in that it enables one to identify the radial location of the apsidal points but not the geometry of the trajectories. For example those trajectories having $\gamma < 1$ and $L = \ell$, say, are bound (see Fig. 10-7) but in general not closed since they do not in general return to their starting point in the spatial coordinates. In order to see this, we have to integrate the equations of motion (10.7.9) once more. The most convenient approach to this problem is the following: from (10.7.9d) and (10.7.9b), putting for convenience $\theta = \pi/2$, $\varrho = 1/u$ and defining $\lambda = \ell/\mu_0 c r_g$, we have:

$$\frac{du}{d\varphi} = -\left(u^3 - u^2 + \frac{u}{\lambda^2} - \frac{\Gamma}{\lambda^2}\right)^{\frac{1}{2}} \tag{10.8.1}$$

where $\Gamma = \gamma^2 - 1$. Differentiating this with respect to φ, namely

$$\frac{d^2u}{d\varphi^2} = -\frac{1}{2}\frac{du}{d\varphi}\left(u^3 - u^2 + \frac{u}{\lambda^2} - \frac{\Gamma}{\lambda^2}\right)^{-\frac{1}{2}}\left(3u^2 - 2u + \frac{1}{\lambda^2}\right)$$

and using (10.8.1) again, we obtain:

$$\frac{d^2u}{d\varphi^2} + u = \frac{3}{2}u^2 + \frac{1}{2\lambda^2}. \tag{10.8.2}$$

Let us now consider the orbit to be a slight deviation from a circular one ($u = u_0$) for which we would have $(d^2u/d\varphi^2)_{u=u_0} = 0$; thus put:

$$\eta = u - u_0 \quad ; \quad \frac{\eta}{u_0} \ll 1 \tag{10.8.3}$$

and from (10.8.2):

$$\frac{d^2\eta}{d\varphi^2} + \eta(1 - 3u_0) = \mathrm{O}(\eta^2). \tag{10.8.4}$$

To first order in η, we finally have:

$$\eta \approx A \cos\left[(1 - 3u_0)^{\frac{1}{2}} \varphi + \delta\right] \tag{10.8.5}$$

with A and δ being integration constants. The apsidal points are now found at

$$(1 - 3u_0)^{\frac{1}{3}} \varphi = n\pi \quad (n = 0, 1, 2 \ldots)$$

and the change in φ needed to cross twice the same apsidal point is:

$$\delta\varphi = \frac{2\pi}{(1 - 3u_0)^{\frac{1}{2}}} \,. \tag{10.8.6}$$

A non-circular orbit therefore does not close on returning to an apsidal value of r; in the case where $u_0 \ll 1$ ($r \gg r_g$), the above equation becomes:

$$\Delta\varphi \approx 2\pi \left(1 + \frac{3}{2} u_0\right) ,$$

hence we define the precession of the apsidal points to be the fractional change in φ after a complete revolution of 2π to reach the same apsidal position:

$$\Delta\hat{\varphi} \approx 3\pi u_0 \quad \text{(radians per revolution)} \,. \tag{10.8.7}$$

From (10.8.3) and (10.8.5) we easily calculate u_0 as:

$$u_0 = \frac{1}{2}(u_+ + u_-) \tag{10.8.8}$$

where $u_\pm$ correspond respectively to the pericenter ($\eta_+ = -A$) and the apocenter ($\eta_- = +A$) of the orbit. To the required accuracy it is sufficient here to take for $u_\pm$ their Newtonian expressions in terms of the semimajor axis a and the eccentricity e, i.e.

$$u_\pm = \frac{r_g}{a(1 \pm e)} \,;$$

hence u_0 becomes:

$$u_0 = \frac{r_g}{(1 - e^2)a} . \tag{10.8.9}$$

In the case of the planet Mercury in the gravitational field of the Sun we have:

$$r_g = \frac{2GM_\odot}{c^2} \approx 2.950 \text{ km} \;\; ; \;\; (1 - e^2)a \approx 55.46 \times 10^6 \text{ km}$$

hence, from (10.8.7):

$$\Delta\varphi \approx 0''.1034 \quad \text{per revolution.}$$

The planet Mercury completes 415 revolutions per century, thus we expect, as predicted by the theory of general relativity, a secular precession of

$$\Delta\varphi \approx 42''.91 \text{ per century.} \tag{10.8.10}$$

When the motion of the solar system is analysed using Newtonian theory and assuming that the Sun is spherically symmetric, one finds, after subtracting the contributions to the precession of Mercury due to the gravitational perturbances of the other planets, an excess in good agreement with (10.8.10) of 42″.7 per century. Unfortunately there is still some doubt as to how spherical the Sun is, although observations of the solar vibrations indicate that the departure from a perfect sphere is very low (Hellings, 1984).

10.9 The plunging-in observer

Ingoing particles endowed with an energy γ larger than the maximum value of the potential barrier which corresponds to their angular-momentum parameters ℓ and L (see Fig. 10-7), move on plunge orbits. They spiral down into the source through its event-horizon, but their motion cannot be followed throughout its whole course by static observers because they fail to be defined on the horizon, the space-time becoming non-static inside the horizon. The best way to see what happens to these particles is to move with them; let us then consider the simplest of the plunging orbits, the radial ones with $\ell = L = 0$, $E/\mu_0 c^2 = 1$ and define a tetrad frame attached to it. The components of the four-velocity read, from (10.7.9):

$$u \equiv \{u^0 = \frac{\varrho}{\varrho - 1} \quad , \quad u^r = -\varrho^{-\frac{1}{2}} \quad , \quad u^\theta = u^\varphi = 0\} \tag{10.9.1}$$

hence a tetrad $\{\lambda_{\hat{a}}\}$ can be defined as:

$$\begin{aligned}
&\lambda_{\hat{0}} = u \\
&\lambda_{\hat{r}} \equiv \{\lambda^0_{\hat{r}} = -\varrho^{-\frac{1}{2}}\left(1-\frac{1}{\varrho}\right)^{-1} \ ; \ \lambda^r_{\hat{r}} = 1 \ ; \ \lambda^\theta_{\hat{r}} = \lambda^\varphi_{\hat{r}} = 0\} \\
&\lambda_{\hat{\theta}} \equiv \{\lambda^0_{\hat{\theta}} = \lambda^r_{\hat{\theta}} = \lambda^\varphi_{\hat{\theta}} = 0 \ ; \ \lambda^\theta_{\hat{\theta}} = \frac{1}{r}\} \\
&\lambda_{\hat{\varphi}} \equiv \{\lambda^0_{\hat{\varphi}} = \lambda^r_{\hat{\varphi}} = \lambda^\theta_{\hat{\varphi}} = 0 \ ; \ \lambda^\varphi_{\hat{\varphi}} = \frac{1}{r\sin\theta}\} \ . \qquad (10.9.2)
\end{aligned}$$

As the particles approach the event horizon, at $r = r_g$, the coordinate time t diverges; from (10.9.1) in fact we have:

$$cdt = \frac{u^0}{u^r}dr = -r_g\frac{\varrho^{\frac{3}{2}}}{\varrho - 1}d\varrho \qquad \varrho = \frac{r}{r_g}$$

which integrated reads:

$$\frac{ct}{r_g} = -\frac{2}{3}\varrho^{\frac{3}{2}} - 2\varrho^{\frac{1}{2}} + \ln\left|\frac{\varrho^{\frac{1}{2}}+1}{\varrho^{\frac{1}{2}}-1}\right| + \text{ const.} \qquad (10.9.3)$$

so as $\varrho \to 1$, we have:

$$\varrho \sim 1 + 4\mathrm{e}^{-\frac{8}{3}}\mathrm{e}^{-(ct/r_g)}. \qquad (10.9.4)$$

On the contrary the proper time of the plunging-in observers remains finite. From (10.9.1) we have in fact:

$$c\delta T = \frac{\delta r}{u^r} = -r_g\varrho^{\frac{1}{2}}\delta\varrho$$

hence, integrating, we obtain:

$$cT = -\frac{2}{3}r_g\varrho^{\frac{3}{2}} + \text{ const.} \qquad (10.9.5)$$

The time the falling-in observers take to reach the event horizon from some initial coordinate radius ϱ_{in}, is given by:

$$\Delta T(\varrho_{\text{in}}, 1) = \frac{2r_g}{3c}\left(\varrho_{\text{in}}^{\frac{3}{2}} - 1\right), \qquad (10.9.6)$$

while the time they spend to reach the singularity at $\varrho = 0$ is:

$$\Delta T(\varrho_{\text{in}}, 0) = \frac{2r_g}{3c}\varrho_{\text{in}}^{\frac{3}{2}} \ . \qquad (10.9.7)$$

For a high-mass black-hole the observers may not be damaged by the increasing tidal forces until after they cross the event horizon, but certainly they will not survive crashing into the singularity at $\varrho = 0$. The time they spend in reaching that point from the event horizon is:

$$\Delta T(1,0) = \frac{2r_g}{3c} = \frac{4GM}{3c^3} \approx 6.55 \times 10^{-6} \frac{M}{M_\odot} \text{ sec.}$$

and approaching the singularity, the observers will suffer in their rest frame unboundedly large tidal stresses. In fact, writing the equation of geodesic deviation in the tetrad frame (10.9.2), we have:

$$\left(\frac{D^2\eta^{\hat{\alpha}}}{D\tau^2}\right) = -R^{\hat{\alpha}}{}_{\hat{0}\hat{\beta}\hat{0}}\eta^{\hat{\beta}} \tag{10.9.8}$$

where η is a space-like vector connecting two radial geodesics and $R^{\hat{\alpha}}{}_{\hat{0}\hat{\beta}\hat{0}}$ are the tetrad components of the curvature tensor relative to $\{\lambda_{\hat{a}}\}$. Some calculations show that:

$$\begin{aligned} R_{\hat{r}\hat{0}\hat{r}\hat{0}} &= -\frac{r_g}{r^3} & R_{\hat{0}\hat{\theta}\hat{0}\hat{\theta}} &= R_{\hat{0}\hat{\varphi}\hat{0}\hat{\varphi}} = \frac{1}{2}\frac{r_g}{r^3} \\ R_{\hat{\theta}\hat{\varphi}\hat{\theta}\hat{\varphi}} &= \frac{r_g}{r^3} & R_{\hat{r}\hat{\theta}\hat{r}\hat{\theta}} &= R_{\hat{r}\hat{\varphi}\hat{r}\hat{\varphi}} = -\frac{1}{2}\frac{r_g}{r^3} \end{aligned} \tag{10.9.9}$$

all the other components being zero. Thus we have:

$$\begin{aligned} \left(\frac{D^2\eta^{\hat{r}}}{D\tau^2}\right) &= -R^{\hat{r}}{}_{\hat{0}\hat{r}\hat{0}}\eta^{\hat{r}} = +\frac{r_g}{r^3}\eta^{\hat{r}} \\ \left(\frac{D^2\eta^{\hat{\theta}}}{D\tau^2}\right) &= -R^{\hat{\theta}}{}_{\hat{0}\hat{\theta}\hat{0}}\eta^{\hat{\theta}} = -\frac{1}{2}\frac{r_g}{r^3}\eta^{\hat{\theta}} \\ \left(\frac{D^2\eta^{\hat{\varphi}}}{D\tau^2}\right) &= -R^{\hat{\varphi}}{}_{\hat{0}\hat{\varphi}\hat{0}}\eta^{\hat{\varphi}} = -\frac{1}{2}\frac{r_g}{r^3}\eta^{\hat{\varphi}} \end{aligned} \tag{10.9.10}$$

and this clearly shows that directions along the motion are infinitely stretched on approaching $r = 0$, the transverse ones on the other hand, being infinitely compressed.

10.10 Null geodesics in the external Schwarzschild solution

The behaviour of the light rays can be deduced from equation (10.7.5) provided we put $\mu_0 = 0$. Since light rays are null geodesics, the constants of motion* $\tilde{E}$ and $\tilde{\ell}$, deducible from the existence of the Killing fields (cf. (10.7.3) and (10.7.4)) retain the same physical significance as for the time-like geodesics. Similarly a separation constant $\tilde{L}$ appears, in addition to $\tilde{E}$ and $\tilde{\ell}$. But unlike the time-like case, to characterize the motion of a light ray, it is now sufficient to know the ratios $(\tilde{E}/\tilde{\ell})$ or $(\tilde{E}/\tilde{L})$. From (10.7.6) and (10.7.7) with $\mu_0 = 0$, we have that the angular equations retain the same form, while the radial equation becomes:

$$P_r^2\left(1-\frac{r_g}{r}\right)-\frac{\tilde{E}^2}{c^2}\left(1-\frac{r_g}{r}\right)^{-1}=-\frac{\tilde{L}^2}{r^2} \tag{10.10.1}$$

where now:

$$P_i=\frac{\partial S}{\partial x^i}=g_{ij}\dot{x}^j$$

and the dot means derivative with respect to some affine parameter on the null geodesics. Hence we have:

$$\dot{r}=\pm\left[\frac{\tilde{E}^2}{c^2}-\frac{\tilde{L}^2}{r^2}\left(1-\frac{r_g}{r}\right)\right]^{\frac{1}{2}} \tag{10.10.2}$$

and so we find a typically non-Newtonian effect of having potential barriers for null rays; these are obtained from the above equation on setting $\dot{r}=0$:

$$\tilde{\gamma}^2=V^2(\varrho)=\frac{\varrho-1}{\varrho^3} \tag{10.10.3}$$

where

$$\tilde{\gamma}=\frac{\tilde{E}}{c\tilde{L}}r_g$$

and $\varrho = r/r_g$. There exists only one potential curve, shown in Fig. 10-9, which has a maximum at $\varrho = (3/2)$ $(r = 3GM/c^2)$ with value: $(\tilde{\gamma})_{\underset{0}{\text{ext}}} = \pm 2/(3\sqrt{3})$.

The maximum in Fig. 10-9 clearly indicates that a light ray is kept there on an unstable circular orbit; meanwhile outgoing light rays emitted at some $1 < \varrho < 3/2$ with $|\tilde{\gamma}| < 2/(3\sqrt{3})$ do not escape the gravitational field.

---

*Parameters with a tilde refer to photons.

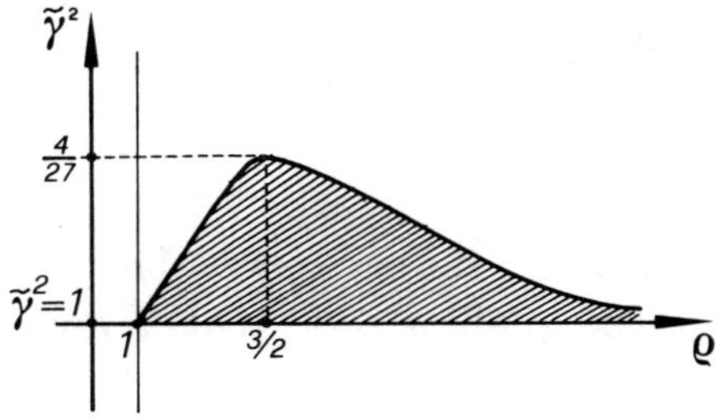

Fig. 10-9 The potential curve for light rays. The shaded region is forbidden to the (classical) motion.

Let us now investigate how a radially infalling observer would see the behaviour of a light ray emitted from within his or her rest frame. This frame is described by the tetrad (10.9.2), hence a null geodesic with tangent vector field k, say, makes an angle with the radial direction given by:

$$\cos\Psi_r = \frac{k^i \lambda_{i\hat{r}}}{(k^{\hat{\alpha}} k_{\hat{\alpha}})^{\frac{1}{2}}} = \frac{\tilde{E}/c + \varrho^{\frac{1}{2}} k^r}{(\tilde{E}/c)\varrho^{\frac{1}{2}} + k^r} \tag{10.10.4}$$

from (9.3.2), (10.9.1) and (10.9.2). Recalling (10.10.2) (with $k^r \equiv \dot{r}$), (10.10.4) becomes:

$$\cos\Psi_{r\pm} = \frac{1 \pm \varrho^{\frac{1}{2}}\left(1 - \tilde{\Lambda}^2(\varrho - 1)/\varrho^3\right)^{\frac{1}{2}}}{\varrho^{\frac{1}{2}} \pm \left(1 - \tilde{\Lambda}^2(\varrho - 1)/\varrho^3\right)^{\frac{1}{2}}} \tag{10.10.5}$$

where

$$\tilde{\Lambda} = \frac{c\tilde{L}}{r_g \tilde{E}}$$

and the upper sign refers to outgoing light rays ($\dot{r} > 0$), the lower sign to ingoing ones ($\dot{r} < 0$). We then see the following features: when $\tilde{\Lambda} = 0$, $\cos\Psi_{r\pm} = \pm 1$ as expected; when $\varrho \to 1$, the angle within which outgoing light rays appear to be confined shrinks to zero, since, whatever $\tilde{\Lambda}^2$:

$$\lim_{\varrho \to 1}(\cos\Psi_{r+}) = 1 \; ; \tag{10.10.6}$$

on the other hand, ingoing light rays can be flashed across the event horizon within a finite angle:

$$\lim_{\varrho \to 1} (\cos \Psi_{r_-}) = -\frac{1 - \tilde{\Lambda}^2}{1 + \tilde{\Lambda}^2}. \qquad (10.10.7)$$

This shows that the event horizon plays the role of a semipermeable membrane to the infalling observers as well.

10.11 The bending of light rays

That light rays are bent by a gravitational field is dramatically shown by the existence of spatially circular orbits which, although unstable, allow a photon to orbit around a collapsed source many times before it plunges into it or escapes to infinity. This phenomenon was predicted by Einstein as the most spectacular consequence of his theory and the experimental confirmation which took place during the solar eclipse in 1918 gave Einstein worldwide fame.

The gravitational strength on the solar surface is small in general relativistic terms, since the ratio

$$\frac{r_g}{R_\odot} = \frac{2GM_\odot}{c^2 R_\odot} \approx 10^{-6}.$$

Nevertheless a light ray from a distant star suffers, when grazing the solar limb, a bending which causes an angular displacement of the star's image (from the position it would have if the Sun were not on the line of sight) of $\theta_\odot = 1''.75$ (Fig. 10-10).

In order to deduce this value let us follow the suggestion first given by Einstein himself, namely that a gravitational field acts on a light ray as a non-uniform optical medium. Since we want to study this effect in the case when the field is weak, as it is at the surface of the Sun, let us write the Schwarzschild line element in isotropic coordinates. Isotropic coordinates allow one to express the metric of any spherically symmetric three-dimensional space as conformal to a flat space-time. In the case of the Schwarzschild solution, the transformation $r = (1 + r_g/4r')^2 r'$ leads to the line

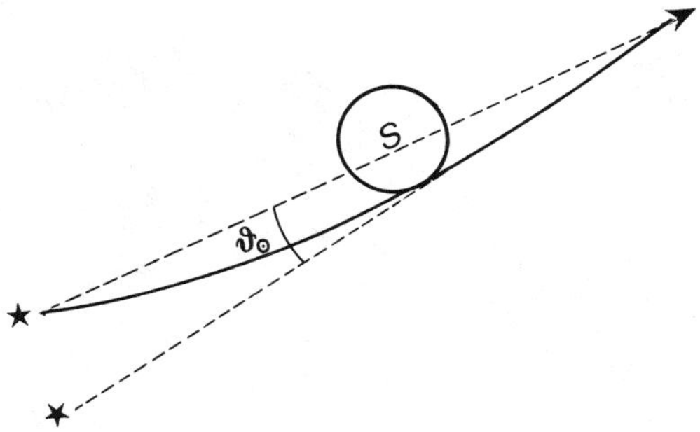

Fig. 10-10 A light ray grazing the Sun suffers a bending of $1''.75$ of arc.

element:

$$ds^2 = -\left(1 - \frac{r_g}{4r'}\right)^2 \left(1 + \frac{r_g}{4r'}\right)^{-2} (dx_0)^2$$
$$+ \left(1 + \frac{r_g}{4r'}\right)^4 [dr'^2 + r'^2(d\theta^2 + \sin^2\theta\, d\varphi^2)] \quad (10.11.1)$$

where now the event-horizon is located at $r' = (1/4)r_g$ and $r' \approx r$ at asymptotic distances. In the limit $r \gg r_g$, (10.11.1) can be written as:

$$ds^2 \approx -c^2dt^2 + \left(1 + \frac{r_g}{r}\right)(dx^2 + dy^2 + dz^2) \; ; \quad \frac{r_g}{r} \ll 1 \; .$$

A light ray then appears to move with a velocity, as measured by an observer at infinity:

$$\frac{dr}{dt} = \frac{c}{1 + r_g/r} \equiv \frac{c}{n}$$

i.e. as if it were moving in a medium with refractive index:

$$n = 1 + \frac{2GM_\odot}{c^2 r} . \quad (10.11.2)$$

Let us then apply the laws of the classical optics in a continuous medium with a three-dimensional Euclidean space-geometry as background (de Felice, 1971). From spherical symmetry, the following formula of Bouguer holds (Born and Wolf, 1965):

$$nr \sin\psi = \text{const.} \quad (10.11.3)$$

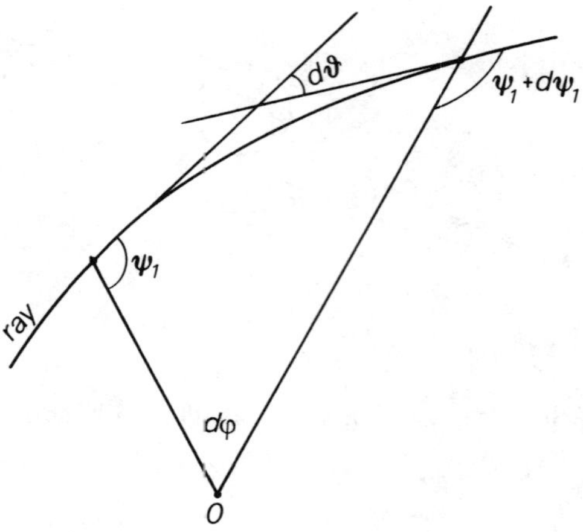

Fig. 10-11 A light ray is bent in a gravitational field as if it were in an optical medium.

which expresses the law of conservation of the axial angular momentum. From the geometrical setting of Fig. 10-11, we have $d\theta = d\psi + d\varphi$, and from (10.11.3) we have:

$$d\psi = -\frac{\ell dy}{y(y^2-\ell^2)^{\frac{1}{2}}} \quad ; \quad d\varphi = \frac{\ell dr}{r(y^2-\ell^2)^{\frac{1}{2}}}, \qquad (10.11.4)$$

where $y = nr$ and ℓ is a constant which measures the impact parameter of the ray with respect to the centre of symmetry of the medium, as measured at large distances.

The angle which measures the bending of the ray is obtained by integrating (10.11.3) over the entire orbit. Due to the symmetry it reads:

$$\theta_\odot = -2\int_\infty^{y(R_\odot)} \frac{\ell}{y(y^2-\ell^2)} dy + 2\int_\infty^{R_\odot} \frac{\ell}{r(y^2-\ell^2)^{\frac{1}{2}}} dr \qquad (10.11.5)$$

where we take the distance of closest approach to the Sun as the Sun's radius. Then also $\ell \approx R_\odot$ so, recalling that $r_g/R_\odot \ll 1$, we have:

$$-2R_\odot \int_\infty^{y(R_\odot)} \frac{dy}{y(y^2-R_\odot^2)^{\frac{1}{2}}} = -2\sec^{-1}\left(\frac{y}{R_\odot}\right)\Bigg|_\infty^{R_\odot} \approx \pi$$

$$2R_{\odot}\int_{\infty}^{R_{\odot}}\frac{dr}{r(r^2+2r_g r-R_{\odot}^2)^{\frac{1}{2}}}\approx-\pi-\frac{2r_g}{R_{\odot}}\,. \qquad (10.11.6)$$

Thus the predicted value is:

$$|\theta_{\odot}|=\frac{4GM_{\odot}}{c^2R_{\odot}}=1''.75\,. \qquad (10.11.7)$$

The observed values, although scattered around the theoretical value by errors which depend on the observational difficulties, do indicate a substantial agreement with Einstein's prediction.

The bending of light rays is related to a change in the frequency of the electromagnetic perturbance. From (9.5.6) we know how two observers compare the frequency of a light ray which is exchanged between them. Let us apply that formula to the radially infalling observer (10.9.1) and to a static observer at infinity with four-velocity

$$u^i_{\infty}=\delta^i_0. \qquad (10.11.8)$$

From the general relation (9.5.6) we have, for outgoing rays:

$$1+z=\frac{\nu_{\rm e}}{\nu_{\rm o}}=\frac{(k_iu^i)_{\rm e}}{(k_iu^i)_{\rm o}}=\frac{\varrho^{\frac{1}{2}}}{\varrho-1}\left[\varrho^{\frac{1}{2}}+\left(1-\tilde{\Lambda}^2\frac{\varrho-1}{\varrho^3}\right)^{\frac{1}{2}}\right]. \qquad (10.11.9)$$

This shows that the red-shift diverges as $\varrho\to 1$ whatever are the values of $\tilde{\Lambda}$. An observer at infinity then, sees the light rays emitted from a freely falling radial emitter, with a frequency which decreases with time as:

$$\frac{\nu_{\rm o}}{\nu_{\rm e}}\approx {\rm e}^{-ct/r_g} \qquad (10.11.10)$$

from (10.9.4).

If the freely falling emitter represents any element at the surface of a collapsing star, then equation (10.11.10) shows how rapid is the decay of the frequency and therefore of the luminosity of the source as measured by an observer at infinity. The e-folding time in fact is of the order of r_g/c; this means for example that a one-solar-mass star which undergoes collapse to a black-hole weakens its luminosity with a decay e-folding time of the order of 10^{-5} sec.

It will then disappear to any available optical device long before it reaches the event horizon.

10.12 The Reissner-Nordström solution

The Schwarzschild solution describes the gravitational field generated by an isolated system which is spherically symmetric and electrically neutral. To know the contribution to the background curvature due to an electric charge, one needs to solve the coupled Einstein-Maxwell equations:

$$\begin{aligned} a) \qquad & R_i^{\ j} - \frac{1}{2}\delta_i^j R = \kappa T_i^{\ j} \\ b) \qquad & F_{[ij,k]} = 0 \\ c) \qquad & F^{ij}_{\ \ ;j} = 4\pi J^i \end{aligned} \tag{10.12.1}$$

where

$$T_i^{\ j} = (T_i^{\ j})_{\rm m} + (T_i^{\ j})_{\rm em}. \tag{10.12.2}$$

Here $(T_i^{\ j})_{\rm m}$ is the energy-momentum tensor of matter and $(T_i^{\ j})_{\rm em}$ is that of the electromagnetic field $((T_i^{\ j})_{\rm em} = \frac{1}{4\pi}(F_i^{\ k}F^j_{\ k} - (1/4)F^{rs}F_{rs}\delta_i^j)$). The condition that (10.12.2) is spherically symmetric assures that the metric retains the form (10.1.12) and the only non-zero component of the electromagnetic tensor F_{ij} is F_{0r}, (see discussion in Sect. 10.1). Since Einstein's equations retain the same form as in (10.2.3), the argument which allowed us to conclude that the metric outside the source is static holds now as well (in fact $(T_0^{\ r})_{\rm em} = 0$, hence outside the source $T_0^{\ r} = 0$), so we have:

$$\lambda = \lambda(r) \quad \nu = \nu(r) \ ; \ F_{ij} = 2E(r)\delta_{0[i}\delta_{j]r}. \tag{10.12.3}$$

Equation (10.12.1b) is identically satisfied, while (10.12.1c) can be written as:

$$\partial_r(\sqrt{-g}F^{0r}) = 4\pi\sqrt{-g}J^0$$

which, from (10.12.3), becomes:

$$\frac{d}{dr}\left(\mathrm{e}^{-\frac{(\nu+\lambda)}{2}}r^2E(r)\right) = -4\pi\mathrm{e}^{\frac{(\nu+\lambda)}{2}}r^2J^0. \tag{10.12.4}$$

This can easily be integrated and yields:

$$e^{-\frac{(\nu+\lambda)}{2}} r^2 E(r) - \lim_{\epsilon \to 0} e^{-\frac{(\nu(\epsilon)+\lambda(\epsilon))}{2}} \epsilon^2 E(\epsilon)$$

$$= -4\pi \int_0^{r_0} e^{\frac{(\nu+\lambda)}{2}} r^2 J^0 dr \qquad r > r_0.$$

Here r_0 is the smallest value of the radial coordinate such that $(T_i^{\ j})_{\mathrm{m}} = 0$ and $J^i = 0$ at $r > r_0$. Symmetry considerations and the assumptions that J^i remains finite as $r \to 0$ and E is continuous, force $E(0) = 0$, hence the last equation becomes:

$$e^{-\frac{(\nu(r)+\lambda(r))}{2}} r^2 E(r) = -4\pi \int_0^{r_0} e^{\frac{(\nu+\lambda)}{2}} r^2 J^0 dr \quad r > r_0. \qquad (10.12.5)$$

The integral in (10.12.5) can be manipulated as follows:

$$-4\pi \int_0^{r_0} e^{\frac{(\nu+\lambda)}{2}} r^2 J^0 dr = -\int_0^{2\pi} d\varphi \int_0^{\pi} d\theta \sin\theta \int_0^{\infty} e^{\frac{(\nu+\lambda)}{2}} r^2 J^0 dr$$

$$= -\int_{\Sigma(t=\mathrm{const.})} J^0 \sqrt{-g}\, d^3x$$

$$= \int_{\Sigma(t=\mathrm{const.})} J^i d\Sigma_i = q \qquad (10.12.6)$$

where $d\Sigma_i$ is the element of the three-surface t =const. and q is the net charge deposited in the metric source. From (7.7.9), however, we know that q is a conserved quantity and therefore it is independent of Σ. Thus, from (10.12.6) and (10.12.5), we have at $r > r_0$:

$$E(r) = e^{\frac{(\nu(r)+\lambda(r))}{2}} \frac{q}{r^2}. \qquad (10.12.7)$$

Let us now consider Einstein's equations:

$$\nu' + \lambda' = \kappa e^{\lambda(r)} r (T_r^{\ r} - T_0^{\ 0}) \qquad (10.12.8)$$

$$e^{-\lambda}\left(\frac{1}{r^2} - \frac{\lambda'}{r}\right) - \frac{1}{r^2} = \kappa T_0^{\ 0}\,. \qquad (10.12.9)$$

From (10.12.7) and (10.12.3), we have:

$$(T_0^{\ 0})_{\mathrm{em}} = (T_r^{\ r})_{\mathrm{e.m}}$$

hence $T_r^{\ r} - T_0^{\ 0} = (T_r^{\ r})_{\mathrm{m}} - (T_0^{\ 0})_{\mathrm{m}}$ and so the considerations which lead to (10.2.9) apply here and give:

$$\lambda(r) = -\nu(r). \qquad (10.12.10)$$

So (10.12.7) at $r > r_0$ takes the *Euclidean* form:

$$E(r) = \frac{q}{r^2}. \tag{10.12.11}$$

Equation (10.12.9) now allows us to determine $\lambda(r)$; its integration leads to:

$$r\mathrm{e}^{-\lambda(r)} - \lim_{\epsilon\to 0} \epsilon \left(\mathrm{e}^{-\lambda(\epsilon)} - 1\right) = r + \kappa \int_0^r r'^2 T_0{}^0 dr' \quad ; \quad r > r_0 \tag{10.12.12}$$

and if $T_0{}^0$ remains finite as $r \to 0$, the second term on the left-hand side vanishes, hence:

$$\begin{aligned} \mathrm{e}^{-\lambda(r)} &= 1 + \frac{\kappa}{r} \int_0^r r'^2 T_0{}^0 dr' \\ &= 1 + \frac{\kappa}{r} \int_0^{r_0} r'^2 (T_0{}^0)_{\mathrm{m}} dr' + \frac{\kappa}{r} \int_0^{r_0} r'^2 (T_0{}^0)_{\mathrm{em}} dr' \\ &\quad + \frac{\kappa}{r} \int_{r_0}^r r'^2 (T_0{}^0)_{\mathrm{em}} dr'. \end{aligned} \tag{10.12.13}$$

From (10.12.10) and the forms of F_{ij} and $(T_i{}^j)_{\mathrm{em}}$, we have:

$$(T_0{}^0)_{\mathrm{em}} = -\frac{E^2}{8\pi}(r) = -\frac{q^2}{8\pi r^4} \quad ; \quad r > r_0 \,. \tag{10.12.14}$$

Thus the last two integrals read:

$$\frac{\kappa}{r}\left(\int_o^{r_0} r'^2 (T_0{}^0)_{\mathrm{em}}(r') dr' - \frac{q^2}{8\pi r_0}\right) + \frac{\kappa q^2}{8\pi r^2} \quad ; \quad r > r_0. \tag{10.12.15}$$

The integral in (10.12.15) clearly measures (w.r.t. infinity) the energy of the electric field inside the charged body, and even if the charge distribution inside it is unknown, symmetry considerations indicate that the integral should be of the form:

$$\int_0^{r_0} r'^2 (T_0{}^0)_{\mathrm{em}}(r') dr' = -\frac{\alpha q^2}{8\pi r_0} \tag{10.12.16}$$

where $\alpha > 0$ is some constant. Finally, the first integral in (10.12.13) measures the total energy of the matter component of the energy-momentum tensor of the source, as it is seen at infinity; though it depends on r_0, in analogy with (10.2.13) we name it:

$$M_s(r_0) = -\frac{4\pi}{c^2} \int_0^{r_0} r'^2 (T_0{}^0)_{\mathrm{m}}(r') dr' \,. \tag{10.12.17}$$

Combining this with (10.12.10) and (10.12.15) we have:

$$e^{-\lambda(r)} = 1 - \frac{2}{r}\left(\frac{GM_s}{c^2} + \beta\frac{Q^2}{r_0}\right) + \frac{Q^2}{r^2} \qquad (10.12.18)$$

where β is a new constant and

$$Q^2 = \frac{G}{c^4}q^2 \qquad (10.12.19)$$

has the dimension of a length squared. Let us define a *generalized mass*

$$M = M_s(r_0) + \frac{\beta}{c^2}\frac{q^2}{r_0}\ ; \qquad (10.12.20)$$

clearly M does not depend on r_0 and, from the Birkoff theorem, it does not depend on t, hence it is constant along the world-tube of the charged source. From (10.12.10), (10.12.18) and (10.12.20) we have the space-time metric outside a charged and spherically symmetric configuration:

$$ds^2 = -\left(1 - \frac{2GM}{c^2 r} + \frac{Q^2}{r^2}\right)dt^2 + \left(1 - \frac{2GM}{c^2 r} + \frac{Q^2}{r^2}\right)^{-1} dr^2 + r^2\left(d\theta^2 + \sin^2\theta\, d\varphi^2\right). \qquad (10.12.21)$$

This is the Reissner-Nordström solution (Reissner, 1916; Nordström, 1913). It reduces to the Schwarzschild solution when $Q = 0$; however its space-time properties are remarkably different.

10.13 The extended Reissner-Nordström solution

The Reissner-Nordström solution in the form (10.12.21) shows divergences in components of the metric or its inverse at $r = 0$ and at

$$r = r_\pm \equiv \frac{GM}{c^2} \pm \left[\left(\frac{GM}{c^2}\right)^2 - Q^2\right]^{\frac{1}{2}}. \qquad (10.13.1)$$

Let us for simplicity write $GM/c^2 = m$; as for the Schwarzschild solution, the singularity at $r = 0$ is a non-removable curvature singularity, while those at $r = r_\pm$ are coordinate singularities which can be removed with a suitable analytical extension of where $r(u', v')$ is the solution of the equations:

$$u'v' = -\mathrm{e}^{\frac{r_+ - r_-}{r_+^2} r} \left(\frac{|r - r_+|}{2m}\right) \left(\frac{|r - r_-|}{2m}\right)^{-(r_-/r_+)^2} \qquad r > r_+ \;\; r < r_-$$

$$u'v' = \mathrm{e}^{\frac{r_+ - r_-}{r_+^2} r} \left(\frac{r_+ - r}{2m}\right) \left(\frac{|r - r_-|}{2m}\right)^{-(r_-/r_+)^2} \qquad r_- < r \le r_+ \; . \tag{10.13.11}$$

In the metric (10.13.10) we have removed the coordinate singularity at $r = r_+$ but not the one at $r = r_-$. A suitable way to extend the metric through $r = r_-$ is to perform the following coordinate transformation:

$$\begin{aligned} \tan \frac{1}{2}(\tilde{t} + \tilde{r}) &= v' \\ \tan \frac{1}{2}(\tilde{t} - \tilde{r}) &= u' \; . \end{aligned} \tag{10.13.12}$$

From (10.13.11) we have:

$$\cos \tilde{t} = \frac{1 + \mathrm{e}^{\frac{r_+ - r_-}{r_+^2} r} \frac{|r - r_+|}{2m} \left(\frac{|r - r_-|}{2m}\right)^{-(r_-/r_+)^2}}{1 - \mathrm{e}^{\frac{r_+ - r_-}{r_+^2} r} \frac{|r - r_+|}{2m} \left(\frac{|r - r_-|}{2m}\right)^{-(r_-/r_+)^2}} \cos \tilde{r} \qquad r > r_+ \, , \; r < r_-$$

$$\sin \tilde{t} = \tanh \left(\frac{r_+ - r_-}{2r_+^2} t\right) \sin \tilde{r} \tag{10.13.13}$$

and

$$\cos \tilde{t} = \frac{1 - \mathrm{e}^{\frac{r_+ - r_-}{r_+^2} r} \frac{(r_+ - r)}{2m} \left(\frac{|r - r_-|}{2m}\right)^{-(r_-/r_+)^2}}{1 + \mathrm{e}^{\frac{r_+ - r_-}{r_+^2} r} \frac{(r_+ - r)}{2m} \left(\frac{|r - r_-|}{2m}\right)^{-(r_-/r_+)^2}} \cos \tilde{r} \qquad r_- \le r \le r_+$$

$$\sin \tilde{t} = \coth \left(\frac{r_+ - r_-}{2r_+^2} t\right) \sin \tilde{r} \; . \tag{10.13.14}$$

The physical space-time initially described in the coordinate patch (t, r) is mapped into the region of the extended manifold $(\tilde{t}, \tilde{r})$ shown in Fig. 10-13.

where $r(u', v')$ is the solution of the equations:

$$u'v' = -\mathrm{e}^{\frac{r_+ - r_-}{r_+^2} r} \left(\frac{|r - r_+|}{2m} \right) \left(\frac{|r - r_-|}{2m} \right)^{-(r_-/r_+)^2} \qquad r > r_+ \quad r < r_-$$

$$u'v' = \mathrm{e}^{\frac{r_+ - r_-}{r_+^2} r} \left(\frac{r_+ - r}{2m} \right) \left(\frac{|r - r_-|}{2m} \right)^{-(r_-/r_+)^2} \qquad r_- < r \le r_+ \,. \tag{10.13.11}$$

In the metric (10.13.10) we have removed the coordinate singularity at $r = r_+$ but not the one at $r = r_-$. A suitable way to extend the metric through $r = r_-$ is to perform the following coordinate transformation:

$$\tan \frac{1}{2}(\tilde{t} + \tilde{r}) = v'$$

$$\tan \frac{1}{2}(\tilde{t} - \tilde{r}) = u' \,. \tag{10.13.12}$$

From (10.13.11) we have:

$$\cos \tilde{t} = \frac{1 + \mathrm{e}^{\frac{r_+ - r_-}{r_+^2} r} \frac{|r - r_+|}{2m} \left(\frac{|r - r_-|}{2m} \right)^{-(r_-/r_+)^2}}{1 - \mathrm{e}^{\frac{r_+ - r_-}{r_+^2} r} \frac{|r - r_+|}{2m} \left(\frac{|r - r_-|}{2m} \right)^{-(r_-/r_+)^2}} \cos \tilde{r} \qquad r > r_+ \,, \; r < r_-$$

$$\sin \tilde{t} = \tanh \left(\frac{r_+ - r_-}{2r_+^2} t \right) \sin \tilde{r} \tag{10.13.13}$$

and

$$\cos \tilde{t} = \frac{1 - \mathrm{e}^{\frac{r_+ - r_-}{r_+^2} r} \frac{(r_+ - r)}{2m} \left(\frac{|r - r_-|}{2m} \right)^{-(r_-/r_+)^2}}{1 + \mathrm{e}^{\frac{r_+ - r_-}{r_+^2} r} \frac{(r_+ - r)}{2m} \left(\frac{|r - r_-|}{2m} \right)^{-(r_-/r_+)^2}} \cos \tilde{r} \qquad r_- \le r \le r_+$$

$$\sin \tilde{t} = \coth \left(\frac{r_+ - r_-}{2r_+^2} t \right) \sin \tilde{r} \,. \tag{10.13.14}$$

The physical space-time initially described in the coordinate patch (t, r) is mapped into the region of the extended manifold $(\tilde{t}, \tilde{r})$ shown in Fig. 10-13.

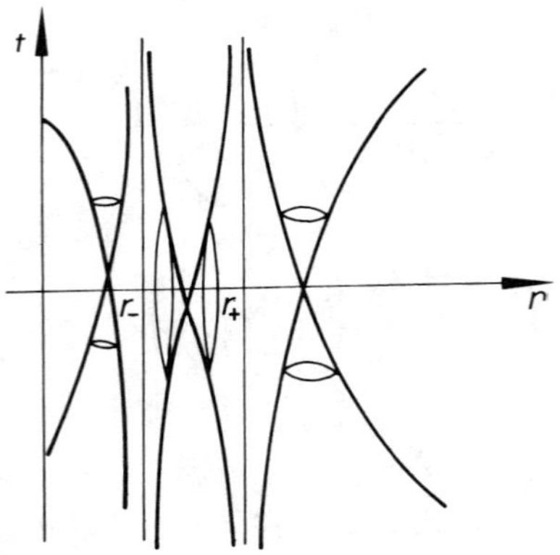

Fig. 10-12 In the (t, r)-plane, the light-cone is singular on both inner and outer horizons.

patches:

$$v' = \exp\left(\frac{r_+ - r_-}{2r_+^2} v\right)$$

$$r > r_+ \ , \ r < r_-$$

$$u' = -\exp\left(-\frac{r_+ - r_-}{2r_+^2} u\right) \qquad (10.13.8)$$

and

$$v' = \exp\left(\frac{r_+ - r_-}{2r_+^2} v\right)$$

$$r_- < r < r_+ \ .$$

$$u' = \exp\left(-\frac{r_+ - r_-}{2r_+^2} u\right) \qquad (10.13.9)$$

These join smoothly at $r = r_+$ and the metric (10.13.5) takes the form:

$$ds^2 = -\frac{16m^2}{r^2}\left(\frac{r_+^2}{r_+ - r_-}\right)^2 e^{-\frac{r_+ - r_-}{r_+^2} r}\left(\frac{|r - r_-|}{2m}\right)^{1+(r_-/r_+)^2} du'dv'$$
$$+ r^2(u', v')(d\theta^2 + \sin^2\theta \, d\varphi^2) \qquad (10.13.10)$$

the metric. In this case however there are three distinct cases: $m^2 > Q^2$, $m^2 < Q^2$, $m^2 = Q^2$. The method here follows the same lines as in Sect. 10.5; radial light rays obey the relation:

$$\frac{dr}{dt} = \pm\left(1 - \frac{2m}{r} + \frac{Q^2}{r^2}\right), \tag{10.13.2}$$

where the $+$ and $-$ signs correspond to the outgoing and ingoing rays respectively. Along them we have:

$$\begin{aligned} u &= t - r^* = \text{const.} \\ v &= t + r^* = \text{const.} \end{aligned} \tag{10.13.3}$$

with

$$r^* \equiv \int \frac{dr}{1 - (2m/r) + (Q^2/r^2)}. \tag{10.13.4}$$

We take u and v as new coordinates (they are null coordinates) so that the metric (10.12.21) becomes:

$$ds^2 = -\left(1 - \frac{2m}{r} + \frac{Q^2}{r^2}\right) dudv + r(u,v)^2(d\theta^2 + \sin^2\theta\, d\varphi^2) \tag{10.13.5}$$

where $r(u,v)$ is a solution of the equation:

$$r^* = \frac{1}{2}(v - u). \tag{10.13.6}$$

Let us now consider the three cases separately.

i) $m^2 > Q^2$. A straightforward integration of (10.13.4) yields:

$$r^* = r + \frac{r_+^2}{r_+ - r_-} \ln \frac{1}{2m}|r - r_+| - \frac{r_-^2}{r_+ - r_-} \ln \frac{1}{2m}|r - r_-|\,. \tag{10.13.7}$$

Hence, noticing that

$$\frac{dt}{dr} = \pm\frac{dr^*}{dr}$$

the light cones in the (t, r)-plane appear as in Fig. 10-12.

As expected, light rays cannot be followed across the surfaces $r = r_\pm$. Let us then introduce the following new coordinate

Combining this with (10.12.10) and (10.12.15) we have:

$$e^{-\lambda(r)} = 1 - \frac{2}{r}\left(\frac{GM_s}{c^2} + \beta\frac{Q^2}{r_0}\right) + \frac{Q^2}{r^2} \tag{10.12.18}$$

where β is a new constant and

$$Q^2 = \frac{G}{c^4}q^2 \tag{10.12.19}$$

has the dimension of a length squared. Let us define a *generalized mass*

$$M = M_s(r_0) + \frac{\beta}{c^2}\frac{q^2}{r_0} \ ; \tag{10.12.20}$$

clearly M does not depend on r_0 and, from the Birkoff theorem, it does not depend on t, hence it is constant along the world-tube of the charged source. From (10.12.10), (10.12.18) and (10.12.20) we have the space-time metric outside a charged and spherically symmetric configuration:

$$ds^2 = -\left(1 - \frac{2GM}{c^2 r} + \frac{Q^2}{r^2}\right)dt^2 + \left(1 - \frac{2GM}{c^2 r} + \frac{Q^2}{r^2}\right)^{-1} dr^2 + r^2\left(d\theta^2 + \sin^2\theta\, d\varphi^2\right). \tag{10.12.21}$$

This is the Reissner-Nordström solution (Reissner, 1916; Nordström, 1913). It reduces to the Schwarzschild solution when $Q = 0$; however its space-time properties are remarkably different.

10.13 The extended Reissner-Nordström solution

The Reissner-Nordström solution in the form (10.12.21) shows divergences in components of the metric or its inverse at $r = 0$ and at

$$r = r_{\pm} \equiv \frac{GM}{c^2} \pm \left[\left(\frac{GM}{c^2}\right)^2 - Q^2\right]^{\frac{1}{2}}. \tag{10.13.1}$$

Let us for simplicity write $GM/c^2 = m$; as for the Schwarzschild solution, the singularity at $r = 0$ is a non-removable curvature singularity, while those at $r = r_{\pm}$ are coordinate singularities which can be removed with a suitable analytical extension of

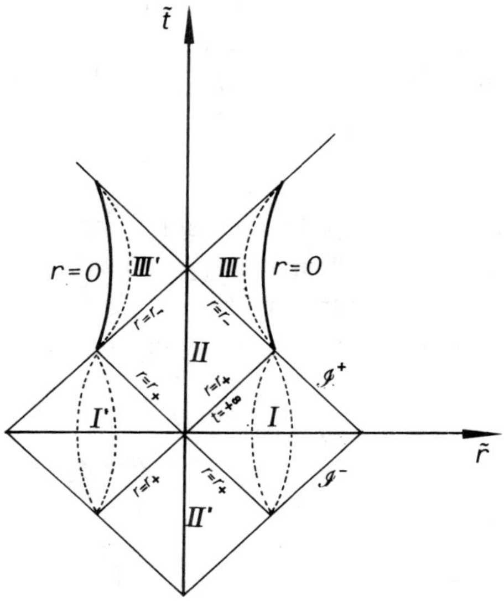

Fig. 10-13 The Penrose diagram of the Reissner-Nordström metric with $Q^2 < m^2$. The patch repeats itself along the $\tilde{t}$-axis indefinitely.

In particular the range $r_+ \leq r < \infty$ is isometric to regions I and I′; the range $r_- \leq r < r_+$ is isometric to region II and $0 \leq r < r_-$ to regions III and III′. The curvature singularity at $r = 0$ is given by the equation:

$$\cos \tilde{t} = \frac{1 + \frac{r_+}{2m}\left(\frac{r_-}{2m}\right)^{-(r_-/r_+)^2}}{1 - \frac{r_+}{2m}\left(\frac{r_-}{2m}\right)^{-(r_-/r_+)^2}} \cos \tilde{r} \tag{10.13.15}$$

and is described by a segment of a time-like curve in the $(\tilde{t}, \tilde{r})$-plane. The pattern in Fig. 10-13, from (10.13.13) and (10.13.14), repeats itself in an infinite series of similar ones which join smoothly.

What we see is that the time-like character of the singularity at $r = 0$ allows the existence of time-like curves which avoid the singularity itself. As in the Schwarzschild space-time the asymptotic regions of the (r, t)-manifold are mapped in the $(\tilde{t}, \tilde{r})$-manifold into neighbourhoods of the null segments $\mathcal{J}^{\pm}$ (Hawking and Ellis, 1973).

ii) $Q^2 = m^2$. In this case, termed *maximally charged*, the Reissner-Nordström metric (10.12.21) is singular at $r = 0$ and $r = r_+ = r_- = m$. Equation (10.13.7) now becomes:

$$r^* = r + m \ln\left(\frac{r}{m} - 1\right)^2 - \frac{rm}{r-m}. \tag{10.13.16}$$

Because we have here a pole divergence at $r = m$ besides the logarithmic divergence, a Kruskal-type coordinate transformation as in (10.13.8) does not suffice to remove the singularity at $r = m$ in the metric. Let us therefore consider the following coordinate transformations:

$$\begin{aligned}\tilde{t} + \tilde{r} &= 2\tan^{-1}\left(\exp\left(\frac{v}{2m}\right)\right)\\ \tilde{t} - \tilde{r} &= 2\tan^{-1}\left(\mp\exp\left(\mp\frac{u}{2m}\right)\right)\end{aligned} \tag{10.13.17}$$

where the "$-$" sign refers to $r > m$ and the "$+$" sign to $r < m$, and $v = t + r^*$, $u = t - r^*$. In terms of these coordinates, the metric takes the form:

$$\begin{aligned}ds^2 = -\frac{(r-m)^2}{r^2} 4m^2 \sin^{-1}(\tilde{t}+\tilde{r})\sin^{-1}(\tilde{t}-\tilde{r})(d\tilde{t}^2 - d\tilde{r}^2)\\ + r^2(d\theta^2 + \sin^2\theta\, d\varphi^2)\end{aligned} \tag{10.13.18}$$

where $r = r(\tilde{t}, \tilde{r})$ is defined implicitly as follows:

$$\begin{aligned}\cos\tilde{t} &= \frac{1 + \left(1 - \frac{r}{m}\right)^2 \mathrm{e}^{r/m}\mathrm{e}^{-r/(r-m)}}{1 - \left(1 - \frac{r}{m}\right)^2 \mathrm{e}^{r/m}\mathrm{e}^{-r/(r-m)}} \cos\tilde{r}\\ & \qquad\qquad\qquad\qquad\qquad\qquad r > m\\ \sin\tilde{t} &= \tanh\left(\frac{t}{2m}\right)\sin\tilde{r}\end{aligned} \tag{10.13.19}$$

and

$$\begin{aligned}\sin\tilde{t} &= -\frac{1 + \left(1 - \frac{r}{m}\right)^2 \mathrm{e}^{r/m}\mathrm{e}^{r/(m-r)}}{1 - \left(1 - \frac{r}{m}\right)^2 \mathrm{e}^{r/m}\mathrm{e}^{r/(m-r)}} \sin\tilde{r}\\ & \qquad\qquad\qquad\qquad\qquad\qquad r < m\,.\\ \cos\tilde{t} &= \tanh\left(\frac{t}{2m}\right)\cos\tilde{r}\end{aligned} \tag{10.13.20}$$

The two patches join smoothly at $r = m$ and the portion of the $(\tilde{t}, \tilde{r})$ plane which is isometric to the original space-time is shown in Fig. 10-14 (regions I and III).

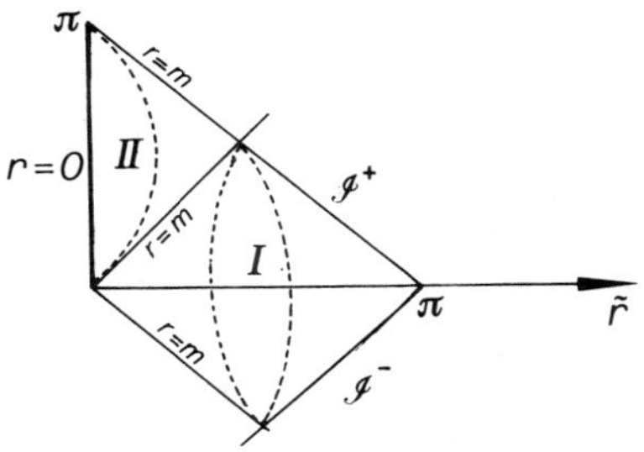

Fig. 10-14 The Penrose diagram of the Reissner-Nordström metric with $Q^2 = m^2$. The patch repeats itself along the $\tilde{t}$-axis indefinitely.

The coalescence of r_+ with r_- leads to the disappearance of the region II; however the multiplicity of the pattern remains (Carter, 1966).

iii) $Q^2 > m^2$. The metric has now only the singularity at $r = 0$; an integration of (10.13.2) gives:

$$r^* = r + m \ln \frac{1}{4m^2}(r^2 - 2mr + Q^2) + \frac{2m^2 - Q^2}{\sqrt{Q^2 - m^2}} \tan^{-1} \frac{r - m}{\sqrt{Q^2 - m^2}}, \tag{10.13.21}$$

and although there is no need for analytical extension since there are no coordinate singularities, we shall, for the sake of completeness, derive the Penrose diagram of the Reissner-Nordström solution also in this case. Let us make the transformation:

$$v' = \mathrm{e}^{v/2m} \qquad u' = -\mathrm{e}^{-u/2m} \tag{10.13.22}$$

so the metric (10.13.5) with (10.13.21) takes the form:

$$ds^2 = -\frac{8m^2}{r^2}\mathrm{e}^{-r/m} \exp\left[-\frac{2m^2 - Q^2}{m\sqrt{Q^2 - m^2}} \tan^{-1} \frac{r - m}{\sqrt{Q^2 - m^2}}\right] du'dv' + r(u', v')(d\theta^2 + \sin^2\theta \, d\varphi^2) \tag{10.13.23}$$

and

$$u'v' = -\frac{1}{4m^2}(r^2 - 2mr + Q^2)\mathrm{e}^{r/m} \times \exp\left[\frac{2m^2 - Q^2}{m\sqrt{Q^2 - m^2}} \tan^{-1} \frac{r - m}{\sqrt{Q^2 - m^2}}\right]. \tag{10.13.24}$$

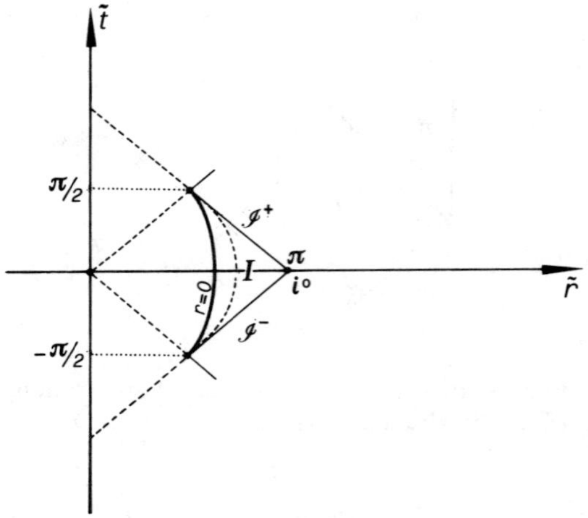

Fig. 10-15 The Penrose diagram of the Reissner-Nordström metric with $Q^2 > m^2$. Here the patch I does not repeat itself in the $(\tilde{t}, \tilde{r})$-plane.

A further transformation:

$$v' = \tan\frac{\tilde{t}+\tilde{r}}{2} \qquad u' = \tan\frac{\tilde{t}-\tilde{r}}{2} \tag{10.13.25}$$

leads to:

$$\cos\tilde{t} = \frac{1+\frac{1}{4}\left[\left(\frac{Q}{m}\right)^2 - 2\frac{r}{m} + \left(\frac{r}{m}\right)^2\right] e^{r/m} \exp\left[\frac{2m^2-Q^2}{m\sqrt{Q^2-m^2}}\tan^{-1}\frac{r-m}{\sqrt{Q^2-m^2}}\right]}{1-\frac{1}{4}\left[\left(\frac{Q}{m}\right)^2 - 2\frac{r}{m} + \left(\frac{r}{m}\right)^2\right] e^{r/m} \exp\left[\frac{2m^2-Q^2}{m\sqrt{Q^2-m^2}}\tan^{-1}\frac{r-m}{\sqrt{Q^2-m^2}}\right]}\cos\tilde{r}$$

$$\sin\tilde{t} = \tanh\left(\frac{t}{2m}\right)\sin\tilde{r}\ . \tag{10.13.26}$$

From the second of (10.13.26) we have that $t = \pm\infty$ implies $\sin\tilde{t} = \pm\sin\tilde{r}$, hence $\tilde{t} = \pm\tilde{r}$. From the first of the above set of equations, we have that $r = \infty$ implies $\cos\tilde{t} = -\cos\tilde{r}$, hence $\tilde{t} = \mp\pi \pm \tilde{r}$. Evidently the condition $r = 0$ implies:

$$\cos\tilde{t} = -\frac{A(0)+1}{A(0)-1}\cos\tilde{r}$$

where $A(r) = -u'v'$. Thus the space-time is now isometric to region I in Fig. 10-15.

10.14 Particle behaviour near the Reissner-Nordström singularity

The conformal (Penrose) diagram of the Reissner-Nordström solution with $m^2 > Q^2$, Fig. 10-13, is made of an infinite set of time-oriented space-time patches, each of them containing two time-like singularities, the left and the right-handed. They are distinguished by having the gradient of the coordinate time t, with respect to a future-oriented proper-time parameter τ, positive and negative respectively. If one wants to trace the future history of a particle on the conformal diagram of Fig. 10-13, an obvious question would be, to which of the two singularities (left or right) will the particle move? Let us consider a charged particle which moves freely in the Reissner-Nordström space-time where the source is collapsed to the singularity at $r = 0$. We neglect here questions concerning the stability of the inner horizon and the feasibility of the space-time structure below the outer horizon as illustrated by the conformal diagram of Fig. 10-13. Our main concern is to learn about the behaviour of the space-time near the time-like singularities from an analysis of the motion of particles.

The equations of motion for a radial infall are given by:

$$\frac{dt}{d\tau} = \frac{r}{\Delta}(\gamma r - eQ) \tag{10.14.1}$$

$$\frac{dr}{d\tau} = -\frac{1}{r}\left[(\gamma r - eQ)^2 - \Delta\right]^{\frac{1}{2}} \tag{10.14.2}$$

where $\Delta = (r - r_+)(r - r_-)$, e is the specific charge of the particle* and $\gamma = E/\mu_0 c^2$ is its specific total energy. The most obvious initial condition is that the particle starts its journey in the patch I of Fig. 10-13. This implies $(dt/d\tau) > 0$ initially and everywhere $r > r_+$. From equation (10.14.1), this condition is fulfilled by any $\gamma > 0$ when $eQ \leq 0$ but only by a $\gamma > eQ/r_+$ when $eQ > 0$. If the particle has a charge with opposite sign to that of the metric source, i.e. $eQ \equiv \beta < 0$, or is neutral ($\beta = 0$), it follows from (10.14.1) that as $r \to 0$, $dt/d\tau > 0$; this shows that the left-hand path is the *only* one compatible with the equations of motion.

---

*Here the charge e is in units of $\mu_0 G^{1/2}$.

The condition $\beta > 0$ is only necessary for the particle to enter the right-hand patch. In order for that to happen, the component $dt/d\tau$ must vanish somewhere between the horizons. This is possible only if $\gamma < \beta/r_-$; if the latter in not satisfied, then a $\beta > 0$ particle also follows a left-hand path. Entering either the left- or the right-handed patches does not imply that the particle will hit the singularity. The motion can be opposed by an infinite potential barrier. Let us in fact calculate, from (10.14.2), the radial acceleration:

$$\frac{d^2r}{d\tau^2} = \frac{\beta\gamma - m}{r^2} + \frac{Q^2 - \beta^2}{r^3}; \tag{10.14.3}$$

regardless of the sign of β, the singularity is repulsive ($d^2r/d\tau^2 \to \infty$ as $r \to 0$) whenever $|\beta| < Q$. This is a remarkable relativistic effect which arises from the way the electric charge contributes to the background curvature. This is particularly evident if we consider a neutral particle ($\beta = 0$) which falls in radially. From (10.14.3) we have:

$$\frac{d^2r}{d\tau^2} = -\frac{1}{r^2}\left(m - \frac{Q^2}{r}\right) \tag{10.14.4}$$

namely the particle *feels* the gravitational field of a varying mass: $m' = m - Q^2/r$, which becomes negative at $r < Q^2/m$; (de la Cruz and Israel, 1967; Cohen and Gautreau, 1979; Hiscock, 1981).

The behaviour of a particle near the Reissner-Nordström singularity is fully understood if we know the potential curves for the radial motion. From (10.14.2) these read:

$$\gamma_\pm = \frac{1}{r}\left(\beta \pm \Delta^{\frac{1}{2}}\right). \tag{10.14.5}$$

This is a one-parameter (β) family of curves which have the following main properties when $\beta > 0$:

$$\lim_{r\to 0} \gamma_+ = +\infty \quad , \quad \lim_{r\to 0} \gamma_- = \begin{cases} +\infty & \beta > Q \\ m/Q & \beta = Q \\ -\infty & \beta < Q \end{cases}. \tag{10.14.6}$$

Here and in what follows the relations for $\beta < 0$ are obtained from those with $\beta > 0$ with the substitution $\gamma_+ \to -\gamma_-$, $\gamma_- \to -\gamma_+$.

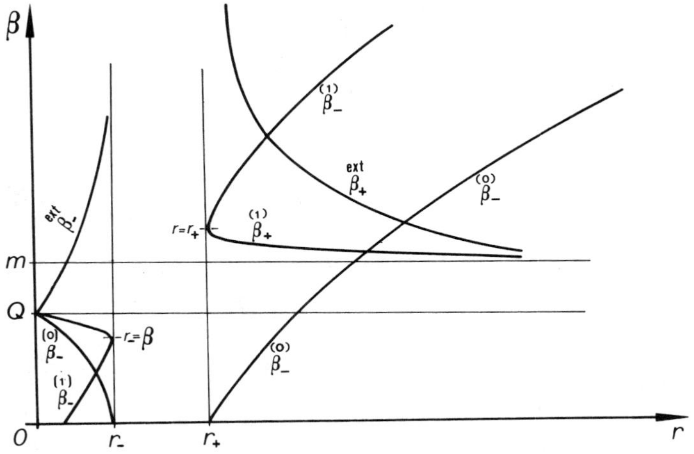

Fig. 10-16 Graphs of the functions $\overset{(1)}{\beta}_{\pm}$, $\overset{(0)}{\beta}_{\pm}$ and $\overset{\text{ext}}{\beta}_{\pm}$ (see text).

Furthermore $\gamma_- = 0$ when $\beta \equiv \overset{0}{\beta}_- = \Delta^{1/2}$, namely at

$$r = m \pm (m^2 + \beta^2 - Q^2)^{\frac{1}{2}} ,$$

(the curve $\overset{0}{\beta}_-$ in Fig. 10-16) and $\gamma_\pm = 1$ when $\beta \equiv \overset{1}{\beta}_\pm = r \mp \Delta^{\frac{1}{2}}$, namely at

$$r = \frac{(\beta^2 - Q^2)}{2(\beta - m)},$$

(the curves $\overset{1}{\beta}_\pm$ in Fig. 10-16).

The extremals of $\gamma_\pm$ occur when

$$\beta \equiv \overset{\text{ext}}{\beta}_{\pm} = \pm \Delta^{-\frac{1}{2}}(mr - Q^2)$$

namely at

$$r'_\pm = \frac{(\beta^2 - Q^2)^{\frac{1}{2}}}{\beta^2 - m^2}\left[m(\beta^2 - Q^2)^{\frac{1}{2}} \pm \beta(m^2 - Q^2)^{\frac{1}{2}}\right] ; \qquad (10.14.7)$$

(the curves $\overset{\text{ext}}{\beta}_{\pm}$ in Fig. 10-16). There, $\gamma_\pm$ take the values:

$$\overset{\max}{\gamma}_+ = \frac{r'_+ - m}{[\Delta(r'_\pm)]^{\frac{1}{2}}} \qquad r'_\pm > r_+$$

$$\overset{\min}{\gamma}_{-} = \frac{m - r'_{-}}{[\Delta(r'_{-})]^{\frac{1}{2}}} \qquad r'_{-} < r_{-} \tag{10.14.8}$$

We can see from here and Fig. 10-16 that if $\beta > m$, $\overset{\min}{\gamma}_{-} > \overset{\max}{\gamma}_{+} > 1$, their difference being:

$$\overset{\min}{\gamma}_{-} - \overset{\max}{\gamma}_{+} = 2(e^2 - 1)^{\frac{1}{2}} \left(\frac{m^2}{Q^2} - 1\right)^{\frac{1}{2}}. \tag{10.14.9}$$

With the aid of Fig. 10-16, for each value of $\beta(> 0)$ one can construct the corresponding potential curve and three cases are shown in Fig. 10-17*.

From these and the above considerations it is possible to conclude that there exists well defined ranges of values for the particle parameters which would lead the particle to hit the singularity, namely:

$$\begin{aligned} &\beta \geq m \begin{cases} \overset{\max}{\gamma}_{+} < \gamma < \overset{\min}{\gamma}_{-} & \text{all} \quad r_{\text{in}} > r_{+} \\ \gamma_{+}(r_{\text{in}}) \leq \gamma < \overset{\max}{\gamma}_{+} & r_{+} < r_{\text{in}} < r'_{+} \end{cases} \\ &Q \leq \beta < m \qquad \gamma_{+}(r_{\text{in}}) \leq \gamma < \overset{\min}{\gamma}_{-} \qquad \text{all} \quad r_{\text{in}} > r_{+} \end{aligned} \tag{10.14.10}$$

In all other cases, the particle is bounced away from the singularity. The fate of this particle after the bounce can be followed in the conformal diagram of Fig. 10.13, where it will emerge in a new asymptotically flat Universe different from the original one, (Boulware, 1973; de Felice and Maeda, 1982).

10.15 Homogeneous and isotropic cosmology

Optical and radio surveys of the sky have, until very recently, indicated that the distribution of galaxies is the same in all directions; and the very accurate uniformity of the microwave background radiation confirms the idea that, on a large enough scale, the Universe is indeed isotropic about our position. Since it is customary to argue that our position in space-time is in no way privileged this suggests that the Universe is isotropic about every space-time

*The figures there are not in scale with Fig. 10-16

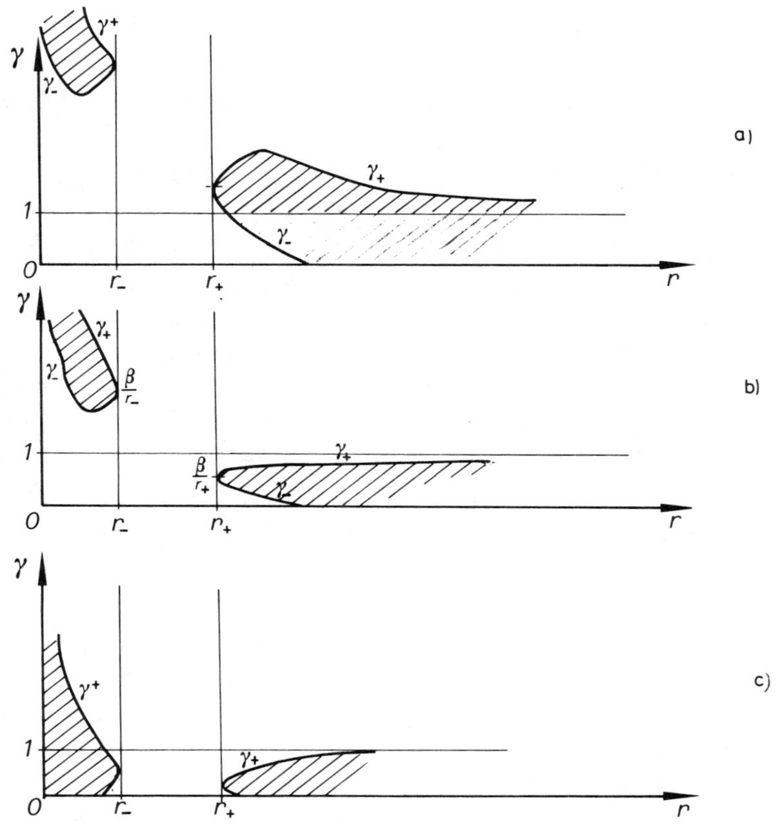

Fig. 10-17 Potential curves in the Reissner-Nordström space-time, when: a) $\beta > m$; b) $m > \beta \geq Q$; c) $< \beta < Q$. The shaded area is forbidden to the (classical) motion.

point. That is, expressed mathematically, for each point p there is an action of the rotation group as a group of isometries of the space-time with two-dimensional orbits in a neighbourhood of p (apart from p itself) and keeping fixed a unit time-like vector $\underset{p}{X}$ at p ($(X|X) = -1$). From this it is easy to argue that space-time must in fact be homogeneous: through every point there is (locally) a space-like surface which is kept fixed by the isotropy groups, and the full group of isometries will contain elements that map any point of the surface into any other point, all points on a *surface of homogeneity* being equivalent.

The assumption that the Universe is homogeneous and isotropic about any of its points is termed the *cosmological principle* (Bondi, 1961). The vectors $\underset{p}{X}$ about which isotropy holds define a corresponding time-like congruence C_X consisting of the world-lines of the observers for whom the Universe appears isotropic. We shall call the field X a *cosmic frame.* The cosmological principle constrains the congruence C_X to be geodesic ($\dot{X} = 0$) and shear and vorticity-free ($\omega_{\hat{\alpha}\hat{\beta}} = \sigma_{\hat{\alpha}\hat{\beta}} = 0$). If any of these parameters, which we refer to some smooth tetrad field $\{\lambda_{\hat{a}}\}$ adapted to the congruence (i.e. $X \equiv \lambda_{\hat{0}}$), were different than zero, the Universe would not appear all the time spatially isotropic, as can be deduced from equations (8.2.5), (8.2.6) and (8.3.3). From Sect. 2.12 and (4.5.9), it follows that C_X is hypersurface-orthogonal; indeed the orthogonal surfaces are simply the surfaces of homogeneity. At least locally there exists a function τ such that these three-dimensional surfaces orthogonal to C_X are given by τ = const. The function τ can be taken as a time coordinate and termed the *standard cosmic time*; the cosmological principle implies naturally that all the physical properties of the matter and energy in the Universe, like density, pressure, temperature, etc., are functions of τ only. Let us choose, on some $\Sigma(\tau_1) = \{p|\tau(p) = \tau_1\}$, a coordinate system $\{\underset{1}{x}^{\alpha}\}_{\alpha=1,2,3}$ and then Lie propagate them along C_X so that $X(x^{\alpha}) = 0$. In this way, the Universe is given a coordinate representation which is termed the *cosmic standard coordinate system.* Any coordinate transformation of the type:

$$t = f(\tau) \qquad x'^{\alpha} = g(x^1, x^2, x^3) \tag{10.15.1}$$

would not change the description of the Universe, hence the sets $\{\tau, x^{\alpha}\}$ and $\{t, x'^{\alpha}\}$ are said to be equivalent. Since, by definition, $X(x^{\alpha}) = 0$, we have: x^{α}=const. along C_X and the observer four-velocity X takes on the form:

$$X^i = X^0(\tau)\delta^i_0 \ . \tag{10.15.2}$$

The homogeneity and the isotropy of the three-surface Σ imply that they admit six independent Killing vector fields; three correspond to the translational invariance of the metric (and of any other non-metric field) along the spatial directions, the remaining three express the invariance relative to the rotation group SO(3). The former three assure the homogeneity, the latter the isotropy of the space. The surfaces Σ are then maximally symmetric (cf. Sect. 3.8), hence they are spaces of constant Gaussian curvature. The general considerations of Sect. 2.12 allow us to conclude that in this case the space-time metric can be written as:

$$ds^2 = g_{00}(\tau)d\tau^2 + f(\tau)ds^2_{\Sigma} \tag{10.15.3}$$

where ds^2_{Σ} is independent of τ, $g_{00}(\tau) < 0$ and $f(\tau) > 0$ are arbitrary metric functions.

The condition that the Σ's have constant Gaussian curvature, K say, determines completely their local geometry. In fact if their curvature is negative, these surfaces are locally isometric to a three-pseudosphere. If their curvature is positive or zero they are locally isometric respectively to a three-sphere or a three-plane. They can be realized as subspaces of flat pseudo-Euclidean or Euclidean spaces defined by

$$\{(x^1, x^2, x^3, z) \mid \epsilon\delta_{\alpha\beta}x^\alpha x^\beta + z^2 = a^2\} \tag{10.15.4}$$

where $a^2 = 1/|K|$ with the (pseudo)-Euclidean space having the metric:

$$ds^2_{\Sigma} = \delta_{\alpha\beta}dx^\alpha dx^\beta + \epsilon(dz)^2 \qquad \epsilon = \text{sign } K \tag{10.15.5}$$

where $\{x^\alpha\}$ are Cartesian coordinates. Differentiating (10.15.4), we have:

$$\epsilon\delta_{\alpha\beta}x^\alpha dx^\beta + zdz = 0 \ , \tag{10.15.6}$$

hence eliminating z between this and (10.15.4) and using the result in (10.15.5), we obtain:

$$ds^2_{\Sigma} = \left(\delta_{\alpha\beta} + \frac{\epsilon x^\alpha x^\beta}{a^2 - \epsilon\delta_{\rho\sigma}x^\rho x^\sigma}\right) dx^\alpha dx^\beta. \tag{10.15.7}$$

Introducing spherical polar coordinates (r, θ, φ), (10.15.7) becomes:

$$ds^2_{\Sigma} = \frac{a^2 dr^2}{a^2 - \epsilon r^2} + r^2(d\theta^2 + \sin^2\theta \, d\varphi^2). \tag{10.15.8}$$

If $K = 0$, this reduces to the metric of a flat space; if $K \neq 0$, it is more convenient to redefine r^2 as $|K|r^2$ so that (10.15.8) becomes:

$$ds^2_{\Sigma} = |K|^{-1}\left[\frac{dr^2}{1-\epsilon r^2} + r^2(d\theta^2 + \sin^2\theta\, d\varphi^2)\right]. \qquad (10.15.9)$$

The arbitrariness in the choice of the cosmic time coordinate allows us to make the following transformation:

$$t = \int^{\tau} (-g_{00}(\tau'))^{\frac{1}{2}} d\tau'$$

hence the metric (10.15.3) takes the form:

$$ds^2 = -dt^2 + R^2(t)\left[\frac{dr^2}{1-\epsilon r^2} + r^2(d\theta^2 + \sin^2\theta\, d\varphi^2)\right] \qquad (10.15.10)$$

where $R(t)$ is a new arbitrary function which is conventionally chosen to be positive. Although in (10.15.10), ϵ is restricted to the values $+1$ or -1, it is convenient to generalize that metric form so to include also the case with $K = 0$. To this end it suffices to substitute ϵ by a new parameter χ, which takes values 0 and ϵ; (10.15.10) can then be written as:

$$ds^2 = -dt^2 + R^2(t)\left[\frac{dr^2}{1-\chi r^2} + r^2(d\theta^2 + \sin^2\theta\, d\varphi^2)\right]. \qquad (10.15.11)$$

The case $\chi = 0$ requires an obvious reinterpretation of the functions $R(t)$ and r. The metric (10.15.11) is the most general form describing a spatially homogeneous and isotropic space-time; it was derived by Robertson (1935, 1936) and Walker (1936) and therefore is known as the Robertson-Walker metric.

To study its dynamical evolution, that is to know what the function $R(t)$ is, one needs to specify the energy-momentum tensor of the matter in the Universe, and solve the Einstein equations. Let us introduce the following field of tetrads adapted to C_X:

$$\begin{aligned}
\lambda^i{}_{\hat{0}} &= X^i = \delta^i_0 \\
\lambda^i{}_{\hat{r}} &= (1-\chi r^2)^{\frac{1}{2}} R^{-1} \delta^i_r \\
\lambda^i{}_{\hat{\theta}} &= (Rr)^{-1} \delta^i_\theta \\
\lambda^i{}_{\hat{\varphi}} &= (Rr\sin\theta)^{-1} \delta^i_\varphi\,.
\end{aligned} \qquad (10.15.12)$$

From (4.5.6) and (4.5.8) we find as expected $\Lambda_{\hat{\alpha}\hat{\beta}} = \omega_{\hat{\alpha}\hat{\beta}} = 0$ and from (4.4.3) $\dot{X}_{\hat{\alpha}} = 0$. Furthermore, from (8.4.9) we find:

$$\Theta_{\hat{\alpha}\hat{\beta}} = \frac{\dot{R}}{R}\delta_{\hat{\alpha}\hat{\beta}} \quad , \quad \Theta = 3\frac{\dot{R}}{R} \quad , \quad \sigma_{\hat{\alpha}\hat{\beta}} = 0. \tag{10.15.13}$$

From these properties we deduce the Einstein tensor; equations (8.4.6) to (8.4.8) become:

$$\begin{aligned} G_{\hat{0}\hat{0}} &= \frac{1}{2}(P + \frac{2}{3}\Theta^2) \\ G_{\hat{0}\hat{\beta}} &= -\frac{2}{3}\nabla_{\hat{\beta}}\Theta \\ G_{\hat{\alpha}\hat{\beta}} &= P_{\hat{\alpha}\hat{\beta}} + \dot{\Theta}_{\hat{\alpha}\hat{\beta}} + \Theta\Theta_{\hat{\alpha}\hat{\beta}} \\ &\quad - \frac{1}{2}\delta_{\hat{\alpha}\hat{\beta}}\left(P + \frac{4}{3}\Theta^2 + 2\dot{\Theta}\right) . \end{aligned} \tag{10.15.14}$$

Here $P_{\hat{\alpha}\hat{\beta}}$ is the Ricci intrinsic curvature of the surfaces Σ and we know from (3.8.17) and (3.8.18) that:

$$P_{\hat{\alpha}\hat{\beta}} = 2\frac{\chi}{R^2}\delta_{\hat{\alpha}\hat{\beta}} \quad , \quad P = \frac{6\chi}{R^2}. \tag{10.15.15}$$

Substituting these and (10.15.13) in (10.15.14), we have finally:

$$\begin{aligned} &a) \quad G_{\hat{0}\hat{0}} = \frac{3}{R^2}(\chi + \dot{R}^2) \\ &b) \quad G_{\hat{0}\hat{\beta}} = 0 \\ &c) \quad G_{\hat{\alpha}\hat{\beta}} = -\left(\chi + 2R\ddot{R} + \dot{R}^2\right)\frac{\delta_{\alpha\beta}}{R^2} . \end{aligned} \tag{10.15.16}$$

Equation (10.15.16b) holds identically, hence we expect that the energy-momentum tensor $T_{\hat{a}\hat{b}}$ satisfies the condition $T_{\hat{0}\hat{\beta}} = 0$. The vanishing of these terms implies, from (9.10.14), that the matter in the Universe is on the average at rest with respect to the standard cosmic frame and there are no dissipative processes due to thermal conduction. Moreover, since the congruence C_X is shear-free, viscous stresses should also be absent, so the matter can be represented by a perfect fluid with an energy-momentum tensor given, from (10.10.12), by:

$$T_{\hat{a}\hat{b}} = (\rho + p)\eta_{\hat{0}\hat{a}}\eta_{\hat{0}\hat{b}} + p\eta_{\hat{a}\hat{b}} \tag{10.15.17}$$

where ρ is the energy density of the fluid, p its uniform pressure. The latter, being uniform over $\Sigma(t)$ (any t), has a vanishing gradient orthogonal to X, hence equation (9.11.7) guarantees the geodesic character of the fluid motion. Moreover the homogeneity of $T_{\hat{a}\hat{b}}$ over $\Sigma(t)$ (any t) assures that the Weyl tensor vanishes identically (cf. (6.4.14) and (8.1.26) to (8.1.29)), while comparison of (10.15.14) and (10.15.15) with (8.4.2) and (8.2.9) shows consistently that the cosmic congruence C_X remains shear-free if it were so initially besides being geodesic and irrotational.

Observations indicate that the Universe is not only filled with matter (galaxies, clusters of galaxies, etc.) which brings us information only about a limited portion of our light cone, but also with radiation, which gives clear evidence that the homogeneity and the isotropy of the Universe extends with a high degree of precision to a much larger portion of our past light cone (deviation from large-scale isotropy is of the order of one part in 10^{-5})*. At the present time the energy density of the background cosmic radiation ($\sim 10^{-34}$ g/cm^3) is negligible compared with the energy density of the observed matter ($\sim 10^{-31}$ g/cm^3); however during some other phases of the Universe, its contribution to $T_{\hat{a}\hat{b}}$ could have been comparable to or even larger than that of matter. Hence ρ should include the contribution from matter and radiation, i.e. $\rho = \rho_{\rm m} + \rho_{\rm r}$ and similarly $p = p_{\rm m} + p_{\rm r}$, where obviously $p_{\rm r} = \frac{1}{3}\rho_{\rm r}$.

We are now in the position to solve Einstein's equations.

10.16 The Friedmann solutions

Combining (10.15.16) with (10.15.17), we write Einstein's equations as

$$\begin{aligned} &a) \quad 3(\chi + \dot{R}^2) = \kappa R^2 \rho \\ &b) \quad \chi + 2R\ddot{R} + \dot{R}^2 = -\kappa R^2 p\,. \end{aligned} \tag{10.16.1}$$

These should be implemented with an equation of state $p = p(\rho)$; a reasonable one is $p_{\rm m} = 0$. In fact, when matter dominates over radiation ($\rho_{\rm r} \approx 0$), the matter is sufficiently diluted to allow us to

---

* See Danese and De Zotti (1977) and De Zotti (1986) for an extensive discussion on this topic.

neglect its pressure. Conversely, when radiation dominates over matter, or even when their energy densities are comparable, the radiation pressure ($\approx \frac{1}{3}\rho_r$) is much larger than the matter pressure, so also in this case the latter can be neglected. Hence in (10.16.1) we shall assume: $\rho = \rho_m + \rho_r$ and $p = p_r = \frac{1}{3}\rho_r$.

An analysis of (10.16.1) leads to two important considerations. The first is of a geometrical nature. Eliminating $\dot{R}^2$ between them yields

$$\ddot{R} = -\frac{\kappa}{6}(\rho + 3p). \tag{10.16.2}$$

Since $R > 0$ and $\rho + 3p > 0$ (this is assured by the strong energy condition), then $\ddot{R} < 0$ always. This implies that

i) $R(t)$ varies with time, hence the Universe is either expanding or contracting;

ii) $R(t) \leq \dot{R}_0(t - t_0) + R_0$, i.e. the curve $R(t)$ is always below the tangent to any general representative point (t_0, R_0). The observations indicate that *now* (at $t = t_0$, say) the Universe is expanding, hence $\dot{R}_0 > 0$.

Choosing the origin of time so that $t_0 = R_0/\dot{R}_0$ (Fig. 10-18), condition ii) implies that $R(t)$ becomes zero at some $\bar{t} > 0$, namely in a finite time interval in our past which is less than $t_0 = R_0/\dot{R}_0$. This agrees with the result already obtained in (8.6.9) (on substituting for Θ from (10.15.13)).

From (10.15.16) we infer that $R = 0$ is a curvature singularity for the cosmological space-time, therefore a Universe modeled by the Robertson-Walker metric has an *origin* which is colloquially termed the "big bang", and may have an *end* which is termed the "big crunch". A deviation from the space-time symmetry induced by the requirements of homogeneity and isotropy does not help to avoid the occurrence of a singularity, hence we talk about the age of the Universe as a time interval which is in general $\leq t_0 = R_0/\dot{R}_0$.

The quantity

$$H_0 \equiv t_0^{-1} = \frac{\dot{R}_0}{R_0} \tag{10.16.3}$$

is known as the Hubble constant. As will be clarified later, H_0 can be deduced from astronomical observations but it is still a

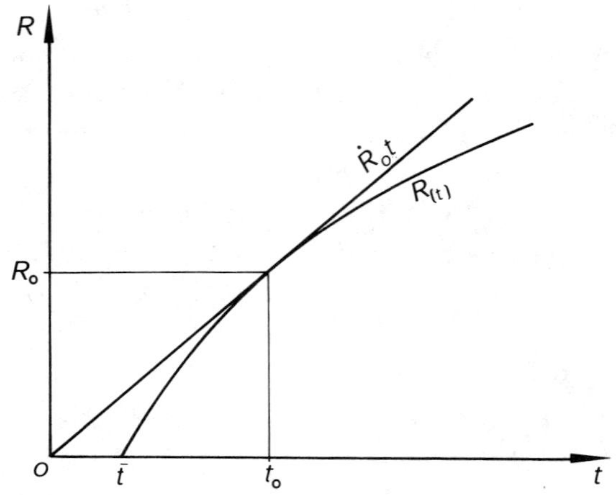

Fig. 10-18 The curve $R(t)$ is always below the tangent to any of its points.

formidable problem of modern cosmology to determine its correct numerical value. At the present time this ranges between 50 and 100 km sec^{-1} Mpc^{-1} (or $1.6-3.2\times10^{-18}$sec^{-1}); an assumed value of 75 km sec^{-1} Mpc^{-1} implies an age of the Universe less than 1.3×10^{10} years (Weinberg, 1972).

The second consideration is of a physical nature. Differentiating (10.16.1a) and eliminating $\ddot{R}$ with (10.16.1b) yields:

$$\dot{\rho} = -(\rho + p)\Theta \tag{10.16.4}$$

which is just equation (9.11.8). If the Universe is matter-dominated ($p = 0,\ \rho = \rho_{\mathrm{m}}$), then (10.16.4) leads to:

$$\rho_{\mathrm{m}} R^3 = \text{const.} \tag{10.16.5}$$

If it is radiation dominated ($\rho = \rho_{\mathrm{r}},\ p = (1/3)\rho_{\mathrm{r}}$), then

$$\rho_{\mathrm{r}} R^4 = \text{const.} \tag{10.16.6}$$

Let us consider the first case; from (10.16.1a), we have:

$$\dot{R}^2 = \frac{\kappa\rho_{0\mathrm{m}}R_0^3}{3R} - \chi\ . \tag{10.16.7}$$

There are three possible dynamical solutions, corresponding to the values of χ and illustrated in Fig. 10-19.

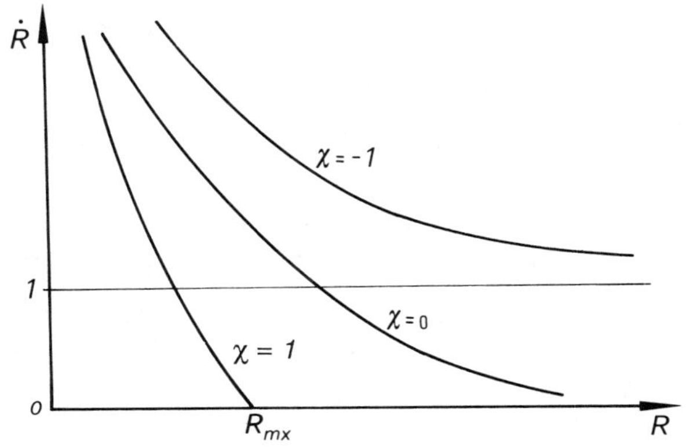

Fig. 10-19 The three possible dynamical evolutions corresponding to the value of χ.

When $\chi = +1$, $\dot{R}$ decreases until it vanishes at

$$R = \frac{\kappa}{3}\rho_{0\mathrm{m}} R_0^3 \equiv R_{\mathrm{max}}. \tag{10.16.8}$$

Hence the expansion of the Universe slows down until it stops and turns into a contraction. Rescaling the coordinate t so that $t = 0$ at $R = 0$, equation (10.16.7) admits a solution:

$$t = \frac{1}{2}R_{\mathrm{max}}\left[\cos^{-1}\left(1 - 2\frac{R}{R_{\mathrm{max}}}\right) - 2\left(\frac{R}{R_{\mathrm{max}}}\right)\left(1 - \frac{R}{R_{\mathrm{max}}}\right)^{\frac{1}{2}}\right]. \tag{10.16.9}$$

This Universe is termed *closed* since it has finite space sections; these have the topology of a three-dimensional sphere with a proper circumference of $L = 2\pi R(t)$ (or of the corresponding projective space).

When $\chi = 0$ or -1, the space sections can be infinite, in which case the universes described by these models are termed *open*. The solutions are respectively

$$\chi = 0 \qquad t = \frac{2}{3}\left(\frac{3}{\kappa\rho_{0\mathrm{m}} R_0^3}\right)^{\frac{1}{2}} R^{\frac{3}{2}} \tag{10.16.10}$$

$$\chi = -1 \qquad t = \left(\frac{\kappa\rho_{0\mathrm{m}} R_0^3}{6}\right)\left\{\left[1 + \left(1 + \frac{6R}{\kappa\rho_{0\mathrm{m}} R_0^3}\right)^2\right]^{\frac{1}{2}}\right.$$

$$- \cosh^{-1}\left(1 + \frac{6R}{\kappa\rho_{0\mathrm{m}}R_0^3}\right)\Bigg\}. \qquad (10.16.11)$$

If we consider a radiation-dominated Universe, then equation (10.16.1a) becomes from (10.16.6):

$$\dot{R}^2 = \frac{\kappa\rho_{0\mathrm{r}}R_0^4}{3R^2} - \chi. \qquad (10.16.12)$$

Comparing this with (10.16.7), we see that the expansion now goes faster than in the previous cases and in any case the law (10.16.12) would prevail over (10.16.7) at an earlier cosmic time, when:

$$\frac{R(t)}{R_0} < \frac{\rho_{0\mathrm{r}}}{\rho_{0\mathrm{m}}}. \qquad (10.16.13)$$

A problem of modern cosmology, which is still unsolved, is which of the above models fits with the observed Universe. To facilitate the confrontation of theory with observations, it is convenient to introduce quantities which can, in principle, be measured. One of these is the Hubble constant already introduced; another is the deceleration parameter:

$$q = -\frac{R\ddot{R}}{\dot{R}^2} = -\frac{\ddot{R}}{RH^2}. \qquad (10.16.14)$$

In terms of these, equations (10.16.1) become:

$$\frac{\chi}{\dot{R}^2} + 1 - 2q = -\frac{\kappa p}{H^2}$$
$$\frac{\chi}{\dot{R}^2} + 1 = \frac{\kappa\rho}{3H^2}. \qquad (10.16.15)$$

At the present time we have $p = 0$, $\rho = \rho_{0\mathrm{m}}$ hence:

$$q_0 = \frac{\kappa\rho_{0\mathrm{m}}}{6H_0^2}$$
$$q_0 = \frac{1}{2} + \frac{\chi}{2H_0^2R_0^2}. \qquad (10.16.16)$$

If $\rho_{0\mathrm{m}}$ and H_0 can be determined by observations, a value of q_0 ($>1/2$, $<1/2$, or $=1/2$), directly deducible from the first of (10.16.16), would imply from the second of (10.16.16) a value of χ and hence fix the model type of our Universe.

All the properties of the cosmological solutions (10.16.9), (10.16.10) and (10.16.11) can be expressed in terms of q_0 and H_0.

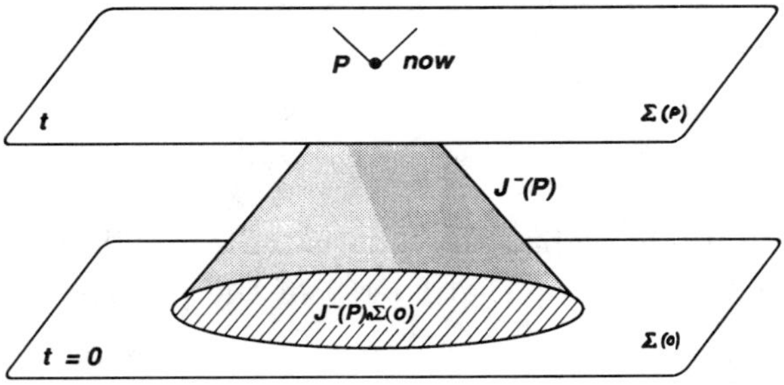

Fig. 10-20 If $J^-(p) \cap \Sigma(0) \neq \Sigma(0)$ then the causal interaction among the various parts of the Universe considered at the time of the event p is inhibited by the particle horizon.

So it is crucial to specify what are the observations which allow for direct information about those parameters: this will be discussed in the following section.

We shall now conclude this section by illustrating a property of the Friedmann solutions known as the *particle horizon* (Rindler, 1956) which sheds doubt on their adequacy as complete models of the real Universe.

To clarify the concept, let us recall that a point p in the Universe at $r = 0$ and $t = t_1$, say, can be causally affected only by events which lie in its causal past, i.e. in $J^-(p)$. The question arises, whether a particle at p can interact with all the other particles in the Universe, if we extend $J^-(p)$ sufficiently far to the past. Since we cannot go beyond the singularity at $t = 0$, the answer would be yes if the intersection of $J^-(p)$ with $\Sigma(0)$ coincided with $\Sigma(0)$* (see Fig. 10-20): i.e.

$$J^-(p) \cap \Sigma(0) = \Sigma(0).$$

If this is not the case, then it follows that the point p can only interact with the particles in the set $J^-(p) \cap \Sigma(0)$; the boundary of this set is termed the particle horizon of p.

*Strictly speaking, $\Sigma(0)$ is not a part of space-time; but the expression has an obvious meaning in the sense of a limit as $t \to 0$.

Its coordinate radius $r_{\rm H}$ is mathematically defined, from (10.15.11), as:

$$\int_0^{r_{\rm H}(t)} \frac{dr}{(1-\chi r^2)^{\frac{1}{2}}} = \int_0^{t_1} \frac{dt}{R(t)} \,. \qquad (10.16.17)$$

The integral on the right-hand side is finite in all the Friedmann solutions. The proper spatial length of the particle horizon of p on $\Sigma(t)$ is:

$$L_{\rm H}(t) = R(t) \int^t \frac{dt'}{R(t')} = R(t) \int^R \frac{dR'}{R'\dot{R}'} \qquad (10.16.18)$$

from (10.16.1a). Solving for the various models, one finds that $L_{\rm H}$ grows steadily in the cases $\chi = 0, -1$ as $t \to \infty$, so the point will ultimately interact with all the Universe. In the case $\chi = 1$, the point will interact with all the Universe only at the time of maximum expansion (Weinberg, 1972).

The existence of the particle horizon, especially in the early phases of the universal expansion, indicates that far apart regions in the Universe cannot interact with each other sufficiently fast to guarantee the homogeneity and isotropy which we observe now. The lack of causal interaction forces one either to impose constraints with no physical justification on the initial conditions and the subsequent evolution of the universe or to abandon the Friedmann-Robertson-Walker type of metric in favour of more general types of solutions.

10.17 Cosmological effects

The most surprising result of Einstein's equations is that the only possible equilibrium between an isotropic and homogeneous distribution of energy and the background geometry is an expanding or contracting Universe. No static solutions exist except at the expense of altering Einstein's equations with an additional term of the type Λg_{ij} where Λ is a constant, termed the cosmological constant. The only static solution (known as the Einstein Universe) allowed by the introduction of the Λ-term is unstable. But there is no need for such a solution (and hence of the cosmological constant) since the Universe *does* show itself to be expanding*.

The light signals which reach us now, at $t = t_0$, from distant galaxies, propagate in space-time along null geodesics. The propagation law is given by the eikonal equation (7.8.8); from (10.15.11), it becomes:

$$-\left(\frac{\partial S}{\partial t}\right)^2 + \frac{1}{R^2(t)}\left[(1-\chi r^2)\left(\frac{\partial S}{\partial r}\right)^2 + \frac{1}{r^2}\left(\frac{\partial S}{\partial \theta}\right)^2 + \frac{1}{r^2 \sin^2\theta}\left(\frac{\partial S}{\partial \varphi}\right)^2\right] = 0. \quad (10.17.1)$$

Here S is the eikonal function; searching for a solution with the method of separation of variables, i.e. $S = S_t(t) + S_r(r) + S_\theta(\theta) + S_\varphi(\varphi)$, we can separate the space-dependent terms from the time-dependent ones which can be written as:

$$R^2(t)\left(\frac{\partial S_t}{\partial t}\right)^2 = \text{const.} \quad (10.17.2)$$

Since t is the proper time of the cosmic observer, then from (9.5.3):

$$-c\frac{\partial S}{\partial t} = \nu(t)$$

is the frequency of the light signal measured by that observer at the time t. Thus, the ratio between the emitted to the observed frequencies reads, from (9.5.6):

$$\frac{\nu(t)}{\nu(t_0)} = 1 + z = \frac{R_0}{R(t)}. \quad (10.17.3)$$

At the time of emission, $R(t) < R_0$. Hence

$$\nu(t_0) = \frac{\nu(t)}{1+z} < \nu(t). \quad (10.17.4)$$

This leads to the most important observational confirmation of the theory: the light from distant galaxies is red-shifted. It is clear that the earlier the time of emission the larger is the ratio

---

*Notwithstanding this, the Λ-term has been reintroduced recently in the context of inflationary world-models (Turner et al., 1984; Peebles, 1984, 1986) to overcome the problem of particle horizons.

R_0/R, hence the larger is the red-shift. It is sometimes convenient, although rather fictitious, to interpret the observed red-shift in terms of a Doppler effect induced by a velocity of recession of the source. Using equation (9.6.4) and taking into account that the isotropy implies that everything is seen (radially) receding with velocity v, say, from whatever point of observation, we have with respect to us:

$$1 + z = \frac{\left(1 - \frac{\mathrm{v}}{c}\right)^{\frac{1}{2}}}{\left(1 + \frac{\mathrm{v}}{c}\right)^{\frac{1}{2}}}. \tag{10.17.5}$$

In the limit of small red-shifts ($|z| \to 0$), the above relation gives:

$$\frac{\mathrm{v}}{c} \approx -z \ . \tag{10.17.6}$$

If we consider only sources nearby to us, that is with $r \ll 1$, the proper distance between us at $r = 0$ and a source located at a point $r = r_1 \ll 1$, both considered at the same cosmic time t since the concept of space-distance is meaningful only if referred to simultaneous events, is from (4.1.12) and (10.15.11):

$$L(t) \approx R(t) r_1. \tag{10.17.7}$$

From (10.17.3) and (10.17.6), the velocity of recession of a nearby source at the time of observation ($t = t_0$) is:

$$\frac{\mathrm{v}}{c} \approx -\frac{\dot{R}_0 r_1}{c} \ ; \tag{10.17.8}$$

hence, from (10.17.7), we obtain a linear relation between red-shift and distances:

$$z \approx \frac{\dot{R}_0 r_1}{c} = \frac{H_0 L_0}{c}. \tag{10.17.9}$$

This relation was observed for the first time by Hubble (1929) and is known as the Hubble law. Since z can be measured directly and L_0 can be independently determined by astronomical techniques (at least within this approximation), one could deduce a value of H_0, which however suffers from the same uncertainties as are inherent in the determination of distances. Conversely, an independent determination of the Hubble constant, together with a knowledge of the red-shift, allows one to deduce distances.

The Hubble law (10.17.9) holds for nearby galaxies only, namely when $z < 1$; when $z \geq 1$, the concept of distance is much less obvious. A particularly useful definition of distance can be deduced from the energy conservation equation applied to the energy-momentum tensor of the radiation emitted from a distant source. From (7.8.24) we have for a beam of radiation:

$$T_{\mathrm{r}}^{ij} = \frac{1}{4\pi}\mu k^i k^j, \tag{10.17.10}$$

where $\mu(t) > 0$ and $k^i = dx^i/d\lambda$ is the tangent to a null ray with parameter λ. The cosmic observer X who receives the light beam at $t = t_0$ measures a power per unit area (apparent luminosity) at its telescope mirror equal to:

$$\ell_0 = \frac{1}{4\pi}\mu_0(k^i X_i)^2 = C\mu_0\nu_0^2 \tag{10.17.11}$$

from (9.5.5), and where C is some constant. From the conservation equation:

$$\nabla_j T_{\mathrm{r}}^{ij} = 0, \tag{10.17.12}$$

we obtain an equation for μ:

$$\frac{d\mu}{d\lambda} = -\mu\Theta_{\mathrm{r}} \qquad \Theta_{\mathrm{r}} = k^i{}_{;i}. \tag{10.17.13}$$

On the other hand, the cross section A to a beam of null rays satisfies the propagation law, from (8.5.15):

$$\frac{dA}{d\lambda} = A\Theta_{\mathrm{r}}, \tag{10.17.14}$$

hence, combining this with (10.17.13), we obtain:

$$\frac{d}{d\lambda}(\mu A) = 0 \tag{10.17.15}$$

that is, μA =const. along the beam. From (10.17.4) and using this last property, equation (10.17.11) can be written as:

$$\ell_0 = \frac{C\mu_0 A_0 \nu_1^2}{A_0(1+z)^2} = \frac{C\mu_1 A_1 \nu_1^2}{A_0(1+z)^2} \tag{10.17.16}$$

where μ_1, $A_1 \ldots$ mean $\mu(t_1)$, $A(t_1) \ldots$ and t_1, say, is the time of emission, and A_0 and A_1 are the cross sections of the beam, respectively, at t_0 and t_1 subtended by a solid angle Ω centred at the source. Relation (10.17.11), when applied to the time of emission t_1, can be written as

$$\ell_1 = C\mu_1\nu_1^2 = \frac{\mathrm{L}}{S_1}$$

where L is the intrinsic absolute luminosiy of the source and S_1 is the area of a two-sphere enclosing the source. If S_1 is sufficiently small in size, the following relation holds:

$$\frac{A_1}{S_1} = \frac{\Omega}{4\pi}\,, \tag{10.17.17}$$

hence (10.17.16) becomes:

$$\ell_0 = \frac{\mathrm{L}}{4\pi(1+z)^2}\,\frac{\Omega}{A_0}. \tag{10.17.18}$$

We then define *luminosity distance* from the source as the quantity:

$$\mathcal{D}_{\mathrm{L}}^2 = \frac{A_0}{\Omega}, \tag{10.17.19}$$

so equation (10.17.18) takes the form:

$$\ell_0 = \frac{\mathrm{L}}{4\pi\mathcal{D}_{\mathrm{L}}^2(1+z)^2}\,. \tag{10.17.20}$$

This equation plays a key role whenever we face a confrontation between observational and theoretical cosmology (Weinberg, 1972).

11

AXIALLY SYMMETRIC SOLUTIONS

11.1 The axially symmetric line element: the canonical form

We shall consider (pseudo)-stationary and axisymmetric space-times, which admit a two-surface $\Sigma_{(2)}$ orthogonal to the two-surface of transitivity of the group $R(1) \otimes SO(2)$ which acts on the manifold. These space-times have been introduced in Sect. 7.10. Since the Weyl tensor of a two-dimensional surface is identically zero, the geometry of $\Sigma_{(2)}$ can be described by a conformally flat metric, hence the space-time metric (7.10.7) can be put in the form:

$$ds^2 = V dt^2 + 2W dt d\varphi + X d\varphi^2 + e^{2\mu}\left((dx^1)^2 + (dx^2)^2\right) \quad (11.1.1)$$

where V, W, X and μ are functions of x^1 and x^2 alone. Any coordinate transformation $x^{A'} = x^{A'}(x^1, x^2)$ ($A' = 1, 2$) in $\Sigma_{(2)}$ leaves invariant the metric functions V, W, and X which behave, in this case, as scalars. Furthermore the metric (11.1.1) is invariant in form under transformations of the type:

$$\varphi' = \alpha\varphi + \beta t \quad , \quad t' = \gamma t + \delta\varphi$$

with α, β, γ, δ constants. The determinant of the metric of the two-dimensional (t, φ)-surface of transitivity of M is

$$-VX + W^2 = \rho^2 \quad . \quad (11.1.2)$$

ρ vanishes on the axis of symmetry where $X = W = 0$; furthermore, comparing (11.1.2) with Sect. 7.10, we see that $\rho^2 = -X\sigma$, hence ρ vanishes on the stationary null surface H, where $\sigma = 0$. The importance of ρ resides in the fact that in the vacuum case, it satisfies the two-dimensional Laplace equation in $\Sigma_{(2)}$:

$$ {}^2\nabla_{A}\partial^{A}\rho = 0 \qquad A = 1, 2 \tag{11.1.3} $$

hence ρ is harmonic in $\Sigma_{(2)}$; moreover it can be shown that ρ has no saddle points there and so it can be taken as a coordinate. Taking z as another coordinate in $\Sigma_{(2)}$ defined so that $\partial_{A} z$ is orthogonal to $\partial_{A}\rho$, the metric on $\Sigma_{(2)}$ becomes:

$$ ds^2_{\Sigma_{(2)}} = e^{2\mu(\rho,z)}\left(d\rho^2 + dz^2\right) \quad . $$

Since from (7.10.17) we deduce that $V < 0$ when $\rho^2 > W^2$, we can put:

$$ W = -hV \quad ; \quad X = -V^{-1}\rho^2 + h^2 V \quad ; \quad e^{2\mu} \equiv -e^{2\gamma}V^{-1} \quad ; \tag{11.1.4} $$

hence (11.1.1) becomes:

$$ ds^2 = -V^{-1}\left(e^{2\gamma}(d\rho^2 + dz^2) + \rho^2 d\varphi^2\right) + V\left(dt - hd\varphi\right)^2 \ . \tag{11.1.5} $$

This is the canonical form of the most general non-singular metric describing a stationary and axisymmetric space-time admitting a two-surface orthogonal to the group orbits; (11.1.5) was found by Lewis (1932) and Papapetrou (1963, 1966).

To derive the vacuum field equations ($R_{ij} = 0$), let us recall that a space-time described by a metric as in (11.1.1) admits a time-like and hypersurface orthogonal congruence of world-lines whose tangent vector field is given by (see Sec. 7.10):

$$ \ell' = Xk - Wm \quad . $$

We can associate to this congruence the tetrad field

$$ \lambda_{\hat{0}} = \frac{1}{\rho\sqrt{X}}(Xk - Wm) $$

$$ \lambda_{\hat{\varphi}} = \frac{1}{\sqrt{X}}\partial_{\varphi} $$

$$ \lambda_{\hat{A}} = e^{-\mu}\partial_{A} \qquad A = 1, 2 $$

where clearly we have reparametrized ℓ' to obtain a unit vector. From the metric symmetries and the property of ℓ' we have $\omega_{\hat{\alpha}\hat{\beta}} = 0$, $\Theta_{\hat{1}\hat{1}} = \Theta_{\hat{1}\hat{2}} = \Theta_{\hat{2}\hat{2}} = \Theta_{\hat{\varphi}\hat{\varphi}} = 0$, and hence $\Theta = 0$ and $\Lambda_{\hat{1}\hat{2}} = 0$. Using these in (8.4.1) to (8.4.3) and carrying out a long algebraic manipulation, one deduces the form of the vacuum field equations. Those relating to V and h become:

$$\begin{aligned} &a) \qquad V\,\nabla^\alpha\,(\partial_\alpha V) = (\partial^\alpha V)(\partial_\alpha V) - \rho^{-2}V^4(\partial_\alpha h)(\partial^\alpha h) \\ &b) \qquad \nabla_\alpha\,(\rho^{-2}V^2\partial^\alpha h) = 0 \qquad\qquad \alpha = \varphi,\ \rho,\ z \end{aligned} \tag{11.1.6}$$

where now the Greek indices and the gradient operator are relative to the fictitious Euclidean metric

$$ds^2_{\mathrm{E}} = \rho^2 d\varphi^2 + d\rho^2 + dz^2 \ .$$

In this metric, equation (11.1.6b) can be written as

$$\partial_\alpha(\rho^{-1}V^2 \times \partial^\alpha h) = 0$$

from (2.10.7a), hence there exists a function Φ', of ρ and z only, such that

$$\rho^{-1}V^2\partial^\alpha h = n_\gamma \epsilon^{\gamma\alpha\beta}\partial_\beta\Phi' \tag{11.1.7}$$

where $n_\gamma = \rho\delta_{\gamma\varphi}$ is the unit normal to the (ρ, z)-plane and $\epsilon^{\gamma\alpha\beta} = \rho^{-1}\delta^{\gamma\alpha\beta}$ is the alternating tensor. If we redefine $\Phi = -\Phi'$, (11.1.7) implies:

$$V^{-2}\partial_\sigma\Phi = -\frac{1}{\rho}\epsilon_{\sigma\pi\alpha}n^\pi\partial^\alpha h \ , \tag{11.1.8}$$

and so equation (11.1.6b) is equivalent to:

$$\nabla^\sigma\,(V^{-2}\partial_\sigma\Phi) = 0 \ . \tag{11.1.9}$$

From this and definitions (11.1.7), equation (11.1.6a) becomes:

$$\nabla^\alpha\left[\frac{\partial_\alpha(V^2+\Phi^2)}{V^2}\right] = 0 \ . \tag{11.1.10}$$

It is now convenient to introduce the complex function ξ:

$$\frac{\xi - 1}{\xi + 1} = -V + \mathrm{i}\Phi \ ; \tag{11.1.11}$$

in terms of this, equations (11.1.9) and (11.1.10) become the single one:

$$(\xi\bar{\xi} - 1)\,\nabla^\alpha\,(\partial_\alpha\xi) = 2\bar{\xi}(\partial_\alpha\xi)(\partial^\alpha\xi) \tag{11.1.12}$$

where the bar means complex conjugation. This is known as the Ernst equation (Ernst, 1968a, b).

A simple class of solutions to this equation are those with constant phase, namely of the type:

$$\xi = \mathrm{e}^{\mathrm{i}\alpha} \coth \psi \tag{11.1.13}$$

where α is constant and ψ is a real function of ρ and z only. With this choice, equation (11.1.12) becomes:

$$\nabla^{\alpha} (\partial_{\alpha} \psi) = 0 \, . \tag{11.1.14}$$

The search for vacuum stationary and axisymmetric solutions is in this case reduced to the integration of the three-dimensional Laplace equation in the function ψ of two variables only. Since equation (11.1.14) is linear, a linear combination of any two solutions ψ_1 and ψ_2 is also a solution of (11.1.14); but this does not apply straightforwardly to the corresponding solutions ξ_1 and ξ_2 of equation (11.1.12), since that is not linear. Thus, for a linear combination of ξ_1 and ξ_2 to be a solution of the Ernst equation, the coefficients need to be constrained. Once V and Φ are known (and thus also h) the metric function γ is deducible from the remaining field equations and satisfies the following relations:

$$\frac{\partial \gamma}{\partial \rho} = \frac{\rho}{(\xi\bar{\xi} - 1)^2} \left(\frac{\partial \xi}{\partial \rho} \frac{\partial \bar{\xi}}{\partial \rho} - \frac{\partial \xi}{\partial z} \frac{\partial \bar{\xi}}{\partial z} \right) \tag{11.1.15a}$$

$$\frac{\partial \gamma}{\partial z} = \frac{2\rho}{(\xi\bar{\xi} - 1)^2} \mathrm{Re} \left(\frac{\partial \xi}{\partial \rho} \frac{\partial \bar{\xi}}{\partial z} \right) . \tag{11.1.15b}$$

11.2 The Kerr solution

The most important vacuum solution of the Einstein field equations for stationary, axisymmetric, asymptotically flat spacetimes, was found by Kerr (1963). This solution is believed to describe the gravitational field of a rotating, isolated system which presumably has collapsed beyond an event horizon. No satisfactory internal solution, matching with the Kerr metric on the outside, has been found so far. In order to deduce this solution from the Ernst equation (11.1.12), it is more convenient to introduce spheroidal coordinates (x, y) which are related to the cylindrical ones (ρ, z) by the relation:

$$\begin{aligned} \rho &= \boldsymbol{k}(x^2-1)^{\frac{1}{2}}(1-y^2)^{\frac{1}{2}} & |y|<1 \\ z &= \boldsymbol{k}xy & |x|>1\ ; \end{aligned} \qquad (11.2.1)$$

$\boldsymbol{k}$ is a constant scale factor. In these coordinates the flat metric in $\Sigma_{(2)}$ becomes:

$$d\rho^2 + dz^2 = \boldsymbol{k}^2(x^2-y^2)\left(\frac{dx^2}{x^2-1} + \frac{dy^2}{1-y^2}\right) \qquad (11.2.2)$$

and the surfaces x =const. and y =const. are orthogonal families of spheroids and hyperboloids, respectively. From (11.2.1) and (11.2.2), the Ernst equation becomes:

$$\begin{aligned} (\xi\bar{\xi}-1)&\left\{\frac{\partial}{\partial x}\left[(x^2-1)\frac{\partial\xi}{\partial x}\right] + \frac{\partial}{\partial y}\left[(1-y^2)\frac{\partial\xi}{\partial y}\right]\right\} \\ &= 2\bar{\xi}\left[(x^2-1)\left(\frac{\partial\xi}{\partial x}\right)^2 + (1-y^2)\left(\frac{\partial\xi}{\partial y}\right)^2\right] . \end{aligned} \qquad (11.2.3)$$

Let us then consider the following solution of (11.1.14):

$$\psi_1 = \frac{1}{2}\ln\left(\frac{x-1}{x+1}\right) \quad ; \quad \psi_2 = \frac{1}{2}\ln\left(\frac{y-1}{y+1}\right) \ ;$$

namely, from (11.1.13):

$$\xi_1 = -e^{i\alpha}x \quad , \quad \xi_2 = -e^{i\alpha}y \ . \qquad (11.2.4)$$

The Kerr solution corresponds to a choice of ξ as*:

$$\xi = px + qy \qquad p, q \in \mathbb{C} \qquad (11.2.5)$$

where, as we mentioned, the constant factors need to be constrained in order that ξ be a solution of (11.2.3). Putting (11.2.5) in there, we obtain:

$$px - qy = (\bar{p}x + \bar{q}y)(p^2 - q^2) \ , \qquad (11.2.6)$$

hence the latter is identically true if:

$$\begin{aligned} p &= \bar{p}(p^2-q^2) \\ q &= -\bar{q}(p^2-q^2) \ . \end{aligned} \qquad (11.2.7)$$

---

*We incorporate the arbitrary phase factor $e^{i\alpha}$ into p and q.

If we express p and q as

$$p = P\mathrm{e}^{\mathrm{i}\alpha} \quad , \quad q = Q\mathrm{e}^{\mathrm{i}\beta}$$

where P and Q are real and α and β are some new phase factors, the condition (11.2.7) amounts to:

$$\mathrm{e}^{2\mathrm{i}\alpha} = -\mathrm{e}^{2\mathrm{i}\beta}$$

which implies

$$\alpha = \beta \pm \frac{\pi}{2} \ . \tag{11.2.8}$$

Choosing the upper sign, and using it in (11.2.7), we have:

$$\begin{aligned} \mathrm{e}^{2\mathrm{i}(\beta+\frac{\pi}{2})} &= P^2\mathrm{e}^{2\mathrm{i}(\beta+\frac{\pi}{2})} - Q^2\mathrm{e}^{2\mathrm{i}\beta} \\ \mathrm{e}^{2\mathrm{i}\beta} &= -P^2\mathrm{e}^{2\mathrm{i}(\beta+\frac{\pi}{2})} + Q^2\mathrm{e}^{2\mathrm{i}\beta} \end{aligned} \tag{11.2.9}$$

which leads to:

$$P^2 + Q^2 = 1 \ . \tag{11.2.10}$$

It is worth noticing here that the Ernst equation (11.1.12) is invariant under a change of the phase α in (11.1.13); hence we can, without loss of generality, choose $\mathrm{e}^{\mathrm{i}\beta} = -\mathrm{i}$, so that $q = -\mathrm{i}Q, \ p = P$.

The required solution (11.2.5), then becomes:

$$\xi = Px - \mathrm{i}Qy \qquad P, Q \in \mathbb{R} \ ; \tag{11.2.11}$$

had we chosen the minus sign in (11.2.8) we would have obtained the complex conjugate of (11.2.11). From this and (11.1.11) we obtain:

$$V = -\frac{P^2x^2 + Q^2y^2 - 1}{(Px+1)^2 + Q^2y^2} \tag{11.2.12}$$

$$\Phi = -\frac{2Qy}{(Px+1)^2 + Q^2y^2} \ . \tag{11.2.13}$$

To obtain the metric functions h and γ, we start by considering (11.1.7). In terms of the x and y coordinates, that equation becomes:

$$\begin{aligned} (x^2-1)^{\frac{1}{2}}\frac{\partial\Phi}{\partial x} &= \frac{V^2}{\rho}(1-y^2)^{\frac{1}{2}}\frac{\partial h}{\partial y} \\ -(1-y^2)^{\frac{1}{2}}\frac{\partial\Phi}{\partial y} &= \frac{V^2}{\rho}(x^2-1)^{\frac{1}{2}}\frac{\partial h}{\partial x} \end{aligned} \tag{11.2.14}$$

hence, from (11.2.12) and (11.2.13), these become:

$$\frac{\partial h}{\partial y} = \frac{4\boldsymbol{k}(x^2-1)PQy(Px+1)}{[P^2x^2-1+Q^2y^2]^2}$$
$$\frac{\partial h}{\partial x} = +\frac{2\boldsymbol{k}(1-y^2)Q[(Px+1)^2-Q^2y^2]}{[P^2x^2-1+Q^2y^2]^2} \ . \qquad (11.2.15)$$

The solution of these equations is:

$$h = -\frac{2\boldsymbol{k}Q}{P}\,\frac{(Px+1)(1-y^2)}{(P^2x^2-1+Q^2y^2)} + C \ ; \qquad (11.2.16)$$

C is an integration constant which we put equal to zero since h vanishes on the axis ($y = \pm 1$).

It remains to evaluate the metric coefficient γ. Let us define a new variable as:

$$e^{2\gamma'} = e^{2\gamma}(x^2-y^2)$$

then in terms of the coordinates (x, y), equations (11.1.15) become:

$$\frac{\partial \gamma'}{\partial y} = \frac{(x^2-1)(1-y^2)}{(x^2-y^2)(\xi\bar{\xi}-1)^2}\left\{x\left(\frac{\partial \xi}{\partial x}\frac{\partial \bar{\xi}}{\partial y} + \frac{\partial \bar{\xi}}{\partial x}\frac{\partial \xi}{\partial y}\right) - y\left[\frac{\partial \xi}{\partial y}\frac{\partial \bar{\xi}}{\partial y}\right.\right.$$
$$\left.\left. - \left(\frac{x^2-1}{1-y^2}\right)\frac{\partial \xi}{\partial x}\frac{\partial \bar{\xi}}{x}\right]\right\} - \frac{y}{(x^2-y^2)}$$
$$\frac{\partial \gamma'}{\partial x} = \frac{(x^2-1)(1-y^2)}{(x^2-y^2)(\xi\bar{\xi}-1)^2}\left\{x\left[\frac{\partial \xi}{\partial x}\frac{\partial \bar{\xi}}{\partial x} - \frac{(1-y^2)}{(x^2-1)}\frac{\partial \xi}{\partial y}\frac{\partial \bar{\xi}}{\partial y}\right]\right.$$
$$\left. -y\left[\frac{\partial \bar{\xi}}{\partial x}\frac{\partial \xi}{\partial y} + \frac{\partial \xi}{\partial x}\frac{\partial \bar{\xi}}{\partial y}\right]\right\} + \frac{x}{x^2-y^2} \ . \qquad (11.2.17)$$

From (11.2.11) however, we have:

$$\frac{\partial \xi}{\partial x}\frac{\partial \bar{\xi}}{\partial y} + \frac{\partial \bar{\xi}}{\partial x}\frac{\partial \xi}{\partial y} = 0 \ , \qquad (11.2.18)$$

hence (11.2.17) become:

$$\frac{\partial \gamma'}{\partial y} = \frac{(x^2-1)y}{(x^2-y^2)(P^2x^2+Q^2y^2-1)} - \frac{y}{x^2-y^2}$$
$$= \frac{Q^2y}{P^2x^2+Q^2y^2-1}$$

$$\frac{\partial\gamma'}{\partial x} = \frac{(1-y^2)x}{(x^2-y^2)(P^2x^2+Q^2y^2-1)} + \frac{x}{x^2-y^2}$$
$$= \frac{P^2x}{P^2x^2+Q^2y^2-1} \ . \tag{11.2.19}$$

The solution of these equations is:

$$\gamma' = \ln\left[C'(P^2x^2+Q^2y^2-1)^{\frac{1}{2}}\right] \tag{11.2.20}$$

where C' is an integration constant to be determined later. Equations (11.2.12), (11.2.16) and (11.2.20) completely determine the line element (11.1.5); from (11.2.1) and (11.2.2), this becomes:

$$ds^2 = -V^{-1}\left[\boldsymbol{k}^2 e^{2\gamma'}\left(\frac{dx^2}{x^2-1}+\frac{dy^2}{1-y^2}\right)+\rho^2 d\varphi^2\right]+V(dt-hd\varphi)^2 \ . \tag{11.2.21}$$

Because of the constraint (11.2.10), the metric depends on two parameters only, hence we introduce the following parameter change:

$$P = \left(1-\frac{a^2}{m^2}\right)^{\frac{1}{2}} \ , \quad Q = \frac{a}{m} \ , \quad \boldsymbol{k} = (m^2-a^2)^{\frac{1}{2}} \ , \quad a < m \ , \tag{11.2.22}$$

with a and m having the dimension of a length.* With these definitions, the metric coefficients become:

$$V = -\frac{\Sigma-2mr}{\Sigma}$$
$$h = -\frac{2arm\sin^2\theta}{\Sigma-2mr}$$
$$\gamma' = \ln\left[\frac{C'}{m}(\Sigma-2mr)^{\frac{1}{2}}\right] \tag{11.2.23}$$

where we define, compatibly with the constraints on x and y in (11.2.1):

$$r = (m^2-a^2)^{\frac{1}{2}}x+m \qquad \theta = \cos^{-1}y \qquad \Sigma = r^2+a^2\cos^2\theta \ . \tag{11.2.24}$$

*The parameter m should not be confused with the Killing vector m which is related to the axial symmetry of the solution.

Finally, if we choose $C' = m/(m^2 - a^2)^{\frac{1}{2}}$, the metric takes the form:

$$ds^2 = -\left(1 - \frac{2mr}{\Sigma}\right) dt^2 - \frac{4amr\sin^2\theta}{\Sigma} dt d\varphi + \frac{\mathcal{A}}{\Sigma}\sin^2\theta\, d\varphi^2 + \frac{\Sigma}{\Delta} dr^2 + \Sigma\, d\theta^2 \quad (11.2.25)$$

where we put:

$$\Delta = r^2 + a^2 - 2mr \quad (11.2.26)$$

$$\mathcal{A} = (r^2 + a^2)^2 - a^2 \Delta \sin^2\theta\ . \quad (11.2.27)$$

The inverse of the metric (11.2.25) is given by:

$$\left(\frac{\partial}{\partial s}\right)^2 = \frac{1}{\Sigma}\left\{\Delta\left(\frac{\partial}{\partial r}\right)^2 + \left(\frac{\partial}{\partial \theta}\right)^2 + \frac{\Delta - a^2\sin^2\theta}{\Delta\sin^2\theta}\left(\frac{\partial}{\partial\varphi}\right)^2 - \frac{4mra}{\Delta}\frac{\partial}{\partial t}\frac{\partial}{\partial\varphi} - \frac{\mathcal{A}}{\Delta}\left(\frac{\partial}{\partial t}\right)^2\right\}. \quad (11.2.28)$$

The Kerr solution reduces to the Schwarzschild solution (10.2.15), when $a = 0$; some components are singular where $\Delta = 0$, namely at:

$$r = r_{\pm} \equiv m \pm (m^2 - a^2)^{\frac{1}{2}} \quad (11.2.29)$$

and at $\Sigma = 0$ namely at $r = 0$ and $\theta = \pi/2$. We shall see in a later section that while (11.2.29) are removable coordinate singularities, $\Sigma = 0$ is a non-removable curvature singularity.

11.3 Physical interpretation of the Kerr metric

The physical significance of the parameters which characterize the solution (11.2.25) can be deduced from an investigation of their behaviour at large values of the cylindrical coordinate ρ. From (11.2.1) and (11.2.24) we have:

$$\rho = \Delta^{\frac{1}{2}}\sin\theta\ , \quad (11.3.1)$$

hence in the limit $r \to \infty$, $\rho \to r\sin\theta$ and, since θ is a latitude, r behaves asymptotically as a radial coordinate. As $r \to \infty$, the metric (11.2.25) tends to the flat Minkowski metric; thus one can exploit this property to apply the conservation equations introduced in Sect. 7.1. However the Kerr metric admits two commuting Killing vector fields, k and m, which in the coordinates used have components:

$$k^i = \delta^i_0 \qquad m^i = \delta^i_\varphi$$

hence a conserved quantity is provided by the Komar integral (7.1.16) (in geometric units):

$$\frac{1}{2}P_{\rm K}\Big|_{\rm A} \equiv I_{\rm A} = \frac{1}{8\pi}\oint_S *\boldsymbol{d\xi}_{\rm A} \quad ; \quad {\rm A} = 0, \varphi \qquad (11.3.2)$$

for each of the Killing one-forms $\boldsymbol{\xi}_{\rm A}$ ($\boldsymbol{\xi}_0 \equiv \boldsymbol{k}$; $\boldsymbol{\xi}_\varphi \equiv \boldsymbol{m}$); $*\boldsymbol{d\xi}_{\rm A}$ is the dual of the two-form $\boldsymbol{d\xi}_{\rm A}$. In (11.3.2) the integral is extended over a closed space-like two-surface S of the background metric.

Let us now consider the orthonormal frame of one-forms (Cohen and de Felice, 1984):

$$\begin{aligned}
\boldsymbol{\lambda}_{\hat{0}} &= -\left(\frac{\Delta\Sigma}{\mathcal{A}}\right)^{\frac{1}{2}} \boldsymbol{dt} \\
\boldsymbol{\lambda}_{\hat{1}} &= \left(\frac{\Sigma}{\Delta}\right)^{\frac{1}{2}} \boldsymbol{dr} \\
\boldsymbol{\lambda}_{\hat{2}} &= \Sigma^{\frac{1}{2}} \boldsymbol{d\theta} \\
\boldsymbol{\lambda}_{\hat{3}} &= \left(\frac{\mathcal{A}\sin^2\theta}{\Sigma}\right)^{\frac{1}{2}} (\boldsymbol{d\varphi} - \omega \boldsymbol{dt})
\end{aligned} \qquad (11.3.3)$$

where:

$$\omega = \frac{2mar}{\mathcal{A}} \quad . \qquad (11.3.4)$$

In terms of these, the stationary Killing one-form $\boldsymbol{\xi}_0 \equiv \boldsymbol{k} = g_{i0}\boldsymbol{dx}^i$ reads:

$$\boldsymbol{\xi}_0 = \left(\frac{\Delta\Sigma}{\mathcal{A}}\right)^{\frac{1}{2}} \boldsymbol{\lambda}_{\hat{0}} - \frac{2mar}{\sqrt{\Sigma\mathcal{A}}}\sin\theta\,\boldsymbol{\lambda}_{\hat{3}} \ . \qquad (11.3.5)$$

Differentiation of (11.3.5) yields after some algebra:

$$\boldsymbol{d\xi}_0 = \tilde{f}\boldsymbol{\lambda}_{\hat{1}} \wedge \boldsymbol{\lambda}_{\hat{0}} + \tilde{g}\boldsymbol{\lambda}_{\hat{2}} \wedge \boldsymbol{\lambda}_{\hat{0}} + \tilde{h}\boldsymbol{\lambda}_{\hat{1}} \wedge \boldsymbol{\lambda}_{\hat{3}} + \tilde{k}\boldsymbol{\lambda}_{\hat{2}} \wedge \boldsymbol{\lambda}_{\hat{3}} \qquad (11.3.6)$$

where:

$$\tilde{f} = -\left(\frac{2\mathcal{A}^{\frac{1}{2}}}{\Sigma^3}\right)(1 - a\omega\sin^2\theta)(M\Sigma - 2mr^2)$$

$$\tilde{g} = -a\omega\left(\frac{\mathcal{A}^3}{\Sigma^6\Delta}\right)^{\frac{1}{2}}\sin 2\theta\left[1 - \frac{r^2+a^2}{a}\omega\right]$$

$$\tilde{h} = -\left(\frac{\Delta}{\mathcal{A}}\right)^{\frac{1}{2}}\left(\frac{2a}{\Sigma^2}\right)(M\Sigma - 2mr^2)\sin\theta$$

$$\tilde{k} = -\frac{2\mathcal{A}^{\frac{1}{2}}\omega\cos\theta}{\Sigma^2}(r^2+a^2)\ . \tag{11.3.7}$$

The dual of (11.3.6) reads*:

$$*\boldsymbol{d\xi}_0 = \tilde{f}\boldsymbol{\lambda}_{\hat{2}}\wedge\boldsymbol{\lambda}_{\hat{3}} - \tilde{g}\boldsymbol{\lambda}_{\hat{1}}\wedge\boldsymbol{\lambda}_{\hat{3}} + \tilde{h}\boldsymbol{\lambda}_{\hat{2}}\wedge\boldsymbol{\lambda}_{\hat{0}} - \tilde{k}\boldsymbol{\lambda}_{\hat{1}}\wedge\boldsymbol{\lambda}_{\hat{0}}. \tag{11.3.8}$$

Expressing (11.3.8) in terms of coordinates, we have:

$$\begin{aligned} *\boldsymbol{d\xi}_0 = &-\left[\tilde{f}\mathcal{A}^{\frac{1}{2}}\omega\sin\theta + \tilde{h}\Sigma\left(\frac{\Delta}{\mathcal{A}}\right)^{\frac{1}{2}}\right]\ \boldsymbol{d\theta}\wedge\boldsymbol{dt} \\ &+\left[\tilde{g}\left(\frac{\mathcal{A}}{\Delta}\right)^{\frac{1}{2}}\omega\sin\theta + \tilde{k}\Sigma\mathcal{A}^{-\frac{1}{2}}\right]\ \boldsymbol{dr}\wedge\boldsymbol{dt} \\ &+\tilde{f}\mathcal{A}^{\frac{1}{2}}\sin\theta\ \boldsymbol{d\theta}\wedge\boldsymbol{d\varphi} - \tilde{g}\left(\frac{\mathcal{A}}{\Delta}\right)^{\frac{1}{2}}\sin\theta\ \boldsymbol{dr}\wedge\boldsymbol{d\varphi}\ ; \end{aligned} \tag{11.3.9}$$

hence integrating over the surface $dt = dr = 0$, which is a surface of simultaneous events relative to the observer who carries the frame (11.3.3), we have from (1.18.18):

$$\oint_S *\boldsymbol{d\xi}_0 = \int\int \tilde{f}\mathcal{A}^{\frac{1}{2}}\sin\theta\ d\theta d\varphi = 8\pi m \tag{11.3.10}$$

thus, from (11.3.2), $I_0 = m$. The parameter m is a conserved quantity associated to the Killing vector field which describes the stationarity of the metric and therefore it is to be interpreted as the total energy of the metric source, in geometric units, relative to an observer at the flat infinity.

The same analysis can be repeated with the Killing one-form $\boldsymbol{\xi}_\varphi \equiv \boldsymbol{m}$ which expresses the axial symmetry. In this case the

*The convention used is $\delta^{\hat{0}\hat{1}\hat{2}\hat{3}} = +1$.

Komar conserved quantity yields the total angular momentum of the metric source (again in units of length) and the result of the calculation is:

$$I_{\varphi} \equiv J = ma \ . \tag{11.3.11}$$

This shows that $a = J/m$ is to be interpreted as the specific angular momentum of the source when measured at the flat infinity (Boyer and Price, 1965; Cohen, 1968).

11.4 The space-time structure

To understand fully the global structure of the Kerr space-time, it is essential to investigate the nature of the metric singularities at $\Delta = 0$ and $\Sigma = 0$. We have mentioned already that those which occur at $r = r_{\pm}$ ($\Delta = 0$) are coordinate (removable) singularities while that at $\Sigma = 0$ is a curvature singularity and hence non-removable. In order to see this, let us compute the components of the Riemann tensor (which in our case coincides with the Weyl tensor). It is simpler to use Cartan algebra and calculate the curvature two-forms:

$$\boldsymbol{\Omega}^{\hat{a}}{}_{\hat{b}} = R^{\hat{a}}{}_{\hat{b}\hat{c}\hat{d}} \boldsymbol{\lambda}^{\hat{c}} \wedge \boldsymbol{\lambda}^{\hat{d}} \ . \tag{11.4.1}$$

However, rather than using the one-forms (11.3.3), it turns out to be more convenient to consider a new frame of one-forms, $\{\hat{\boldsymbol{\lambda}}_{\hat{a}}\}$, first introduced by Carter (Carter, 1968a), and related to (11.3.3) by:

$$\begin{aligned}
\hat{\boldsymbol{\lambda}}^{\hat{0}} &= \frac{\mathcal{A}^{\frac{1}{2}}}{\Sigma}(1 - a\omega \sin^2\theta)\boldsymbol{\lambda}^{\hat{0}} - \left(\frac{\Delta}{\mathcal{A}}\right)^{\frac{1}{2}} a \sin\theta \boldsymbol{\lambda}^{\hat{3}} \\
\hat{\boldsymbol{\lambda}}^{\hat{1}} &= \boldsymbol{\lambda}^{\hat{1}} \\
\hat{\boldsymbol{\lambda}}^{\hat{2}} &= \boldsymbol{\lambda}^{\hat{2}} \\
\hat{\boldsymbol{\lambda}}^{\hat{3}} &= \left(\frac{\Delta}{\mathcal{A}}\right)^{\frac{1}{2}} a \sin\theta \boldsymbol{\lambda}^{\hat{0}} - \frac{(r^2 + a^2)}{\mathcal{A}^{\frac{1}{2}}} \boldsymbol{\lambda}^{\hat{3}} \ .
\end{aligned} \tag{11.4.2}$$

With respect to this, the curvature two-forms (11.4.1) become (Marck, 1983):

$$\begin{aligned}
\hat{\boldsymbol{\Omega}}^{\hat{1}}{}_{\hat{2}} &= -I_1\hat{\boldsymbol{\lambda}}^{\hat{1}} \wedge \hat{\boldsymbol{\lambda}}^{\hat{2}} + I_2\hat{\boldsymbol{\lambda}}^{\hat{0}} \wedge \hat{\boldsymbol{\lambda}}^{\hat{3}} \\
\hat{\boldsymbol{\Omega}}^{\hat{0}}{}_{\hat{3}} &= -I_1\hat{\boldsymbol{\lambda}}^{\hat{0}} \wedge \hat{\boldsymbol{\lambda}}^{\hat{3}} - I_2\hat{\boldsymbol{\lambda}}^{\hat{1}} \wedge \hat{\boldsymbol{\lambda}}^{\hat{2}} \\
\hat{\boldsymbol{\Omega}}^{\hat{0}}{}_{\hat{1}} &= -2I_1\hat{\boldsymbol{\lambda}}^{\hat{1}} \wedge \hat{\boldsymbol{\lambda}}^{\hat{0}} + 2I_2\hat{\boldsymbol{\lambda}}^{\hat{2}} \wedge \hat{\boldsymbol{\lambda}}^{\hat{3}} \\
\hat{\boldsymbol{\Omega}}^{\hat{3}}{}_{\hat{2}} &= -2I_1\hat{\boldsymbol{\lambda}}^{\hat{2}} \wedge \hat{\boldsymbol{\lambda}}^{\hat{3}} - 2I_2\hat{\boldsymbol{\lambda}}^{\hat{1}} \wedge \hat{\boldsymbol{\lambda}}^{\hat{0}} \\
\hat{\boldsymbol{\Omega}}^{\hat{0}}{}_{\hat{2}} &= I_1\hat{\boldsymbol{\lambda}}^{\hat{2}} \wedge \hat{\boldsymbol{\lambda}}^{\hat{0}} + I_2\hat{\boldsymbol{\lambda}}^{\hat{1}} \wedge \hat{\boldsymbol{\lambda}}^{\hat{3}} \\
\hat{\boldsymbol{\Omega}}^{\hat{3}}{}_{\hat{1}} &= I_1\hat{\boldsymbol{\lambda}}^{\hat{1}} \wedge \hat{\boldsymbol{\lambda}}^{\hat{3}} - I_2\hat{\boldsymbol{\lambda}}^{\hat{2}} \wedge \hat{\boldsymbol{\lambda}}^{\hat{0}}
\end{aligned} \qquad (11.4.3a)$$

where:

$$I_1 = \frac{mr}{\Sigma^3}(r^2 - 3a^2\cos^2\theta) \quad ; \quad I_2 = \frac{ma\cos\theta}{\Sigma^3}(3r^2 - a^2\cos^2\theta) \ . \qquad (11.4.3b)$$

From (11.4.3a and b) the curvature invariant $I = R_{ijkl}R^{ijkl}$ becomes*:

$$I(r,\theta) = \frac{48m^2}{\Sigma^6}(r^2 - a^2\cos^2\theta)(r^4 - 14a^2r^2\cos^2\theta + a^4\cos^4\theta) \ . \qquad (11.4.4)$$

This reduces to (10.5.1) when $a = 0$; it diverges at $\Sigma = 0$ but remains finite and smooth at $\Delta = 0$. This suggests that one can find a coordinate transformation which removes the singularity at $\Delta = 0$ in the metric (11.2.25). In fact with the following:

$$\begin{aligned}
r &= \tilde{r} \\
\theta &= \tilde{\theta} \\
d\varphi &= d\tilde{\varphi} - a\frac{dr}{\Delta} \\
dt &= d\tilde{t} - 2mr\frac{dr}{\Delta}
\end{aligned} \qquad (11.4.5)$$

the metric (11.2.25) becomes:

$$\begin{aligned}
ds^2 = dr^2 - 2a\sin^2\theta dr d\tilde{\varphi} + (r^2 + a^2)\sin^2\theta d\tilde{\varphi}^2 + \Sigma d\theta^2 - d\tilde{t}^2 \\
+ \frac{2mr}{\Sigma}(dr - a\sin^2\theta \, d\tilde{\varphi} + d\tilde{t})^2 \ . \qquad (11.4.6)
\end{aligned}$$

*For a discussion on the behaviour of (11.4.4) see de Felice and Bradley (1988).

This form of the metric was first deduced by Boyer and Lindquist who recognized the usefulness of the spheroidal coordinates to evince the symmetries of Kerr space-time as compared with the flat Cartesian coordinates in which the Kerr metric was originally found. In fact, applying to (11.4.6) the following coordinate transformation:

$$
\begin{aligned}
\mathrm{x} &= (r^2+a^2)^{\frac{1}{2}} \sin\theta \, \cos\left[\varphi - \tan^{-1}\left(\frac{a}{r}\right)\right] \\
\mathrm{y} &= (r^2+a^2)^{\frac{1}{2}} \sin\theta \, \sin\left[\varphi - \tan^{-1}\left(\frac{a}{r}\right)\right] \\
\mathrm{z} &= r\cos\theta \\
t &= t
\end{aligned}
\tag{11.4.7}
$$

we obtain:

$$
ds^2 = \left(\eta_{ij} + \frac{2mr^3}{r^4 + a^2\mathrm{z}^2} k_i k_j\right) dx^i dx^j \quad ; \quad \{x^i\} \equiv \{\mathrm{x}, \mathrm{y}, \mathrm{z}, t\}
\tag{11.4.8}
$$

where the one-form:

$$
\boldsymbol{k} \equiv \left\{\frac{r\mathrm{x} + a\mathrm{y}}{r^2+a^2} \; ; \; \frac{r\mathrm{y} - a\mathrm{x}}{r^2+a^2} \; ; \; \frac{\mathrm{z}}{r} \; ; \; 1\right\}
\tag{11.4.8'}
$$

is null and r is determined implicitly in terms of x, y and z as:

$$
\frac{\mathrm{x}^2 + \mathrm{y}^2}{r^2+a^2} + \frac{\mathrm{z}^2}{r^2} - 1 = 0 \ .
\tag{11.4.9}
$$

From (11.4.9) we deduce that each point in the $(\mathrm{x}, \mathrm{y}, \mathrm{z})$-space, corresponds to two points in the (r, θ, φ)-space having a value of r with opposite sign and given by:

$$
r = \pm\left\{\frac{\mathrm{x}^2+\mathrm{y}^2+\mathrm{z}^2-a^2}{2} + \left[\left(\frac{\mathrm{x}^2+\mathrm{y}^2+\mathrm{z}^2-a^2}{2}\right)^2 + \mathrm{z}^2a^2\right]^{\frac{1}{2}}\right\}^{\frac{1}{2}} .
\tag{11.4.10}
$$

In the plane $\mathrm{z} = 0$, all the points with $\mathrm{x}^2 + \mathrm{y}^2 \leq a^2$ have $r = 0$, hence in that plane, r measures the coordinate distances from the rim of a disk with radius a which will be described as "the $r = 0$-disk" (see Fig. 11-1a); moreover the points with $r =$ const. form circles of radius $R = (r^2+a^2)^{\frac{1}{2}}$. Off the plane $\mathrm{z} = 0$, the surfaces $r =$ const. are oblate spheroids, which cross the z-axis at $\mathrm{z} = r$ (see Fig. 11-1b).

a)

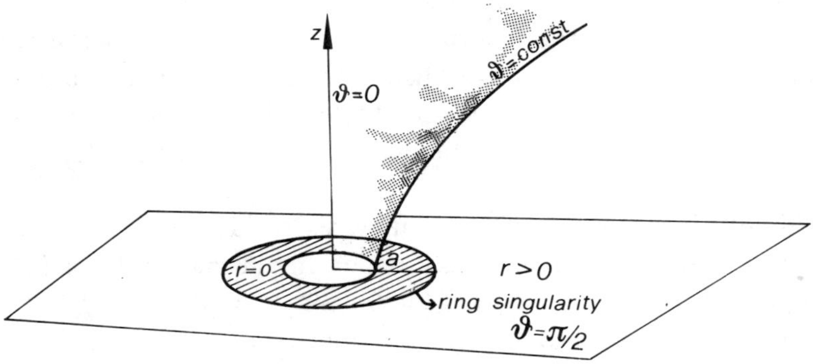

b)

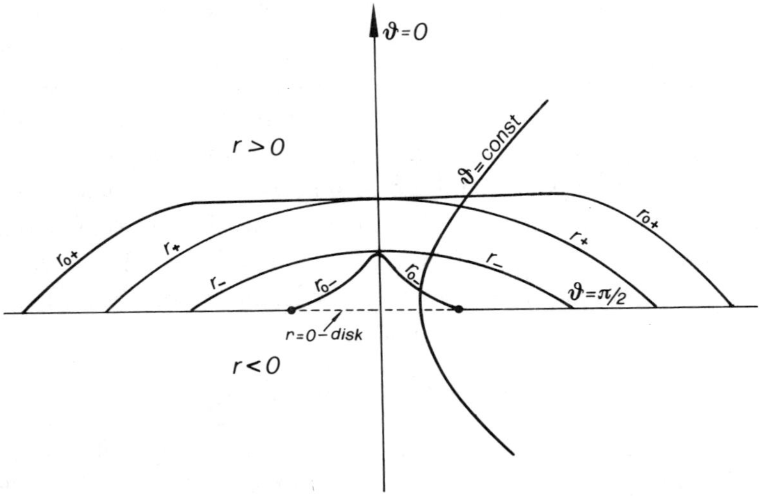

Fig. 11-1 The geometrical structure of Kerr space-time. a) The surfaces $\theta =$ const. are hyperboloids which cross the $r = 0$-disk. b) The $r = 0$-disk connects continuously the $r > 0$ to the $r < 0$ sheets. The horizons and the ergospheres are only in the $r > 0$ sheet.

More interesting are the surfaces θ = const. From (11.4.7) we recognize that these surfaces are hyperboloids which cross the plane z = 0 inside the $r = 0$-disk with intercepts which are circles of radius $R(\theta) = a \sin\theta$ (see Fig. 11-1). The surface $\theta = \pi/2$, hereafter called the equatorial plane, coincides with the plane z = 0, *without* the $r = 0$-disk.

The spheroidal coordinates allow us to identify more precisely the location of the singularity which in the metric form (11.4.8) appeared to be the entire $r = 0$-disk ($r = 0$, z = 0). In the form (11.4.6) or (11.2.25) the singularity is located at $\Sigma = r^2 + a^2 \cos^2\theta = 0$, that is at $r = 0$, $\theta = \pi/2$, thus the metric is singular only on the rim of the $r = 0$-disk; the rim will be named the *ring singularity.* This remarkable property of the Kerr metric, together with the form of (11.4.6), implies that the space-time is regular inside the $r = 0$-disk, thus permitting a continuous transition from $r > 0$ to $r < 0$ along paths having $\theta < \pi/2$. The $r > 0$ and $r < 0$ parts of the Kerr metric describe space-times which are asymptotically flat but non-isometric to each other. While, for example, the function Δ vanishes only in the $r > 0$ region, implying the existence of event horizons (provided $a < m$), the function $\mathcal{A}$ vanishes only in the $r < 0$ part; this can be clearly seen if we write it as

$$\mathcal{A} = (r^2 + a^2)\Sigma + 2mra^2 \sin^2\theta \ .$$

The vanishing of $\mathcal{A}$ implies the existence of regions where the axial Killing vector ∂_φ becomes time-like, hence assuring a local causality violation.

Finally the surfaces φ =const. can easily be visualized by noting from (11.4.7) that the surfaces

$$\psi \equiv \varphi - \tan^{-1}\left(\frac{a}{r}\right) = \text{const.}$$

are planes crossing the axis of symmetry. Then in order to keep $\varphi = \psi + \tan^{-1}(a/r)$ constant, the angle ψ, which coincides with φ at large r, has to increase when $r \to 0$ up to a maximum amount of $\pi/2$, when $r = 0$. Thus, in the spheroidal coordinate system used in (11.4.7), a point Q with coordinates (r, θ, φ), is mapped along

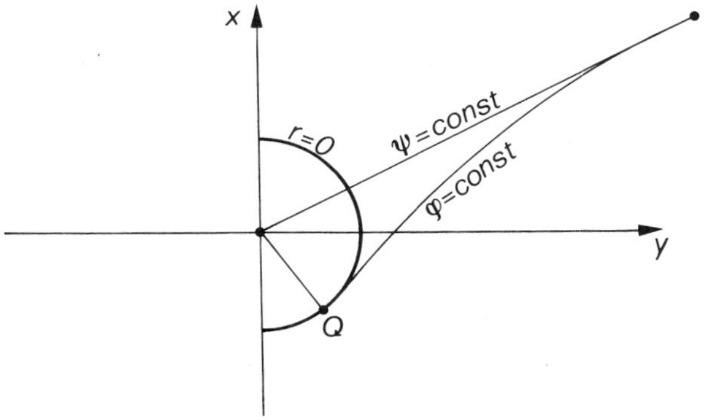

Fig. 11-2 In the spheroidal coordinates, each point in space is mapped into a corresponding point inside the $r = 0$-disk.

the θ =const., φ =const. surfaces, to a point Q′ in the $r = 0$-disk with coordinates $\{\mathrm{Z} = 0,\ \mathrm{X} = -a\sin\theta\sin\varphi,\ \mathrm{Y} = a\sin\theta\cos\varphi\}$, (see Fig. 11-2).

The point Q′ is the *origin* of the coordinates relative to all points with the same θ and φ; when $a = 0$ (the static case) the $r = 0$-disk shrinks to a point ($\mathrm{X} = \mathrm{Y} = \mathrm{Z} = 0$) which is the coordinate origin relative to all points. This property of the spheroidal coordinates will play a role in the interpretation of the constants of the motion.

Let us conclude this section by reconsidering the metric form (11.4.8). As mentioned already, Kerr and Schild (1967) proved that the Kerr solution is a special example of the vacuum solutions whose metrics have the form:

$$g_{ij} = \eta_{ij} + \mathcal{H}k_i k_j \quad ; \quad g^{ij} = \eta^{ij} - \mathcal{H}k^i k^j \qquad (11.4.11)$$

where η_{ij} is the Minkowski metric, k^i is tangent to a null geodesic and $\mathcal{H}$ is some function, positive and at least C^2. A direct calculation provides:

$$R^i{}_{jkl}k^j k^l = -\ddot{\mathcal{H}}k^i k_k \qquad (11.4.12)$$

where $\dot{\mathcal{H}} = \mathcal{H}_{,i}k^i$, which implies that the condition:

$$k_{[m}R_{i]jkl}k^j k^l = 0 \qquad (11.4.13)$$

is identically satisfied. Since in the vacuum ($R_{ij} = 0$), the Riemann tensor coincides with the Weyl tensor, (11.4.13) shows that the solutions (11.4.11), and therefore also the Kerr solution, are algebraically special of type II according to the Petrov classification (see Sect. 5.10). Thus the vector field k is a repeated principal null direction of the Weyl tensor and, according to a theorem due to Goldberg and Sachs (1962), it is shear-free (see also Newman and Penrose (1962)).

11.5 Time-like geodesics

The importance of the Kerr metric resides in its interpretation as the space-time of a rotating collapsed body. It is of interest now to outline the properties of the geodesic motion.

Consider a point particle with rest mass μ_0, moving on a geodesic. The Killing vector fields $k \equiv \partial_0$ and $m \equiv \partial_\varphi$ imply from Sect. 7.5 the following set of equations:

$$g_{00}\dot{t} + g_{0\varphi}\dot{\varphi} = -\frac{E}{\mu_0 c^2} \equiv -\gamma$$
$$g_{\varphi 0}\dot{t} + g_{\varphi\varphi}\dot{\varphi} = \frac{\ell}{\mu_0 c} \equiv \lambda \tag{11.5.1}$$

where $\dot{x}^i = (dx^i/d\tau)$, τ being the proper time on the geodesic, and E and ℓ are constants of the motion which, from considerations similar to those made in the Schwarzschild case, can be interpreted as the particle total energy and the azimuthal angular momentum both measured by a static observer at infinity. From (11.5.1) and (11.2.25) we have:

$$\dot{t} = \frac{\mathcal{A}}{\Delta\Sigma}(\gamma - \lambda\omega) \tag{11.5.2}$$
$$\dot{\varphi} = \frac{\mathcal{A}}{\Delta\Sigma}\left[\left(\frac{\Sigma - 2mr}{\mathcal{A}}\right)\frac{\lambda}{\sin^2\theta} + \omega\gamma\right] \tag{11.5.3}$$

where, we remember, ω is the angular velocity of the gravitational dragging, introduced in (7.10.18).

In order to solve for the other two components, $\dot{\theta}$ and $\dot{r}$, of the geometric tangent to the geodesic, we need to consider the

Hamilton-Jacobi equation (7.4.11) (Carter, 1968b). From (7.4.11) and applying the method of separation of variables as in Sect. 10.7, we have:

$$\frac{\Delta}{\Sigma}P_r^2 + \frac{1}{\Sigma}P_\theta^2 + \frac{\Delta - a^2\sin^2\theta}{\Sigma\Delta\sin^2\theta}P_\varphi^2 - \frac{4mra}{\Sigma\Delta}P_\varphi P_0 - \frac{\mathcal{A}}{\Sigma\Delta}P_0^2 + \mu_0^2c^2 = 0 \tag{11.5.4}$$

where $P_i = \partial S_{|i|}/\partial x^i$. Since P_φ = const. = $\mu_0 c\lambda$, and P_0 = const. = $-\mu_0 c\gamma$, equation (11.5.4) can be separated with respect to r and θ. Multiplying by $\Sigma/\mu_0^2c^2$ that equation can be rewritten as:

$$\Delta\frac{P_r^2}{\mu_0^2c^2} - \frac{a^2}{\Delta}\lambda^2 + \frac{4mra}{\Delta}\gamma\lambda + \frac{(r^2+a^2)^2}{\Delta}\gamma^2 + r^2 + \frac{P_\theta^2}{\mu_0^2c^2} + \frac{\lambda^2}{\sin^2\theta} + \gamma^2a^2\sin^2\theta + a^2\cos^2\theta = 0\ , \tag{11.5.5}$$

thus there exists a (separation) constant $\boldsymbol{K}$ such that:

$$\Delta\frac{P_r^2}{\mu_0^2c^2} - \frac{a^2\lambda^2}{\Delta} + \frac{4mra\lambda\gamma}{\Delta} + \frac{(r^2+a^2)^2}{\Delta}\gamma^2 + r^2 = -\boldsymbol{K}$$

$$\frac{P_\theta^2}{\mu_0^2c^2} + \frac{\lambda^2}{\sin^2\theta} - (\gamma^2-1)a^2\cos^2\theta + \gamma^2a^2 = \boldsymbol{K}\ . \tag{11.5.6}$$

It is more convenient to introduce a new constant:

$$\boldsymbol{L} = \boldsymbol{K} - a^2\gamma^2\ , \tag{11.5.7}$$

so that one can write the remaining components of the tangent vector as follows:

$$\Sigma^2\dot\theta^2 = \boldsymbol{L} - \frac{\lambda^2}{\sin^2\theta} + (\gamma^2-1)a^2\cos^2\theta \tag{11.5.8}$$

$$\Sigma^2\dot r^2 = (2mr\gamma - a\lambda)^2 + \Delta\left[r^2(\gamma^2-1) + 2mr\gamma^2 - \boldsymbol{L}\right]\ . \tag{11.5.9}$$

The separation constant $\boldsymbol{L}$ (or $\boldsymbol{K}$) cannot be interpreted straightforwardly as the square of a total angular momentum as it would be in the $a = 0$ case since now $\boldsymbol{L}$ (or $\boldsymbol{K}$) can take negative values. However from (11.5.8) we see that the quantity:

$$F(\theta) = \boldsymbol{L} + (\gamma^2-1)a^2\cos^2\theta \tag{11.5.10}$$

is always positive and indeed is the square of the total angular momentum of a particle at infinity, with angular coordinate θ, and calculated with respect to the *centre* of the $r = 0$-disk. But this latter point has angular coordinate $\theta = 0$, thus one obviously expects that the coordinate θ appears explicitly in the expression of the angular momentum, as it does in (11.5.10). This dependence disappears if one calculates the total angular momentum of a particle, located at some point Q at infinity and with angular coordinates (θ, φ), with respect to the point Q' in the $r = 0$-disk having the same angular coordinates as the point Q; (see Fig.11-2 and discussion in Sect. 11.4). In this case the separation constant appears explicitly related to the square of that angular momentum (see de Felice, 1980).

11.6 Rotationally induced effects

Gravitational dragging is a general relativistic effect which enters into any physical process taking place in the gravitational field of a rotating source. In the Kerr metric this effect is measured by the angular velocity $\omega = -g_{0\varphi}/g_{\varphi\varphi}$, but it is often more convenient to subtract the dragging effects when performing physical measurements, and thus a suitable frame of reference at each point is given by the following tetrad:

$$\lambda_{\hat{0}} = \left(\frac{\mathcal{A}}{\Sigma\Delta}\right)^{\frac{1}{2}} (\partial_0 + \omega\partial_\varphi) \quad ; \quad \lambda_{\hat{r}} = \left(\frac{\Delta}{\Sigma}\right)^{\frac{1}{2}} \partial_r$$
$$\lambda_{\hat{\varphi}} = \left(\frac{\Sigma}{\mathcal{A}\sin^2\theta}\right)^{\frac{1}{2}} \partial_\varphi \quad ; \quad \lambda_{\hat{\theta}} = \Sigma^{-\frac{1}{2}}\partial_\theta \; . \tag{11.6.1}$$

It is easy to verify that this is the orthonormal tangent base dual to the base of one-forms (11.3.3). The observers defined by the integral curves of $\lambda_{\hat{0}}$ in (11.6.1), have zero azimuthal angular momentum with respect to infinity; in fact from (7.10.8) and (7.10.18) we have identically:

$$(\partial_\varphi|\lambda_{\hat{0}}) = \left(\frac{\mathcal{A}}{\Sigma\Delta}\right)^{\frac{1}{2}} (W + \omega X) = 0 \; . \tag{11.6.2}$$

The frame (11.6.1) is termed *locally non-rotating* and generalizes the concept of a non-rotating observer; on the other hand a static (zero angular velocity) observer in a stationary space-time (namely when his or her four-velocity is given by $u = (-g_{00})^{-\frac{1}{2}}\partial_0$) has a non-zero angular momentum with respect to infinity:

$$(\partial_\varphi|u) = \frac{g_{0\varphi}}{\sqrt{-g_{00}}} = -\frac{\omega\mathcal{A}}{\Sigma^{\frac{1}{2}}(\Sigma - 2mr)^{\frac{1}{2}}}\sin^2\theta \ .$$

A particle moving on a geodesic has an energy measured in units of $\mu_0 c^2$ relative to $\{\lambda_{\hat{a}}\}$ equal to:

$$\frac{\mathcal{E}}{\mu_0 c^2} = -\dot{x}_i\lambda^i_{\hat{0}} = (\gamma - \lambda\omega)\left(\frac{\mathcal{A}}{\Sigma\Delta}\right)^{\frac{1}{2}} \qquad (11.6.3)$$

and since one has necessarily $\mathcal{E} > 0$, the geodesic motion must take place with $\gamma > \lambda\omega$. A particle with positive local energy may have negative energy with respect to infinity, i.e. $\gamma < 0$, but, from the definition (11.5.1) and the discussion at the end of Sect. 1.15, this may occur only below the surface $V = 0$ where the time Killing vector k becomes space-like. When that happens, it is always possible to choose a time-like vector whose scalar product with k is positive, which then defines a geodesic with $\gamma < 0$. From the above relation, particles with negative γ's can only have negative λ's; furthermore from the general discussion of Sect. 7.10 we deduce that in this region these particles can only have $\Omega = \dot{\varphi}/\dot{t} > 0$; hence since $\dot{t} > 0$ it follows from (11.5.2) that $\dot{\varphi} > 0$.

The existence of particles with negative energy with respect to infinity has important implications. Suppose that a particle coming in from infinity with a given initial energy γ_0 ($\gamma_0 > 0$) crosses the surface $V = 0$ entering the region where k is space-like ($V > 0$) and decays into two particles with energies $\gamma_1 < 0$ and $\gamma_2 > \gamma_0$. While the negative-energy particle can only plunge into the event horizon, the positive-energy particle can escape to infinity bringing with it more energy than it had initially. An observer at infinity then experiences a process of energy extraction which presumably takes place at the expense of the energy of the metric source (Penrose and Floyd, 1971).

The region confined between the horizon $\Delta = 0$ and the surface $V = 0$ is termed the *ergoregion*; the surface $V = 0$ is termed the *ergosurface*. The above process, known as the *Penrose process of energy extraction*, has an important wave mechanical analogue, which bears the name of *superradiance* (Starobinskii, 1973; Starobinskii and Churilov, 1974). It consists of the amplification of a low-frequency wave upon scattering by the Kerr metric source.

Let us consider a scalar field Φ, which satisfies the massless Klein-Gordon wave equation:

$$g^{ij}\nabla_i\nabla_j\Phi = 0 \ . \tag{11.6.4}$$

Once written in Boyer–Lindquist coordinates, this equation is separable, hence, putting $\Phi = R(r)P(\theta)\mathrm{e}^{\mathrm{i}n\varphi-\omega t}$, n being an azimuth quantum number and ω the frequency, the radial part is given by:

$$\Delta\frac{d}{dr}\left(\Delta\frac{dR}{dr}\right) + \left[\omega^2(r^2+a^2) - 4amnr\omega + a^2n^2 - \lambda\Delta\right]R = 0 \tag{11.6.5}$$

where λ is a separation constant and n is, as stated, the quantum number of the spheroidal harmonics which solve the angular part of (11.6.4). Defining a new coordinate z and a function $f(r)$ as:

$$\begin{aligned} \frac{dr}{dz} &= \frac{\Delta}{r^2+a^2} \qquad -\infty < z < +\infty \\ R(r) &= \frac{f(r)}{(r^2+a^2)^{\frac{1}{2}}} \end{aligned} \tag{11.6.6}$$

equation (11.6.5) becomes:

$$\frac{d^2f}{dz^2} + F^2(z)f = 0 \tag{11.6.7}$$

where:

$$\begin{aligned} F^2(z) &= (\omega - n\omega_{\mathrm{H}})^2 - \frac{\Delta}{(r^2+a^2)^2} \\ &\quad \times \left\{\lambda - 2an\omega + (r^2+a^2)^{\frac{1}{2}}\frac{d}{dr}\left[\frac{\Delta r}{(r^2+a^2)^{\frac{3}{2}}}\right]\right\} \end{aligned} \tag{11.6.8}$$

and

$$r = r(z) \ , \ \omega_{\mathrm{H}} = \frac{a}{r_+^2 + a^2}.$$

When $z \to +\infty$ $(r \to +\infty)$, the function $F^2(z) \to \omega^2$, while when $z \to -\infty$ $(r \to r_+)$, $F^2(z) \to (\omega - n\omega_{\mathrm{H}})^2$. The boundary condition on the horizon $(r \to r_+,\ z \to -\infty)$ is that of an ingoing wave only, namely:

$$f \sim \exp[-\mathrm{i}(\omega - n\omega_{\mathrm{H}})z], \tag{11.6.9}$$

while at infinity ($r \to +\infty$, $z \to +\infty$) we have a mixture of ingoing and outgoing waves as:

$$f = A\mathrm{e}^{-\mathrm{i}\omega z} + B\mathrm{e}^{\mathrm{i}\omega z} . \tag{11.6.10}$$

The Wronskian of (11.6.7) tells us that:

$$|A|^2 - |B|^2 \sim \mathrm{sign}\,(\omega - n\omega_{\mathrm{H}}) \tag{11.6.11}$$

hence we have an amplification of the scattered wave if $|B|^2 > |A|^2$, namely when $\omega < n\omega_{\mathrm{H}}$. The frequency $\omega_c = n\omega_{\mathrm{H}}$ is termed the *critical frequency of superradiance.*

An obvious question now arises; by how much can the energy of the metric source be decreased? The previous arguments show that a particle with negative γ has negative λ, therefore any decrease in energy is also accompanied by a decrease in angular momentum. Now relation (11.6.3) shows that the minimum energy which can be poured into the source is given, in the limit of $\Delta \to 0$, by:

$$\gamma_{\min} = \lambda\omega_{\mathrm{H}} = \frac{\lambda a}{r_+^2 + a^2} ; \tag{11.6.12}$$

thus for each unit of energy accreted, the source will suffer a change of parameters:

$$\delta m = \gamma \qquad ; \qquad \delta J = \lambda \tag{11.6.13}$$

related by:

$$\delta m \geq \frac{a\delta J}{r_+^2 + a^2} \tag{11.6.14}$$

from (11.6.12). Recalling that $J = ma$ and that to the changes (11.6.13), there corresponds a change in the radial coordinate of the (outer) horizon:

$$\delta r_+ = \frac{r_+\delta m - a\delta a}{r_+ - m} , \tag{11.6.15}$$

relation (11.6.14) can be written more conveniently as:

$$\frac{r_+ - m}{2(r_+^2 + a^2)}\delta(r_+^2 + a^2) \geq 0 \ . \tag{11.6.16}$$

Hence, insofar as $a < m$ ($r_+ > m$), accretion of matter into the source would never decrease the quantity $(r_+^2 + a^2)$. From (11.4.7) moreover, we see that $\mathcal{S}_{\rm H} \equiv 4\pi(r_+^2 + a^2)$ is the area of the (outer) event horizon, hence (11.6.16) establishes an important fact which has relevant consequences in the development of the black-hole theory: no transformation of the metric due to the injection of matter through the event horizon can ever decrease its area from the point of view of an observer at infinity, i.e. $\delta\mathcal{S}_{\rm H} \geq 0$. (This is known as the area theorem and can be proven rigorously.) The transformations which leave the area unchanged are termed *reversible*.

An alternative interpretation of the area theorem is given directly by the definition of r_+; in fact writing:

$$(r_+^2 + a^2) = 2mr_+ = 2m^2 + 2m(m^2 - a^2)^{\frac{1}{2}}$$

and solving with respect to m, we have:

$$m^2 = \frac{(r_+^2 + a^2)}{4} + \frac{J^2}{(r_+^2 + a^2)} \tag{11.6.17}$$

hence if we define:

$$m_{\rm irr} \equiv \frac{1}{2}(r_+^2 + a^2)^{\frac{1}{2}} \ , \tag{11.6.18}$$

(11.6.17) becomes (Christodoulou, 1970):

$$m^2 = m_{\rm irr}^2 + \frac{J^2}{4m_{\rm irr}^2} \ . \tag{11.6.19}$$

This relation indicates that the total energy m of the source is contributed by its rotational energy and by an irreducible part, termed *irreducible mass*, which in fact would never decrease whatever changes we might subject the metric to.

These considerations and relation (11.6.19) show that the Penrose processes of energy extraction would only reduce the rotational component of the energy of the metric source.

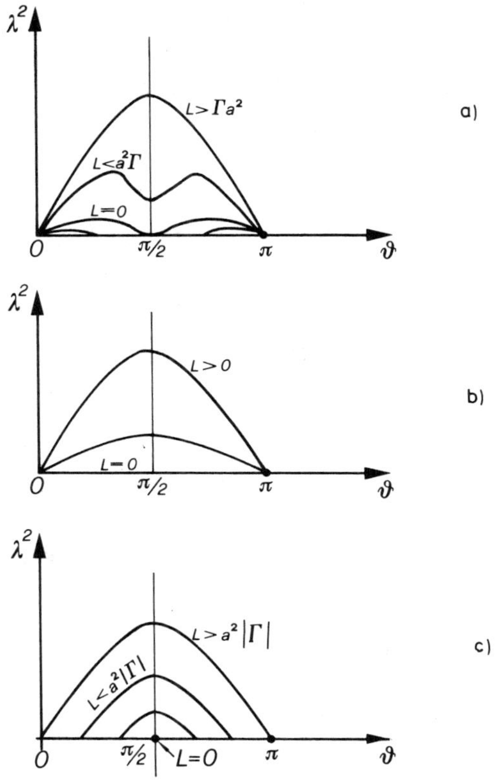

Fig. 11-3 Behaviour of the coordinate θ; a) $\Gamma > 0$; b) $\Gamma = 0$; c) $\Gamma < 0$.

11.7 The angular geodesic equation

The most important consequence of the rotational symmetry of the Kerr metric is that the geodesic motion is not a plane motion. This can be recognized from the explicit dependence of the angular equations (11.5.3) and (11.5.8) on the particle energy which always couples with the rotational parameter a. Let us consider (11.5.8); imposing $\dot{\theta} = 0$ and solving with respect to λ^2 we have:

$$\lambda^2 = \boldsymbol{L} \sin^2 \theta + \frac{1}{4} \Gamma a^2 \sin^2(2\theta) \qquad (11.7.1)$$

where $\Gamma = \gamma^2 - 1$. This describes a two-parameter family of curves which are shown in Fig. 11-3 (de Felice and Calvani, 1972).

The motion in the equatorial plane ($\theta = \pi/2$) is characterized by $\boldsymbol{L} = \lambda^2$, however it is stable when $\boldsymbol{L} > \epsilon a^2 \Gamma$ where ϵ is the sign of Γ; it is unstable when $\Gamma > 0$ and $\boldsymbol{L} < a^2\Gamma$ (see Fig. 11-3a).

Stable non-equatorial motion with θ =const. is possible if:

$$\Gamma \geq 0 \qquad \boldsymbol{L} = -\Gamma a^2 \cos 2\theta \qquad \lambda = \pm \Gamma^{\frac{1}{2}} a \sin^2\theta \ . \tag{11.7.2}$$

Particles with $\boldsymbol{L} < 0$ or

$$\boldsymbol{L} < \lambda^2 \leq \frac{(\boldsymbol{L} + a^2\Gamma)^2}{4a^2\Gamma}$$

would never cross the equatorial plane $\theta = \pi/2$. These are in fact the only particles which could cross the $r = 0$-disk without hitting the ring singularity and connect the $r > 0$ part to the $r < 0$ part of the Kerr space-time.

11.8 The equatorial circular geodesics

Physically, the most relevant properties of the geodesic motion are contained in the radial equation (11.5.9). Since this has been extensively analysed in the specialized literature (de Felice, 1968; Chrandrasekhar, 1983) we shall limit ourselves to the equatorial time-like (spatially) circular geodesics which are the most important from the astrophysical point of view.

If we write equation (11.5.9) as:

$$\Sigma^2 \dot{r}^2 = V(r) \tag{11.8.1}$$

the condition for equatorial circular motion is:

$$V(r) = 0 \quad ; \quad \frac{d}{dr}V(r) = 0 \quad ; \quad \lambda^2 = \boldsymbol{L} \ . \tag{11.8.2}$$

Solving with respect to γ and λ, we obtain after some algebra (Bardeen, 1972):

$$\underset{c}{\gamma}_{\pm} = \frac{r^{\frac{3}{2}} - 2mr^{\frac{1}{2}} \pm am^{\frac{1}{2}}}{r^{\frac{3}{4}}\left(r^{\frac{3}{2}} - 3mr^{\frac{1}{2}} + 2am^{\frac{1}{2}}\right)^{\frac{1}{2}}} \tag{11.8.3}$$

$$\underset{c}{\lambda}_{\pm} = \frac{m^{\frac{1}{2}}\left(r^2 \pm 2am^{\frac{1}{2}}r^{\frac{1}{2}} + a^2\right)}{r^{\frac{3}{4}}\left(r^{\frac{3}{2}} - 3mr^{\frac{1}{2}} + 2am^{\frac{1}{2}}\right)^{\frac{1}{2}}} \ . \tag{11.8.4}$$

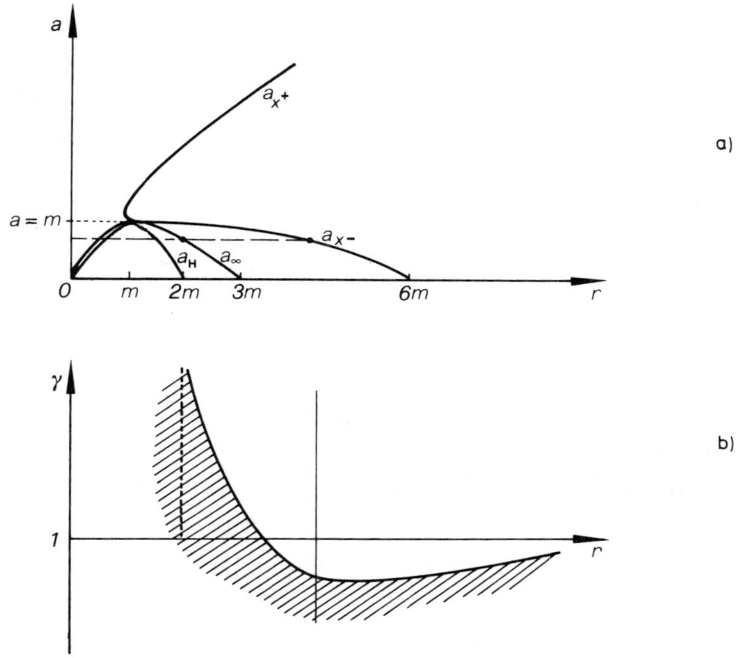

Fig. 11-4 Particle behaviour in the Kerr metric; a) graphs of the functions a_∞, $a_{x\pm}$, a_{H} (see text); b) potential curve for corotating orbits in the equatorial plane of the Kerr metric. The dashed area is forbidden to the (classical) motion.

The upper sign refers to the corotating orbits, the lower sign to the counterrotating ones. The behaviour of γ_{c+} (corotating orbits only) as a function of r is shown in Fig. 11.4b, for $a < m$. It diverges where:

$$a = a_\infty \equiv \frac{m}{2}\left(\frac{r}{m}\right)^{\frac{1}{2}}\left(3 - \frac{r}{m}\right) \tag{11.8.5}$$

and has extreme points where:

$$a = a_{x\pm} \equiv \frac{m}{3}\left(\frac{r}{m}\right)^{\frac{1}{2}}\left[4 \pm \sqrt{\frac{3r}{m} - 2}\right] . \tag{11.8.6}$$

These functions are shown in Fig. 11-4a (the curves a_∞ and a_x).

From the latter in particular we see that the minimum in the curve γ_{c+}, which for each value of a characterizes the innermost stable circular orbit, also known as *marginally stable*, extends deeply into the field, to $r = m$, in the limit of $a = m$. In this case however, since when $a = m$ the outer and inner horizons coincide at $r_+ = r_- = m$ (see Fig. 11-4a, curve $a_{\rm H}$), one might erroneously think that a stationary time-like (circular) orbit can be found on a null surface. This apparent paradox is due to the peculiarity of the Boyer–Lindquist coordinates; in the limit $a = m$ in fact the radial coordinates of the horizons, that of the marginally stable orbit as well as of the divergence limit of $\gamma_{c\,+}$, coincide at $r = m$ as a result of a coordinate degeneracy. However these regions remain physically distinct in space-time (Bardeen et al., 1972) the only significant characterization being that of their energy and angular momentum.

We shall identify the values of the parameters corresponding to the marginally stable time-like circular orbits with the subscript m.s. Then the energy which corresponds to such an orbit at $r = m$, is

$$\gamma_{\rm m.s.\,+}\Big|_{r=m\;\; a=m} = \frac{1}{\sqrt{3}}$$

from (11.8.6) and (11.8.3). Thus, if an accretion disk around an extreme Kerr black-hole ($a = m$) allows a particle to spiral down from the outer part of the disk ($r \gg m,\ \gamma \approx 1$) to the innermost stable circular orbit ($r = m\ ,\ \gamma \approx 1/\sqrt{3}$) due to some kind of friction, the loss of energy which contributes either to the thermal energy of the disk or to radiation is:

$$1 - \gamma_{\rm m.s.\,+}\Big|_{r=m} = 0.422\ . \tag{11.8.7}$$

Hence, a rotating black-hole allows a gravitational energy conversion up to $\sim 42\%$ of the particle rest energy; this efficiency is higher than we would have in the Schwarzschild case or any other known physical process (Bardeen, 1970). For this reason, the Kerr metric has recently received much attention and played a key role in the recent development of black-hole astrophysics.

Let us now consider the effects on the metric source by the accretion of matter through the event horizon. After having left the marginally stable orbit, the particle carries into the source the energy and the angular momentum which corresponded to the marginally stable orbit itself, hence it contributes a change in the source parameters given by:

$$\delta m = \gamma_{\mathrm{m.s.}} \qquad \delta J = \lambda_{\mathrm{m.s.}} \tag{11.8.8}$$

per unit of the accreted mass. Since $\gamma_{\mathrm{m.s.}}$ and $\lambda_{\mathrm{m.s.}}$ are both positive, accretion would increase the mass and the angular momentum of the metric source. Moreover the condition for the existence of the event horizon is, from (11.2.29):

$$\frac{a}{m} \leq 1 \tag{11.8.9}$$

hence the change in that ratio due to an increment of the mass and angular momentum given by (11.8.3) is, recalling (11.3.11):

$$\frac{\delta(a/m)}{\delta m} = \frac{1}{m}\left[\frac{1}{m}\frac{\delta J}{\delta m} - 2\frac{a}{m}\right] . \tag{11.8.10}$$

The radius of the marginally stable orbit can be expressed in terms of the a/m ratio as follows (Bardeen et al., 1972):

$$r\Big|_{\mathrm{m.s.}} = mf\left(\frac{a}{m}\right) \tag{11.8.11}$$

where:

$$\begin{aligned} f\left(\frac{a}{m}\right) &= 3 + Z_2 - [(3 - Z_1)(3 + Z_1 + 2Z_2)]^{\frac{1}{2}} \\ Z_1 &= 1 + \left[1 - \left(\frac{a}{m}\right)^2\right]^{\frac{1}{3}}\left[\left(1 + \frac{a}{m}\right)^{\frac{1}{3}} + \left(1 - \frac{a}{m}\right)^{\frac{1}{3}}\right] \\ Z_2 &= \left[3\left(\frac{a}{m}\right)^2 + Z_1^2\right]^{\frac{1}{2}} . \end{aligned} \tag{11.8.12}$$

Using these expressions in (11.8.3) and (11.8.4), we obtain after some algebra, (de Felice and Yu, 1986):

$$\frac{1}{m}\frac{\delta J}{\delta m} = \frac{1}{m}\left(\frac{\lambda}{\gamma}\right)_{\mathrm{m.s.}} = \frac{f^2 - 2\frac{a}{m}f^{\frac{1}{2}} + \left(\frac{a}{m}\right)^2}{f^{\frac{3}{2}} - 2f^{\frac{1}{2}} + \frac{a}{m}} \equiv g\left(\frac{a}{m}\right) . \tag{11.8.13}$$

Using finally (11.8.13) to (11.8.10) and taking the limit to infinitesimals, we obtain a differential equation in a/m:

$$\frac{d}{dm}\left(\frac{a}{m}\right) = \frac{1}{m}\left[g\left(\frac{a}{m}\right) - 2\left(\frac{a}{m}\right)\right] . \tag{11.8.14}$$

We can notice from (11.8.13) and the various definitions that:

$$\begin{array}{llll} i) & g\left(\frac{a}{m}\right) > 2\frac{a}{m} & \text{for} & 0 \leq \frac{a}{m} < 1 \\ ii) & g\left(\frac{a}{m}\right) \rightarrow 2\frac{a}{m} & \text{when} & \frac{a}{m} \rightarrow 1 \ ; \end{array}$$

thus a/m is a monotonic increasing function of m and may reach one after accretion of a finite amount of mass-energy:

$$\Delta m = m_0 \left[\exp \int_{\left(\frac{a}{m}\right)_0}^{1} \frac{d\alpha}{g(\alpha) - 2\alpha} - 1 \right] \qquad (11.8.15)$$

where the subscript $_0$ refers to the initial value.

Once the mass $m_1 = m_0 + \Delta m$ is attained, namely when $a/m = 1$, the rate of change of a/m with the mass accreted vanishes, i.e. $\frac{d}{dm}(a/m) = 0$ so any further accretion will not increase the ratio a/m above one. In a real situation, matter and radiation will accrete onto the source through the event horizon, following orbits different from the marginally stable circular orbits we have considered. While the latter contribute most favorably to the ratio a/m of the source in the sense of bringing into it more angular momentum than mass so that the ratio increases, all the other orbits contribute less efficiently, inhibiting the increase of a/m which, in real situations, is found to remain always less than one, (Thorne, 1974).

11.9 Null geodesics

The null geodesics in the Kerr metric differ from the time-like ones only in their $\theta-$ and r-components, the time and φ-components remaining unaltered. From these, namely (11.5.2) and (11.5.3), and from (11.5.4) where we put $\mu = 0$, we obtain*:

$$\Sigma^2 \dot{\theta}^2 = \tilde{\boldsymbol{L}} - \frac{\tilde{\lambda}^2}{\sin^2\theta} + \tilde{\gamma}^2 a^2 \cos^2\theta \qquad (11.9.1)$$

$$\Sigma^2 \dot{r}^2 = (2mr\tilde{\gamma} - a\tilde{\lambda})^2 + \Delta\left[\tilde{\gamma}^2 r(r+2m) - \tilde{\boldsymbol{L}}\right] \qquad (11.9.2)$$

*Parameters with tildes refer to photons.

$$\dot{t} = \frac{\mathcal{A}}{\Delta\Sigma}(\tilde{\gamma} - \tilde{\lambda}\omega) \tag{11.9.3}$$

$$\dot{\varphi} = \frac{\mathcal{A}}{\Delta\Sigma}\left[\left(\frac{\Sigma - 2mr}{\mathcal{A}}\right)\frac{\tilde{\lambda}}{\sin^2\theta} + \omega\tilde{\gamma}\right] . \tag{11.9.4}$$

The analysis of the null trajectories follows the same lines as those for the time-like ones; we shall consider here two important examples, the equatorial null geodesics and the integral curves of the principal null directions of the Weyl tensor which, as we said in Sec. 11.4, are null geodesics.

In the equatorial plane ($\theta = \pi/2$, $\tilde{\lambda}^2 = \tilde{\boldsymbol{L}}$), the behaviour of the null geodesics with respect to the radial coordinate, is deduced by imposing $\dot{r} = 0$ in (11.9.2) and solving with respect to the ratio $\tilde{\lambda}/\tilde{\gamma} \equiv \tilde{\Lambda}$; namely (de Felice, 1968; Boyer and Lindquist, 1967):

$$\tilde{\Lambda}_{\pm} = \frac{2ma \pm r\Delta^{\frac{1}{2}}}{2m - r} . \tag{11.9.5}$$

This describes a family of curves parametrized by a; each of these curves is completely determined by the locus of the extreme points for $\tilde{\Lambda}_{\pm}$, namely the curve of equation:

$$a^2 = \frac{r}{4m}(r - 3m)^2 \equiv a^2_{\text{ph}} \tag{11.9.6}$$

and by the limits:

$$\lim_{r \to +\infty} \tilde{\Lambda}_{\mp} = \pm\infty \quad , \qquad \lim_{r \to 2m} \tilde{\Lambda}_{-} = +\infty$$

$$\lim_{r \to r_+} \tilde{\Lambda}_{\pm} = \tilde{\Lambda}_{\pm}(r_{\pm}) \quad , \qquad \lim_{r \to 2m} \tilde{\Lambda}_{+} = \frac{a^2 + 2m^2}{2am} .$$

Curve (11.9.6) is shown in Fig. 11-5a and a general curve of the family (11.9.5) is shown in Fig. 11-5b.

The points of (11.9.6), which are on the branch $m \le r \le 3m$, represent unstable spatially circular null trajectories corotating with the metric source, the others are unstable counterrotating ones. As expected, when $a = 0$, namely in the Schwarzschild case, these orbits are found only at $r = 3m$. It is worth noticing that the net rotational asymmetry between the corotating ($\tilde{\Lambda} > 0$) and counterrotating ($\tilde{\Lambda} < 0$) light rays, indicate that the latter are preferentially captured by the metric source, hence inhibiting whatever increases its angular momentum.

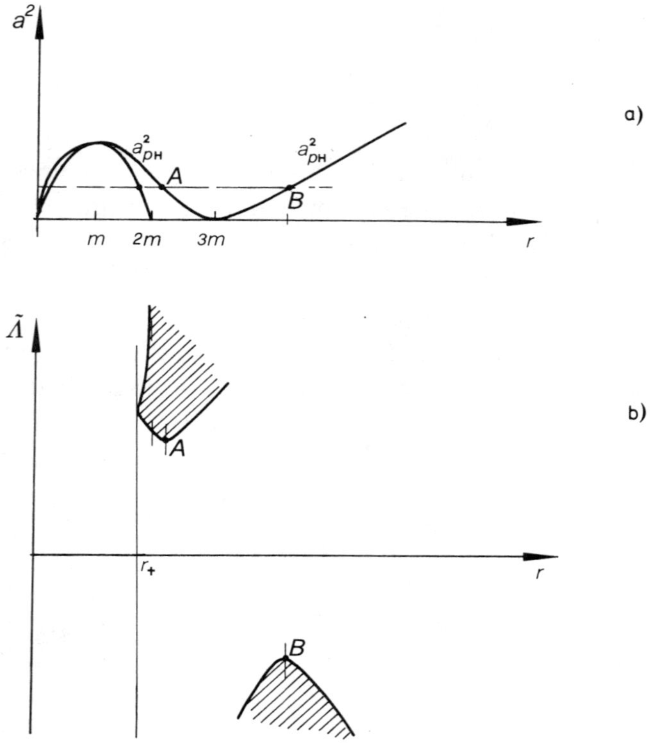

Fig. 11-5 Behaviour of the light rays in the equatorial plane of the Kerr metric. a) Graphs of the functions $a^2_{\rm ph}$, $a^2_{\rm H}$. b) Potential curves for light rays corresponding to a value of $a < m$ (dashed line in a)). The shaded area is forbidden to the (classical) motion.

Off the equatorial plane, the most interesting null geodesics are those which constitute the principal null directions of the Kerr solution; they form a shear-free congruence and their characterization was given in Cartesian coordinates by (11.4.8′). In spheroidal coordinates, they turn out to be trajectories confined on the hyperboloids $\theta =$ const., since their constants of the motion satisfy the conditions (11.7.2) (in our case $\Gamma \equiv \tilde{\gamma}^2 > 0$) and in addition they are corotating with the metric source since $\tilde{\lambda} > 0$ (de Felice and Calvani, 1972). Hence using:

$$\tilde{\boldsymbol{L}} = -\tilde{\gamma}^2 a^2 \cos(2\theta) \qquad \tilde{\lambda} = \tilde{\gamma} a \sin^2\theta \tag{11.9.7}$$

in (11.9.1) to (11.9.4) we have for the principal null directions:

$$\dot{t} = \frac{(r^2 + a^2)\tilde{\gamma}}{\Delta} \qquad \dot{r} = \pm\tilde{\gamma}$$
$$\dot{\varphi} = \frac{a\tilde{\gamma}}{\Delta} \qquad \dot{\theta} = 0 \, . \tag{11.9.8}$$

Obviously the divergence on $\Delta = 0$ is due to the coordinates in use; if we refer to the coordinates $\tilde{\varphi}$, $\tilde{t}$ introduced with the transformation (11.4.5), the equations for ingoing principal null directions take a very simple form:

$$\dot{\tilde{t}} = -\dot{r} = \tilde{\gamma} \quad ; \quad \dot{\tilde{\varphi}} = \dot{\theta} = 0 \, . \tag{11.9.9}$$

Hence they can go across the horizon at $\Delta = 0$ with continuity. The outgoing null congruence however shows the singularity at $\Delta = 0$ as in (11.9.8). From these, we see that the only geodesic which can *stay* on the horizon is an outgoing principal null direction with $\tilde{\gamma} = 0$; it is dragged around with an angular velocity equal to that of the gravitational dragging:

$$\omega_{\rm H} = \left.\frac{\dot{\varphi}}{\dot{t}}\right|_{\rm H} = \frac{a}{r_+^2 + a^2} \, . \tag{11.9.10}$$

These geodesics are said to be the *generators* of the horizon.

11.10 The Kerr-Newman solution

The metric form (11.2.5) is manifestly invariant under the simultaneous inversion of the axial and the stationary Killing vectors; this invariance property was shown by Papapetrou (1966) to be a property of any connected stationary axisymmetric solution of Einstein vacuum equations. This result can be generalized, as shown by Carter (1970), to non-empty space-times, provided the energy-momentum tensor is invariant under this transformation or else if the contribution to it comes only from a source-free electromagnetic field. Thus the arguments of Sect. 11.1 apply to a charged generalization of the Kerr solution found by Newman and collaborators (Newman et al., 1965).

Written in spherical coordinates, the new exact solution of Einstein-Maxwell field equations, known as the Kerr-Newman solution, reads:

$$ds^2 = -\left(\frac{\Delta'\Sigma}{\mathcal{A}'}\right) dt^2 + \frac{\mathcal{A}'\sin^2\theta}{\Sigma}(d\varphi - \omega' dt)^2 + \left(\frac{\Sigma}{\Delta'}\right) dr^2 + \Sigma\, d\theta^2 \qquad (11.10.1)$$

$$\boldsymbol{A} = -\frac{Qr}{\Sigma}(\boldsymbol{dt} - a\sin^2\theta\, \boldsymbol{d\varphi}) \qquad (11.10.2)$$

where $\boldsymbol{A}$ is the electromagnetic potential one-form, Q is the charge in geometrized units (see (10.12.19)) and:

$$\omega' = \frac{a}{\mathcal{A}'}(2mr - Q^2)$$
$$\Delta' = r^2 + a^2 - 2mr + Q^2$$
$$\mathcal{A}' = (r^2 + a^2)^2 - a^2\Delta'\sin^2\theta \ . \qquad (11.10.3)$$

Here the source of the space-time metric is assumed to be all contained in the ring singularity at $\Sigma = 0$ and in the electromagnetic field outside it.

The Kerr-Newman metric (11.10.1) reduces to the Kerr metric when $Q = 0$, and to the Reissner-Nordström metric when $a = 0$, hence it combines the properties of both solutions. The condition for the existence of an event horizon at $\Delta' = 0$ is now, from (11.10.3):

$$m^2 \geq a^2 + Q^2 \ . \qquad (11.10.4)$$

Also in this case, the equality sign is only marginally attainable by means of physical processes, since any contribution to the specific angular momentum and/or to the charge of the metric source is also accompanied by a contribution to the mass so that the inequality in (11.10.4) is always preserved.

In the Kerr-Newman metric the motion of charged particles (with rest mass μ_0) is somewhat complicated by the presence of the electromagnetic interaction. Although they do not move on geodesics, their equations of motion:

$$u^i\nabla_i u^j = eF^j{}_k u^k \qquad (11.10.5)$$

where e is the charge of the particles (in units of $\mu_0 G^{1/2}$) and $u^i \equiv \dot{x}^i$ are the components of their four-velocity, can be integrated and yield:

$$\begin{aligned}
\Sigma\dot{t} &= -a(a\gamma\sin^2\theta - \lambda) + (r^2+a^2)\frac{\boldsymbol{P}}{\Delta'} \\
\Sigma\dot{r} &= \pm\left\{\boldsymbol{P}^2 - \Delta'\left[r^2 + (\lambda - a\gamma)^2 + (\boldsymbol{L} - \lambda^2)\right]\right\}^{\frac{1}{2}} \\
\Sigma\dot{\theta} &= \pm\left\{(\boldsymbol{L} - \lambda^2) - \cos^2\theta\left[a^2(1-\gamma^2) + \frac{\lambda^2}{\sin^2\theta}\right]\right\}^{\frac{1}{2}} \\
\Sigma\dot{\varphi} &= -\left(a\gamma - \frac{\lambda}{\sin^2\theta}\right) + \frac{a\boldsymbol{P}}{\Delta'}
\end{aligned} \tag{11.10.6}$$

where

$$\boldsymbol{P} = \gamma(r^2+a^2) - a\lambda - eQr\ .$$

Here $\boldsymbol{L}$ is a separation constant and γ and λ are constants of the motion which have the same physical meaning as in the Kerr metric, namely that of the total energy (in units of $\mu_0 c^2$) and azimuthal angular momentum (in units of $\mu_0 c$) of the particle as measured by a static observer at infinity. These parameters however are defined differently from the Kerr case; in fact the presence of the electromagnetic potential leads to a generalized four-momentum given by:

$$\pi_i = P_i + e\mu_0 c A_i \tag{11.10.7}$$

where $P_i = \mu_0 c u_i$, thus it can be shown that in this case:

$$\frac{1}{\mu_0 c}\pi_i k^i = u_0 + eA_0 = -\gamma \tag{11.10.8}$$

$$\frac{1}{\mu_0 c}\pi_i m^i = u_\varphi + eA_\varphi = \lambda \tag{11.10.9}$$

with k and m being the stationary and axial Killing vectors of the metric.

Since a family of zero angular-momentum observers can be defined as in (11.6.1), with the obvious substitutions $\mathcal{A} \to \mathcal{A}'$, $\omega \to \omega'$, $\Delta \to \Delta'$, the energy of the particles in this frame is given by:

$$\mathcal{E}_\lambda = \left(\frac{\mathcal{A}'}{\Sigma\Delta'}\right)^{\frac{1}{2}}\left[\gamma - \lambda\omega' - \frac{eQr}{\mathcal{A}'}(r^2+a^2)\right]\ . \tag{11.10.10}$$

Hence the argument of Sect. 11.6 generalizes here easily; the minimum energy which can be poured into the metric source is now:

$$\gamma_{\text{min}} = \frac{\lambda a + eQr_+}{r_+^2 + a^2} \tag{11.10.11}$$

where $r_+ = m + [m^2 - a^2 - Q^2]^{1/2}$, and therefore equation (11.6.19) becomes now:

$$m^2 = \left(m_{\text{irr}} + \frac{Q^2}{4m_{\text{irr}}}\right)^2 + \frac{J^2}{4m_{\text{irr}}} \tag{11.10.12}$$

which explicitly includes the electromagnetic contribution to the total mass of the source.

The importance of the Kerr-Newman three-parameter family of solutions resides in a general theorem conjectured by Carter (1970b) and proved by Robinson (1975) according to which it is the *only* class of electrovac solutions which describe a space-time admitting an event horizon and being asymptotically flat, stationary and axisymmetric. Each of these solutions is described only by the three parameters m, a and Q; thus any two sources with the same values of $(m,\ a,\ Q)$, and collapsed below an event horizon, will generate space-times completely indistinguishable, regardless of the many different microstates which would locally distinguish the two systems. As a consequence, a black-hole is eventually unperturbed by the physical nature of the matter that produced it, and so its shape is completely smooth: *a black-hole has no hair.* The corresponding theorem for the non-rotating charged case, i.e. the Reissner-Nordström solution, was proved by Israel (Israel, 1967, 1968).

11.11 The Weyl and the T-S solutions

Both the Kerr and the Kerr-Newman solutions describe space-times generated by a rotating isolated system, but the Ernst equation (11.1.12) admits a class of *simpler* solutions which describe the space-times of non-rotating (static), axisymmetric and in general deformed systems given by the Weyl solutions (Weyl, 1917;

Kramer et al., 1980):

$$\xi = \frac{(x+1)^\delta + (x-1)^\delta}{(x+1)^\delta - (x-1)^\delta} \ . \tag{11.11.1}$$

Here x is the spheroidal coordinate introduced in (11.2.1) and δ is a positive parameter which measures the degree of deformation of the metric source. If we call m the mass of the source, its quadrupole moment, defined asymptotically, is given by:

$$\boldsymbol{Q} = m^3 \left(\frac{\delta^2 - 1}{3\delta^2} \right) \ . \tag{11.11.2}$$

The case with $\delta = 1$ corresponds to the Schwarzschild metric; those with $\delta > 1$ ($\delta < 1$) represent the exterior gravitational fields of oblate (prolate) axisymmetric masses.

A generalization of (11.11.1) to a stationary case with a generic δ is given by a solution of the type:

$$\xi = \frac{\alpha}{\beta} \tag{11.11.3}$$

where α and β are complex polynomials of x and y.

The Kerr metric corresponds to $\delta = 1$, $\beta = 1$ and

$$\alpha = Px - \mathrm{i}Qy \tag{11.11.4}$$

as we know from (11.2.11). The solutions of (11.11.3) for discrete $\delta = 2,\ 3,\ 4$, were found by Tomimatsu and Sato and are known as T-S metrics (Tomimatsu and Sato, 1972, 1973). The complexity of the polynomials α and β increases considerably with δ, hence we shall consider only the T-S solutions with $\delta = 2$. In this case we have:

$$\begin{aligned} \alpha &= P^2x^4 + Q^2y^4 - 1 - 2\mathrm{i}PQxy(x^2 + y^2) \\ \beta &= 2Px(x^2 - 1) - 2\mathrm{i}Qy(1 - y^2) \end{aligned} \tag{11.11.5}$$

where

$$P^2 + Q^2 = 1 \ .$$

At large distances from the source, namely for $Px \ll 1$, the metric coefficients V and h from (11.1.1) are given by (cf. with

(11.2.12) and (11.2.16)):

$$V \approx 1 - \frac{2\delta}{Px} + \frac{2\delta^2}{P^2x^2} + \frac{2\delta^3}{P^3x^3}\left(Q^2y^2 - 1 + \frac{\delta^2-1}{2\delta^2}P^2\right)$$
$$h \approx 2mQ(y^2-1)\frac{\delta}{Px} \tag{11.11.6}$$

where in our case $\delta = 2$, but a similar behaviour is found in the T-S metrics with $\delta = 3$ and $\delta = 4$. The quadrupole moment reads:

$$\boldsymbol{Q} = m^3\left(\frac{\delta^2-1}{3\delta^2}P^2 + Q^2\right) \tag{11.11.7}$$

and clearly it exceeds the value with $\delta = 1$ (Kerr case) stressing the significance of a deformation attributed to the parameter δ.

Despite their complexity, the T-S solutions have been shown to play an important role in the problem of δ colliding Kerr black-holes as the limit of δ-superimposed Kerr metrics.

11.12 Hawking radiation: an overview*

The classical feature of a gravitational collapse which leads to the appearance of an event horizon is a progressive suppression of any emission to the far regions. This however appears to be false from an analysis of the quantum processes which take place near the event horizon. In 1974 Hawking (Hawking, 1974) announced the result that an incipient black-hole emits radiation with a thermal spectrum, the energy which is so emitted being at the expense of the black-hole mass. In the case of the collapse of a spherically symmetric star, the spectrum of the massless particles which would be detected at infinity is given by

$$\nu_{\omega\ell n} = \frac{\mathcal{T}_{\omega\ell n}}{2\pi}\left[e^{\frac{2\pi\omega}{k}} + \epsilon\right]^{-1} \qquad \epsilon = \pm 1 \tag{11.12.1}$$

where $\nu_{\omega\ell n}$ is the number of particles (fermions if $\epsilon = +1$, bosons if $\epsilon = -1$) emitted per unit time, unit frequency ω and with

*In this section all the physical quantities are expressed in Planckian units, namely mass in units of the Planck mass: $M^* = (\hbar c/G)^{1/2} = 2.177 \times 10^{-5}$ g., time in units of the Planck time: $t^* = \left(\hbar G/c^5\right)^{1/2} = 5.391 \times 10^{-44}$ sec. and lengths in units of the Planck length: $L^* = \left(\hbar G/c^3\right)^{1/2} = 1.616 \times 10^{-33}$ cm.

the quantum numbers ℓ and n of the spherical harmonics which solve the angular part of the wave equation in the Schwarzschild solution. In (11.12.1), k is the surface gravity, which characterizes a stationary (in this case static) event horizon and is given by (7.11.10); in the Schwarzschild metric it is equal to

$$\mathrm{k} = \frac{1}{4m} \quad ; \qquad (11.12.2)$$

$\tau_{\omega\ell n}$ is the absorption coefficient which measures the fraction of the particles which, being produced in the $(\omega,\ \ell,\ n)$ mode, are captured by the black-hole itself. The existence of the Planck factor in (11.12.1) indicates that the boson particles could be revealed at large distances from the source with an effective temperature $T = \frac{\mathrm{k}}{2\pi}$. In conventional units, if k_b is the Boltzmann constant, we have:

$$\mathcal{T} = T\frac{\hbar M^*}{k_b t^*} = \frac{\hbar c^3}{8\pi k_b GM} \approx 6 \times 10^{-8}\frac{M_\odot}{M}\ {}^0\mathrm{K}\ . \qquad (11.12.3)$$

The presence of the absorption coefficient $\tau_{\omega\ell n}$ is essentially due to the background curvature and since it depends on ω, the spectrum (11.12.1) is not strictly Planckian. Nevertheless one can prove that, despite the formal complexity needed to evaluate $\tau_{\omega\ell n}$, the relation (11.12.1) still describes a thermal spectrum at temperature $\mathrm{k}/2\pi$, so an incipient black-hole emits like a black-body at that temperature. From that spectrum one deduces that the peak frequency of the emitted radiation is $\omega \sim \mathrm{k} \sim m^{-1}$; hence, since m is the size of the black-hole itself, it is impossible to localize the process of emission. Thus a thorough analysis of the Hawking process requires a rigorous quantum mechanical approach (Birrel and Davis, 1982); we shall not embark on this here, however we give an order of magnitude bird's eye view of the main properties of this effect. Let us calculate the luminosity of the black-hole due to the Hawking emission.

From (11.12.1), the energy (per unit time) emitted in the mode (ℓn) and the frequency range $(\omega,\ \omega + d\omega)$ is:

$$dW = \frac{\tau_{\omega\ell n}}{2\pi}\frac{\omega d\omega}{e^{\frac{2\pi\omega}{\mathrm{k}}} - 1} \qquad (11.12.4)$$

hence the total energy (per unit time) over all modes and frequencies (the luminosity) reads:

$$L = \frac{1}{2\pi} \sum_{\ell=0}^{\infty} \sum_{n=-\ell}^{+\ell} \int_0^{\infty} \frac{\tau_{\omega \ell n} \, \omega d\omega}{e^{\frac{2\pi\omega}{k}} - 1} \, . \tag{11.12.5}$$

Evidently this expression depends on $\tau_{\omega \ell n}$, however the order of magnitude of L is easily calculated. Treating the black-hole of mass m as a black-body with temperature $T \propto m^{-1}$, it is

$$L \propto \text{Area} \times T^4 \propto m^2 T^4 \propto m^{-2}.$$

A more accurate determination is due to Page (Page, 1976) who takes into account several species of particles and finds:

$$L \approx 8.6 \times 10^{-21} \left(\frac{M_\odot}{M} \right)^2 \quad \text{ergs sec}^{-1} \tag{11.12.6}$$

where $M_\odot$ is the solar mass ($\approx 1.9 \times 10^{33}$ g). As for the temperature, also the luminosity decreases with the mass of the black-hole; on the other hand the latter decreases with time (proper time of an observer at infinity) as:

$$\frac{dm}{dt} \propto -L \propto -m^{-2}. \tag{11.12.7}$$

Assuming a quasi-stationary emission, so that the metric is a slowly varying function of time, (11.12.7) leads, from (11.12.6), to a time scale for a complete *evaporation* of a black-hole given by:

$$t \approx 5 \times 10^3 m^3 \approx 5 \times 10^3 G^2 \hbar^{-1} c^{-4} M^3 \approx 6.7 \times 10^{66} \left(\frac{M}{M_\odot} \right)^3 \quad \text{years.} \tag{11.12.8}$$

The assumption of quasi-stationary emission fails when the mass approaches zero, since the process is self-accelerating. In fact the time for the emission of a quantum of energy ω is, from (11.12.7) $\omega/\tau \propto m^{-2}$, but since $\omega \leq m^{-1}$, it turns out that: $\tau \leq 5 \times 10^3 (GM/c^3) \approx 2.4 \times 10^{-2} (M/M_\odot)$ sec. Thus the assumption of quasi-stationary emission holds whenever $t \gg \tau$ or alternatively $M \gg\sim 10^{-5}$g.

From (11.12.8) one deduces that black-holes of stellar size ($M \approx M_\odot$), and a fortiori also more massive ones, evaporate in a time much longer than the present age of the Universe ($\approx 10^{10}$ years). When the mass of the hole becomes smaller than the threshold ($\approx 10^{-5}$ g.) for quasi-stationary emission, the process of evaporation proceeds very quickly and is expected to culminate in the emission of a burst of energy. Nevertheless, we would not be able to witness *now* any of these events unless black-holes had been formed with mass initially less than about 10^{15} g. (see (11.12.8)). Black-holes of mass $\leq 10^{15}$ g., termed *mini black-holes*, could only have been formed at the beginning of the Universe, since only then were there energies high enough to compress so small a mass into a black-hole.

If the collapsing system is rotating and leads to a Kerr black-hole, the Hawking arguments can be generalized to the axisymmetric case with only an increase in complexity of form (Hawking, 1975); however at large distances from the source, the effect of the rotation on the Hawking radiation is that of substituting ω by $\omega - n\omega_{\text{H}}$ in (11.12.1), where ω_{H} is the angular velocity of the event horizon (see Sect. 7.10); n is now the azimuth quantum number of the spheroidal harmonics which solve the angular wave equation in the Kerr metric. The observed radiation will then have a spectrum equal to:

$$\nu_{\omega\ell n} = \frac{\tau_{\omega\ell n}}{2\pi} \left[e^{\frac{2\pi}{k}(\omega - n\omega_{\text{H}})} + \epsilon \right]^{-1} \quad ; \quad \epsilon = \pm 1 \,. \tag{11.12.9}$$

The consequence of this correction is that the probability (11.12.9) depends on the azimuth number n, thus the emission will be asymmetric around the black-hole; quanta with positive angular momentum ($n > 0$) will be favoured with respect to those with negative n, hence the Hawking radiation will slow down the black-hole rotation. In the boson case ($\epsilon = -1$), when $\omega < n\omega_{\text{H}}$, the absorption coefficient $\tau_{\omega\ell n}$ becomes negative; this manifests the quantum mechanical analogue of the superradiance effect discussed in Sect. 11.6. Here it is more appropriate to identify this behaviour as the result of a stimulated emission while, when $\omega > n\omega_{\text{H}}$, we would talk of spontaneous emission. It is worth noticing that the latter vanishes in the limit of large masses ($m \to \infty$, $T \to 0$), but the former, when $\omega < n\omega_{\rm H}$, being induced by rotation, persists also in that limit, and gives

$$\nu_{\omega\ell n} \to \frac{|\tau_{\omega\ell n}|}{2\pi} .$$

The effect of rotation is to decrease the temperature of the Hawking radiation; in fact in the Kerr metric, the surface gravity (7.11.10) can be written as:

$$\mathrm{k} = \frac{(m^2 - a^2)^{\frac{1}{2}}}{r_+^2 + a^2} , \qquad (11.12.10)$$

hence it vanishes when $a \to m$. The same situation is met in the more general case when the outcome of collapse is a Kerr-Newman metric, thus an extreme black-hole is equivalent, from the point of view of the Hawking radiation, to a black-body at absolute zero temperature.

NOTATION

M : background space.
$\boldsymbol{t}$: topology on M.
$(M, \boldsymbol{t})$: topological space.
U : open sets of M.
$\mathbb{R}$: the real line.
$\mathbb{R}^n$: the space of real n-tuples.
$(U_\alpha,\ \varphi_\alpha : U \to \mathbb{R}^n)$: a chart (coordinate neighbourhood) on M.
$x = \varphi_\alpha(p)$: local coordinates of p in U_α.
$f : M \to \mathbb{R}$: a function on M.
$\tilde{f}_\alpha = f \circ \varphi_\alpha^{-1}(x)$: local coordinate representation of f.
$T_p(M)$: tangent space at p.
$\overset{*}{T}_p(M)$: cotangent space at p.
$\Phi : M \to N$: a map of M into N.
$\Phi_* : T_p(M) \to T_{\Phi(p)}(N)$: the tangent map.
$\Phi^* : \overset{*}{T}_{\Phi(p)}(N) \to \overset{*}{T}_p(M)$: the dual tangent map.
$\underset{p}{u}\,,\ \underset{p}{v} \in T_p(M)$: vectors at a point.
o : zero vector.
$\underset{p}{\boldsymbol{u}}\,,\ \underset{p}{\boldsymbol{v}} \in \overset{*}{T}_p(M)$: covectors at a point.
$\boldsymbol{o}$: zero covector.
$\{e_i\}_p\,,\ \{\partial_i\}_p$: basis at a point.
$\{\boldsymbol{e}\}_p\,,\ \{\boldsymbol{dx}^i\}_p$: dual basis at a point.
$\gamma : t \to \mathbb{R}$: curve (γ is also used to mean *energy* per unit rest-mass energy of a particle; Chap. 10).
$\gamma(t)$: point on a curve.
$\dot{\gamma}$: tangent vector to a curve γ at p.
$\overset{p}{\boldsymbol{\gamma}}$: geodesic.
$\boldsymbol{\gamma}(t)$: point on a geodesic.
$\dot{\boldsymbol{\gamma}}(p)$: tangent to a geodesic at p.

X , Y , Z : vector fields.
$\boldsymbol{\omega}$, $\boldsymbol{\alpha}$, $\boldsymbol{X}$, $\boldsymbol{Y}$: covector fields or one-forms.
X_p : value of a vector field at p.
$\boldsymbol{\omega}_p$: value of a one-form at p.
$\{e_i\}$, $\{\boldsymbol{e}^i\}$, $\{\partial_i\}$, $\{\boldsymbol{dx}^i\}$: basis fields.
$\{\lambda_{\hat{a}}\}$, $\{\boldsymbol{\lambda}^{\hat{a}}\}$: orthonormal basis (tetrads).
$\dot{\gamma}(\tau)$: tangent field.
$\otimes$: tensor product.
$\wedge$: exterior product.
$\overset{*}{\boldsymbol{g}}$: covariant metric tensor.
$g = \text{Det}\left[g_{ij}\right]$: determinant of the metric.
$(u|v) = \overset{*}{\boldsymbol{g}}(u,v) = g_{ij}u^iv^j$: scalar product.
δ^i_j : Kronecker delta.
δ_{ijkl} : Levi–Civita alternating tensor.
$\delta^{ijk\cdots}_{rst\cdots}$: generalized Kronecker delta.
C_X : congruence of integral curves to a vector field X.
∂_i , $()_{,i}$: partial derivative.
$\left(\pounds_X Y\right)^i \equiv \pounds_X Y^i \equiv [X,Y]^i$; Lie drivative
$\Gamma(p,q;\gamma)$: connector on γ.
$\Gamma^i{}_{jk}$: connection coefficients.
$\nabla_{e_i} \equiv \nabla_i$, $()_{;i}$: covariant derivative.
$\left(\nabla_X Y\right)^i \equiv \nabla_X Y^i$: i-component of the covariant derivative of Y with respect to X.
$\frac{DY}{Dt} \equiv \nabla_{\dot{\gamma}} Y$: absolute derivative of Y along the curve γ.
$C^i{}_{jk}$; $C^{\hat{a}}{}_{\hat{b}\hat{c}}$: structure coefficients.
$R^i{}_{jkl}$: Riemann tensor.
$R^i{}_j$: Ricci tensor.
$R = R^i{}_i = R^i{}_{ji}{}^j$: curvature scalar.
$C^i{}_{jkl}$: Weyl tensor.
$G^i{}_j = R^i{}_j - \frac{1}{2}\delta^i_j R$: Einstein tensor.
$U_{\mathcal{N}}(p)$: normal neighbourhood of p.
$\Omega(p_0,p_1): U_{\mathcal{N}} \times U_{\mathcal{N}} \to \mathbb{R}$: world function.
$\Gamma_{\hat{a}\hat{b}\hat{c}}$: Ricci rotation coefficient.
$\Lambda_{\hat{\alpha}\hat{\beta}}$: Fermi rotation coefficients.
$\partial_{\hat{a}} = \lambda^i_{\hat{a}}\partial_i$: tetrad component of a partial derivative.

$h: h^i{}_j = \delta^i_j + u^i u_j$: orthogonal projecting operator relative to the vector u at a point.

$\pi: \pi^i{}_j = -u^i u_j$: parallel projecting operator relative to a vector u at a point.

$T_{\perp p}(M)$: space of tangent vectors orthogonal to u at p.

$T_{\parallel p}(M)$: space of tangent vectors parallel to u at p.

$\boldsymbol{L}$: the Lorentz group.

$\mathcal{L}$: the *proper* and *orthochronous* Lorentz group.

$\tilde{\mathcal{L}}$: the Lorentz internal rotation group.

$\Box_\eta = \eta^{ij}\partial_i\partial_j$: d'Alembertian operator in Minkowski metric.

$\overset{*}{F}{}^{ij} = \frac{1}{2}\eta^{ijrs}F_{[rs]}$: dual of $F_{[ij]}$.

$\overset{*}{R}{}^{ijk}{}_l$: left-dual of $R^{ijk}{}_l$.

$R^i{}_j{}^{\overset{*}{kl}}$: right-dual of $R^i{}_{jkl}$.

$\overset{*\,*}{R}{}^{ijkl}$: double-dual of R_{ijkl}.

Exp_p: $T_p(M) \to U(p)$: exponential mapping.

REFERENCES

Anile A.M. (1989) *Relativistic fluids and magnetofluids* Cambridge Univ. Press, Cambridge U.K.

Arnowitt R., Deser S. and Misner C.W. (1962) "Dynamical structure and definition of energy in General Relativity" *Phys. Rev.* **116**, 1322

Ashtekhar A. and Hansen R.O. (1978) "A unified treatment of null and spatial infinity in general relativity I" *J. Math. Phys.* **19**, 1542

Bardeen J.M. (1970) "Kerr metric black-holes" *Nature*, **226**, 64

Bardeen J.M., Press W.H. and Teukolsky S.A. (1972) "Rotating black-holes: locally non rotating frames, energy extraction and scalar synchroton radiation" *Ap. J.* **178**, 347

Beiglböck W. (1967) "The center of mass in Einstein theory of gravitation" *Comm. Math. Phys.* **5**, 106

Birrel N.D. and Davis P.C.W. (1982) *Quantum fields in curved space* Cambridge Univ. Press, Cambridge U.K.

Birkoff G.D. (1923) *Relativity and modern physics* Harvard Univ. Press, Cambridge, Mass.

Bondi H. (1961) *Cosmology* Cambridge Univ. Press, Cambridge U.K.

Born M. and Wolf E. (1965) *Principles of optics* Pergamon Press.

Boyer R.H. and Price T.G. (1965) "An interpretation of the Kerr metric in General Relativity" *Proc. Camb. Phil. Soc.* **61**, 531

Boyer R.H. and Lindquist R.W. (1967) "Maximal analytical extension of the Kerr metric" *J. Math. Phys.* **8**, 265

Boyer R.H. (1969) "Geodesic Killing orbits and bifurcate Killing horizons" *Proc. Roy. Soc.* A**311**, 245

Bourbaki N. (1948) *Topologie Générale* Hermann, Paris.

Boulware D.G. (1973) "Naked singularities, Thin shells, and Reissner-Nordström metric" *Phys. Rev.* D **8**, 2363

Braginsky V.B. and Panov V.I. (1971) "Verification of the equivalence of inertial and gravitational mass" *Zh. Eksp. & Teor. Fiz.* **61**, 873 (also *Sov. Phys.-JETP* **34**, 464)

Buchdahl H.A. (1959) "General relativistic fluid spheres" *Phys. Rev.* **116**, 1027

Buchdahl H.A. (1967) "General relativistic fluid spheres II. A: static gaseous model" *Ap. J.* **147**, 310

Cartan E. (1945) *Les système differentiels extérieurs et leur applications géométriques* Hermann, Paris, N. 994.

Carter B. (1966) "The complete analytical extension of the Reissner-Nordström metric in the special case $e^2 = m^2$" *Phys. Lett.* **21**, 423

Carter B. (1968a) "Hamilton-Jacobi and Schrödinger separable solutions of Einstein's equations" *Comm. Math. Phys.* **10**, 208

Carter B. (1968b) "Global structure of the Kerr family of gravitational fields" *Phys. Rev.* **174**, 1559
Carter B. (1969) "Killing horizons and orthogonal transitive groups in space-time" *J. Math. Phys.* **10**, 70
Carter B. (1970a) "The commutation property of a stationary axisymmetric system" *Comm. Math. Phys.* **17**, 233
Carter B. (1970b) "An axisymmetric black-hole has only two degrees of freedom" *Phys. Rev. Lett.* **26**, 331
Chandrasekhar S. (1965) "The post-newtonian equations of hydrodynamics in general relativity" *Ap. J.* **142**, 1488
Chandrasekhar S. and Esposito F.P. (1970) "The $2\frac{1}{2}$-post-newtonian equations of hydrodynamics and radiation reaction in general relativity" *Ap. J.* **160**, 153
Chandrasekhar S. (1983) *The mathematical theory of black-holes* Clarendon Press, Oxford U.K.
Chevalley C. (1946) *Theory of Lie groups* Princeton Univ. Press, Princeton
Choquet-Bruhat Y., De Witt C. and Dillard-Bleick M. (1977) *Analysis, manifolds and physics* North-Holland, Amsterdam
Christodoulou D. (1970) "Reversible and irreversible transformations in black-hole physics" *Phys. Rev.* Lett. **25**, 1596
Clarke C.J.S. (1987) "Spherical symmetry does not imply a direct product" *Class. Quantum Grav.* **4**, L37
Clarke C.J.S. (1988) "A condition for forming trapped surfaces" *Class. Quantum Grav.* **5**, 1029.
Clarke C.J.S. (1990) *Singularities in space-time* (in preparation)
Cohen J.M. (1968) "Angular momentum and the Kerr metric" *J. Math. Phys.* **9**, 905
Cohen J.M. and de Felice F. (1984) "The total effective mass in Kerr-Newman metric" *J. Math. Phys.* **25**, 992
Cohen J.M. and Gautreau D. (1979) "Naked singularities, event horizon, and charged particles" *Phys. Rev.* D **19**, 2273
Cornish F.H.J. (1964a) "Energy and momentum in general relativity I. The 4-momentum expressed in terms of four invariants when space-time is asymptotically flat" *Proc. Roy. Soc.* A **282**, 358
Cornish F.H.J. (1964b) "Energy and momentum in general relativity II. The total energy and momentum of an isolated axisymmetric system generating gravitational waves" *Proc. Roy. Soc.* A **282**, 372
Danese L. and De Zotti G. (1977) "The relic radiation spectrum and the thermal history of the Universe" *Rivista del Nuovo Cimento* **7**, 277
de Felice F. (1968) "Equatorial geodesic motion in the gravitational field of a rotating source" *Nuovo Cimento* **57B**, 351
de Felice F. (1971) "On the gravitational field acting as an optical medium" *Gen. Rel. & Gravit.* **2**, 347
de Felice F. and Calvani M. (1972) "Orbital and vortical motion in the Kerr metric" *Nuovo Cimento* **10B**, 447
de Felice F. and Maeda K. (1982) "Topology of collapse in conformal diagrams: the Reissner-Nordström case" *Prog. Theor. Phys.* **68**, 1967
de Felice F. (1979) "On the velocity composition law in general relativity" Lett. *Nuovo Cimento* **25**, 531
de Felice F. (1980) "Angular momentum and separation constant in the Kerr metric" J. Phys. A: Math. & Gen. **13**, 1701

de Felice F. and Yu Yunqiang (1986) "Behaviour of the ratio a/m for rotating bodies during accretion and gravitational collapse" *Proc. IV Marcell Grossman*; Ed. R. Ruffini

de Felice F. and Bradley M. (1988) "Rotational anisotropy and repulsive effects in the Kerr metric" *Class. Quantum Gravit.* **5**, 1577

de la Cruz V. and Israel W. (1967) "Gravitational bounce" *Nuovo Cimento* **51**, 744

De Zotti G. (1986) "The spectrum of the microwave background as a probe of the early universe" *Progr. in Particle and Nuclear Phys.* **Vol. 17**, 117

De Witt B.S. and Brehme R.W. (1960) "Radiation damping in a gravitational field" *Ann. Phys.* **9**, 220

De Witt C.M. and De Witt B.S. (1964) "Falling charges" *Physics* **Vol. I**, 3

De Rham G. (1955) *Varietés differentiables* Hermann. Paris, N. 1222

Dixon W.G. (1970a) "Dynamics of extended bodies in general relativity I. Momentum and angular momentum" *Proc. Roy. Soc. Lond.* A **314**, 499

Dixon W.G. (1970b) "Dynamics of extended bodies in general relativity II. Moments of charged-current vectors" *Proc. Roy. Soc. Lond.* A **319**, 509

Dixon W.G. (1979) "Extended bodies in general relativity: Their description and motion" *Course 67 of E. Fermi School* (Varenna) Ed. J. Ehlers. North Holland, Amsterdam

Eddington A.S. (1922) *The mathematical theory of relativity* Cambridge Univ. Press, Cambridge, U.K.

Eckart C. (1940) "The thermodynamics of irreversible processes III. Relativistic theory of the simple fluid" *Phys. Rev.* **58**, 919

Ehlers J. and Rudolf E. (1977) "Dynamics of extended bodies in general relativity. Center of mass description and quasi rigidity" *Gen. Rel. & Gravit.* **8**, 197

Einstein A. (1915a) "Zür allgemeinen Relativistätstheorie" *Preuss. Akad. Wiss.* Berlin, Sitzber 778

Einstein A. (1915b) "Zür allgemeinen Relativistätstheorie" *Preuss. Akad. Wiss.* Berlin, Sitzber 799

Einstein A., Infeld L. and Hoffmann B. (1938) "Gravitational equations and the problem of motion" *Ann. Math.* **39**, 65

Eisenhart L.P. (1926) *Riemannian geometry* Princeton Univ. Press, Princeton

Ellis G.F.R. (1967) "Dynamics of pressure free matter in general relativity" *J. Math. Phys.* **8**, 1171

Ellis G.F.R. (1971) *Relativity and cosmology* Ed. Sachs Academic Press New York

Ernst F.J. (1968a) "New formulation of the axially symmetric gravitational field problem" *Phys. Rev.* **167**, 1175

Ernst F.J. (1968b) "New formulation of the axially symmetric gravitational field problem II" *Phys. Rev.* **168**, 1415

Fermi E. (1922) "Sopra i fenomeni che avvengono in vicinanza di una linea oraria" *Atti Acc. Naz. Lincei Cl. Sci. Fis. Mat. & Nat.* **31**, 184, 306

Finkelstein D. (1958) "Past-future asymmetry of the gravitational field of a point particle" *Phys. Rev.* **110**, 965

Fisher A.E. and Marsden J.E. (1972) "The Einstein equation of evolution. A geometrical approach" *J. Math. Phys.* **13**, 546

Flanders H. (1963) *Differential forms* Academic Press, New York

Freud P. (1939) "Über die Ausdrücke der Gesamtenergie und des Gesamtimpulses eines Materiellen Systems in der Allgemeinen Relativistätstheorie" *Ann. Math.* **40**, 417
Friedlander F.G. (1975) *The wave equation on a curved space-time* Cambridge Univ. Press, Cambridge, U.K.
Friedmann A. (1922) "Über die Krümmung des Raumes" *Z. Phys.* **10**, 377
Frobenius G. (1911) "Über die unzerlegbaren diskreten Bewegungsgruppen" *Sitzungsberichte König. Preuss. Akad. Wiss.* Berlin **29**, 654
Geroch R. (1968) "Spinor structure of space-times in general relativity I" *J. Math. Phys.* **9**, 1739
Geroch R. (1985) *Mathematical physics* The Univ. of Chicago Press, Chicago
Goldberg J.N. and Sachs R.K. (1962) "A theorem on Petrov types" *Acta Phys. Pol. Suppl.* **22**, 13
Greenberg P.J. (1972) "The algebra of the Riemann curvature tensor in general relativity: Preliminaries" *Studies in Applied Maths.* Vol II. M.I.T.
Hawking S.W. and Ellis G.F.R. (1973) *The large scale structure of space-time* Cambridge Univ. Press, Cambridge, U.K.
Hawking S.W. (1974) "Black-hole explosions?" *Nature* (London) **248**, 30
Hawking S.W. (1975) "Particle creation by black-holes" *Comm. Math. Phys.* **43**, 199
Helgason S. (1962) *Differential geometry and symmetric spaces* Academic press, New York
Hellings R.W. (1983) "Testing relativity with solar system dynamics" *General relativity and gravitation* Ed. Bertotti, de Felice and Pascolini, Reidel
Hilbert D. (1917) "Die Gründlagen der Physik" *König. Gesell. d. Wiss. Göttingen Nachr. Math. Phys.* Kl. 53
Hiscock W.A. (1981) "On the topology of charged spherical collapse" *J. Math. Phys.* **22**, 215
Hobbs J.M. (1968) "A vierbein formalism of radiation damping" *Ann. of Phys.* **47**, 141
Hubble E.P. (1929) "A relation between distances and radial velocity among extragalactic nebulae" *Proc. Nat. Acad. Sci. U.S.A.* **15**, 169
Israel W. (1963) "Relativistic kinetic theory of a simple gas" *J.Math. Phys.* **4**, 1163
Israel W. (1967) "Event horizons in static vacuum space-times" *Phys. Rev.* **164**, 1776
Israel W. (1968) "Event horizons in static electrovac space-times" *Comm. Math. Phys.* **8**, 245
Israel W. (1970) "Differential forms in general relativity" *Comm. of Dublin Inst. Adv. Stud.* Sez. A N. 19
Israel W. (1976) "Non stationary irreversible thermodynamics: A causal relativistic theory" *Ann. Phys.* **100**, 310
Israel W. and Stewart J.M. (1979) "Transient relativistic thermodynamics and kinetic theory" *Ann. Phys.* **118**, 341
Jordan P., Ehlers J. and Sachs R. (1961) "Exact solutions of the field equations in general relativity II. Contribution to the theory of pure gravitational radiation" *Akad. Wiss. Lit. Mainz Abh. Math. Nat. Kl.* **1**, 3
Kerr R.P. (1963) "Gravitational field of a spinning mass as an example of algebraically special metric" *Phys. Rev. Lett.* **11**, 237
Kerr R.P. and Shild A. (1967) "A new class of vacuum solutions of the Einstein field equations" *Atti del Convegno sulla Rel. Gen.* Firenze 222

Kobayashi S. and Nomizu K. (1969) *Foundations of differential geometry* Interscience, New York

Komar A. (1959) "Covariant conservation laws in general relativity" *Phys. Rev.* **113**, 934

Kramer D., Stefani H., MacCallum M. and Herlt E. (1980) *Exact solutions of Einstein's field equations* VEB Deutchen Verlag der vissenshaften, Berlin

Kruskal M.D. (1960) "Maximal extension of Schwarzschild metric" *Phys. Rev.* **119**, 1743

Lake K. (1978) "White holes" *Nature* **272**, 599

Landau L.P. and Lifschitz E.M. (1959) *Fluid mechanics* Pergamon Press, London

Landau L.P. and Lifschitz E.M. (1962) *The classical theory of fields* Pergamon Press, London

Lathrop J.D. (1973) "Covariant description of motion in general relativity" Ann. of Phys. **79**, 580

Lewis T. (1932) "Some special solutions of the equations of axially symmetric gravitational fields" *Proc. Roy. Soc. London* A **136**, 176

Levi-Civita T. (1918) "La teoria di Einstein ed il principio di Fermat" *Nuovo Cimento* **16**, 105

Lichnerowicz A. (1967) *Relativistic hydrodynamics and magnetohydrodynamics* W.A. Benjamin Inc., New York

Marck J.A. (1983) "Solution to the equations of parallel transport in Kerr geometry; tidal tensor" *Proc. Roy. Soc. London* A **385**, 431

Martin C.F., Torrence M.H. and Misner C.W. (1985) "Relativistic effects on an Earth orbiting satellite in the baricenter coordinate system" *J. Geophys. Research* **90**, 9403

Mathisson M. (1937) "Neue Mechanik materieller Systeme" *Acta Phys. Pol.* **6**, 163

Misner C.W., Thorne K.S. and Wheeler J.A. (1973) *Gravitation* Freeman & Co., San Francisco

Møller C. (1958) "On the localization of the energy of a physical system in the general theory of relativity" *Ann. Phys.* **4**, 347

Møller C. (1961) "Further remarks on the localization of the energy in the general theory of relativity" *Ann. Phys.* **12**, 118

Newman E.T. and Penrose R. (1962) "An approach to gravitational radiation by a method of spin coefficients" *J. Math. Phys.* **3**, 566

Newman E.T., Couch E., Chinnapared K., Exton A., Prakash A. and Torrence R. (1965) "Metric of radiating charged mass" *J. Math. Phys.* **6**, 918

Nordström G. (1913) "Zür Theorie der Gravitation vom Standpunkt des Relativistätsprinzipe" *Ann. Phys.* (Germany) **42**, 533

Novikov I.D. and Thorne K.S. (1972) in *Black-holes* Ed. C. and B.S. De Witt, Gordon & Breach

Page D.N. (1976) "Particle emission rates from a black-hole: massless particles from an uncharged, non rotating hole" *Phys. Rev.* D**13**, 198

Palatini A. (1919) "Deduzione invariantiva delle equazioni gravitazionali dal principio di Hamilton" *Rend. Circ. Mat. Palermo* **43**, 203

Papapetrou A. (1951) "Spinning test particles in general relativity I" *Proc. Roy. Soc.* A **209**, 248

Papapetrou A. (1963) "Quelques remarques sur les champs gravitationnels stationnaires" *C. R. Acad. Sci.* (Paris) **257**, 2797

Papapetrou A. (1966) "Champs gravitationnels stationaires à symmetrie axiale" *Ann. Inst. H. Poincaré* A **4**, 83
Peebles P.J.E. (1984) "Tests of cosmological models constrained by inflation" *Ap. J.* **284**, 439
Peebles P.J.E. (1986) "The mean mass density of the Universe" *Nature* **321**, 27
Penrose R. (1960) "A spinor approach to general relativity" *Ann. of Phys.* **10**, 171
Penrose R. (1964) "Conformal treatment of infinity" in *Relativity groups and topology* Ed. C. and B. De Witt; Les Houches Summer School 1963, Gordon & Breach
Penrose R. and Floyd R.M. (1971) "Extraction of rotational energy from a black-hole" *Nature* **229**, 177
Penrose R. and Rindler W. (1984) *Spinors and space-time I. Spinor calculus and relativistic fields* Cambridge Univ. Press, Cambridge, U.K.
Petrov A.Z. (1969) *Einstein spaces* Pergamon Press, New York
Pirani F.A.E. (1956a) "On the physical significance of the Riemann tensor" *Acta Phys. Pol.* **15**, 389
Pirani F.A.E. (1956b) "Tetrad formulation of general relativity theory" *Helv. Phys. Acta Suppl.* **4**, 198
Pirani F.A.E. (1965) in *Lectures on General Relativity* Brandeis Summer Institute, Vol. I, Prentice Hall
Pound R.V. and Rebka G.A. (1960) "Apparent weight of Photons" *Phys. Rev. Lett.* **4**, 337
Raychaudhuri A. (1955) "Relativistic cosmology" *Phys. Rev.* **98**, 1123
Reissner H. (1916) "Über die Eigengravitation des elektrischen Feldes nach der Einsteinschen Theorie" *Ann. Phys.* (Germany) **50**, 106
Rindler W. (1956) "Visual horizons in world-models" *Mon. Not. Roy. Astr. Soc.* **116**, 663
Robertson H.P. (1935) "Kinematics and world structure" *Ap.J.* **82**. 284
Robertson H.P. (1936) "Kinematics and world structure" *Ap.J.* **83**, 187
Robinson D.C. (1975) "Uniqueness of the Kerr black-hole" *Phys. Rev. Lett.* **34**, 905
Roll P.G., Krotkov R. and Dicke R.H. (1964) "The equivalence of inertial and passive gravitational mass" *Ann. Phys.* (U.S.A.) **26**, 442
Ruse H.S. (1931) "Taylor's theorem in the tensor calculus" *Proc. London Math. Soc.* **32**, 87
Sachs R.K. (1962) "Gravitational waves in general relativity VIII. Waves in asymptotically flat space-time" *Proc.Roy.Soc.* A **270**, 103
Schoen R. and Yau S.T. (1983) "The existence of a black-hole due to condensation of matter" *Comm. Math. Phys.* **90**, 575
Schouten J.A. (1954) *Ricci calculus* Springer, Berlin
Schrödinger E. (1963) *Space-time structure* Cambridge Univ. Press, Cambridge, U.K.
Schutz, B.F. (1970) "Perfect fluids in general relativity: velocity potentials and a variational approach" *Phys. Rev.* D**2**, 2762
Schwarzschild K. (1916a) "Über das Gravitationsfeld einer Massenpunktes nach der Einsteinschen Theorie" *Sitz. Preuss. Akad. Wiss.* **189**
Schwarzschild K. (1916b) "Über das Gravitationsfeld einer Kugel aus inkompressibles Flussigkeit nach der Einsteinschen Theorie" *Sitz. Preuss. Akad. Wiss.* **424**

Starobinskii A.A. (1973) "Amplification of waves during reflection from a rotating black-hole" *Soviet Phys.-JETP* **37**, 28

Starobinskii A.A. (1974) "Amplification of electromagnetic and gravitational waves scattered by a rotating black-hole" *Sov. Phys.-JETP* **38**, 1

Stewart J.M. and Ellis G.F.R. (1968) "Solutions of Einstein's equations for a fluid which exhibits local rotational symmetry" *J. Math. Phys.* **9**, 1072

Synge J.L. (1960) *Relativity The general theory* North Holland Amsterdam.

Thorne K.S. (1969) "Non radial pulsation of general relativistic stellar models IV. The weak field limit" *Ap. J.* **158**, 997

Thorne K.S. (1974) "Disk accretion onto a black-hole II. Evolution of the hole" *Ap.J.* **191**, 507

Thorne K.S. (1980) "Gravitational wave research: current status and future prospects" *Rev. Mod. Phys.* **52**, 285

Tod K.P.. de Felice F. and Calvani M. (1976) "Spinning test particles in the field of a black-hole" *Nuovo Cimento* **34B**, 365

Tomimatsu A. and Sato H. (1972) "New exact solution for the gravitational field of a spinning mass" *Phys. Rev.* Lett. **29**, 1344

Tomimatsu A. and Sato H. (1973) "New series of exact solutions for gravitational fields of spinning masses" *Progr. Theor. Phys.* **50**, 95

Trautman A. (1965) in *Lectures on General Relativity* Brandeis Summer Institute, Vol. I, Prentice Hall

Tulczyjew W.M. (1957) "On the energy momentum density for single pole particles" *Bull. Acad. Polon. Sci. Cl.III* **5**, 279

Tulczyjew W.M. (1959) "Motion of multipole particles in general relativity theory" *Acta Phys. Polon.* **18**, 393

Tulczyjew W.M. (1977) "The Legendre transformations" *Ann. Inst. H. Poincaré* **27**, 101

Turner M.S., Steigman G. and Krauss L.M. (1984) "Flatness of the Universe: reconciling theoretical prejudices with observational data" *Phys. Rev. Lett.* **52**, 1090

Veblen O. (1962) *Invariants of quadratic differential forms* Cambridge Univ. Press, Cambridge, U.K.

Vishveshwara C.V. (1968) "Generalization of the Schwarzschild surface to arbitrary static and stationay metrics" *J. Math. Phys.* **9**, 1319

Wald R.M. (1984) *General Relativity* The University of Chicago Press

Walker A.G. (1932) "Relative coordinates" *Proc. Roy. Soc. Edinburgh* **52**, 345

Walker A.G. (1963) "On Milne's theory of world structure" *Proc. Lond. Math. Soc. (2)* **42**, 90

Weinberg S. (1972) *Gravitation and cosmology* Wiley, New York

Weyl H. (1917) "Zür Gravitationstheorie" *Ann. Phys.* **54**, 117

Weyl H. (1922) *Space, time, matter* Dover

Williams J.G., Dicke R.H., Bender P.L., Alley C.O., Carter W.E., Currie D.G., Eckhardt D.H., Faller J.E., Kaula W.M., Mulholland J.D., Plotkin H.H., Poultney S.K., Shelus P.J., Silverberg E.C., Sinclair W.S., Slade M.A. and Wilkinson D.T. (1978) "New test of the equivalence principle from lunar laser ranging" *Phys. Rev.* Lett. **36**, 551

Witteborn F.C. and Fairbank W.H. (1967) "Experimental comparison of the gravitational force on freely falling electrons and metallic electrons" *Phys. Rev. Lett.* **19**, 1049

INDEX

Abelian ring, 19
Absolute space, 5, 6
  time, 6
Acceleration, 4, 6
  deceleration parameter in cosmology, 382
  four-acceleration, 138, 253, 255, 257, 259
  nearby a singularity, 370
  relative acceleration of nearby geodesics, 117, 118, 257, 258
  uniform, 9
Achronal sets, 238
Action
  Hilbert, 192
  of a field, 188, 189
  of a gravitational field, 191
Affine connection, 78, 108
  geodesic, 76
  parameter, 76, 107
  transformation 78
Age of the Universe, 380, 429
Algebra of bivectors, 155
  exterior, 51
  Lie, 151
  of tensors, 34, 57
Algebraic classification of the Weyl tensor, 180*ff*
Alternating tensors, 48
Analytical extension, 330, 331
Angle, 281
  definition of, 281
  variation induced by curvature, 285
Angular momentum (*see* Momentum)
Antisymmetrization of tensors, 38, 39
Apsidal points, 347, 348
Area theorem, 412
Aristotle, 1
Asymptotic flatness, 207*ff*
  properties of space-time, 338*ff*, 361*ff*
Atlas, 17
Axial symmetry, 239, 389

Basis
  coordinate (natural), 25
  in cotangent space, 29
  in tangent space, 24
  orthonormal (*see also* Tetrad), 43, 110
Bending of light rays, 354
Bianchi identities, 114, 116, 254
Bifurcation point, 247
Binding energy of marginally stable orbits, 347, 416
Birkoff theorem, 324, 342
Bitensors, 77
Bivector space, 107
Black holes, 324, 338, 357
  energy extraction from, 409, 410, 411, 412
  evaporation of, 428
  mini black holes, 429
  no hair theorem, 424
Boltzman, 427
Bouguer, 355
Boyer, 247, 402, 416

Carter, 400, 421, 424
Cauchy (*see* Initial value problem)
Causal structure of space-time, 236*ff*
  curve, 236

Center of mass, 216*ff*
Chart, 13, 17
Christoffel symbols, 85
Circular orbits: in Schwarzschild, 345; in Kerr, 414
Class of a manifold, 19
  of a function, 19
Codazzi, 253
Commutator (*see* Lie brackets)
  of matrices, 169
Compact spaces (*see* Spaces)
Complex
  Einstein, 203
  Freud, 204
  Landau and Lifschitz, 205
Conformal compactification, 341
  factor, 114, 174*ff*
  invariance of the Weyl tensor, 114
  spinors, 174*ff*
Conformal transformations, 114
Conformal weight, 175*ff*
Congruence, 57*ff*, 60, 62
  hypersurface orthogonal, 141, 143, 145, 260, 374, 390
  geometry of, 250*ff*
  null, 261, 271
Conjugate points, 118
  to a surface, 272, 273
Connecting vector, 117
Connection coefficients, 77
  compatibility with metric, 84
  spinor, 166*ff*
  vanishing of, 83, 84
Connector, 68*ff*
Conservation laws, 186, 206*ff*, 387
Contraction, 36, 47, 58
Contravariant components, 26, 34
Constant curvature spaces, 116, 117, 127, 128
Coordinate system, 17*ff*
  Boyer–Lindquist, 402, 416
  local, 17
  normal, 83, 107, 108
  right- and left-handed, 45, 48
  standard in cosmology, 374
Cosmic radiation, 378
Cotangent space, 28, 29
Covariant components, 31, 35, 36
  derivative (*see* Derivative)
Covector, 29
Cosmic time, 374
Cosmology, 372*ff*
Cosmological constant, 384
Cosmological principle, 373
Curvature tensor, 104*ff*, 143
    number of components, 107
    symmetry properties, 106
  Gaussian, 108, 116, 128, 322, 375
  intrinsic, 252
  invariants, 332, 401
  scalar, 112
  sectional, 108
  small curvature limit, 306*ff*
  spatially projected, 143
  spinor (*see* Spinors)
Current density, 229
Curve, 21
  integral, 62

D'alambertian, 228, 313
Deceleration parameter, 382
Decomposable r-forms, 53
Degree of freedom (*see* Gauge transformations)
De Rham, 228, 231
Determinant of a $n \times n$ matrix, 49, 53
  of a k-rowed minor, 49
Derivative
  absolute, 78, 79
  convective, 63, 68
  covariant of vectors 79
    application of, 86*ff*
    of one-forms, 79, 80
    of spinors, 176
    of tensors, 81
  exterior, 88*ff*
  Lie, 62*ff*
Diffeomorphism, 20, 60, 330
Differential forms, 28*ff*, 57
Dirac delta, 199
Discontinuities of an electromagnetic field, 232*ff*
Distance
  spatial, 132, 275, 277
  in cosmology (luminosity distance), 388
Dragging (gravitational), 243, 406, 408

Dual (Hodge), 50
right- left-dual, 50
map, 55
self-dual and anti-self-dual, 157, 171, 181

Eddington, 334, 335
Einstein, 11, 116, 191, 193, 260, 354, 384
complex, 203, 204
equations, 191, 193
in axial symmetry, 390, 391
in spherical symmetry, 323, 378
in tetrad forms, 260*ff*
in the weak field limit, 309, 310, 312, 313
Maxwell's equations, 358, 421
tensor, 116
Universe, 385
Eikonal equation, 235, 286, 385
Electrodynamics, 230, 231
Electromagnetic fields
free, 227*ff*
measure of, 298
null, 235
polarization of, 236
radiation field, 231*ff*
with charges and currents, 228*ff*
Energy of a particle, 297, 343, 344
Energy conditions, 269
Energy density of a fluid, 185, 301*ff*
of an electromagnetic field, 304
Energy extraction form a black-hole, 409
Energy momentum tensor
for Newtonian fluids, 187
in special relativity, 187, 188
of a perfect fluid, 195, 197
of a single particle, 199, 200, 222
of an electromagnetic field, 200
pseudotensor, 201, 204, 205
Entropy, 195, 305, 306
Equivalence (*see* Principle of)
Ergoregion-ergosphere, 409
Ernst, 391, 424
Euclidean geometry, 1, 2
Euler equation, 186, 306
Eulerian coordinates, 195
Evaporation of a black-hole, 428
Event horizon, 245, 338, 424
future and past, 337, 338
Expansion, 140, 225, 256
in the weak field limit, 310
isotropic (volumetric), 257
Exponential map, 108
Exterior product, 52, 53
derivative, 88*ff*
Extrinsic curvature, 252, 260

Fermi (Walker) transport, 138, 139, 221
rotation coefficients, 138, 259
Fields (of vectors and tensors), 57*ff*
basis, 58
Finkelstein, 334, 335
Fluids, 185*ff*
Newtonian, 186, 187
perfect, 195*ff*
Forms, 51, 52
integrable, 94
irrotational, 96
one-forms, 57
r-forms, 51; simple r-forms, 52
Frames
freely falling, 10
inertial, 6
locally inertial, 11
locally non-rotating, 408
non-inertial, 11
tetrads (*see* Tetrads)
spin-entangled, 160
Fraser, 7
Frequency, 286, 385, 427
Doppler shift, 287, 386
red-shift in cosmology, 385, 386
shifts, 286, 287, 357
Freud, 204
Friedmann, 320, 378*ff*, 384
Frobenius, 90*ff*, 241, 245

Galileo, 4*ff*
transformations, 6, 7

Gauge transformations
in electromagnetic fields, 200, 227
in the weak gravitational field limit, 308*ff*
Lorentz in gravitational radiation theory, 314
transverse traceless in gravitational radiation theory, 314
Gauss equation, 143, 252
Gaussian curvature (*see* Curvature)
General covariance, 11
General relativity, 7, 11
Geodesics, 72, 75
affine, 76
deviation equation, 117, 258, 351
null in the Schwarzschild metric, 351
null in the Kerr metric, 418
time-like in the Schwarzschild metric, 342*ff*
time-like and circular in the Kerr metric, 406, 414
Geometrical optics, 235
Geroch, 271
Globally pseudo-Euclidean coordinates, 110, 111
Goldberg–Sachs theorem, 406
Gravitation
equations of the field, 191*ff*
surface gravity, 249, 427
weak gravitational field, 306*ff*
Gravitational radiation, 314
Gravitational red-shift, 357
Green's function, 119, 231
Group
Lorentz, 135*ff*, 146*ff*
rotation SO(3), 320
SL(2,**C**), 146

Hamilton-Jacobi equation, 225, 342, 407
Harmonic function, 315, 390
Hausdorff, 14, 129, 331
Hawking, 271, 426
radiation, 426*ff*
Hermitian matrices, 147
Hilbert, 192
Hypersurface, 21
orthogonal (*see* Congruence)
Homeomorphism, 17, 21
Homogeneity of the Universe, 373*ff*
Homomorphism, 42, 150, 154
Homotopy, 109
Horizon, 245
bifurcated, 247, 248
event, 245, 338, 339, 354, 363, 403, 404, 426*ff*
particle, 383, 385
Hubble, 380, 386, 387

Immersion, 21
Inertial observer, 6*ff*
forces, 6, 9
frames, 6
Inflationary Universe, 385
Initial value problem, 261
Integral curve, 62
Irreducible mass (*see* Mass)
Isometries, 99*ff*
Isomorphism, 42, 132, 154, 162
Isotropy of the Universe, 373*ff*
Isotropic coordinates, 354

Jacobian, 48
Jacobi field, 118
Jacobi identity, 65, 151, 251

Kant, 1
Kennedy-Thorndyke, 7
Kerr, 392*ff*
horizons, 397, 410, 411, 416, 421
Killing one-forms, 398
Newman solution, 421
solution, 397
Killing fields, 99, 125, 126, 225
equations, 99
one-forms in conservation laws, 209
Klein-Gordon, 410
Komar, 210, 328
Kronecker delta, 24
generalized delta, 48
Kruskal coordinates, 336

Lagrangian coordinates, 195
Lagrangian formulation, 189*ff*

Lagrangian of a perfect fluid, 196, 197
of an electromagnetic field, 200, 227
of the background geometry, 192
Landau and Lifschitz, 204, 205
Left-handed coordinates (*see* Coordinate system)
Length
measure of, 2, 55, 275
Lewis, 390
Lie, 62*ff*
algebra, 151*ff*
brackets, 65, 91, 102
group, 151
Light-cone, 236
Light rays, 233
bending of, 354
velocity, 7, 187, 297, 303
Lindquist, 402, 416
Line element, 56
Lipschitz, 19, 61
Locally non rotating frame, 408, 409
Lorentz
force, 298, 300
group, 135, 150
internal rotation group, 136
transformations, 7
proper and orthocronous, 135
Lorentzian metric, 45
Lowering of a tensor (*see* Contraction)

Manifold, 19*ff*
differentiable, 19
imbedded, 21
Map, 15
bilinear, 33
exponential, 108
linear, 22
of manifolds, 20
the tangent map, 27
the dual map, 31
Marginally stable orbits in the Kerr metric, 416
Mass
generalized mass in the Reissner-Nordström metric, 361
gravitational, 8
in geometrized units, 361
in the Kerr metric, 399
inertial, 8
irreducible, 412
of an extended body, 217
total in Schwarzschild, 325
Matrix algebra, 49, 50
Laplace expantion of, 50
non-singular, 26
Maxwell equations, 227, 232, 233
(Einstein)-Maxwell, 358, 421
Measure
of length, 2, 55*ff*
theory of, 274*ff*
Metric tensor, 39
Michaelson-Morley, 7
Module, 57
Modulus of a vector, 40
Momentum
angular, 209, 214, 343, 406, 407, 408, 417, 423
in the Kerr metric, 400
linear of a particle, 297
of a fluid, 301
proper, 213
Motion
equations of motion for a particle, 221*ff*
for extended bodies, 211*ff*
for a fluid, 304*ff*
constants of, 213, 225, 343, 352, 406, 407, 420
slow motion approximation, 307*ff*

Neighbourhood, 14
coordinate, 17
normal, 83
Newman, 421
Newtonian limit of the gravitational field, 313
Nordström (*see* Reissner-Nordström solution)

Null
  cone (*see* Light-cone)
  curve, 45 (*see* also Geodesics)
  geodesic (*see* Geodesics)
  principle directions in the Kerr metric, 420, 421
  subspace, 45
  tetrad, 263

Observer, 130
  definition of, 130
  in plunging orbits, 349
Open sets, 14
Operator
  projecting parallel and transverse, 130
Optical scalars, 266
Orbits (*see* Geodesics)
  marginally stable, 416
Orientability of a manifold, 45
Orthogonality, 40
Orthonormal basis (*see also* Tetrad), 43

Palatini identity, 192
Papapetrou, 215, 390, 421
Paracompactness, 15, 129
Parallel vector, 85
Parallelism, 85*ff*
Parameter on a curve, 21
  on a congruence, 62
  transformation to an affine parameter, 75, 76
Particle horizon, 383, 385
Penrose, 166, 271, 338, 409, 412
  diagrams, 338*ff*
  flag, 166
  process of energy extraction, 409
Perfect fluid (*see* fluids)
Petrov classification, 146, 177*ff*
Planck, 426, 427
Plato, 1
Poincaré, 2, 3
Pole-dipole approximation, 220
Potential curves for time-like geodesics, 345, 414
  in general relativity, 192
  Newtonian, 255, 313
  one-form in electromagnetism, 200, 227*ff*, 422, 423
  velocity potentials, 198
Precession of the apsidal points, 347
Principal null directions, 180, 405, 421
Principle of equivalence, 10
  of general covariance, 11
  of relativity, 297
Pseudotensor, 201, 205

Quadrupole moment
  of extended bodies, 215, 218
  of gravitationally radiating sources, 318

Radiation
  cosmic, 378
  electromagnetic, 231*ff*
  gravitational, 314*ff*
Raising (of a tensor) (*see* Contraction)
Rank (of a tensor), 36
Raychauduri equation, 257, 267, 272
Red-shift in cosmology, 385
Reissner-Nordström
  extended solution, 361
  singularity, 369
  solution, 358
Retarded potentials, 231, 315
Reversible processes in black-holes, 412
Ricci
  rotation coefficients, 140, 251
  tensor, 111
  tetrad decomposition of, 259, 260
Riemannian manifold, 44, 85
Riemann
  integral, 54
  tensor, 104
    tetrad decomposition of, 250*ff*
Right-handed coordinates, 45
Robertson, 320, 376, 384
Robinson, 424

Rotational effects, 243
in the Kerr metric, 408
Rotation group, 320

Sachs (*see* Goldberg–Sachs theorem)
Sato, 425
Scalar
curvature, 112
product, 40
Schild, 405
Schwarzschild
external solution, 322*ff*
extended external solution, 332
internal solution, 326
radius, 332, 333
Segré symbols, 183, 184
Shear, 256, 257, 266, 278, 302, 305, 319
free congruence, 374, 376, 420
Shoen, 272
Signature, 43
Simultaneity, 277
Singularity, 267*ff*
apparent, 332, 361, 401
crushing, 270
in cosmology, 380
in Kerr, 400
in Reissner-Nordström, 369
in Schwarzschild, 332
Space
compact, 15
concept of space in theory of measurements, 275*ff*
connected, 14, 129
maximally symmetric, 125
of r-forms, 51
paracompact, 15, 129
separated (Hausdorff), 14, 129
topological, 14, 129
Space-like curve, 45
subspace, 45
Space-time, 4, 129
Special relativity, 4, 187
Spherically symmetric line element, 322
Spherical symmetry, 320, 329
Spin-curvature interaction, 214*ff*
Spin entanglement, 160
Spin frame, 160
vector, 219, 221
Spinors, 146, 159
conformal, 174
connection, 166
curvature, 168
Static space-times, 239
Stationary space-times, 240
limit of stationarity, 244
null surfaces, 244*ff*
Stokes, 111
Stress-tensor (*see* Energy-momentum tensor)
Strong energy condition, 269
Structure coefficients, 65
Summation convention, 23
Superpotential, 204, 205
Superradiance, 409, 410
Surface gravity, 246, 249, 427*ff*
Symmetrization of tensors, 38, 39
Symmetry
axial, 239, 389
operations, 37
spherical, 320
symmetric spaces (*see* spaces)

Tangent space, 12, 21
vectors, 22
Temperature: effective for black-holes, 427
Tensors, 11, 32*ff*
alternating, 48
function of two points, 77, 120
product, 33
Tetrad, 129*ff*
Fermi transported, 138, 139, 145, 286
null, 263
Thermal spectrum, 426
Thermodynamic laws, 186, 197, 304, 305
Thomas precession, 221
Thorndyke (*see* Kennedy-Thorndyke)
Tidal stresses, 254, 350, 351
Time
absolute, 6
measure of, 275*ff*
proper, 224, 277
standard cosmic time, 374

Time-like curve, 45 (*see also* Geodesics)
  subspace, 45
Time orientability, 45
Tomimatsu, 425
Topology, 14
  induced, 14
Torque, 214
Torsion, 82, 172
Transformations
  Galileo, 6, 7
  Lorentz, 7, 135
  of alternating tensor, 48
  of contravariant indices, 26, 28
  of covariant indices, 31, 32
  of spinors, 164
  of tensor components, 34, 36
  of the connector, 77*ff*
Transverse-traceless (*see* Gauge)
Trapped surface, 271, 272

Units
  geometrized for mass, 361, 396
    for angular momentum, 400
    for electric charge, 361
  planckian, 426
Universe, 372*ff*
  closed, 381
  open, 381

Valence, 36
Vector
  cotangent, 28
  null, 40
  space, 22*ff*
  space-like, 45
  tangent, 21
  time-like, 45
Velocity
  composition law, 295*ff*
  Doppler velocity, 289, 386
  measure of relative velocities, 287*ff*
  recession velocity, 288, 289, 386
Vorticity, 140, 253, 259
  free congruence, 140, 268
  in the weak field limit, 310
Walker, 320, 376, 384
Weak energy condition, 269
Weak field limit, 306
Wedge product (*see* Exterior product)
Weyl
  solutions, 424
  spinor, 177
  tensor, 111*ff*, 180, 181, 254
    classification of, 180, 181
    electric and magnetic parts, 254, 255
White-holes, 338
World-function, 55*ff*
  covariant derivatives of, 120
  in theory of measurements, 275*ff*

Yau, 272